Aerospace Avionics Systems

Aerospace Avionics Systems

A Modern Synthesis

George M. Siouris

Air Force Institute of Technology
Department of Electrical and Computer Engineering
Wright-Patterson Air Force Base, Ohio

Academic Press, Inc.
Harcourt Brace & Company

San Diego   New York   Boston   London   Sydney   Tokyo   Toronto

This book is printed on acid-free paper. ∞

Copyright © 1993 by ACADEMIC PRESS, INC.
All Rights Reserved.
No part of this publication may be reproduced or transmitted in any form or by any means, electronic or mechanical, including photocopy, recording, or any information storage and retrieval system, without permission in writing from the publisher.

Academic Press, Inc.
1250 Sixth Avenue, San Diego, California 92101-4311

United Kingdom Edition published by
Academic Press Limited
24–28 Oval Road, London NW1 7DX

Library of Congress Cataloging-in-Publication Data

Aerospace Avionics Systems: a modern synthesis / George M. Siouris.
    p. cm.
  Includes bibliographical references and index.
  ISBN 0-12-646890-7
  1. Inertial navigation (Aeronautics)   2. Navigation (Aeronautics)
I. Title.
TL588.5.S56   1993
625.135'1-dc20   92-33766
                                                                                     CIP

PRINTED IN THE UNITED STATES OF AMERICA
93 94 95 96 97 98   MM   9 8 7 6 5 4 3 2 1

To the memory of my parents

Contents

Preface xi

1 Introduction

1.1 Organization of the Text 5

2 Coordinate Systems and Transformations

2.1 Coordinate Systems 7
    2.1.1 True Inertial Frame (I) 9
    2.1.2 Earth-Centered Inertial (ECI) Frame (i) 9
    2.1.3 Earth-Centered Earth-Fixed (ECEF) Frame (e) 10
    2.1.4 Navigation Frame (n) 10
    2.1.5 Body Frame (b) 10
    2.1.6 Wander–Azimuth Frame (c) 11
    2.1.7 Platform, Accelerometer, and Gyroscope Frames 12
2.2 Coordinate Transformations 13
    2.2.1 The Direction Cosine Matrix (DCM) 14
    2.2.2 The Direction Cosine Matrix Differential Equation 17
    2.2.3 Euler's Angles 20
    2.2.4 Transformation of Angular Velocities 24
    2.2.5 Earth Sidereal Rotation Rate 26
    2.2.6 Earth-Centered Inertial (ECI) to Earth-Fixed 27
    2.2.7 Earth-Fixed to Navigation 31
    2.2.8 Earth-Fixed to Wander–Azimuth 32

vii

- 2.2.9 Navigation Frame to Wander–Azimuth Frame 33
- 2.2.10 Body Frame to Navigation Frame 33
- 2.2.11 Body Frame to Fixed Line-of-Sight (LOS) Transformation 37
- 2.2.12 Doppler Velocity Transformation from Aircraft Axes to Local-Level Axes 39
- 2.2.13 Geodetic to Earth-Centered Earth-Fixed (ECEF) Coordinate Transformation 40
- 2.3 Coordinate Frames for INS Mechanization 41
  - 2.3.1 North-Slaved System 45
  - 2.3.2 Unipolar System 45
  - 2.3.3 Free-Azimuth System 45
  - 2.3.4 Wander–Azimuth System 46
- 2.4 Platform Misalignment 48
- 2.5 Quaternion Representation in Coordinate Transformations 51
- 2.6 The Universal Transverse Mercator Grid Reference System 68
  - 2.6.1 UTM Grid Zone Description 69
  - 2.6.2 The UTM Equations 74
  - 2.6.3 Computation of Convergence 77
- 2.7 Rotation Vector 79
  - References 82

3 Inertial Sensors

- 3.1 Introduction 83
- 3.2 The Ring Laser Gyro 84
  - 3.2.1 Introduction 84
  - 3.2.2 Laser Gyro Fundamentals 86
  - 3.2.3 Theoretical Background 87
- 3.3 The Ring Laser Gyro Principle 101
  - 3.3.1 Laser Gyro Error Sources 107
  - 3.3.2 Multioscillator Ring Laser Gyros 113
  - 3.3.3 Alternate RLG Designs: Passive Laser Gyros 117
- 3.4 Fiber-Optic Gyros 118
- 3.5 The Ring Laser Gyro Error Model 123
- 3.6 Conclusion 130
  - References 132

4 Kinematic Compensation Equations

- 4.1 Introduction 135
- 4.2 Rotating Coordinates; General Relative Motion Equations 137
- 4.3 General Navigation Equations 140

Contents   ix

    4.4 Standard Mechanization Equations   155
        4.4.1   Latitude–Longitude Mechanization   157
        4.4.2   Wander–Azimuth Mechanization   161
        4.4.3   Space-Stabilized Mechanization   187
    4.5 The Vertical Channel   197
        4.5.1   Physics of the Atmosphere   199
        4.5.2   The Central Air Data Computer (CADC)   205
        4.5.3   Vertical Channel Damping   209
        4.5.4   Altitude Divergence   223
        4.5.5   Altitude Calibration   227
        References   227

5 Error Analysis

    5.1 Introduction   229
    5.2 Development of the Basic INS System Error Equations   231
    5.3 General INS Error Equations   241
        5.3.1   Position Error Equations   242
        5.3.2   Velocity Error Equations   245
        5.3.3   Attitude Error Equations   249
        References   268

6 Externally Aided Inertial Navigation Systems

    6.1 Introduction   269
    6.2 Navaid Sensor Subsystems   272
        6.2.1   Radar   274
        6.2.2   Tactical Air Navigation (TACAN)   277
        6.2.3   Long Range Navigation (Loran)   281
        6.2.4   Omega   290
        6.2.5   The Navstar–Global Positioning System (GPS)   296
        6.2.6   Very-High-Frequency Omnidirectional Ranging (VOR)   337
        6.2.7   Distance Measurement Equipment (DME)   339
        6.2.8   VOR/TACAN (VORTAC)   341
        6.2.9   Joint Tactical Information Distribution System (JTIDS) Relative Navigation (RelNav)   341
        6.2.10   Visual Flyover   345
        6.2.11   Forward-Looking Infrared (FLIR) Updates   346
        6.2.12   Terrain Contour Matching (TERCOM)   348
        6.2.13   Star Sightings   357
        6.2.14   Doppler Radar   360
        6.2.15   Indicated Airspeed (IAS)   366

      6.2.16 EM-Log or Speedlog   368
6.3 The Updating Process   369
      References   373

7 Steering, Special Navigation Systems, and Modern Avionics Systems

7.1 Introduction   375
7.2 Steering   376
      7.2.1   Great Circle Steering   385
      7.2.2   Rhumb-Line Navigation   399
7.3 Special Navigation Systems   406
      7.3.1   Dead-Reckoning Navigation   406
      7.3.2   Area Navigation   413
      7.3.3   Attitude-and-Heading Reference Systems   415
      7.3.4   Electronic Combat Systems and Techniques   422
7.4 Modern Avionics Systems   423
      7.4.1   Flight Instruments and Displays   424
      References   441

Appendix A: System Performance Criteria

A.1 Circular Error Probable   443
A.2 Spherical Error Probable   452
A.3 Radial Position Error   453
      References   456

Appendix B: The World Geodetic System

B.1 Summary of Selected Formulas   459
      Reference   460

Index   461

Preface

The purpose of an inertial navigation and guidance system is to determine the geographical position of a vehicle (e.g., aircraft, spacecraft, ship, missile, or helicopter) and to guide the vehicle from one point to another. The basic difference between inertial navigation and other guidance systems, such as radar, sonar, and artificial satellites, is that inertial navigation is completely self-contained and independent of the environment. By self-contained we mean that the system measurement, correctional, and computational elements are an integral part of the system. Thus, the system is free of external influences such as weather, magnetic disturbances, electronic jamming, and distortion. Specifically, this freedom from external influences is achieved by utilizing precision gyroscopes and accelerometers to sense all motion of the vehicle, both linear and rotational, and by combining and operating on the accelerometer outputs with an onboard computer to determine velocity, position, attitude, and correctional data. This information is then delivered to a variety of electromechanical devices in the vehicle to accomplish changes in the vehicle's velocity, position, and attitude. For instance, each time an aircraft changes speed or the direction of flight, it experiences acceleration of some magnitude and in some direction. The function of the accelerometers is to sense and measure the specific force and to produce an acceleration signal for use by the onboard navigation computer. Consequently, the different values of accelerations are integrated with time to produce a continuously corrected velocity signal. The computer also processes the velocity signal to obtain earth reference information, distance traveled, and position data.

Many excellent texts have been written on the subject of inertial navigation and guidance systems. This book has evolved from classroom lecture notes (as do most texts) of a series of graduate courses on inertial navigation systems given by the author at the Air Force Institute of Technology (AFIT), Department of Electrical and Computer Engineering. In the course of this teaching, I have tried to get a feeling of what students do and do not know and to incorporate improvements suggested by many colleagues at AFIT and in industry. In gathering material I have drawn on many sources, but the organization and method of presentation are my own. I have added original material, from my many years of experience (both in industry and government) in the field of inertial navigation and guidance, and many new or revised illustrations and plots. The largest amount of space is devoted to topics that I believe will be useful to most readers or that are not adequately or clearly treated elsewhere in the technical literature. In this book I have tried to present a balanced combination of theory and up-to-date practice and application. Above all, my intent is to serve a broad spectrum of users, from senior undergraduate or graduate students to experienced engineers and avionics systems specialists in industry, the airlines, and government.

The book is organized into seven chapters and two appendices. Chapter 1 is an introduction to inertial navigation, emphasizing fundamental concepts and ideas while at the same time motivating the reader to study these systems. Chapter 2 considers the various coordinate systems and transformations that are necessary in the design and analysis of inertial navigation systems. In particular, this chapter discusses such coordinate systems as the inertial frame, navigation frame, body frame, Euler's angles, coordinate transformations, platform misalignment, quaternions, the universal transverse mercator (UTM) grid reference system, and the rotation vector concept. Chapter 3, titled "Inertial Sensors," briefly covers the theory and physics of the ring laser gyro, the fiber-optic gyro, and the ring laser gyro error model. The conventional inertial sensors (e.g., spinning mass gyroscopes and accelerometers) have been omitted since the literature is replete with discussions of these sensors. Chapter 4 presents the basic kinematic equations, which form the heart of all inertial navigation systems. The attention here is focused on the general navigation equations, standard mechanization equations, wander-azimuth and space-stable mechanization, a discussion of the vertical channel, and altitude divergence. Chapter 5 may be considered as a natural extension of Chapter 4, dealing with the error analysis of inertial systems. Chapter 6 deals with externally aided inertial navigation systems. The navigation aids (or navaids) considered here are the following: radar, TACAN, Loran, Omega, the Navstar-GPS system, VOR and DME, terrain contour matching (TERCOM), the joint tactical information distribution system (JTIDS), Doppler radar, EM-Log, and star trackers. All these navaids are treated in some detail. In the last chapter, Chapter 7, the reader will find useful information on steering, special navigation systems, and modern avionics systems. Of particular importance here are great circle and rhumb-line navigation, dead-reckoning navigation, area navigation, and modern cockpit flight instruments. The book concludes with two appendices. Appendix A provides the reader with methods of evaluating system performance, while Appendix B summarizes the world geodetic system (WGS), its parameters, and its values. References are provided at the end of each chapter. Where necessary, a computer program listing is provided to facilitate the engineer's analysis of his or her particular system design.

It is a pleasure for the author to acknowledge his many colleagues and students who throughout the years have contributed directly or indirectly to this work. In particular, I am grateful to Professor John J. D'Azzo, Head, Department of Electrical and Computer Engineering, Air Force Institute of Technology, for giving me the opportunity to be associated with this fine institution of higher learning, and for the friendship and encouragement that he provided me in all these years. To Professor Cornelius T. Leondes of the University of California, San Diego, I am indebted for the many valuable suggestions in the content and organization of the text. Also, I would like to thank Dr. Daniel J. Biezad, Associate Professor, Department of Aeronautical Engineering, California Polytechnic State University, San Luis Obispo; Dr. Randall N. Paschall, Assistant Professor of Electrical and Computer Engineering at AFIT; and Dr. Brian D. Brumback of General Dynamics Corporation, Fort Worth, Texas, for reading the manuscript and making many invaluable suggestions and corrections. To the editorial staff of Academic Press, in particular Messrs. Dean Irey and William LaDue, the author expresses his appreciation for their unfailing cooperation and for the high standards that they have set and maintained. In every instance, the author has received encouragement and complete cooperation. Finally, but perhaps most importantly, the author wishes to acknowledge an invaluable debt of time to his supportive wife, Karin, and their daughter, Andrea. Their patience throughout this endeavor is deeply appreciated.

Introduction

The navigation problem has existed since humans first began to travel. Seen from a broad viewpoint, going "from here to there" is the overall role of navigation. Navigation may be considered as the art of directing the movement of a vehicle from one place on the earth (or space) to another. In order to calculate course and distance for the next leg of the trip, the navigator must determine the present position either periodically or continuously. Historically, three basically independent types of navigation exist: (1) *Celestial*—present position is computed by measuring the elevation angles, or altitudes, of stars and noting the time of observation; (2) *dead reckoning*—course and distance traveled from the point of departure are maintained by plotting on a chart, or by continuous computation of north–south and east–west components from heading and speed of the vehicle; and (3) *piloting*—landmarks or beacons and the visual pattern on the earth's surface are used, as for taking a ship into port. Traditionally, in celestial navigation the observation consisted of the navigator measuring the star altitude with a sextant. A vertical reference must be available for measuring the altitude; and since the earth rotates, a time reference is required to correlate the observations with known information about the star. Celestial navigation supplements conventional dead reckoning by providing an accurate "fix" periodically to bring the dead reckoning up to date.

With the advent of faster and more advanced aircraft, the solution to the navigation problem requires greater and greater accuracy. In essence, then, the solution to the navigation problem can be resolved by considering the following questions: (1) Where am I? (i.e., present position); (2) How fast am I going? (i.e., velocity); (3) In which direction is my destination?

1

(i.e., relative bearing); and (4) How far is it to my destination? [i.e., distance-to-go (remaining)]. Traditionally, the earliest aircraft were low-speed, low-altitude, fair-weather vehicles. As mentioned above, navigation was mainly accomplished using celestial fixes or by following visual landmarks. However, today's aircraft, flying at altitudes in excess of 12,192 m (40,000 ft) and at speeds over Mach 2, require some other means of navigation. Radio and radar navigation systems are useful for providing position, direction, and average-velocity information. These systems, however, are susceptible to weather conditions and electromagnetic interference. An inertial navigation system (INS), on the other hand, overcomes these difficulties, since it does not require any external device or signal in order to compute solutions to the navigation problem. The INS provides instantaneous outputs of present position, ground speed, true heading, distance-to-go, and relative bearing to the destination.

Inertial navigation systems have undergone a remarkable evolution since the 1940s and early 1950s, progressing from simple dead-reckoning and celestial methods to modern sophisticated techniques such as the global positioning system, stellar–inertial, Doppler radar, Loran, and Omega. That is, the widening scope of INS applications has frequently resulted in the integration of pure inertial systems with additional noninertial navigation equipment or sensors. Thus, the inertial navigation system is used for a precise reference on a continuous basis with external data being used to periodically update the system. The requirement for aircraft inertial navigation systems with long-range flights and global capability has become commonplace for both commercial and military applications. Today, inertial navigation systems are self-contained (i.e., autonomous) and are among the most accurate systems in the field of instrumentation, and are capable of operating for extended periods of time and in all weather conditions. The purpose of an inertial navigation system is therefore to provide, through self-contained equipment, the velocity, position, and attitude of the carrying vehicle in a convenient coordinate system. The system outputs of velocity, position, and attitude, however, are not exact for a number of reasons. The primary reason is the presence of inertial sensor (i.e., gyroscopes and accelerometers) inaccuracies, such as gyroscope drift rates and accelerometer errors. Another important reason is the error due to gravity modeling. As a result, these system error sources drive the system errors, causing unbounded velocity, position, and attitude errors. Within a short period of time, the navigation system errors become excessive for the majority of missions. Specifically, INS errors grow with time in the so-called free inertial mode, but are predictable from an estimate at a given point in time. Consequently, in order to avoid this system error growth, in aircraft or ships operating over extended periods of time, the inertial navigation system is provided with auxiliary information from noninertial external sensors (also known as *NAVAIDs*), such as Doppler radar, Loran, Omega, star trackers, and navigation satellites. In general, the errors in the noninertial sensors do not increase with time. Therefore, system errors can be bounded and reduced by use of (1) high-quality inertial sensors, (2) a good gravity model, and (3) noninertial (i.e., redundant) sensors and the mixing of information from these sensors with the INS information in order to obtain the combination of short-term accuracy of the inertial instruments and the long-term accuracy of noninertial (NAVAID) sensors.

Generally, the accuracy of any inertial navigation system depends on how well the system is preconditioned and aligned. Furthermore, since the navigation system is dependent on its initial inputs (or initial conditions), it follows that the more accurately the present position coordinates and heading are known and inserted in the onboard navigation computer, the more accurate the navigation performance will be. The modern concept of inertial navigation and guidance systems has been made possible by the development of the precision gyroscope. The gyroscope remains the heart of an inertial navigation system, with system performance being directly related to the ability of the gyroscope to provide a precision inertial reference frame, in which to reference the accelerometers used to measure the forces acting on the carrying vehicle, as the vehicle moves from an initial position A to a position B. More specifically, accelerometers measure the components of specific force (acceleration plus gravity) in a reference frame defined by the gyroscopes. Thus, inertial navigation systems have the ability to maintain a reference frame in which the combined effects of inertial acceleration and gravity are resolved. The initial orientation of the reference coordinate frame, initial position, and velocity are required for the navigation system to determine the future orientation, velocity, or position. A priori knowledge of the gravitational field allows the navigator to deduce the kinematic inertial acceleration, which, when integrated once, provides velocity. A second integration determines the position of the vehicle. In essence, the onboard navigation digital computer converts the incremental accelerometer outputs into actual velocity and position, maintains a set of steering laws, and generates attitude error signals used to control the vehicle.

Modern inertial navigation systems for aircraft, marine applications, missiles, helicopters, and spacecraft applications consist of an inertial measurement unit (IMU), a digital processor, and navigation algorithms. Gimbaled inertial navigation systems use a stable platform with three or more gimbals, which is kept either nonrotating with respect to an inertial frame, or establishes a reference frame which precesses from the inertial frame at a known rate. Moreover, the typical stable platform consists of three single-degree-of-freedom gyroscopes, or two 2-degree-of-freedom gyroscopes, three linear accelerometers, and their associated electronics. The gyroscope input axes establish a coordinate frame on the platform. These gyroscopes detect any angular deviation of the platform with respect to inertial space. The gyroscope outputs are used to drive the gimbal torque motors in such a way as to maintain the platform in alignment with the inertial coordinate frame, regardless of the vehicle orientation. Thus, the accelerometer outputs are coordinatized in that frame, and navigation information can easily be generated by performing computations in the instrumented coordinate frame. Furthermore, the vehicle orientation with respect to that coordinate frame can be determined by observing the angles formed between the various gimbals. If desired, an earth-fixed reference frame can be maintained by providing either a bias signal to the gyroscope torque generators to maintain a known precession rate of the stable platform with respect to the inertial frame or additional gimbaled freedom to maintain both the inertial and earth-fixed frames simultaneously. A gimbal readout system therefore gives the vehicle attitude with respect to the desired reference frame. The gimbals isolate the stable platform from vehicle rotations and tend to damp the vibrational environment to which the inertial sensors are exposed, except at the mechanical gimbal resonant frequencies (typically 100 to 200 Hz). The complete inertial measurement unit is generally mounted on a navigation base that further isolates vibrational inputs from the inertial sensors.

Computer processing power and throughput has increased with every generation of equipment. In the not-too-distant future, embedded computers will provide vital intelligence for military weapons systems. Smart and brilliant weapons* systems currently being developed require new computer architectures in order to provide robust and efficient computing capability for use with next-generation languages. Furthermore, these new weapons systems must have the capability for dealing effectively with the complexities and uncertainties of vast amounts of information in the real world. Artificial-intelligence techniques promise to aid this task considerably. In the field of communications, it is safe to say that requirements between elements of a system including the sensors, computer, human interface, weapons, navigation, electronic warfare, and intelligence, also tend to increase rapidly with technological advancements.

Standardization of inertial navigation systems is another important area that is often addressed. Standardization of avionics subsystems between the military services has been the goal of the Department of Defense (DoD). Cost savings from a single procurement specification, the availability of multiple competitive sources, and the advantages of volume production, logistics, and maintenance are obvious and have long been recognized. Commercial airlines standardize for the same reasons. In the mid-1970s, the USAF Standardization Program (*Form*, *Fit*, and *Function*) was initiated in response to Congressional tasking in order to reduce the proliferation of

---

*For example, Strategic Defense Initiative (SDI) weapons systems.

system types and derive maximum benefit from such systems. In addition, the idea behind the standardization was to improve reliability and reduce the acquisition and life cycle of inertial navigation systems. Therefore, cost of procurement, flexibility, maintainability and reliability, and life-cycle cost (LCC) are some of the driving factors influencing this trend. The next generation of precision INS will undoubtedly be highly accurate and more reliable and also will have shorter reaction times and reduced LCC, power, volume, and weight. This will come about by the use of highly accurate inertial sensors such as ring laser gyros and fiber-optic gyros, better gravity models, and efficient navigation error models and mechanization algorithms. Very-high-speed integrated circuits (VHSICs), parallel processing, and fault-tolerant systems will also play an important role in the next generation of inertial navigation system design. Current generation of INSs are of high quality and performance, typically better than 1 km/h circular error probable (CEP) rate.

1.1 ORGANIZATION OF THE TEXT

The material of the book is divided into seven chapters and two appendices. After a short introduction of basic concepts, Chapter 1 discusses inertial navigation and its evolution from a simple dead-reckoning system. Chapter 2 is devoted to the various coordinate systems and coordinate transformations that are available to the inertial system designer, as well as a discussion of quaternions and the Universal Transverse Mercator grid system. Chapter 3 presents an in-depth discussion of such state-of-the-art inertial sensors as the ring laser and fiber-optic gyros. Since the literature is replete with the theory and design of mechanical gyroscopes and accelerometers, their treatment has been omitted here. Chapter 4 is concerned with the inertial system kinematic equations. Among the topics treated in this chapter are rotating coordinates, the general navigation equations, standard mechanization equations, the vertical channel, and altitude divergence. Chapter 5 may be considered as a natural extension of Chapter 4, and deals with the error analysis of inertial systems. Error analysis is an integral part of any design and/or analysis process. Chapter 6, entitled "Externally Aided Inertial Navigation Systems," treats in-depth the various aids available in augmenting and thence improving the performance of inertial systems. The last chapter, Chapter 7, is devoted to steering and special navigation systems such as area navigation and attitude and heading reference systems. Modern, state-of-the-art as well as future flight instruments are also discussed. Appendix A discusses system performance criteria, namely, the circular error probable, spherical error probable, and radial position error. These are important concepts in assessing and/or evaluating the inertial navigation system's performance. Finally, Appendix B presents the World Geodetic System (WGS-84) parameters. The treatment of this book is at the senior-undergraduate or first-year graduate level. Also, this book is intended for engineers who need to design and analyze modern avionics systems, as well as those engaged in the theory and conceptual studies of aircraft and missile avionic systems. The mathematical background expected of the reader is a year course in calculus, ordinary differential equations, vector and matrix analysis, and some familiarity with modern control and estimation theory.

2. Coordinate Systems and Transformations

2.1. COORDINATE SYSTEMS

Several reference coordinate systems (or frames) are available to the inertial navigation system designer for use in navigation computations of position, velocity, and attitude. The choice of the appropriate coordinate frame will depend, to a large degree, on the mission requirements, ease of implementation, computer storage and speed, and navigation equation complexity. Expressing acceleration, velocity, and position as vector quantities allows the use of vector operations to express relationships between system variables. These vectors have no meaning, however, unless expressed with respect to some preselected reference coordinate system. For example, when determining a vehicle's motion, it becomes necessary to relate the solution to the moving earth. This can be accomplished by first defining a convenient inertial coordinate system, with respect to the earth, and then determining the motions of both the vehicle and the earth with respect to this reference frame. Thus, the definition of a suitable inertial coordinate system requires a knowledge of the motion of the earth. The initial orientation of the reference coordinate frame, initial position, and initial velocity are therefore required for the navigation system in order to determine any future orientation, position, or velocity.

The choice of a coordinate system for a specific mechanization depends on many considerations. For example, in many inertial navigation systems latitude ϕ, longitude λ, and altitude h are the desired outputs, and consequently the system should be mechanized to yield these outputs directly. There are generally six fundamental coordinate frames of interest for expressing motion relative to some frame of reference: (1) true inertial, (2) earth-centered inertial, (3) earth-centered earth-fixed, (4) navigation, (5) body, and (6) wander–azimuth frames. Unless specifically defined to the contrary, these coordinate frames are orthogonal, right-handed, Cartesian frames and differ in the location of the origin, the relative orientation of the axes, and the relative motion between the frames. The characteristics and relationship of these coordinate frames will now be defined. The interrelationship between the various coordinate systems is illustrated in Fig. 2-1.

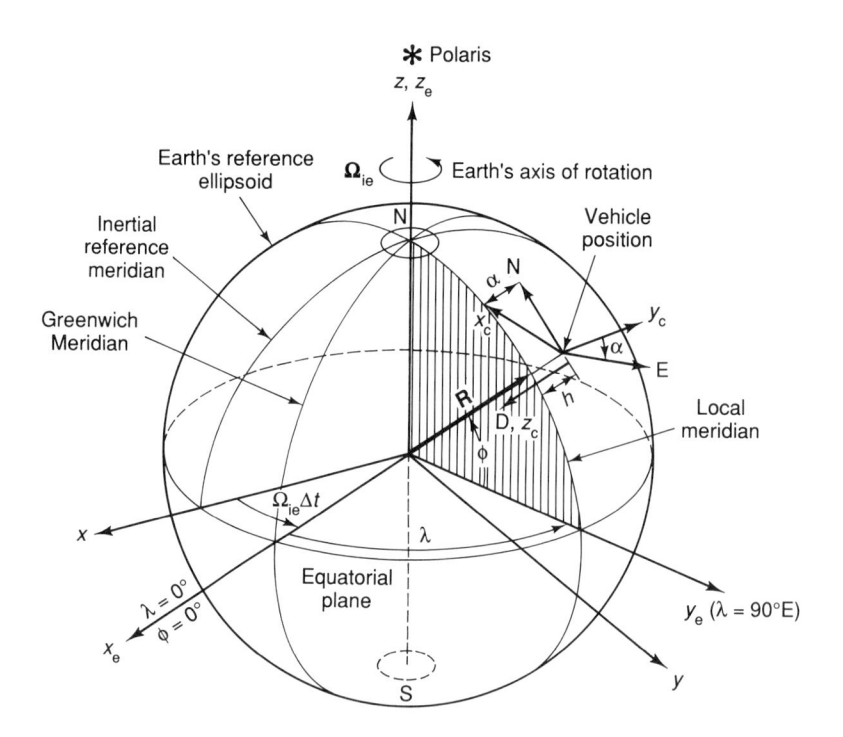

x_i, y_i, z_i or x, y, z = inertial

x_e, y_e, z_e = earth-fixed

N, E, D or x_n, y_n, z_n = geographic (or navigational frame)

x_c, y_c, z_c = wander–azimuth (or computational frame)

FIGURE 2-1

Coordinate-frame geometry.

2.1.1. True Inertial Frame (I)

The true inertial frame is the only reference frame in which Newton's laws of motion are valid. Newton's laws are also valid in Galilean frames (i.e., those that do not rotate with respect to one another and which are uniformly translating in space). This is a larger class than one whose absolute motion is zero; however, they too are impractical for use as practical reference frames. Newton assumed that there was a frame of reference whose absolute motion was zero. Furthermore, Newton considered such an inertial frame (fixed relative to the stars) to be one of absolute zero motion, and his laws of motion to be valid when referred to such a reference frame. Since Newton's time, controversies regarding the existence of such a reference frame of absolute zero motion led to the formulation of the theories of relativity for which Newtonian mechanics is a special case. The true inertial frame consists of a set of mutually perpendicular axes that neither accelerate nor rotate with respect to inertial space. The true inertial frame is not a practical reference frame; it is used only for visualization of other reference frames.

2.1.2. Earth-Centered Inertial (ECI) Frame (i)

This basic frame has its origin at the earth's center of mass and is nonrotating relative to the inertial space (i.e., the "fixed stars"). However, this frame accelerates with respect to inertial space since it moves with the earth. As the earth rotates and moves about the sun, the inertial frame appears to an earth-fixed observer to be rotating at a rate that is the combination of the earth's rotational rate (Ω) and the earth's position about the sun (Julian Day). At the start of the navigation mode, the x_i–y_i axes of this frame lie in the earth's equatorial plane with the x_i axis typically defined toward a star and the z_i axis is aligned with the earth's spin axis. For this reason, it is also called earth-centered inertial (ECI) frame. It should be pointed out that this frame does not rotate with the earth. Theoretically, axes that are fixed to the earth are not inertial axes per se, because of the various modes of motion that the earth exhibits relative to the "fixed space." The most important of these noninertial influences are (1) daily rotation of the earth about its polar axis, (2) monthly rotation of the earth–moon system about its common mass or barycenter, (3) precession of the earth's polar axis about a line fixed in space, (4) motion of the sun with respect to the galaxy, and (5) irregularities in the polar precession (or nutation). The inertial frame is important in that Newton's laws are approximately correct in this frame, and that for vehicles navigating in the vicinity of the earth, computations of specific force are performed in this frame.

2.1.3. Earth-Centered Earth-Fixed (ECEF) Frame (e)

The earth frame (also known as the *geocentric frame*), like the inertial frame, has its origin at the earth's center of mass. This frame, however, rotates with the earth, and coincides with the inertial frame once every 24 h (actually once every sidereal day). The rotation of the earth with respect to the ECI frame, which is $\Omega_{ie}\Delta t$, is about the same axis and in the same sense as the longitude. The z_e axis is directed north along the polar axis, while the x_e, y_e axes are in the equatorial plane with the x_e axis directed through the Greenwich Meridian (0° latitude, 0° longitude), and the y_e axis is directed through 90° east longitude.

2.1.4. Navigation Frame (n)

The navigation frame (also called the *geographic frame*)* has its origin at the location of the inertial navigation system. This is a local-level frame with its $x_n - y_n$ axes in a plane tangent to the reference ellipsoid (see Appendix B) and the z_n axis perpendicular to that ellipsoid. Typically, the x_n axis will point north, the y_n axis east, and the z_n axis down (or up) depending on the coordinate convention the system designer selects. It should be pointed out that the largest class of inertial navigation systems is the local-level type, where the stable platform is constrained with two axes in the horizontal plane. In the past, many INSs have been built using the local-level mechanization, mainly due to the error compensation simplifications of maintaining constant platform alignment to the gravity vector. Historically, the early local-level INSs were of the north-slaved type, but use of this conventional geographic set of axes leads to both hardware and computational difficulties in operation at the polar regions. As we shall see in Section 2.3, different ways or mechanizations have been devised for avoiding the singularity occurring in the polar regions. Each mechanization corresponds to a different platform azimuth control.

2.1.5. Body Frame (b)

This coordinate frame has its origin at the vehicle (e.g., aircraft, ship) center of mass. This is a convenient coordinate system for developing the equations of motion of a vehicle, since the vehicle equations of motion are normally written in the body-axes coordinate frame. In aircraft applications, the convention is to choose the x_b axis pointing along the aircraft's longitudinal axis (the roll axis), the y_b axis out to the right wing (the pitch axis), and the z_b axis pointing down (the yaw axis). The body axis is typically used in strapdown (or analytic) systems.

---

*Some authors refer to the navigation frame as the "vehicle carried" vertical frame.

2.1.6. Wander–Azimuth Frame (c)

This coordinate frame is a special case of the navigation frame, also known as the *computational frame,* has its origin at the system's location, and is coincident with the origin of the navigation frame. The horizontal axes of this (x_c, y_c, z_c) local-level geodetic wander–azimuth frame lie in a plane tangent to the local vertical. Furthermore, this frame is maintained "locally level" and is defined with respect to the earth frame by three successive Eulerian (see Section 2.2.3) angle rotations (longitude λ, geodetic latitude ϕ, and wander angle α). Latitude is defined to be positive in the northern hemisphere, and the wander angle α is defined to be positive west of true north and measured in the geodetic horizon plane. If the wander angle α is zero, this frame will be aligned with the navigational (or geographic) frame. Also, it will be noted that when the geodetic latitude, longitude, and wander angle are zero, the (x_c, y_c, z_c) axes will be aligned with the (x_e, y_e, z_e) axes of the earth-fixed frame. Because of its importance, we have chosen to define this coordinate frame since many of the present-day inertial navigation systems are mechanized in this frame.

It should be pointed out that in most work in inertial navigation an ellipsoidal (ellipsoid of revolution) earth is assumed. Therefore, a set of geodetic (or geographic) coordinates must be defined. Geodetic coordinates are earth-fixed parameters which are defined in terms of the earth reference ellipsoid as follows: (1) the geodetic longitude is positive east of the Greenwich Meridian ($\lambda = 0°$), measured in the reference equatorial plane, (2) the geodetic latitude is positive north measured from the reference equatorial plane to the ellipsoidal surface passing through the point of interest (usually present position), and (3) the altitude h above the reference ellipsoid measured along the normal passing through the point of interest (in case of a surface vehicle, such as a ship, h will, of course, be zero). Pilots and/or navigators most readily interpret positional information in terms of geographic latitude ϕ, geographic longitude λ, and altitude h.

The local-level terrestrial navigation reference frame was defined as one having one axis along the radial direction, with the other two axes being perpendicular to the first, defining the level plane. Specifically, this system instruments the local geographic frame. Furthermore, in this mechanization the inertial navigation unit's (INU) platform axes are commanded into alignment with a right-handed, local north-east-down (NED) coordinate system. The precessional rate of the local-level frame about two level axes is completely defined by the requirement that these axes remain at all times horizontal. Moreover, the precessional rate about the vertical axis, which is normal to the reference ellipsoid, is dependent on the particular navigation equation implementation and/or mechanization. For example, in a conventional NED mechanization, the vertical axis is precessed at a rate which keeps the two level axes pointing north and east at all times. However, this leads to a problem if one of the earth's poles is traversed, in which case the required vertical precessional rate becomes infinitely large. As a practical matter, this means that the north–east navigator is limited to latitudes at which the allowable precessional rate of its vertical axis will not be exceeded and, therefore, does not have an all-earth navigation capability. In spite of this fact, and as mentioned in Section 2.1.4, this type of local-level system has been used in a great number of inertial navigation systems. At this point, it is very important to note that there is no uniformity among inertial navigation system designers in choosing one coordinate frame over another. The selection of one frame over another depends on such factors as onboard navigation computer storage and throughput, mission requirements, etc. For example, one designer might choose a local-level navigation frame with the x axis pointing north, the y axis pointing east, and the z axis pointing up, while another designer may select a coordinate frame in which the x axis points north, the y axis west, and the z axis up. Still another operational system uses a coordinate frame with the z axis pointing north, the y axis east, and the x axis up. In subsequent sections we will be using these coordinate frames interchangeably. The transformations relating the various coordinate systems will be considered in Section 2.2.

2.1.7. Platform, Accelerometer, and Gyroscope Frames

In addition to the preceding coordinate frames, three more coordinate frames are often used for measurements made by the inertial instruments with regard to the above defined frames: (1) platform, (2) accelerometer, and (3) gyroscope frames. These frames are then defined as follows.

2.1.7.1. Platform Frame (p)

This right-handed, orthogonal coordinate frame is defined by the input axes of inertial sensors (typically, the input axes of the gyroscopes) and has its origin at the system location (INS), with its orientation in space being fixed. Moreover, this reference coordinate frame is a function of the configuration and mechanization of the particular inertial navigator under design. In fact, as we shall see in Section 2.3, the three classical mechanizations of an inertial navigation system (INS) are based on the platform frame coinciding with one of the fundamental coordinate frames discussed above. For example, if the platform frame coincides with the body frame, we have a strapdown system, whereas if the platform frame coincides with the inertial frame, we have a space-stable system.

2.1.7.2. Accelerometer Frame (a)

This frame is a nonorthogonal frame defined by the input or sensitive axes of the instruments mounted on the inertial platform.

2.1.7.3. Gyroscope Frame (g)

This frame, like the accelerometer frame, is a nonorthogonal frame defined by the input or sensitive axes of the instruments mounted on the inertial platform.

2.2. COORDINATE TRANSFORMATIONS

The description of the position and orientation of a rigid body with respect to a right-handed Cartesian reference frame, say, $\{R\} = \{O_R; x_R, y_R, z_R\}$, is the subject of kinematics. Consequently, in order to describe the motion of a rigid body, a Cartesian frame, say, $\{E\} = \{O_E; x_E, y_E, z_E\}$, is attached to it as illustrated in Fig. 2-2. At any instant of time, the following apply: (1) the position of the rigid body is defined by a 3×1 position vector $(O_R O_E) = [p_x \; p_y \; p_z]^T$ of the origin O_E of the rigid body and (2) the orientation of the rigid body is defined by a 3×3 rotation matrix, whose unit vectors $1_{xE}, 1_{yE}, 1_{zE}$ describe the axes of the rigid-body frame.

Here we assume that $O_E \equiv O_R \Leftrightarrow p_x = p_y = p_z = 0$. The rotation matrix approach utilizes nine parameters, which obey the orthogonality and unit length constraints, to describe the orientation of the rigid body. Since a rigid body possesses three rotational degrees of freedom, three independent parameters are sufficient to characterize completely and unambiguously its orientation. For this reason, three-parameter representations are popular in engineering because they minimize the dimensionality of the rigid-body control problem. This section is concerned with several methods of describing

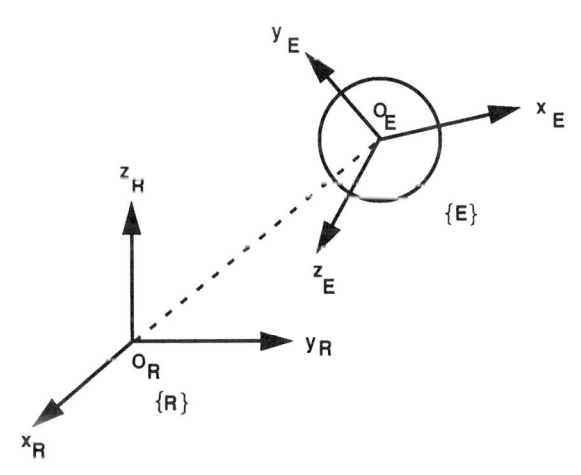

FIGURE 2-2
Definition of Cartesian and rigid-body reference frames.

the orientation of a rigid body in space. The transformation of coordinate axes is an important aspect of inertial navigation systems. Moreover, the transformation of coordinate axes is an important necessity in resolving angular positions and angular rates from one coordinate system to another. The transformation between the various coordinate systems can be expressed in terms of the relative angular velocity between them. Stated another way, the transformation from one frame to another can be accomplished through an intermediate frame or frames. Specifically, coordinate frames are related to each other by coordinate transformations, which are the direction cosines of one coordinate frame with respect to the other, arranged as an orthogonal matrix. Mathematically, the orientation of a right-handed Cartesian coordinate frame with reference to another one is characterized by the coordinate transformation matrix, which maps the three components of a vector, resolved in one frame, into the same vector's components resolved into the other frame. There are several methods to perform a required transformation. Among these methods, the following are noteworthy: (1) direction cosine matrix, (2) Euler angles, (3) rotation vector, (4) 4×4 robotic method of rotation and displacement, and (5) quaternions. The direction cosine matrix and the quaternions are classified as four-parameter transformations, while the Euler angles and the rotation vector are classified as three-parameter transformations. The significance of the three versus four parameters is that all known three-parameter transformations contain singularities at certain orientations, while the four-parameter transformations have no such limitations. In certain applications, it is necessary or advantageous to use more than one transformation. This becomes necessary in view of the fact that in many applications the inertial system designer must use more than one coordinate frame.

2.2.1. The Direction Cosine Matrix (DCM)

The direction cosine matrix (DCM) is widely used in aerospace applications, and plays an important role in the design of inertial navigation systems. As a result, the most commonly employed method for generating the coordinate transformation matrix is based on the computation of direction cosines between each axis of one frame and every axis of another one. This is done by calculating the vector dot products between the axes. In order to demonstrate this fact, consider two unit vectors $\mathbf{1}_i^a$ and $\mathbf{1}_j^b$. From vector analysis, it can be shown that the projection of two unit vectors is simply a scalar trigonometric function of the angle α_{ij} between these two vectors. Thus

$$\mathbf{1}_i^a \cdot \mathbf{1}_j^b = |\mathbf{1}_i^a||\mathbf{1}_j^b| \cos \alpha_{ij} = \cos \alpha_{ij}$$

Specifically, the angle between two vectors $\mathbf{r}_1$ and $\mathbf{r}_2$ can be obtained from the following diagram. From trigonometry, the cosine of θ can be determined

2.2. Coordinate Transformations

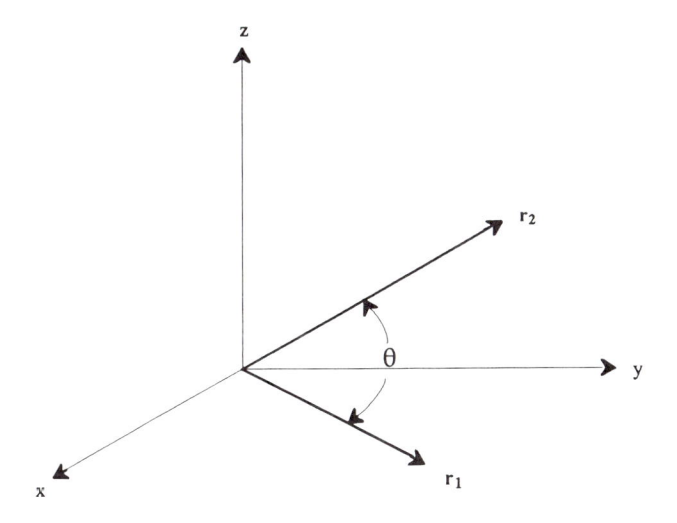

from the projection of $\mathbf{r}_2$ on $\mathbf{r}_1$, divided by the magnitude of $\mathbf{r}_2$, or the projection of $\mathbf{r}_1$ on $\mathbf{r}_2$, divided by the magnitude of $\mathbf{r}_1$. Using the former definition, the projection of $\mathbf{r}_2$ on $\mathbf{r}_1$ is given by the expression

$$\frac{\mathbf{r}_2^T \mathbf{r}_1}{|\mathbf{r}_1|} = \frac{\mathbf{r}_2^T \mathbf{r}_1}{(\mathbf{r}_1^T \mathbf{r}_1)^{1/2}}$$

and the angle between the two vectors is

$$\theta = \cos^{-1}\left[\frac{\mathbf{r}_2^T \mathbf{r}_1}{(\mathbf{r}_1^T \mathbf{r}_1)^{1/2}(\mathbf{r}_2^T \mathbf{r}_2)^{1/2}}\right]$$

Note that any point in space can be located by the position vector $\mathbf{r} = [x \ y \ z]^T$, while its magnitude $|\mathbf{r}|$ is given by

$$|\mathbf{r}| = (\mathbf{r}^T \mathbf{r})^{1/2} = \left\{ [x \ y \ z] \begin{bmatrix} x \\ y \\ z \end{bmatrix} \right\}^{1/2}$$

$$= (x^2 + y^2 + z^2)^{1/2}$$

Finally, we note that the scalar measure of length (or metric) is called a *Euclidean* or *quadratic norm*; thus, the n-dimensional vector $\mathbf{x}$ can be expressed as

$$\|\mathbf{x}\| = (\mathbf{x}^T \mathbf{x})^{1/2} = (x_1^2 + x_2^2 + \cdots + x_n^2)^{1/2}$$

Next, the transformation matrix between two frames, say, frames a and b, results in an array of nine direction cosines as follows:

$$C_a^b = \begin{bmatrix} C_{11} & C_{12} & C_{13} \\ C_{21} & C_{22} & C_{23} \\ C_{31} & C_{32} & C_{33} \end{bmatrix}$$

Therefore, the direction cosine matrix C_a^b transforms (or rotates) a vector in R^3 from one frame into another. For example, the transformation from the inertial (i-frame) to the earth-centered earth-fixed (e-frame) can be denoted mathematically by the relation [9]*

$$\mathbf{R}^e = C_i^e \mathbf{R}^i \qquad (2.1)$$

where C_i^e is a 3×3 direction cosine matrix that transforms a mathematical vector coordinatized in the e-frame to the equivalent vector coordinatized in the e-frame. Each element C_{ij} of the DCM represents the cosine of the angle or a projection between the ith axis of the i-frame and the jth axis of the e-frame. An important property of the DCM is

$$\mathrm{Det}(C_a^b) \equiv |C_a^b| = 1$$

Also, for orthogonal systems, the orthogonality property of a matrix is satisfied by the identity

$$(C_a^b)^{-1} = (C_a^b)^T = C_b^a \qquad (2.2)$$

Specifically, in order to illustrate the transformation from one coordinate frame to another, consider two points having orthogonal coordinates (x, y, z) and (X, Y, Z), respectively; the former coordinate is in frame a and the latter in frame b. Mathematically, these two coordinate systems are related by the equation [4]

$$\begin{bmatrix} x \\ y \\ z \end{bmatrix}^a = C_b^a \begin{bmatrix} X \\ Y \\ Z \end{bmatrix}^b \qquad (2.3)$$

or as indicated by Eq. (2.1), we can write this equation in a more compact form as

$$\mathbf{R}^a = C_b^a \mathbf{R}^b$$

where, as before, the elements of C_b^a are the direction cosines and $\mathbf{R}$ is any vector. The matrix C_b^a, also known as a *rotation matrix*, represents the direction cosines of the b-system axes with respect to the a-system axes. In other words, the matrix C_b^a may be regarded as a coordinate transformation matrix that contains the relative orientation between the two coordinate frames. Furthermore, the DCM projects the vector $\mathbf{R}^b$ into the a-system.

*Bracketed numbers correspond to entries in References list.

2.2. Coordinate Transformations

Let us now consider the geometric interpretation of the direction cosine matrix for two coordinate frames a and b. The DCM is, by definition, the projection of each a-frame unit vector onto each unit vector of the b-frame. To this end, let a vector **R** coordinatized in the reference frame a be denoted $\mathbf{R}^a$. Next, let another vector **R** coordinatized in the b-frame be denoted $\mathbf{R}^b$. These two vectors can then be expressed as

$$\mathbf{R}^a = (\mathbf{R}^T \mathbf{1}_x^a)\mathbf{1}_x^a + (\mathbf{R}^T \mathbf{1}_y^a)\mathbf{1}_y^a + (\mathbf{R}^T \mathbf{1}_z^a)\mathbf{1}_z^a$$

$$\mathbf{R}^b = (\mathbf{R}^T \mathbf{1}_x^b)\mathbf{1}_x^b + (\mathbf{R}^T \mathbf{1}_y^b)\mathbf{1}_y^b + (\mathbf{R}^T \mathbf{1}_z^b)\mathbf{1}_z^b$$

where for $i = x, y, z$, each $\mathbf{R}^T \mathbf{1}_i^a$ is the scalar component of the vector **R** projected (dot product) along the ith a-frame coordinate direction. A similar analogy exists for the $\mathbf{R}^b$ vector in the b-frame. Now we can relate the unit vectors $\mathbf{1}_i^a$ to the unit vectors $\mathbf{1}_j^b$ ($i, j = x, y, z$) as follows [9]:

$$\mathbf{1}_i^b = (\mathbf{1}_i^{bT} \mathbf{1}_x^a)\mathbf{1}_x^a + (\mathbf{1}_i^{bT} \mathbf{1}_y^a)\mathbf{1}_y^a + (\mathbf{1}_i^{bT} \mathbf{1}_z^a)\mathbf{1}_z^a$$

Thus, the ith component of $\mathbf{R}^b$, being $\mathbf{R}^T \mathbf{1}_i^b$, can be expressed as

$$\mathbf{R}^T \mathbf{1}_i^b = \mathbf{R}^T[(\mathbf{1}_i^{bT} \mathbf{1}_x^a)\mathbf{1}_x^a + (\mathbf{1}_i^{bT} \mathbf{1}_y^a)\mathbf{1}_y^a + (\mathbf{1}_i^{bT} \mathbf{1}_z^a)\mathbf{1}_z^a]$$

Distributing $\mathbf{R}^T$ inside the brackets and using the scalar property of projections, $A^T B = B^T A$, each ith component of $\mathbf{R}^b$ becomes

$$\mathbf{R}^T \mathbf{1}_i^b = (\mathbf{1}_i^{bT} \mathbf{1}_x^a)\mathbf{R}^T \mathbf{1}_x^a + (\mathbf{1}_i^{bT} \mathbf{1}_y^a)\mathbf{R}^T \mathbf{1}_y^a + (\mathbf{1}_i^{bT} \mathbf{1}_z^a)\mathbf{R}^T \mathbf{1}_z^a$$

Therefore, the entire vector $\mathbf{R}^b$ assembled from its components is

$$\mathbf{R}^b = \begin{bmatrix} \mathbf{R}^T \mathbf{1}_x^b \\ \mathbf{R}^T \mathbf{1}_y^b \\ \mathbf{R}^T \mathbf{1}_z^b \end{bmatrix} = \begin{bmatrix} \mathbf{1}_x^{bT} \mathbf{1}_x^a & \mathbf{1}_x^{bT} \mathbf{1}_y^a & \mathbf{1}_x^{bT} \mathbf{1}_z^a \\ \mathbf{1}_y^{bT} \mathbf{1}_x^a & \mathbf{1}_y^{bT} \mathbf{1}_y^a & \mathbf{1}_y^{bT} \mathbf{1}_z^a \\ \mathbf{1}_z^{bT} \mathbf{1}_x^a & \mathbf{1}_z^{bT} \mathbf{1}_y^a & \mathbf{1}_z^{bT} \mathbf{1}_z^a \end{bmatrix} \mathbf{R}^a = \begin{bmatrix} C_{x_a}^{x_b} & C_{y_a}^{x_b} & C_{z_a}^{x_b} \\ C_{x_a}^{y_b} & C_{y_a}^{y_b} & C_{z_a}^{y_b} \\ C_{x_a}^{z_b} & C_{y_a}^{z_b} & C_{z_a}^{z_b} \end{bmatrix} \mathbf{R}^a$$

$$= C_a^b \mathbf{R}^a$$

$$= [C_{ij}] \mathbf{R}^a$$

which is Eq. (2.1). Figure 2-3 illustrates the coordinatization of the vector **R** in the b-frame [9].

Finally, each element of $C_a^b = [C_{ij}]$ is, as we have seen above, the cosine of the angle between the unit vectors $\mathbf{1}_i^a$ and $\mathbf{1}_j^b$, which gives rise to the name *direction cosine matrix*.

2.2.2. The Direction Cosine Matrix Differential Equation

In navigation and tracking systems, it is necessary to maintain knowledge of the orientation of a rigid body relative to inertial space. This may be done by propagating C, a 3×3 matrix of direction cosines. Consider now the relative rotational motion of the two right-handed Cartesian coordinate

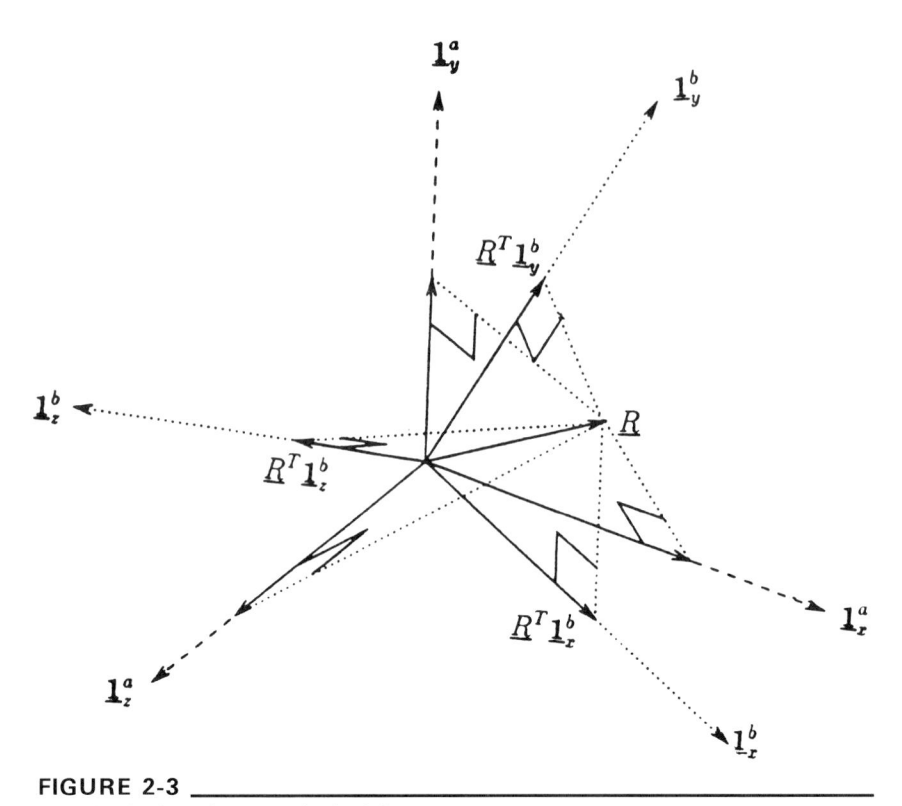

FIGURE 2-3
Coordinatization of vector **R** in the b-frame.

frames a and b discussed in the previous section. For the general rotation of the b-frame (body frame) with respect to the a-frame (the inertial frame) at an angular velocity $\omega_{ab}^{b}(t)$ vector, the relative orientation of these two frames can be described by the direction cosine matrix differential equation. That is, at time t, the a- and b-frames are related via the DCM, $C_b^a(t)$. During the next instant of time Δt, frame b rotates to a new orientation such that the DCM at $t+\Delta t$ is given by $C_b^a(t+\Delta t)$. Thus, by definition, the time rate of change of $C_b^a(t)$ is given by

$$\frac{dC_b^a(t)}{dt} = \dot{C}_b^a(t) = \lim_{\Delta t \to 0} \frac{\Delta C_b^a}{\Delta t} = \lim_{\Delta t \to 0} \frac{C_b^a(t+\Delta t) - C_b^a(t)}{\Delta t} \qquad (2.4)$$

The DCM differential equation is a nonhomogeneous, linear matrix differential equation, forced by the angular velocity vector ω in its skew–symmetric matrix form $\Omega_{ab}^{b} = \Omega_{ab}^{b}[\omega_{ab}^{b}(t)]$. Thus, the direction cosine matrix differential equation is related to the angular velocity matrix, and can be

2.2. Coordinate Transformations

expressed mathematically as [3, 4, 9]

$$\frac{dC_b^a}{dt} = C_b^a \Omega_{ab}^b = C_b^a \begin{bmatrix} 0 & -\omega_3 & \omega_2 \\ \omega_3 & 0 & -\omega_1 \\ -\omega_2 & \omega_1 & 0 \end{bmatrix} \quad (2.5)$$

where the skew-symmetric form of ω_{ab}^b is denoted by Ω_{ab}^b and

$$\omega_{ab}^b = [\omega_1 \quad \omega_2 \quad \omega_3]^T = -\omega_{ba}^b$$

The designation Ω_{ab}^b denotes the angular velocity of the b-frame relative to the a-frame, coordinatized in the b-frame. Equation (2.5) is a linear, matrix differential equation, which can be represented by nine scalar, linear, coupled differential equations, forced by the scalar components of the angular velocity vector ω_{ab}^b. Furthermore, this equation is easily integrated with the initial conditions $C_b^a(t_0)$, where the initial conditions represent the initial orientation of the a-frame with respect to the b-frame. Consequently, the nine scalar differential equations can be written in the form [9]

$$\dot{C}_{i,j} = C_{i,j+1}\omega_{j+2} - C_{i,j+2}\omega_{j+1} \qquad i,j = 1, 2, 3 \quad (2.6)$$

where the second subscript is modulo 3 (if $j+1>3$, then $j+1=j+1-3$; similarly for $j+2$).

In general, Eq. (2.5) can be solved by assuming a solution of the form

$$C(t) = C(t_0) \exp(\Omega t) \quad (2.7)$$

Utilizing the Caley–Hamilton* theorem [6] to solve for the C-matrix as an explicit time function, we have

$$C(t) = C(t_0)\left[I + \left(\frac{\sin \omega t}{\omega}\right)\Omega + \left(\frac{1-\cos \omega t}{\omega^2}\right)\Omega^2\right] \quad (2.8)$$

where I = identity matrix

$$\omega = (\omega_x^2 + \omega_y^2 + \omega_z^2)^{1/2}$$

Equation (2.7) is an important equation used in strapdown inertial navigation systems. The matrix $C(t)$ can also be updated at time intervals ΔT, for $t_k < t < t_k + \Delta T$, using a Taylor series in the form (second-order expansion)

$$C(t_k + \Delta T) = \{I + \Omega(t_k)\,\Delta T + [\Omega^2(t_k) + \dot{\Omega}(t_k)]\,\Delta T^2/2\}C(t_k)$$

Note that for a simple, first-order expansion, the matrix $C(t_k)$ can be updated by the expression

$$C(t_k + \Delta T) = \{I + \Omega(t_k)\,\Delta T\}C(t_k)$$

The Caley–Hamilton theorem: Associated with a matrix **A** is a characteristic equation, namely, $\phi(\lambda) = \det[\lambda \mathbf{I} - \mathbf{A}] = 0$. The Caley–Hamilton theorem states that "any square matrix satisfies its own characteristic equation," that is, $\phi(\mathbf{A}) = 0$.

2.2.3. Euler's Angles

One of the most used methods of specifying the angular orientation of one coordinate system with respect to another is the use of the three Euler angles. A point on a rigid body can be defined in terms of body axes (x, y, z). In the Eulerian representation, a series of three ordered right-handed rotations is needed to coalesce the reference frame with the rigid-body frame. In order to determine the orientation of the body itself, we now introduce Euler's angles (ϕ, θ, ψ), which are three independent quantities capable of defining the position of the (x, y, z) body axes relative to the inertial (X, Y, Z) axes, as depicted in Fig. 2-4. The Euler angles (ϕ, θ, ψ) correspond to the conventional roll–pitch–yaw angles. For example, the heading (yaw) of an aircraft is displayed to the pilot on a heading indicator, such as the heading situation indicator (HSI), while the pitch and roll of the aircraft are indicated on an attitude indicator, such as the attitude direction indicator (ADI).

It should be pointed out, however, that the Euler angles are not uniquely defined, since there is an infinite set of choices. Unfortunately, there are no

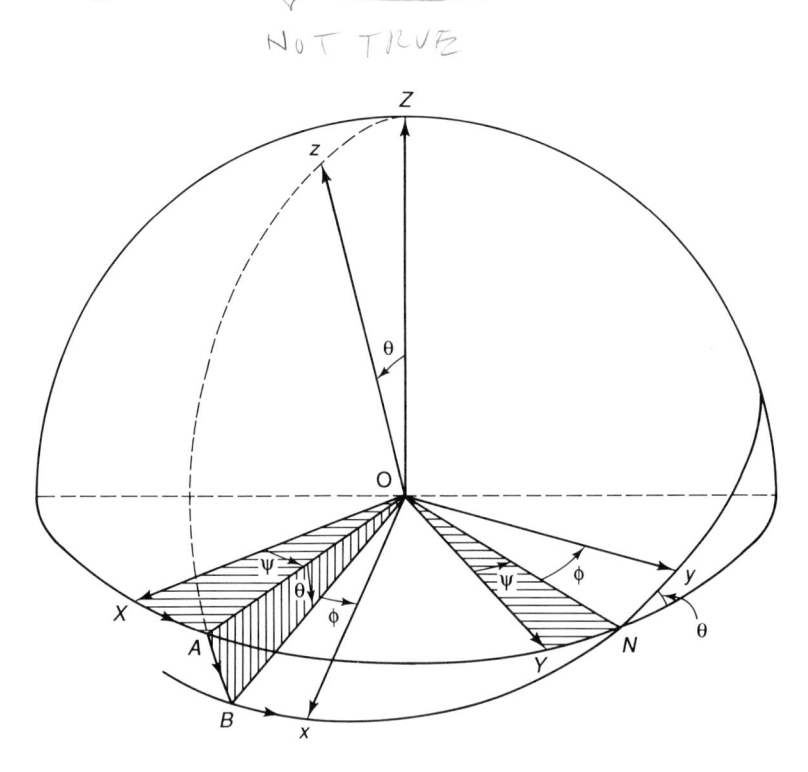

FIGURE 2-4
Body axes defined relative to inertial axes by Euler's angles ϕ, θ, ψ.

2.2. Coordinate Transformations

standardized definitions of the Euler angles. For a particular choice of Euler angles, the rotation order selected and/or defined must be used consistently. That is, if the order of rotation is interchanged, a different Euler angle representation is defined.

As stated above, the position of the body axes can be arrived at by a series of three ordered right-handed rotations. With the (x, y, z) axes coinciding with the (X, Y, Z) axes, a rotation is made of the (X, Y, Z) system about the Z axis through an angle ψ, the yaw or heading angle. This rotation and the subsequent rotations are made in the positive, that is, anticlockwise sense, when looking down the axis of rotation toward the origin. The rotation about Z of ψ results in a new set of axes (X', Y', Z') as shown below. The coordinates of a point, or equivalently a position vector, in the new system can be expressed in terms of the original coordinates by using the rotation matrix [7, 13]

$$\begin{bmatrix} X' \\ Y' \\ Z' \end{bmatrix} = [A] \begin{bmatrix} X \\ Y \\ Z \end{bmatrix} = \begin{bmatrix} \cos\psi & \sin\psi & 0 \\ -\sin\psi & \cos\psi & 0 \\ 0 & 0 & 1 \end{bmatrix} \begin{bmatrix} X \\ Y \\ Z \end{bmatrix} \quad (2.9)$$

We next allow a second rotation to be made by rotating the (X', Y', Z') system about the Y' axis by an angle θ, the pitch angle. The coordinate transformation from the (X', Y', Z')-axis system to the new (X'', Y'', Z'') system is then given by

$$\begin{bmatrix} X'' \\ Y'' \\ Z'' \end{bmatrix} = [B] \begin{bmatrix} X' \\ Y' \\ Z' \end{bmatrix} = \begin{bmatrix} \cos\theta & 0 & -\sin\theta \\ 0 & 1 & 0 \\ \sin\theta & 0 & \cos\theta \end{bmatrix} \begin{bmatrix} X' \\ Y' \\ Z' \end{bmatrix} \quad (2.10)$$

Finally, the third rotation about the X'' axis by the angle ϕ, the roll or bank angle, moves the (X'', Y'', Z'') system to the (x, y, z) system. The transformation equations are

$$\begin{bmatrix} x \\ y \\ z \end{bmatrix} = [D] \begin{bmatrix} X'' \\ Y'' \\ Z'' \end{bmatrix} = \begin{bmatrix} 1 & 0 & 0 \\ 0 & \cos\phi & \sin\phi \\ 0 & -\sin\phi & \cos\phi \end{bmatrix} \begin{bmatrix} X'' \\ Y'' \\ Z'' \end{bmatrix} \quad (2.11)$$

Consequently, when performed in the order ψ, θ, ϕ, (yaw, pitch, roll), the rotation of the (X, Y, Z) system by the Euler angles brings that system to the orientation of the body or (x, y, z) system. The total rotation matrix for expressing the components of a vector in the body system in terms of the components of a vector of the inertial system can be formed by substitution of the above rotations. Thus

$$\begin{bmatrix} x \\ y \\ z \end{bmatrix} = [D][B][A] \begin{bmatrix} X \\ Y \\ Z \end{bmatrix} = [C] \begin{bmatrix} X \\ Y \\ Z \end{bmatrix}$$

$$= \begin{bmatrix} \cos\theta\cos\psi & \cos\theta\sin\psi & -\sin\theta \\ \cos\psi\sin\theta\sin\phi - \sin\psi\cos\phi & \cos\psi\cos\phi + \sin\psi\sin\theta\sin\phi & \cos\theta\sin\phi \\ \cos\psi\sin\theta\cos\phi + \sin\psi\sin\phi & \sin\psi\sin\theta\cos\phi - \cos\psi\sin\phi & \cos\theta\cos\phi \end{bmatrix} \begin{bmatrix} X \\ Y \\ Z \end{bmatrix} \quad (2.12)$$

where the matrix $[C]$ is the product of $[D]$, $[B]$, and $[A]$ in that order in terms of the angles ϕ, θ, ψ. In other words, we have chosen the notation $[C]$ to represent this Euler angle transformation matrix. The orientation of the rigid body is thus completely specified by the three roll–pitch–yaw angles ϕ, θ, ψ.

In the sequel, we will have the occasion to denote the matrix $[C]$ as simply C_a^b. Ranges for the Euler angles are $-\pi \leq \psi \leq +\pi$, $-\pi/2 \leq \theta \leq \pi/2$ and $-\pi \leq \phi \leq +\pi$. The Euler angles can be physically instrumented by a set of gimbal angles in an inertial platform. Figure 2-5 depicts the natural progression of the coordinate-frame transformations. This is intended only as a guide; the inertial navigation system designer must select the coordinate frames which are best suited to the particular mission.

We will now illustrate the various coordinate transformations and their characteristics. In all transformations, the vehicle will be assumed to be

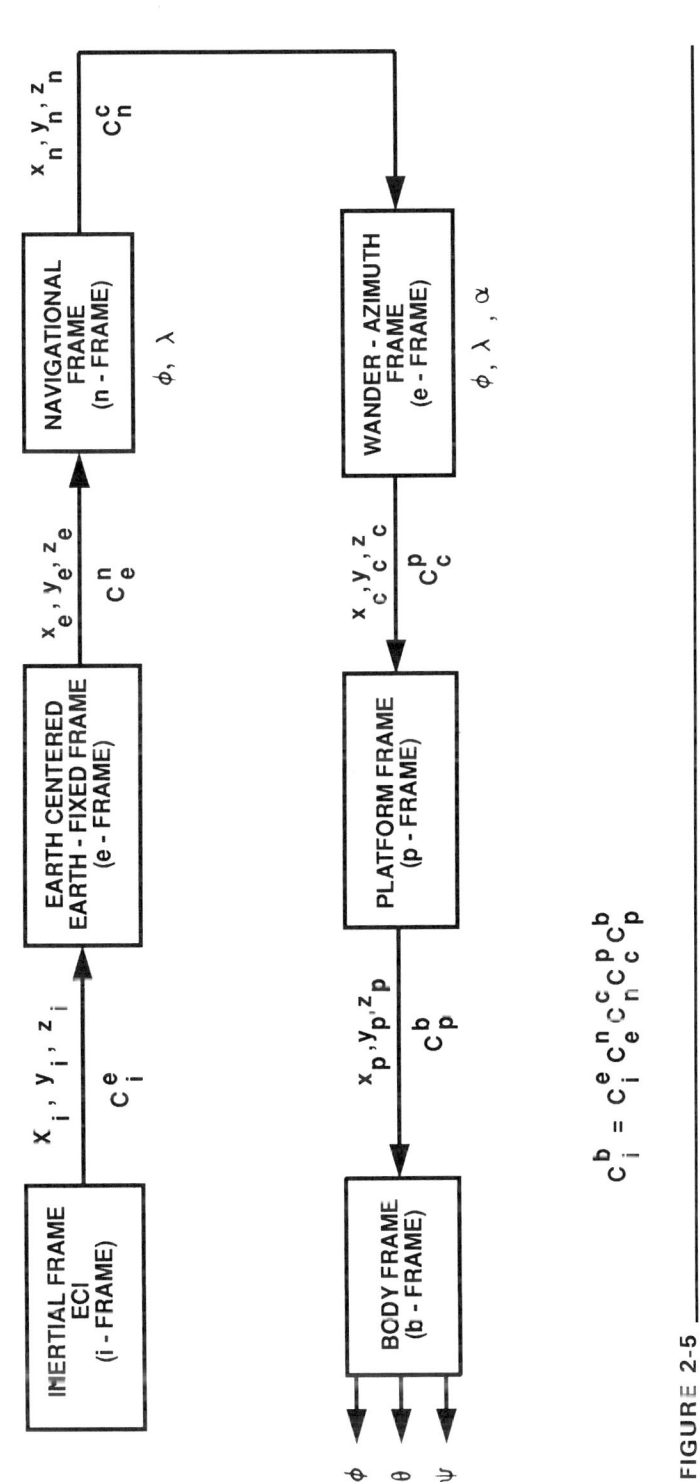

FIGURE 2-5
Transformation sequence.

located on the surface of the earth; that is, the altitude h will be taken to be zero ($h=0$).

2.2.4. Transformation of Angular Velocities

Frequently we need to express the angular velocities $\omega_x, \omega_y, \omega_z$ about the body axes (x, y, z) in terms of Euler angles ψ, θ, ϕ. Since an aircraft normally has some rotational motion, the values of ψ, θ, and ϕ are changing with time. That is, the Euler angles, just as the DCM orientation parameters, vary with time when an input angular velocity vector is applied between the two reference frames. Therefore, it is desirable to express the time derivatives of the Euler angles in terms of the components of the angular velocity of the body-fixed axes. The angular velocity vector ω, in the body-fixed coordinate system, has components p in the x direction, q in the y direction, and r in the z direction. In order to express $\dot\psi, \dot\theta$, and $\dot\phi$ in terms of (p, q, r), we consider each derivative of an Euler angle as the magnitude of the angular velocity vector in the coordinate system in which the angle is defined. Thus, $\dot\psi$ is the magnitude of $\dot{\boldsymbol{\psi}}$ that lies along the Z axis of the earth-fixed coordinate system. The values of the components of $\dot{\boldsymbol{\psi}}$ are $(0, 0, \dot\psi)$. If we symbolize the components of $\dot{\boldsymbol{\psi}}$ in the (x, y, z) system as $(\dot\psi_x, \dot\psi_y, \dot\psi_z)$, the transformation equations are [13]

$$\dot{\boldsymbol{\psi}} = \begin{bmatrix} \dot\psi_x \\ \dot\psi_y \\ \dot\psi_z \end{bmatrix} = [C] \begin{bmatrix} 0 \\ 0 \\ \dot\psi \end{bmatrix} \tag{2.13}$$

Using Eq. (2.12), we may write these equations as follows:

$$\dot\psi_x = -\dot\psi \sin \theta$$

$$\dot\psi_y = \dot\psi \cos \theta \sin \phi$$

$$\dot\psi_z = \dot\psi \cos \theta \cos \phi \tag{2.14}$$

The vector $\dot{\boldsymbol{\theta}}$ is defined in the $(X'Y'Z')$ system where its components have the values $(0, \dot\theta, 0)$, so the transformation to the body-fixed coordinate system is

2.2. Coordinate Transformations

given by

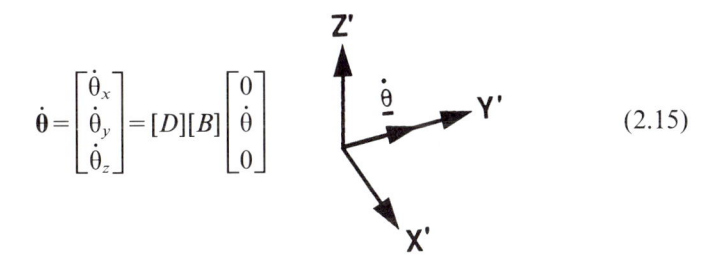

$$\dot{\boldsymbol{\theta}} = \begin{bmatrix} \dot{\theta}_x \\ \dot{\theta}_y \\ \dot{\theta}_z \end{bmatrix} = [D][B] \begin{bmatrix} 0 \\ \dot{\theta} \\ 0 \end{bmatrix} \qquad (2.15)$$

or using Eqs. (2.10) and (2.11),

$$\begin{aligned} \dot{\theta}_x &= 0 \\ \dot{\theta}_y &= \dot{\theta} \cos \phi \\ \dot{\theta}_z &= -\dot{\theta} \sin \phi \end{aligned} \qquad (2.16)$$

where $\dot{\theta}_x$, $\dot{\theta}_y$, $\dot{\theta}_z$ are the (x, y, z) components of $\dot{\boldsymbol{\theta}}$. Finally, the components of $\dot{\boldsymbol{\phi}}$ in the $(X''Y'Z'')$ system have the values $(\dot{\phi}, 0, 0)$, so that we make the transformation as

$$\dot{\boldsymbol{\phi}} = \begin{bmatrix} \dot{\phi}_x \\ \dot{\phi}_y \\ \dot{\phi}_z \end{bmatrix} = [D] \begin{bmatrix} \dot{\phi} \\ 0 \\ 0 \end{bmatrix} \qquad (2.17)$$

or from Eq. (2.11)

$$\begin{aligned} \dot{\phi}_x &= \dot{\phi} \\ \dot{\phi}_y &= 0 \\ \dot{\phi}_z &= 0 \end{aligned} \qquad (2.18)$$

The components of each of the derivatives of the Euler angles must add vectorially along a given axis of the body-fixed coordinate system to the components of $\boldsymbol{\omega}$ in that system since [6]

$$\boldsymbol{\omega} = \dot{\boldsymbol{\psi}} + \dot{\boldsymbol{\theta}} + \dot{\boldsymbol{\phi}} \qquad (2.19)$$

in any coordinate system. Therefore, we may write

$$\omega_x = p = \dot{\psi}_x + \dot{\theta}_x + \dot{\phi}_x = \dot{\phi} - \dot{\psi} \sin \theta$$
$$\omega_y = q = \dot{\psi}_y + \dot{\theta}_y + \dot{\phi}_y = \dot{\psi} \cos \theta \sin \phi + \dot{\theta} \cos \phi \qquad (2.20)$$
$$\omega_z = r = \dot{\psi}_z + \dot{\theta}_z + \dot{\phi}_z = \dot{\psi} \cos \theta \cos \phi - \dot{\theta} \sin \phi$$

In order to express $\dot{\psi}$, $\dot{\theta}$, and $\dot{\phi}$ in terms of (p, q, r), we may solve (2.20) using determinants, yielding

$$\dot{\psi} = (1/\cos \theta)(q \sin \phi + r \cos \phi)$$
$$\dot{\theta} = q \cos \phi - r \sin \phi \qquad (2.21)$$
$$\dot{\phi} = p + \tan \theta (q \sin \phi + r \cos \phi).$$

Equations (2.21) relate the components of the angular velocity in the aircraft reference system to the rates of change of the Euler angles. These are very useful relationships, since the equations of rotational motion of the aircraft are more simply expressed in the body-fixed coordinate system than in the earth-fixed coordinate system. However, Eqs. (2.21) also reveal a serious difficulty in the use of Euler angles for expressing orientations. For example, as the pitch angle θ approaches $\pm 90°$, the roll rate $\dot{\phi}$, and yaw rate $\dot{\psi}$, become undefined because of the terms $(\cos \theta)^{-1}$ and $\tan \theta$. This is the so-called gimbal-lock problem, which if not solved would obviously severely limit the usefulness of any gimbaled inertial navigation system. An alternative method of specifying angular position and angular velocity that avoids the gimbal-lock problem is the quaternion representation, which is the subject of Section 2.5.

2.2.5. Earth Sidereal Rotation Rate

In certain coordinate transformations, the earth sidereal rotation rate is used in these transformations as well as in the mechanization (or error) equations. Because of its importance, the earth rate vector will be derived for a north–east–up coordinate frame. The derivation will be carried out with reference to Fig. 2-6.

In Fig. 2-6, λ is the longitude and ϕ is the geographic latitude. Let **i, j, k** be unit vectors along the (x, y, z) axes, respectively. Furthermore, we can write the earth rotation rate in vector form as [11]

$$\mathbf{\Omega}_{ie} = 0 \cdot \mathbf{i} + \Omega_y \mathbf{j} + \Omega_z \mathbf{k}$$

where

$$\Omega_y = \Omega_{ie} \cos \phi$$
$$\Omega_z = \Omega_{ie} \sin \phi$$

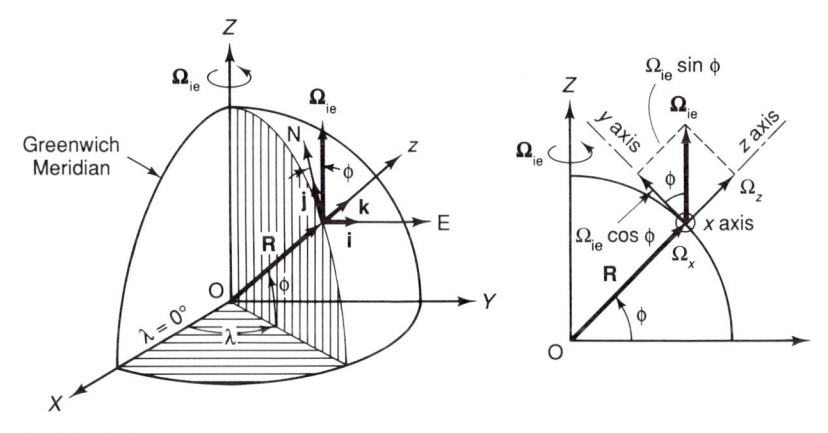

2.2. Coordinate Transformations

FIGURE 2-6
Definition of coordinate frames.

and Ω_{ie} is the magnitude of the earth rate vector. Thus, the earth rate vector consists of two components (the third component being zero) as follows:

$$\Omega_{ie} = \Omega_{ie} \cos \phi \cdot \mathbf{j} + \Omega_{ie} \sin \phi \cdot \mathbf{k} \tag{2.22}$$

2.2.6. Earth-Centered Inertial (ECI) to Earth-Fixed

Origin Both coordinate frames have their respective origin at the center of the earth.

Orientation At the current time of interest t, the inertial coordinate frame (x_i, y_i, z_i) has the x_i axis pointing toward the true equinox* of date at time t_0, z_i axis along the earth's rotational axis, and the y_i axis completes the right-handed Cartesian orthogonal system. The earth-fixed coordinate frame (x_e, y_e, z_e) is related to the inertial frame by a single positive rotation about the z_{ie} axis of $\Omega_{ie} \Delta t$, where Ω_{ie} is the earth's sidereal rotation rate. That is, the coordinates appear to rotate at a rate of 360° per sidereal day (23 h, 56 min, 4.09 s) [11]. Therefore

$$\Omega_{ie} = \frac{360°}{23 + [56 + (4.09/60)]/60} = 15.04106874 \text{ deg/h}$$

$$= 4.178074648 \times 10^{-3} \text{ deg/s}$$

$$= 7.2921159 \times 10^{-5} \text{ rad/s}$$

---

*The "true" place of a star is defined with the sun at the origin and with respect to the "true equator" and "true equinox" at the instant of observation.

and $\Delta t = t - t_0$ (this transformation is illustrated in Fig. 2-7); $\Omega_{ie} \Delta t$ is known as the "sidereal hour angle," and Δt is the time of year in hours. Strictly speaking, the earth's spin axis of date does not lie exactly along the geocentric polar axis of the reference ellipsoid. However, this displacement is varying slowly (polar precession and nutation) so that it can be considered constant over a period of days or weeks.

In addition to the sidereal time, two other time bases are defined. The first is the *Coordinated Universal Time* (UTC), standardized by the International Astronomical Union. Coordinated Universal Time is the mean solar time determined from the rotation of the earth by using astronomical observations and is referred to as UT1. Since the raw observations of time are referred to as UT0, they must be corrected for the polar motion of the earth in order to obtain UT1. UTC uses the atomic second as its time base, and is adjusted in epoch so as to remain close to UT2 (UT2 is obtained from an empirical correction added to UT1 to take into account the annual changes in speed of rotation). The Bureau International de L'Heure (BIH) in Paris, France, maintains a UT1 time scale. UTC forms the basis for civil time in most countries and is commonly referred to as Greenwich Mean Time (GMT). The other time base is the Ephemeris Time (ET). Ephemeris

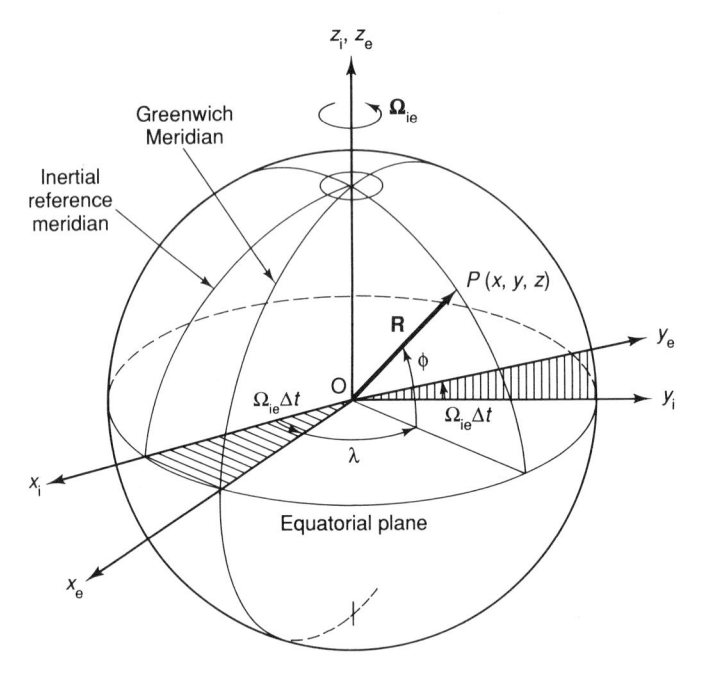

FIGURE 2-7
Earth-centered inertial to earth-fixed coordinate transformation.

2.2. Coordinate Transformations

Time is obtained from the orbital motion of any planet or satellite. In practice, the moon is generally used to determine ET because of its rapid orbital motion.

For most applications and to within sufficient accuracy, the vector $\mathbf{\Omega}_{ie}$ is expressed with respect to the ECEF frame as

$$\mathbf{\Omega}_{ie} = \begin{bmatrix} 0 \\ 0 \\ \dfrac{2\pi}{86,164} \end{bmatrix} = \begin{bmatrix} 0 \\ 0 \\ 7.292115 \times 10^{-5} \text{ rad/s} \end{bmatrix}$$

From Fig. 2-7 it can be seen that the i-frame is related to the e-frame by the earth rate and the time of the year (Julian Day). This relationship is defined by the sidereal hour angle

$$\begin{bmatrix} x_e \\ y_e \\ z_e \end{bmatrix} = \begin{bmatrix} \cos \Omega_{ie} \Delta t & \sin \Omega_{ie} \Delta t & 0 \\ -\sin \Omega_{ie} \Delta t & \cos \Omega_{ie} \Delta t & 0 \\ 0 & 0 & 1 \end{bmatrix} \begin{bmatrix} x_i \\ y_i \\ z_i \end{bmatrix} \quad (2.23)$$

or $\mathbf{r}^e = C_i^e \mathbf{r}^i$.

In astronomical work, the inertial axes are conveniently chosen so that the x_i axis points along the vernal Equinox (equinoctial coordinates), also called the "first point of Aries" and denoted by the symbol ϒ. The vernal Equinox is the point of intersection of the earth's equator and the plane of the earth's motion through which the sun crosses the equator from south to north. Consequently, the angle Λ by which the earth-fixed geocentric axes are separated from the inertial axes depends only on the time that has elapsed from the vernal Equinox and the earth's rotation rate Ω_{ie}. Mathematically, this can be expressed as

$$\Lambda = \Omega_{ie}(t - t_0) - 2n\pi$$

where n is an integer chosen so that $0 \leq \Lambda \leq 2\pi$ is satisfied. In this case, the inertial coordinates can be obtained by solving Eq. (2.23) for the (x_i, y_i, z_i) vector and taking the transpose of the transformation matrix and substituting Λ for $\Omega_{ie} \Delta t$ in the argument. Thus

$$\begin{bmatrix} x_i \\ y_i \\ z_i \end{bmatrix} = \begin{bmatrix} \cos \Lambda & -\sin \Lambda & 0 \\ \sin \Lambda & \cos \Lambda & 0 \\ 0 & 0 & 1 \end{bmatrix} \begin{bmatrix} x_e \\ y_e \\ z_e \end{bmatrix} \quad (2.24)$$

The rectangular, earth-fixed coordinates of a point $P(x, y, z)$ may be found from the geocentric latitude and longitude from the relation

$$x_e = R \cos \phi \cos \lambda$$
$$y_e = R \cos \phi \sin \lambda \quad (2.25)$$
$$z_e = R \sin \phi$$

A similar expression can be written for the inertial coordinates of point $P(x, y, z)$ with the exception that the angle Λ must be added to the longitude λ.

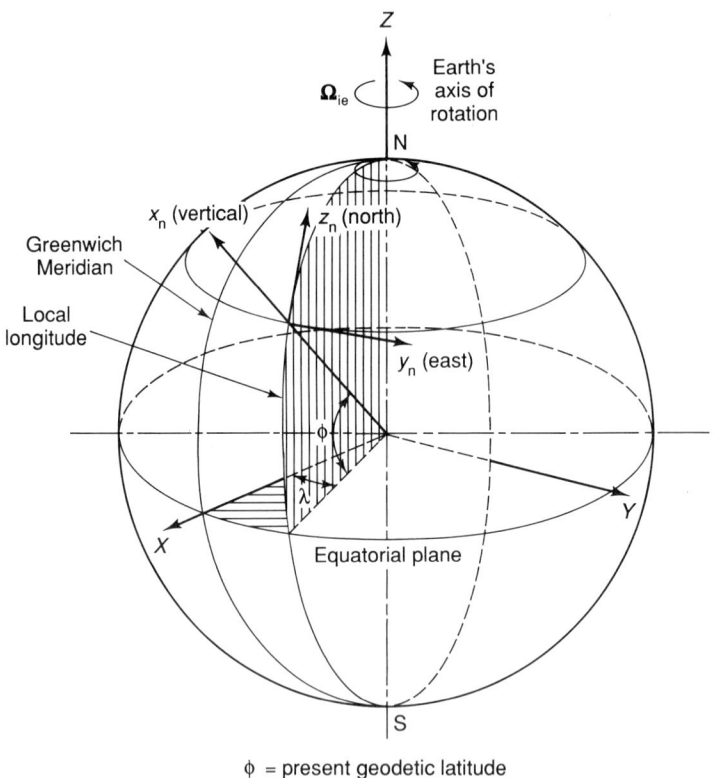

FIGURE 2-8
Earth-fixed to local-level navigation frames.

2.2.7. Earth-Fixed to Navigation

Origin System location (INS).

Orientation The navigation axes (x_n, y_n, z_n) are commonly aligned with the north, east, up (or down) directions. For the present transformation, we will assume that the x_n axis points in the up direction, the y_n axis points east, and the z_n axis points north (see Fig. 2-8). This transformation, C_e^n, is realized by two rotations: one through the angle λ about the z_e axis, and the other through the angle ϕ about the y_e axis.

1. Rotation about the z_e axis through the angle λ

$$\begin{bmatrix} x_e' \\ y_e' \\ z_e' \end{bmatrix} = \begin{bmatrix} \cos\lambda & \sin\lambda & 0 \\ -\sin\lambda & \cos\lambda & 0 \\ 0 & 0 & 1 \end{bmatrix} \begin{bmatrix} x_e \\ y_e \\ z_e \end{bmatrix}$$

2. Rotation about the y_e axis through the angle ϕ

$$\begin{bmatrix} x_e'' \\ y_e'' \\ z_e'' \end{bmatrix} = \begin{bmatrix} \cos\phi & 0 & \sin\phi \\ 0 & 1 & 0 \\ -\sin\phi & 0 & \cos\phi \end{bmatrix} \begin{bmatrix} x_e' \\ y_e' \\ z_e' \end{bmatrix}$$

From these transformations we obtain

$$\begin{bmatrix} x_n \\ y_n \\ z_n \end{bmatrix} = \begin{bmatrix} \cos\phi & 0 & \sin\phi \\ 0 & 1 & 0 \\ -\sin\phi & 0 & \cos\phi \end{bmatrix} \begin{bmatrix} \cos\lambda & \sin\lambda & 0 \\ -\sin\lambda & \cos\lambda & 0 \\ 0 & 0 & 1 \end{bmatrix} \begin{bmatrix} x_e \\ y_e \\ z_e \end{bmatrix}$$

$$\begin{bmatrix} x_n \\ y_n \\ z_n \end{bmatrix} = \begin{bmatrix} \cos\phi\cos\lambda & \cos\phi\sin\lambda & \sin\phi \\ -\sin\lambda & \cos\lambda & 0 \\ -\sin\phi\cos\lambda & -\sin\phi\sin\lambda & \cos\phi \end{bmatrix} \begin{bmatrix} x_e \\ y_e \\ z_e \end{bmatrix} \qquad (2.26)$$

or

$$\begin{bmatrix} x_n \\ y_n \\ z_n \end{bmatrix} = C_e^n \begin{bmatrix} x_e \\ y_e \\ z_e \end{bmatrix}$$

$$\mathbf{r}^n = C_e^n \mathbf{r}^e$$

where $x_n \equiv U$, $y_n \equiv E$, $z_n \equiv N$ (U=up, E=east, N=north). In certain applications, the coordinate transformation from the inertial (x_i, y_i, z_i) axes to the local vertical (UEN) is desired. This transformation is exactly the same as the transformation from the earth-fixed to the local vertical navigation frame, except that the longitude λ must be replaced by $\Lambda = \Omega_{ie}t + (\lambda - \lambda_0)$ where Ω_{ie} is the earth's sidereal rate and λ_0 is the longitude at time $t=0$. That is, the z axis rotates at a rate of 360° per sidereal day. Therefore, in the above transformation, the longitude will be replaced by Λ. As an example, consider a NEU coordinate system similar to those illustrated in Figs. 2-6 and 2-8. The unit vector representation of this local vertical coordinate system is as follows:

$$1_E = \frac{\Omega_{ie} \times R}{|\Omega_{ie} \times R|} \qquad (2.27)$$

$$1_N = \frac{R \times (\Omega_{ie} \times R)}{|R \times (\Omega_{ie} \times R)|} \qquad (2.28)$$

$$1_U = \frac{R}{|R|} \qquad (2.29)$$

2.2.8. Earth-Fixed to Wander–Azimuth

Origin Point of interest.

Orientation The wander–azimuth frame (x_c, y_c, z_c) (sometimes referred to as the computational frame) has its x_c axis pointing α degrees from north, the y_c axis pointing east, and the z_c axis is perpendicular to the surface of the reference ellipsoid (if $h=0$). The horizontal axes x_c, y_c are displaced from the east and north axes by the wander angle α. The wander angle is taken to be positive west of true north.

When the latitude (ϕ), longitude (λ), and wander angle (α) are zero, the (x_c, y_c, z_c) axes are aligned with (x_e, y_e, z_e) of the earth-fixed frame. Referring to Fig. 2-1, the sequence of rotations is shown below. Order of rotation: (1) a positive rotation of λ about y_c, (2) a negative rotation of ϕ about the displaced x_c axis, and (3) a positive rotation of α about the displaced z_c axis.

$$\begin{bmatrix} x_e \\ y_e \\ z_e \end{bmatrix} = \begin{bmatrix} 1 & 0 & 0 \\ 0 & \cos\alpha & \sin\alpha \\ 0 & -\sin\alpha & \cos\alpha \end{bmatrix} \overset{\text{Latitude}}{\begin{bmatrix} \cos\phi & 0 & \sin\phi \\ 0 & 1 & 0 \\ -\sin\phi & 0 & \cos\phi \end{bmatrix}} \overset{\text{Longitude}}{\begin{bmatrix} \cos\lambda & \sin\lambda & 0 \\ -\sin\lambda & \cos\lambda & 0 \\ 0 & 0 & 1 \end{bmatrix}} \begin{bmatrix} x_c \\ y_c \\ z_c \end{bmatrix}$$

$$\begin{bmatrix} x_e \\ y_e \\ z_e \end{bmatrix} = \begin{bmatrix} \cos\phi\cos\lambda & \cos\phi\sin\lambda & \sin\phi \\ -\cos\alpha\sin\lambda - \sin\alpha\sin\phi\cos\lambda & \cos\alpha\cos\lambda - \sin\alpha\sin\phi\sin\lambda & \sin\alpha\cos\phi \\ \sin\alpha\sin\lambda - \cos\alpha\sin\phi\cos\lambda & -\sin\alpha\cos\lambda - \cos\alpha\sin\phi\sin\lambda & \cos\alpha\cos\phi \end{bmatrix} \begin{bmatrix} x_c \\ y_c \\ z_c \end{bmatrix}$$

$$\mathbf{r}^e = C_c^e \mathbf{r}^c \qquad (2.30)$$

The wander angle (α), latitude (ϕ), and longitude (λ) are determined

2.2. Coordinate Transformations

in terms of the direction cosine elements as follows:

$$\alpha = \tan^{-1}(C_{zy}/C_{zz}) \tag{2.31}$$

$$\phi = \tan^{-1}\left(\frac{C_{zx}}{\sqrt{C_{xx}^2 + C_{yx}^2}}\right) \tag{2.32}$$

$$\lambda = \tan^{-1}(C_{yx}/C_{xx}) \tag{2.33}$$

2.2.9. Navigation Frame to Wander–Azimuth Frame

Origin System location.

Orientation The axes of these two coordinate frames is shown in Fig. 2-1. This transformation relates the north–east–up (NEU) local-vertical north-pointing frame to the ideal local-level wander–azimuth frame. It is a positive, single-axis rotation about the vertical (U, z) axis through the wander angle α.

$$\begin{bmatrix} x_c \\ y_c \\ z_c \end{bmatrix} = \begin{bmatrix} \cos\alpha & -\sin\alpha & 0 \\ \sin\alpha & \cos\alpha & 0 \\ 0 & 0 & 1 \end{bmatrix} \begin{bmatrix} x_n \\ y_n \\ z_n \end{bmatrix} \tag{2.34}$$

2.2.10. Body Frame to Navigation Frame

Origin Aircraft center of mass.

Orientation This transformation involves three successive single-axis rotations through the ordinary Euler angles of roll (ϕ), pitch (θ), and yaw or heading (ψ), which yield the aircraft attitude. The aircraft body axes are defined as follows. The x_b axis points in the forward (longitudinal) direction, the y_b axis points out the right wing, and the z_b axis nominally pointing down, when the x_b and y_b axes are nominally in the horizontal plane. These axes correspond to the vehicle roll, pitch, and yaw axes, respectively. Furthermore, the sign convention will be taken to be such that the roll angle ϕ is defined to be positive when the right wing dips below the horizontal plane, the pitch angle θ is defined positive when the nose of the aircraft is elevated above the horizontal plane, and the yaw (or heading) angle ψ is defined positive when the aircraft nose is rotating from north to east (i.e., positive clockwise looking down). Heading is the angle between x_n and the projection of the body axis on the horizontal plane. Pitch angle is the angle between the body axis and the body axis projection on the horizontal plane. Roll angle is the negative of the rotation about x_b that would bring y_b into the horizontal plane. In this transformation, it will be assumed that the navigation frame is defined by the local earth axes with the x_n axis pointing north, the y_n axis east, and the z_n axis down.

With this arrangement of navigation coordinates, we have a right-handed north–east–down (NED) system. The geometric configuration of the aircraft body axes is illustrated in Fig. 2-9. Also, this figure shows the order or rotation from the navigation axes to the body axes.

The direction cosine matrix linking the body (or aircraft) coordinates and the navigation coordinates will be denoted by C_n^b, where

$$\begin{bmatrix} x_b \\ y_b \\ z_b \end{bmatrix}_{aircr} = C_n^b \begin{bmatrix} x_n \\ y_n \\ z_n \end{bmatrix}_{nav} \qquad \begin{matrix} x_n \equiv N \\ y_n \equiv E \\ z_n \equiv D \end{matrix}$$

Therefore, one can go from the navigation coordinates to the body

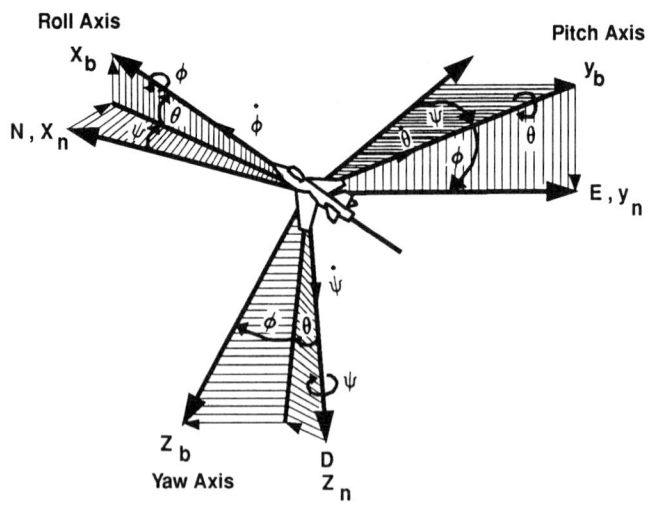

FIGURE 2-9
Euler angles and their relationships between body and navigation coordinates.

2.2. Coordinate Transformations

coordinates by successive rotations of ψ, θ, and ϕ. The order of rotation is as follows: (1) the first rotation is through the angle ψ about the down (D) or "z_n-axis," (2) the second rotation is through the angle θ about the displaced y_n axis called y'_n axis, and (3) the third rotation is through the angle ϕ about the new x_n axis called x''_n axis, which is in the desired aircraft or body (b) frame. These rotations are shown below.

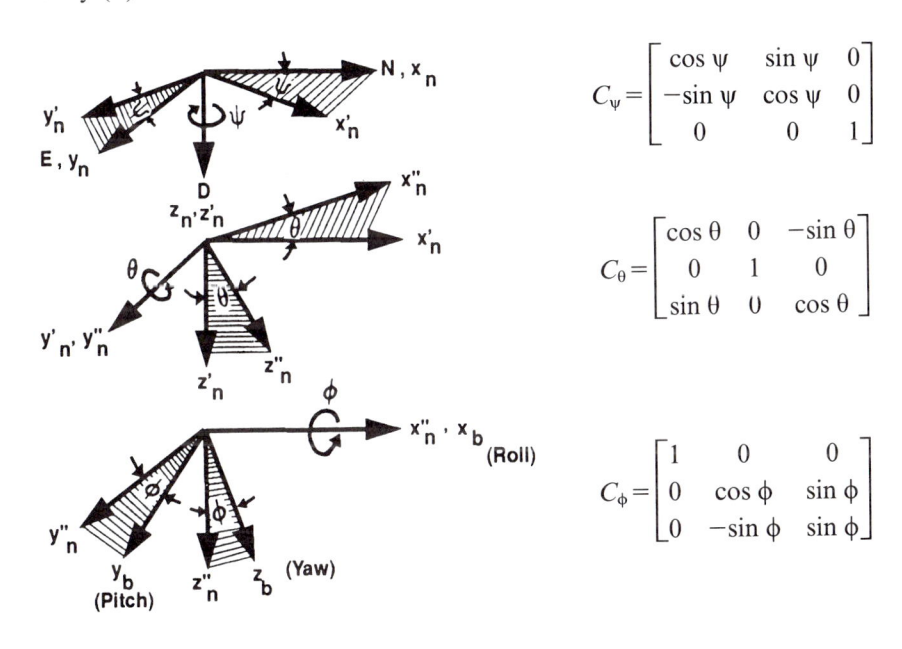

$$C_\psi = \begin{bmatrix} \cos\psi & \sin\psi & 0 \\ -\sin\psi & \cos\psi & 0 \\ 0 & 0 & 1 \end{bmatrix}$$

$$C_\theta = \begin{bmatrix} \cos\theta & 0 & -\sin\theta \\ 0 & 1 & 0 \\ \sin\theta & 0 & \cos\theta \end{bmatrix}$$

$$C_\phi = \begin{bmatrix} 1 & 0 & 0 \\ 0 & \cos\phi & \sin\phi \\ 0 & -\sin\phi & \sin\phi \end{bmatrix}$$

Then C_n^b can be defined as follows:

$$C_n^b = C_\psi C_\theta C_\phi$$

$$C_n^b = \begin{bmatrix} \cos\psi & \sin\psi & 0 \\ -\sin\psi & \cos\psi & 0 \\ 0 & 0 & 1 \end{bmatrix} \begin{bmatrix} \cos\theta & 0 & -\sin\theta \\ 0 & 1 & 0 \\ \sin\theta & 0 & \cos\theta \end{bmatrix} \begin{bmatrix} 1 & 0 & 0 \\ 0 & \cos\phi & \sin\phi \\ 0 & -\sin\phi & \cos\phi \end{bmatrix}$$

(Yaw) ──────────── (Pitch) ──────────── (Roll)

$$C_n^b = \begin{bmatrix} \cos\theta\cos\psi & \cos\theta\sin\psi & -\sin\theta \\ \sin\phi\sin\theta\cos\psi - \cos\phi\sin\psi & \sin\phi\sin\theta\sin\psi + \cos\phi\cos\psi & \sin\phi\cos\theta \\ \cos\phi\sin\theta\cos\psi + \sin\phi\sin\psi & \cos\phi\sin\theta\sin\psi - \sin\phi\cos\psi & \cos\phi\sin\theta \end{bmatrix} \quad (2.35)$$

This transformation is very important in strapdown* inertial navigation system error analyis. It should be pointed out that the Euler angles ϕ, θ, and ψ of the aircraft are measured by the INS with respect to the platform coordinate system (x_p, y_p, z_p) whose orientation is fixed in space. Now, if the INS output attitude angles ϕ, θ, and ψ denote respectively roll, pitch, and yaw, one can go from the platform frame to the body frame by successive rotation of these angles. Moreover, since the origin of the aircraft body coordinate system coincides with the origin of the platform coordinate system, the direction cosine linking these two coordinate system is exactly the same as the direction cosine C_n^b. Therefore, $C_n^b \equiv C_b^p$:

$$C_n^b = \begin{bmatrix} C_{11} & C_{12} & C_{13} \\ C_{21} & C_{22} & C_{23} \\ C_{31} & C_{32} & C_{33} \end{bmatrix}$$

Consequently, the elements of the new matrix C_b^n can be obtained by taking the transpose of C_n^b as follows:

$$(C_n^b)^T = C_b^n = \begin{bmatrix} C_{11} & C_{21} & C_{31} \\ C_{12} & C_{22} & C_{32} \\ C_{13} & C_{23} & C_{33} \end{bmatrix}$$

From Eq. (2.35), the direction cosine elements are as follows:

$C_{11} = \cos\theta \cos\psi$
$C_{12} = \cos\theta \sin\psi$
$C_{13} = -\sin\theta$
$C_{21} = \sin\phi \sin\theta \cos\psi - \cos\phi \sin\psi$
$C_{22} = \sin\phi \sin\theta \sin\psi + \cos\phi \cos\psi$
$C_{23} = \sin\phi \cos\theta$
$C_{31} = \cos\phi \sin\theta \cos\psi + \sin\phi \sin\psi$
$C_{32} = \cos\phi \sin\theta \sin\psi - \sin\phi \cos\psi$
$C_{33} = \cos\phi \cos\theta$

---

*A strapdown (also referred to as "analytic") INS differs from the conventional gimbaled INS in that the inertial sensors (i.e., gyroscopes and accelerometers) are mounted directly onto the host vehicle's (e.g., aircraft, ship, or missile) frame instead of on the stable platform. More specifically, the onboard navigation computer keeps track of the vehicle's attitude with respect to some preselected reference frame, based on information from the gyroscopes. As a result, the computer is able to provide the coordinate transformation necessary to coordinatize the accelerometer outputs in a computational reference frame. That is, the transformation is done analytically in the computer. Thus, the stable platform of the gimbaled system is replaced by two computer functions: (1) establish an attitude reference based on the gyroscope outputs, and (2) transform the accelerometer outputs into the reference computational frame. A strapdown INS provides vehicle rate information directly. Furthermore, the usual gimbals, resolvers, sliprings, torquers, pickoffs, synchros, and electronics associated with gimbaled systems are eliminated.

2.2. Coordinate Transformations

By examining the direction cosine matrix, we note that the Euler angles can be obtained from C_n^b. Thus

$$\tan \phi = \frac{C_{23}}{C_{33}} = \frac{\sin \phi \cos \theta}{\cos \phi \cos \theta} = \frac{\sin \phi}{\cos \phi} \quad \text{or} \quad \phi = \tan^{-1}\left(\frac{C_{23}}{C_{33}}\right)$$

$$\tan \psi = \frac{C_{12}}{C_{11}} = \frac{\cos \theta \sin \psi}{\cos \theta \cos \psi} = \frac{\sin \psi}{\cos \psi} \quad \text{or} \quad \psi = \tan^{-1}\left(\frac{C_{12}}{C_{11}}\right)$$

$$-\tan \theta = \frac{C_{13}}{\sqrt{1-C_{13}^2}} = \frac{-\sin \theta}{\sqrt{1-\sin^2 \theta}}$$

$$= \frac{-\sin \theta}{\cos \theta} \quad \text{or} \quad \theta = \tan^{-1}\left[\frac{-C_{13}}{\sqrt{1-C_{13}^2}}\right]$$

In the general case, if one desires the transformaton from the wander-azimuth frame to the body frame, the sine and cosine terms of the C_ψ rotation matrix must include the wander angle α; that is, these elements become $\sin(\psi + \alpha)$ and $\cos(\psi + \alpha)$.

We have seen that the Euler angle method is based on the use of one rotation about each of three axes in an ordered sequence. Therefore, if the rotations are made in an ordered sequence, a transformation matrix will result that exactly describes the orientation of one coordinate frame with respect to another.

2.2.11. Body Frame to Fixed Line-of-Sight (LOS) Transformation

Origin Vehicle center of mass

Orientation The orientation of the body or vehicle axes for this transformation is similar to the navigation body coordinate transformation. However, the ensemble LOS vectors do not, in general, form a set of orthogonal axes. Here, the x_b axis is directed along the aircraft or missile longitudinal axis, the y_b axis out to the right, and the z_b axis down. These axes correspond to the vehicle roll, pitch, and yaw axes, respectively. The transformation from the body axes to the LOS is realized by two rotations as follows: (1) a positive (or clockwise) rotation of the x_b axis about the z_b axis through the azimuth angle of magnitude ψ and (2) a positive rotation of the new y_b axis through the elevation angle θ. Figure 2-10 illustrates this transformation.

This transformation can be accomplished by considering an arbitrary vector, say, **r**. Then

$$\mathbf{r}^{\text{LOS}} = C_b^{\text{LOS}} \mathbf{r}^b$$

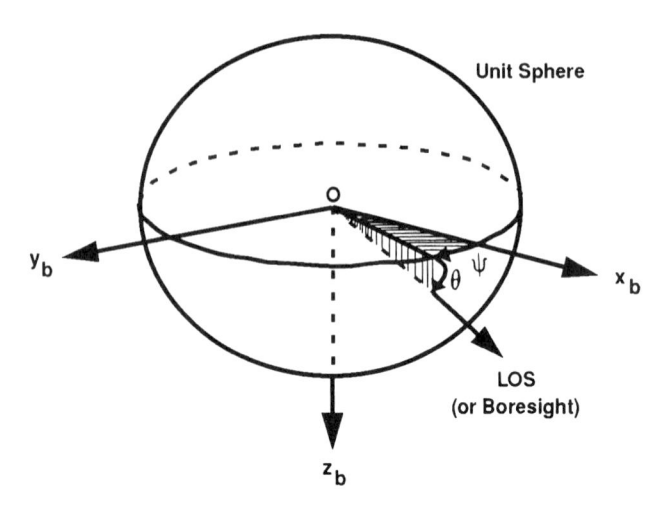

FIGURE 2-10
Body-fixed to LOS coordinate transformation.

where the transformation matrix C_b^{LOS} is given by

$$C_b^{LOS} = \begin{bmatrix} \cos\theta & 0 & \sin\theta \\ 0 & 1 & 0 \\ -\sin\theta & 0 & \cos\theta \end{bmatrix} \begin{bmatrix} \cos\psi & \sin\psi & 0 \\ -\sin\psi & \cos\psi & 0 \\ 0 & 0 & 1 \end{bmatrix}$$

Performing the indicated operation, we can therefore write

$$\mathbf{r}^{LOS} = \begin{bmatrix} \cos\theta\cos\psi & \cos\theta\sin\psi & \sin\theta \\ -\sin\psi & \cos\psi & 0 \\ -\sin\theta\cos\psi & -\sin\theta\sin\psi & \cos\theta \end{bmatrix} \mathbf{r}^b \qquad (2.36)$$

where $(\mathbf{r}^b)^T = (x_b \; y_b \; z_b)$. The above transformation is useful in determining the pointing angle error for, say, a radar laser sensor mounted on a stable platform or forward-looking infrared (FLIR) sensor. For this particular application, Fig. 2-11 will be of interest.

Measurements can also be made in spherical coordinates as follows (see Fig. 2-11):

$$x_b = R\cos\theta\cos\psi$$
$$y_b = R\cos\theta\sin\psi$$
$$z_b = R\sin\theta$$

where R = slant range identified as the LOS (or boresight) to the target or point in question
ψ = azimuth angle
θ = elevation angle

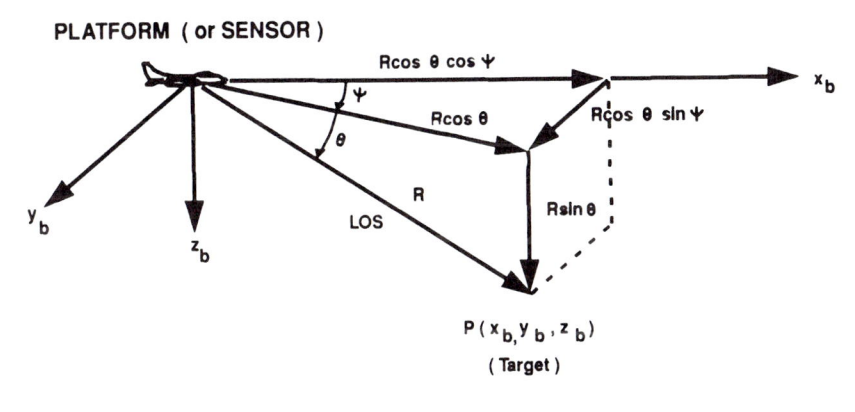

2.2. Coordinate Transformations

FIGURE 2-11
Sensor measurement coordinate transformation.

In the ideal case, no errors will be present in the pointing direction of the LOS. That is, the real platform axes and sensor axes will coincide. However, it must be pointed out that platform misalignment errors will cause the object or target location to deviate or be displaced. These platform pointing errors must be considered in any sensitivity error analysis. With the above equations, a digital simulation program can be used to investigate the effects of sensor measurement and platform errors for either an aircraft or missile offset aim point. The LOS coordinates find extensive use when considering GPS satellite positions with respect to the user.

2.2.12. Doppler Velocity Transformation from Aircraft Axes to Local-Level Axes

Origin Aircraft center of mass.

Orientation As we shall discuss later, the majority of aircraft use Doppler radar as a source for reference velocity. Doppler radar measurements are used to provide velocity damping for aircraft inertial navigation systems. Specifically, Doppler velocity is measured in aircraft body axis coordinates (V_H, V_D, V_V). The Doppler heading velocity V_H is measured along the aircraft longitudinal or x_b-body axis, the Doppler drift velocity V_D is measured out the right wing or along the y_b-body axis, and the Doppler vertical velocity V_V is measured up along the z_b-body axis. These components are transformed through roll (ϕ), pitch (θ), and yaw or heading (ψ) into local-level north-east-up coordinates. Following the transformation procedure given in Section 2.2.10, we have

$$\begin{bmatrix} V_N \\ V_E \\ \dot{H} \end{bmatrix} = \begin{bmatrix} \cos\theta\cos\psi & \sin\phi\sin\theta\cos\psi - \cos\phi\sin\psi & -\cos\phi\sin\theta\cos\psi - \sin\phi\sin\psi \\ \cos\theta\sin\psi & \sin\phi\sin\theta\sin\psi + \cos\phi\cos\psi & -\cos\phi\sin\theta\sin\psi + \sin\phi\cos\psi \\ \sin\theta & -\sin\phi\cos\theta & \cos\phi\cos\theta \end{bmatrix} \begin{bmatrix} V_H \\ V_D \\ V_V \end{bmatrix}$$

where $\dot{H}$ is the altitude rate.

2.2.13. Geodetic to Earth-Centered Earth-Fixed (ECEF) Coordinate Transformation

Origin Vehicle center of mass.

Orientation As we have seen in Sections 2.1.2 and 2.1.5, the geodetic coordinates are defined by the parameters λ, ϕ, and h, while in the ECEF Cartesian coordinate system the z_e axis is directed along the ellipsoidal polar axis, the x_e axis passes through the Greenwich Meridian, and the y_e axis is 90° to the right of the x_e axis, forming a right-handed rotation. Both the x_e and the y_e axes are in the equatorial plane. Therefore, in the general case, where $h \neq 0$, the transformation from the geodetic coordinates to the ECEF coordinates can be computed as follows:

$$x_e = (R_\lambda + h)\cos\phi\cos\lambda$$
$$y_e = (R_\lambda + h)\cos\phi\sin\lambda$$
$$z_e = [(1-\varepsilon^2)R_\lambda + h]\sin\phi$$

where R_λ = the radius of curvature in the east–west direction
$\qquad\quad = R_e/(1-\varepsilon^2\sin^2\phi)^{1/2}$
$\qquad R_e$ = equatorial radius of the reference ellipsoid
$\qquad \varepsilon^2$ = reference ellipsoid eccentricity squared $= 1 - b^2/a^2 \cong 2f$ [where a = semimajor (equatorial) axis of the ellipsoid; b = semiminor (polar) axis of the ellipsoid; f = flattening]
$\qquad h$ = altitude above the reference ellipsoid and normal to the surface
$\qquad \phi$ = geodetic latitude
$\qquad \lambda$ = geodetic longitude

Thus, in typical applications, the position and velocity states are mechanized in the earth-centered earth-fixed frame, while attitude is defined as the error in the tranformation between the body frame (b) and the e-frame, coordinatized in the e-frame. In these applications, the e-frame is defined as before. That is, its origin is at the center of the World Geodetic System (WGS-84) reference ellipsoid, with its z_e axis aligned with the earth's angular velocity vector; the x_e axis points in the direction of the intersection of the equatorial plane and the Greenwich Meridian, and the y_e axis completes the right-hand set. Efficient algorithms have been developed for the transformation between the various coordinate systems. Computationally, the earth-fixed Cartesian coordinate system is faster than either the inertial coordinates or local-level coordinates [5, 10].

2.3. COORDINATE FRAMES FOR INS MECHANIZATION

In Section 2.1 it was stated that, in designing an inertial navigation system, the designer must define from the onset a coordinate system where the velocity and position integrations must be performed. That is, the particular INS coordinate frame selected will be used in the onboard navigation computer in order to keep track of the vehicle velocity, position, and attitude estimates. Three types of INS configurations are commonly employed, depending on the application: (1) local-level, (2) space-stable, and (3) strapdown mechanizations. Most of the coordinate systems in use are of the local-level type. In the ideal case, the level axes of the platform coordinate system will lie in a plane tangent to the local vertical at the position defined by the direction cosines and longitude. As we shall see in Chapter 4, the navigation system equations can be mechanized in any of these frames. For convenience, the representative coordinate configurations are listed as follows:

1. Local-level mechanization
    - North-slaved (or north-pointing)
    - Unipolar
    - Free azimuth
    - Wander azimuth
2. Space-stabilized mechanization
3. Strapdown (gimballess or analytic) mechanization

The inertial navigation system computations of velocity, position, and attitude may be performed in any of the above frames. The criteria for the selection of the particular frame will depend on worldwide navigation capability, onboard computer complexity, and interface with other avionic subsystems (e.g., flight control, weapon delivery). Specifically, terrestrial measurements are usually made with respect to a coordinate system fixed in the earth, which therefore rotates uniformly with a constant angular velocity relative to the inertial system. The word "fixed" is generally used with reference to terrestrial objects to denote invariable position relative to the surface of the earth at the location considered. In what follows, a derivation of the

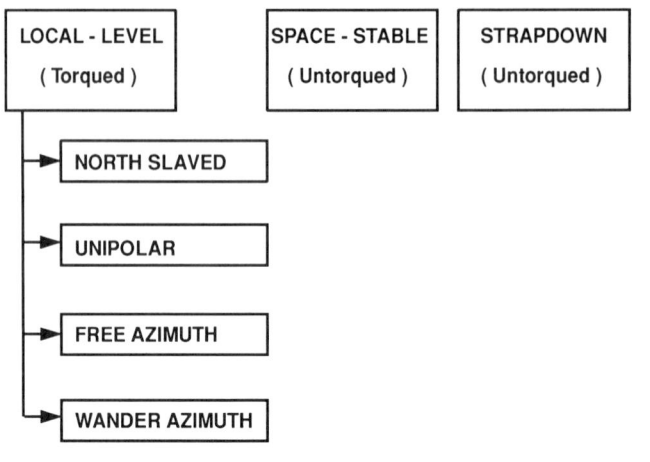

```
┌─────────────────┐    ┌─────────────────┐    ┌─────────────────┐
│  LOCAL - LEVEL  │    │  SPACE - STABLE │    │   STRAPDOWN     │
│   ( Torqued )   │    │   ( Untorqued ) │    │   ( Untorqued ) │
└────────┬────────┘    └─────────────────┘    └─────────────────┘
         │
         │    ┌─────────────────┐
         ├───▶│  NORTH SLAVED   │
         │    └─────────────────┘
         │
         │    ┌─────────────────┐
         ├───▶│    UNIPOLAR     │
         │    └─────────────────┘
         │
         │    ┌─────────────────┐
         ├───▶│  FREE AZIMUTH   │
         │    └─────────────────┘
         │
         │    ┌─────────────────┐
         └───▶│ WANDER AZIMUTH  │
              └─────────────────┘
```

FIGURE 2-12
Coordinate system mechanizations.

various local-level systems will be given, and the advantages and disadvantages of using one system or the other will be summarized. Figure 2-12 illustrates the various coordinate system mechanizations.

Before the various local-level systems are derived, certain preliminaries are in order. Local-level systems are normally expressed in the navigation frame, and measured with respect to an earth-centered inertial (ECI) reference frame. The distinction between the various local-level mechanization categories is in the azimuth torque rate. These systems maintain two of the accelerometer input axes in the horizontal plane. Specifically, the accelerometer platform is required to align with the unit vectors of a local-level coordinate frame. Therefore, in order for the platform to remain locally level, the horizontal gyroscopes on the platform must be driven with the components along their input axes of the desired inertial angular velocity. The earth rate vector, as we have seen in Section 2.2.5, is a function of latitude and consists of two components as follows [Eq. (2.22)]:

$$\mathbf{\Omega}_{ie} = \Omega_{ie} \cos \phi \cdot \mathbf{j} + \Omega_{ie} \sin \phi \cdot \mathbf{k}$$

Now consider the geographic (latitude–longitude) coordinate system of Fig. 2-13, where the x_g axis points east, the y_g axis points north, and the z_g axis points up. Thus, position will be expressed in terms of latitude (ϕ) and longitude (λ).

The origin of the system coincides with the vehicle as it moves on the surface of the earth. The angular velocity $\boldsymbol{\omega}$ of the geographic frame in inertial space is the sum of the angular velocity $\boldsymbol{\rho}$ of the geographic frame with respect to the rotating earth plus the angular velocity $\mathbf{\Omega}_{ie}$ of the earth

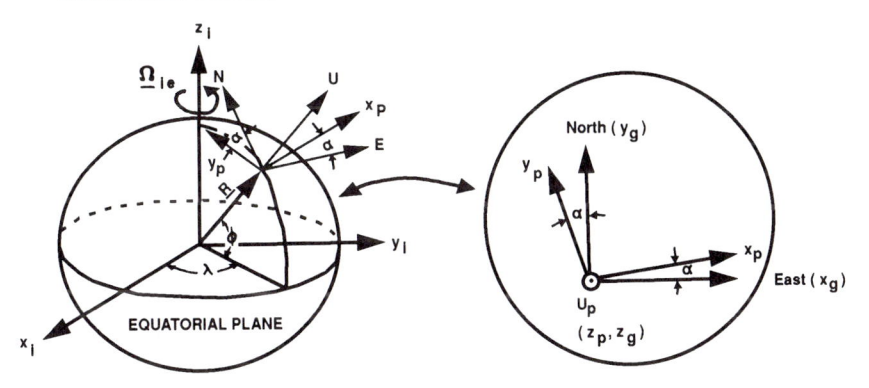

2.3. Coordinate Frames for INS Mechanization

FIGURE 2-13
Definition of coordinate frames.

with respect to inertial space and can be expressed as

$$\omega_E = \rho_E$$
$$\omega_N = \rho_N + \Omega_N = \rho_N + \Omega_{ie} \cos \phi$$
$$\omega_Z = \rho_Z + \Omega_Z = \rho_Z + \Omega_{ie} \sin \phi$$

Furthermore, the components of ρ are given by

$$\rho_E = -\frac{v_N}{R_\phi + h} \tag{2.37}$$

$$\rho_N = \frac{v_E}{R_\lambda + h} \tag{2.38}$$

$$\rho_Z = \rho_N \tan \phi = \left[\frac{v_E}{R_\lambda + h}\right] \tan \phi \tag{2.39}$$

where v_N and v_E are the north and east velocity components with respect to the earth; R_ϕ and R_λ are the radii of curvature of the reference ellipsoid in the north–south (or constant longitude) and east–west (or constant latitude) directions, respectively; and h is the altitude above the reference ellipsoid. In most inertial navigation system applications, the radii of curvature of the earth reference ellipsoid are taken to be

$$R_\phi = \frac{R_e(1-\varepsilon^2)}{(1-\varepsilon^2 \sin^2 \phi)^{3/2}} \tag{2.40}$$

$$R_\lambda = \frac{R_e}{(1-\varepsilon^2 \sin^2 \phi)^{1/2}} \tag{2.41}$$

where R_e is the equatorial radius of the earth and ε is the eccentricity given by $\varepsilon^2 = 1 - b^2/a^2$ (a is the semimajor axis of the reference ellipsoid and b is the semiminor axis).

Now assume that, as the vehicle moves over the surface of the earth, the platform coordinate system diverges by an angle α, called the *wander angle*, from the geographic or navigation frame. That is, the platform coordinate system is described by a rotation about the z axis of the geographic frame. Specifically, the divergence rate $\dot\alpha$ is due to the vertical components of (1) the earth's rotation and (2) vehicle motion. Therefore, the analysis of the local-level system considers only this vertical component of the earth rate and vehicle motion.

Now let $\boldsymbol{\omega}_{gt}$ be the total angular velocity of the geographic coordinate system with respect to inertial space. In this system, we will assume that the x_g axis points east, the y_g axis north, and the z_g axis up. Therefore, it can be shown that $\boldsymbol{\omega}_{gt}$ is given by

$$\boldsymbol{\omega}_{gt} = -\dot\phi \cdot \mathbf{i} + (\dot\lambda + \Omega_{ie})\cos\phi \cdot \mathbf{j} + (\dot\lambda + \Omega_{ie})\sin\phi \cdot \mathbf{k} \qquad (2.42)$$

where $\mathbf{i}, \mathbf{j}, \mathbf{k}$ are unit vectors along the (x_g, y_g, z_g) directions, respectively, and Ω_{ie} is the earth's sidereal rotation rate. From Eq. (2.42) it is seen that the vertical component of the geographic frame is

$$\omega_{zg} = (\dot\lambda + \Omega_{ie})\sin\phi \qquad (2.43)$$

where ϕ is the geographic latitude and λ is the longitude. As we have seen above, the wander angle α is the amount that the platform frame has rotated out of the geographic frame. This rotation is described by the relation

$$\mathbf{1}_p = C_g^p \mathbf{1}_g$$

where

$$C_g^p = \begin{bmatrix} \cos\alpha & -\sin\alpha & 0 \\ \sin\alpha & \cos\alpha & 0 \\ 0 & 0 & 1 \end{bmatrix}$$

The rate of change of α is defined as follows:

$$\dot\alpha = \omega_{zp} - \omega_{zg} \qquad (2.44)$$

where ω_{zp} describes the motion of the platform relative to the geographic coordinate system. Furthermore, the platform motion ω_{zp} is equivalent to the commanded rate ω_{zc} ($\omega_{zp} \equiv \omega_{zc}$). That is, ω_{zc} results from the input torque applied to the platform by the system. It is the azimuth gyroscope commanded precession rate. Consequently, the value of ω_{zc} determines which of the four systems is being used. Having done the preliminary analysis, these four local-level mechanizations will now be described.

2.3. Coordinate Frames for INS Mechanization

2.3.1. North-Slaved System

The vertical platform axis of this system is torqued to maintain alignment with the geographic (navigation) axes. That is, the platform must stay aligned with the geographic axes. In order for this to be true, the wander angle α must be zero (i.e., no rotation out of the geographic axes) as the vehicle moves over the surface of the earth. Thus

$$\dot{\alpha} = 0 = \omega_{zc} - \omega_{zg} \qquad (2.45)$$
$$\omega_{zc} = \omega_{zp}$$

Therefore,

$$\omega_{zc} = \Omega_{ie} \sin \phi + \dot{\lambda} \sin \phi = (\dot{\lambda} + \Omega_{ie}) \sin \phi \qquad (2.46)$$

2.3.2. Unipolar System

In this system, the platform vertical axis is torqued to maintain the wander angle equal to the longitude angle. Therefore, the platform must be torqued to (1) cancel the precession caused by the earth rate and vehicle rate and (2) maintain the wander angle equal to the longitude. Mathematically, this is expressed as

$$\dot{\alpha} = \omega_{zc} - \omega_{zg} \qquad (2.47)$$
$$\dot{\alpha} = \pm \dot{\lambda} \qquad (2.48)$$

where

$$\omega_{zc} = \omega_{zg} \pm \dot{\lambda}$$

Therefore

$$\omega_{zc} = \Omega_{ie} \sin \phi + \dot{\lambda} \sin \phi \pm \dot{\lambda}$$

or

$$\omega_{zc} = (\dot{\lambda} + \Omega_{ie}) \sin \phi \pm \dot{\lambda} \qquad (2.49)$$

The double sign indicates that the direction of the wander angle is reversed with respect to the longitude rate when crossing the equator. Specifically, above the equator, the negative sign is used. This is necessary in order to eliminate the singularity at the North Pole (i.e., heading is undefined at the pole and/or the $\lambda \sin \phi$ term as mentioned in Section 2.3.4).

2.3.3. Free-Azimuth System

The vertical platform axis in this system is not torqued. That is, it is inertially nonrotating along the vertical or z axis, which in effect eliminates the torquing error associated with vertical gyroscope. Therefore, $\omega_{zc} = 0$. As a result, the platform axes then diverge in azimuth from the geographic axes in accordance with

$$\dot{\alpha} = \omega_{zc} = \omega_{zg}$$

and since $\omega_{zc} = 0$, then

$$\dot{\alpha} = -\Omega_{ie} \sin \phi \quad \dot{\lambda} \sin \phi = -(\dot{\lambda} + \Omega_{ie}) \sin \phi \qquad (2.50)$$

The fact that the vertical gyroscope is not torqued can be considered as an advantage, since normally the vertical gyroscope exhibits poorer drift rate characteristics than either of the horizontal gyroscopes. Free-azimuth coordinates would ordinarily be used in aircraft inertial navigation systems where high platform angular rates about the vertical axis are required.

2.3.4. Wander–Azimuth System

In the wander–azimuth system, the vertical platform axis is torqued to compensate only for the vertical component of the earth rate

$$\omega_{zc} = \Omega_{ie} \sin \phi \qquad (2.51)$$

and not the vertical component of the aircraft transport rate. That is, no attempt is made to maintain the level axes in a preferred azimuth direction, such as north; the level axes are allowed to rotate freely about the vertical axis. As is the case with all local-level systems, the inertial platform is initially aligned by first driving it with torquers until the outputs of the two-level accelerometers are zero. This concept will be discussed in more detail in the section on alignment and gyrocompassing. This system finds extensive use in many of today's aircraft navigation system mechanizations, since no singularity exists at the pole:

$$\dot{\alpha} = \omega_{zc} - \omega_{zp}$$

$$\dot{\alpha} = \Omega_{ie} \sin \phi - \Omega_{ie} \sin \phi - \dot{\lambda} \sin \phi \qquad (2.52)$$

Thus $\quad \dot{\alpha} = -\dot{\lambda} \sin \phi$

For each of the four local-level mechanizations derived, the azimuth gyro commanded precession rate ω_{zc} and wander–azimuth angle rate $\dot{\alpha}$ are summarized in Table 2-1.

Table 2-2 lists the advantages of these local-level systems.

Next, we shall discuss briefly the space-stable and strapdown (analytic) coordinate configurations. In the space-stable configuration, the inertial platform maintains a constant orientation with respect to the inertial space. Moreover, in space-stable systems, the platform frame coincides with the inertial frame. Space-stable mechanizations, like local-level mechanizations, are associated with space-stable navigation. The primary input to the space-

2.3. Coordinate Frames for INS Mechanization

TABLE 2-1
Local-Level Systems

System	Commanded ω_{zc}	Parameter $\dot{\alpha}$
North-slaved	$(\Omega_{ie} + \dot{\lambda}) \sin \phi$	0
Unipolar	$(\Omega_{ie} + \dot{\lambda}) \sin \phi \pm \dot{\lambda}$	$\pm \dot{\lambda}$
Free azimuth	0	$-(\Omega_{ie} + \dot{\lambda}) \sin \phi$
Wander azimuth	$\Omega_{ie} \sin \phi$	$-\dot{\lambda} \sin \phi$

stable navigation is sensed acceleration in inertial reference (space-stable) coordinates. That is, the velocity increment data when divided by the computation interval can be viewed as an equivalent average sensed acceleration. In this system, the Z axis is parallel to the earth's polar axis, while the Y axis is defined as being parallel to east at the start of navigation. This coordinate frame is used as the reference for the inertial integrations. Therefore, the resulting values of (X, Y, Z) are the coordinates of vehicle position relative to the center of the earth. The acceleration due to vehicle motion with respect to inertial space is integrated twice in inertial reference coordinates. Consequently, the output of the first integrator is inertial relative velocity, while the output of the second integrator yields position relative to the center of the earth in Cartesian coordinates. Standard position outputs, such as geodetic latitude and longitude, are derived by performing inverse trigonometric functions on the inertial Cartesian components. These angles are also used for the transformation, which relate the local-level north coordinates to the inertial reference coordinates. We have noted earlier that inertial

TABLE 2-2
Mechanization Advantages

Mechanization	Advantages
North-slaved	Latitude and longitude coordinates can be read directly from the platform azimuth gimbal
Unipolar	Platform azimuth is maintained equal to longitude; can obtain azimuth and longitude with a single computation
Free azimuth	Needs no accurate torque electronics for the vertical or z-gyroscope
Wander azimuth	No singularity at the poles; the coordinate transformations are simplified considerably

navigation depends on the integration of acceleration with respect to a Newtonian reference frame. These conventional systems physically maintain such a reference frame by means of three or more gimbals.

In strapdown configurations, the inertial sensors are mounted directly on the host vehicle so that the transformation from the sensor axes to the inertial reference frame is computed rather than mechanized. Therefore, for all practical purposes, in a strapdown system the platform frame coincides with the body frame. Consequently, the accelerometers rotate (with respect to the inertial space) in the same way as the vehicle.

We can summarize the results of Section 2.3 by noting that the largest class of inertial navigation systems are of the local-level type, where the stable platform is constrained with two axes in the horizontal plane. Several thousands of inertial navigation systems have been built over the years using local-level mechanization because of the error compensation simplifications of maintaining constant platform alignment to the gravity vector. Historically, the first local-level inertial navigation systems were of the north-slaved type, but as we have seen earlier, use of this conventional geographic set of axes leads to both hardware and computational difficulties when operating at or in the vicinity of the polar regions. Naturally, different designers have devised different ways of avoiding the singularities occurring in the polar regions. The distinction between the various categories discussed is in the azimuth torque rate. Moreover, different inertial navigation system designers favor different platform azimuth drive implementations. For example, the free azimuth requires no accurate torque electronics for the z-gyroscope. The wander–azimuth design (or mechanization) eliminates variation of platform azimuth with time, simplifying the necessary coordinate transformations. The north-slaved design is not used much, since the $\dot{\lambda} \sin \phi$ term has singularities at both the North and South Poles. Finally, the unipolar system design is computationally economic, as the platform azimuth is maintained equal to longitude (with an arbitrary initial condition), and thus a single computation serves for both azimuth and longitude drives. Mechanization of the unipolar system is relatively simple. As a singularity in azimuith rate still exists at one pole, the direction of the azimuth drive with respect to longitude is reversed at the equator.

2.4. PLATFORM MISALIGNMENT

Manufacturing imperfections in mounting the accelerometer orthogonal triad onto a stable platform will inadvertently create axis misalignment. That is, the sensitive axes of the accelerometer triad will not be perfectly aligned with the platform axes. Platform or instrument misalignment causes the accelerometers to sense a specific force component due to gravity. This axis

2.4. Platform Misalignment

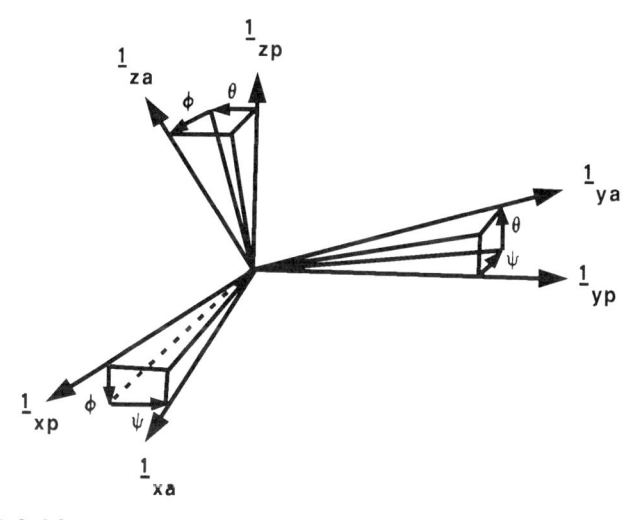

FIGURE 2-14
Platform–accelerometer misalignment angles.

misalignment must be compensated mathematically. In order to get a physical insight into this misalignment, consider Fig. 2-14.

Let the platform coordinates consist of an orthogonal triad defined by the unit vectors $\mathbf{1}_{xp}, \mathbf{1}_{yp}, \mathbf{1}_{zp}$, and the accelerometer triad defined by the unit vectors $\mathbf{1}_{xa}, \mathbf{1}_{ya}, \mathbf{1}_{za}$. Since the misalignment angles are usually small, it will be assumed that $\sin\theta \approx \theta$ and $\cos\theta \approx 1$. The order of rotation will be taken as follows:

1st rotation: rotate θ about $\mathbf{1}_{xp}$
2nd rotation: rotate ϕ about $\mathbf{1}_{yp}$
3rd rotation: rotate ψ about $\mathbf{1}_{zp}$

Following the above rotation convention, we have

$$\begin{aligned}
\mathbf{1}_{xa} &= \mathbf{1}_{xp} + \cos\left(\frac{\pi}{2} - \psi\right)\mathbf{1}_{yp} + \cos\left(\frac{\pi}{2} + \phi\right)\mathbf{1}_{zp} \\
&= \mathbf{1}_{xp} + \psi\mathbf{1}_{yp} - \phi\mathbf{1}_{zp} \\
\mathbf{1}_{ya} &= \cos\left(\frac{\pi}{2} + \psi\right)\mathbf{1}_{xp} + \mathbf{1}_{yp} + \cos\left(\frac{\pi}{2} - \theta\right)\mathbf{1}_{zp} \\
&= -\psi\mathbf{1}_{xp} + \mathbf{1}_{yp} + \theta\mathbf{1}_{zp} \\
\mathbf{1}_{za} &= \cos\left(\frac{\pi}{2} - \phi\right)\mathbf{1}_{xp} + \cos\left(\frac{\pi}{2} + \theta\right)\mathbf{1}_{yp} + \mathbf{1}_{zp} \\
&= \phi\mathbf{1}_{xp} - \theta\mathbf{1}_{yp} + \mathbf{1}_{zp}
\end{aligned} \qquad (2.53)$$

Therefore, the matrix that relates the two triads can be written as follows:

$$\begin{bmatrix} \mathbf{1}_{xa} \\ \mathbf{1}_{ya} \\ \mathbf{1}_{za} \end{bmatrix} = \begin{bmatrix} 1 & \psi & -\phi \\ -\psi & 1 & \theta \\ \phi & -\theta & 1 \end{bmatrix} \begin{bmatrix} \mathbf{1}_{xp} \\ \mathbf{1}_{yp} \\ \mathbf{1}_{zp} \end{bmatrix} \quad (2.54)$$

or in compact form

$$\mathbf{1}_a = C_a^p \mathbf{1}_p \quad (2.55)$$

In the case of interest, the misalignment orientation matrix has main diagonal elements, which are near unity, and off-diagonal elements, which are small compared with unity. This means that the two coordinate frames are in near coincidence. In general, transformation between two frames can be accomplished by use of the direction cosine matrix as indicated by Eq. (2.55). Now consider the transformation from the platform frame p into the navigation frame n. Specifically, let $\hat{C}_p^n$ be the estimated direction cosine matrix. This matrix, as we shall see later, is used by the inertial navigation system to convert the specific force measurements from the platform frame into the navigation frame. Since this direction cosine matrix is only an estimate, it contains errors that manifest themselves as attitude errors. Furthermore, since these misalignment errors are usually small, they can be treated as vectors. That is, small angles can be considered as vectors [5]. Therefore, the estimated direction cosine matrix $\hat{C}_p^n$ is related to the true direction cosine matrix C_p^n by the equation [3]

$$\hat{C}_p^n = \{I - [M]^{pn}\} C_p^n \quad (2.56)$$

where I is the unit matrix and $[M]^{pn}$ is the skew–symmetric (or antisymmetric) matrix given by

$$[M]^{pn} = \begin{bmatrix} 0 & -M_z & M_y \\ M_z & 0 & -M_x \\ -M_y & M_x & 0 \end{bmatrix} \quad (2.57)$$

which consists of six independent angles. The angles M_x, M_y, and M_z are, in general, very small and are defined in a conventional manner; that is, they will normally correspond to the angles θ, ϕ, and ψ. Equation (2.57) constitutes an orthogonal transformation, that is

$$[\hat{C}_p^n]^T = \hat{C}_n^p = C_n^p \{I - [M]^{pn}\} \quad (2.58)$$

Therefore, the estimated (or computed) transformation $\hat{C}_p^n$ will contain errors caused by

1. Initial misalignment.
2. Errors in the measured angles $\hat{\theta}$, $\hat{\phi}$, and $\hat{\psi}$ which include such errors as

synchro errors, analog-to-digital converter noise, and platform orientation errors.
3. Elastic deformation of the aircraft such as bending and twisting.

The matrix $[M]^{pn}$ consists of the errors above listed. Equation (2.56) is also valid if the misalignment between the platform and inertial coordinate system is desired.

In preprocessing of INS data, the accelerometer-derived velocity increments must be compensated for bias, scale factor, and nonorthogonality calibration values and transformed to inertial reference coordinates. In aircraft applications using gimbaled inertial platforms, aircraft attitude is computed from inertial platform gimbal angle data. As we shall see later, a resolver mounted on each gimbal axis encodes the angular position of each gimbal about its respective axis. Calibration terms are normally applied to correct for resolver zero biases and inertial measurement unit (IMU) misalignments. As alignment corrections are applied to the platform during ground alignment, the cross-coupling effects due to the initial physical misalignments are reduced.

Now consider the platform–sensor arrangement discussed in Section 2.2.11 and illustrated in Fig. 2-11. To a first-order approximation, the transformation from an ideal platform to the actual platform can be realized by

$$\mathbf{X} = R[\cos\theta \cos\phi + E_z \sin\theta \cos\phi - E_y \sin\phi]\boldsymbol{\rho}_x$$
$$\mathbf{Y} = R[-E_z \cos\theta \cos\phi + \sin\theta \cos\phi + E_x \sin\phi]\boldsymbol{\rho}_y$$
$$\mathbf{Z} = R[E_y \cos\theta \cos\phi - E_x \sin\theta \cos\phi + \sin\phi]\boldsymbol{\rho}_z$$

where $\Delta R, \Delta\theta, \Delta\phi$ = sensor measurement error
E_x, E_y, E_x = platform pointing error
$\boldsymbol{\rho}_x, \boldsymbol{\rho}_y, \boldsymbol{\rho}_z$ = ideal platform unit vectors

2.5. QUATERNION REPRESENTATION IN COORDINATE TRANSFORMATIONS

As mentioned in Section 2.2, the orientation of a right-handed Cartesian coordinate frame with reference to another one is characterized by the coordinate transformation matrix that maps the three components of a vector resolved in one frame into the same vector's components resolved into the other coordinate frame. The coordinate transformation matrix can be represented by three Euler angles or by a four-parameter quaternion. The Euler angles are a set of three rotations taken in a specified order to generate the desired orientation. The same orientation can be uniquely described by means of the nine direction cosines that exist between the unit vectors of the two frames. Whereas there is no redundancy in the Euler angle description, there are six constraints on the direction cosines (e.g., the sums of the squares of elements in a single row or column equal unity). More specifically, in order to describe the inertial coordinate frame with respect to a body frame, a set of three direction cosine elements is employed to determine the orientation of each inertial axis with respect to the three axes of the body frame. Thus, a set of nine direction cosine elements is employed to specify the orientation of the three inertial axes with respect to the body frame. For example, the gyroscopes in a strapdown inertial navigation system measure angular increments of the vehicle body with respect to inertial space. If it is desired to obtain the attitude of the vehicle with respect to an earth-referenced coordinate frame, then the angular increments that the reference coordinate frame makes with respect to inertial space must be subtracted from the body angular increments. The "inertial" reference frame in the strapdown computer must be rotated at appropriate rates in order to yield an "earth" reference frame in the computer. Moreover, in order to effect a proper subtraction from the body angular increments, which are in body axis coordinates, the coordinate frame rates must first be transformed from earth-reference coordinates to body-axes coordinates. In this section we will discuss the quaternion method and its associated algebra. The quaternion is the basis for all of the four-parameter transformations. Essentially, a quaternion is a means of describing angular orientation with four parameters, the minimum redundancy that removes the indeterminate points of three-parameter (Euler angles) descriptions. Thus, a rotation of one coordinate frame relative to another can be expressed uniquely in terms of four parameters, three to define the axis of rotation and the fourth to specify the amount of rotation. Because of the amount of computation involved with higher-order integration methods, in certain applications it is more efficient to propagate four quaternion parameters instead of nine direction cosines, and then calculate the matrix entries algebraically.

Quaternion algebra is an extension of three-dimensional algebra developed as a "new kind of algebraic formalism" by Sir William Rowan Hamilton in 1843. A quaternion $[Q]$ is a quadruple of real numbers, written in a definite order, as a three-dimensional vector. That is, Hamilton's quaternion is defined as a hypercomplex number of the form [2, 16]

$$[Q] = q_0 + q_1\mathbf{i} + q_2\mathbf{j} + q_3\mathbf{k} = (q_0, q_1, q_2, q_3) = (q_0, \mathbf{q}) \qquad (2.59)$$

where q_0, q_1, q_2, q_3 are real numbers, and the set $\{1, \mathbf{i}, \mathbf{j}, \mathbf{k}\}$ forms a basis for a quaternion vector space. Specifically, quaternions span the space of real and imaginary numbers and, since $\mathbf{i}, \mathbf{j}, \mathbf{k}$ can serve as an orthogonal basis, quaternion algebra includes scalar and vector algebra. The quaternion will have its tensor equal to unity. Addition, subtraction, and multiplication of a quaternion by a scalar are done in the same manner as in vector algebra:

2.5. Quaternion Representation in Coordinate Transformations

Addition The sum of two quaternions $[Q]$ and $[S]$ is as follows:

$$[Q]+[S] = (q_0 + q_1\mathbf{i} + q_2\mathbf{j} + q_3\mathbf{k}) + (s_0 + s_1\mathbf{i} + s_2\mathbf{j} + s_3\mathbf{k})$$
$$= (q_0+s_0) + (q_1+s_1)\mathbf{i} + (q_2+s_2)\mathbf{j} + (q_3+s_3)\mathbf{k}$$

Subtraction Quaternion subtraction is merely the addition of a negative quaternion:

$$-[Q] = (-1)[Q]$$

or

$$[Q]-[S] = Q + (-1)S = (q_0 - s_0, q_1 - s_1, q_2 - s_2, q_3 - s_3)$$

Multiplication The product of two quaternions $[Q]$ and $[S]$ is

$$[Q][S] = (q_0 + q_1\mathbf{i} + q_2\mathbf{j} + q_3\mathbf{k})(s_0 + s_1\mathbf{i} + s_2\mathbf{j} + s_3\mathbf{k})$$
$$= (q_0 s_0 - q_1 s_1 - q_2 s_2 - q_3 s_3) + (q_0 s_1 + q_1 s_0 + q_2 s_3 - q_3 s_2)\mathbf{i}$$
$$+ (q_0 s_2 - q_1 s_3 + q_2 s_0 + q_3 s_1)\mathbf{j}$$
$$+ (q_0 s_3 + q_1 s_2 - q_2 s_1 + q_3 s_0)\mathbf{k}$$

In general, quaternion multiplication is not commutative; that is, $[Q][S] \neq [S][Q]$. Multiplication by a scalar λ yields

$$\lambda[Q] = \lambda q_0 + \lambda q_1\mathbf{i} + \lambda q_2\mathbf{j} + \lambda q_3\mathbf{k}$$

Quaternion multiplication is defined by using the distributive law on the elements as in ordinary algebra, except that the order of the unit vectors must be preserved. The bases $\mathbf{i}$, $\mathbf{j}$, $\mathbf{k}$ are generalizations of $\sqrt{-1}$, which satisfy the relationships

$$\mathbf{i}^2 = \mathbf{j}^2 = \mathbf{k}^2 = -1$$

and by cyclic symmetry

$$\mathbf{ij} = -\mathbf{ji} = \mathbf{k}$$
$$\mathbf{jk} = -\mathbf{kj} = \mathbf{i}$$
$$\mathbf{ki} = -\mathbf{ik} = \mathbf{j}$$

Conjugate The conjugate $[Q]^*$ of a quaternion is defined as follows:

$$[Q]^* = q_0 - q_1\mathbf{i} - q_2\mathbf{j} - q_3\mathbf{k}$$

Moreover, the norm (or length) of a quaternion $N(Q)$ is a scalar defined as the product of a quaternion and its conjugate. Thus,

$$N(Q) = [Q][Q]^* = [Q]^*[Q]$$
$$= q_0^2 + q_1^2 + q_2^2 + q_3^2$$

Using the isomorphism we have $[Q][Q]^* = \det(Q)$. Conventionally, $\det(Q)$ is called the *norm*, denoted by $N(Q)$, so we have $[Q][Q]^* = N(Q)$. Similarly, $[Q]^*[Q] = N(Q)$, or $[Q][Q]^* = N(Q) = [Q]^*[Q]$. The norm of the product of two quaternions is equal to the product of their norms. Hence $N(QS) = NQ \cdot N(S)$ and consequently, $|[Q[[S]]| = |[Q]||[S]|$. By mathematical induction, the product of n quaternion factors is

$$N(Q_1 Q_2 \cdots Q_n) = NQ_1 NQ_2 \cdots NQ_n$$

Inverse If $[Q]$ is not zero, then the inverse quaternion $[Q]^{-1}$ is defined as

$$[Q][Q]^{-1} = [Q]^{-1}[Q] = 1$$

Using the norm concept, we obtain

$$[Q]^{-1} = [Q]^*/N(Q)$$

where $N(Q) \neq 0$. Thus, we can also write

$$[Q][Q]^{-1} = [Q][Q]^*/N(Q) = [Q]^{-1}[Q] = 1$$

Identities

1. A zero quaternion $[Q_0]$ is a quaternion with zero scalar and zero vector:

$$[0] = 0 + 0\mathbf{i} + 0\mathbf{j} + 0\mathbf{k}$$

2. A unit quaternion is defined as any quaternion whose norm is 1. Thus

$$[1] = 1 + 0\mathbf{i} + 0\mathbf{j} + 0\mathbf{k}$$

Since there is a single redundancy in a four-parameter description of coordinate rotations, as opposed to six for the direction cosines, the quaternion parameters satisfy a single constraint equation. The constraint equation on a unit quaternion then is

$$q_0^2 + q_1^2 + q_2^2 + q_3^2 = [Q][Q]^* = 1$$

Computationally, $(q_0^2 + q_1^2 + q_2^2 + q_3^2)^{1/2}$ can be used as a normalizing factor for each parameter. This is analogous to the periodic orthogonalization procedure used in conjunction with the direction cosine propagations.

Equality The equality of two quaternions $[Q]$ and $[S]$ is defined when their scalars are equal and their vectors are equal. Thus, $[Q] = [S]$ if and only if

$$q_0 = s_0, \quad q_1 = s_1, \quad q_2 = s_2, \quad q_3 = s_3, \quad \text{or} \quad [Q]_i = [S]_i, \quad i = 0, 1, 2, 3$$

2.5. Quaternion Representation in Coordinate Transformations

From the preceding discussion, the unit quaternion can now be defined in a similarity transformation in which a vector **R** in the inertial coordinate frame is transformed into the body frame as [2]

$$\mathbf{R}^i = [Q]\mathbf{R}^b[Q]^*$$
$$N(Q) = 1 \tag{2.60}$$

Now consider $\{\mathbf{i}, \mathbf{j}, \mathbf{k}\}$ as the orthogonal basis for the real-vector space R^3. Then, any vector $\mathbf{x} = x_1\mathbf{i} + x_2\mathbf{j} + x_3\mathbf{k}$ in R^3 can be expressed in quaternion form with a zero scalar term. Similarly, from our previous definition of a quaternion, any quaternion

$$[Q] = q_0 + q_1\mathbf{i} + q_2\mathbf{j} + q_3\mathbf{k}$$

can be expressed as the sum of a scalar and a vector ($[Q] = q_0 + \mathbf{q}$). Figure 2-15 depicts the spatial orientation of a quaternion.

Next, we will discuss how the quaternions are used in coordinate transformations. Consider Fig. 2-16, illustrating two arbitrary coordinate frames $a(X, Y, Z)$ and $b(X', Y', Z')$. Furthermore, the quaternion that describes the rotation of coordinate system a into coordinate system b will be denoted by Q_a^b.

As in the case of vectors, there are equivalence classes of quaternions. For example, a quaternion $[Q]$ represents a transformation [2]

$$\mathbf{x}' = [Q]^{-1}\mathbf{x}[Q] \tag{2.61}$$

Let us now recapitulate some of the previous results.

$$[Q] = q_0 + q_1\mathbf{i} + q_2\mathbf{j} + q_3\mathbf{k}$$
$$[Q]^* = q_0 - q_1\mathbf{i} - q_2\mathbf{j} - q_3\mathbf{k}$$
$$[Q][Q]^* = N(Q) = q_0^2 + q_1^2 + q_2^2 + q_3^2 = [Q]^*[Q]$$
$$[Q]^1 = [Q]^*/N(Q)$$

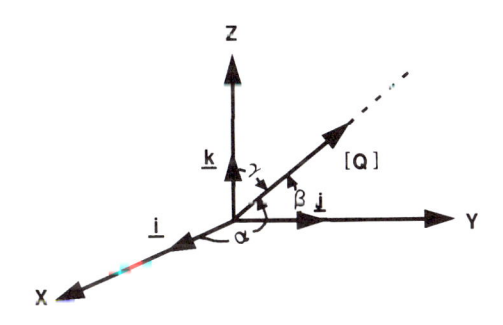

FIGURE 2-15

Spatial representation of the quaternion.

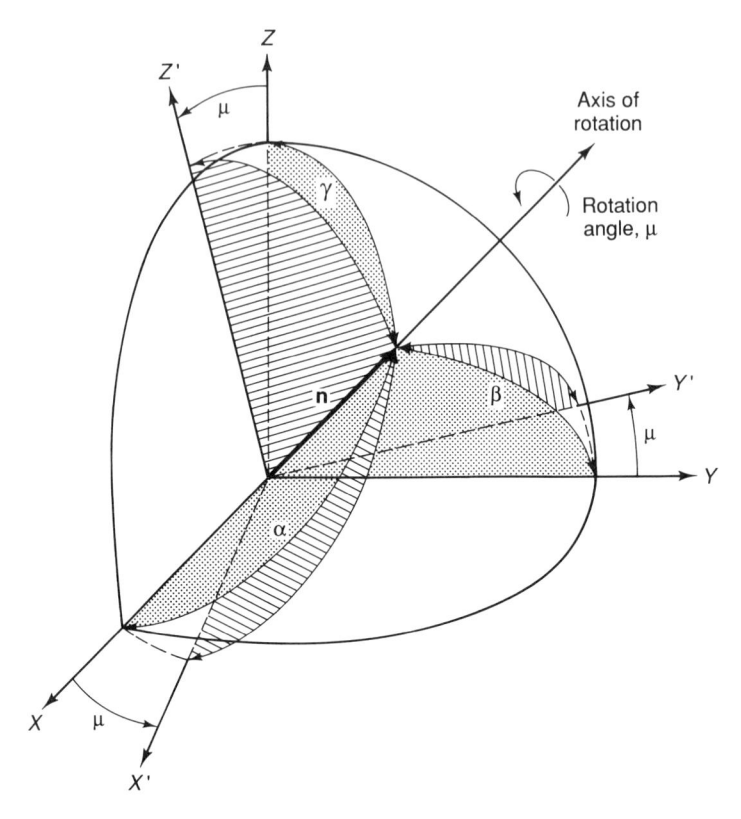

FIGURE 2-16

Rotation axis of "vector" about which system a is rotated in order to coordinate with system b.

(*Note*: If $N(Q) = 1$, then $[Q]^{-1} = [Q]$.) Similar to Eq. (2.61), we can write

$$[Q]^* \mathbf{p}[Q] = \mathbf{p}' \tag{2.62}$$

Performing the indicated operation, we have

$$(q_0 - q_1\mathbf{i} - q_2\mathbf{j} - q_3\mathbf{k})(X\mathbf{i} + Y\mathbf{j} + Z\mathbf{k})(q_0 + q_1\mathbf{i} + q_2\mathbf{j} + q_3\mathbf{k}) = \mathbf{p}'$$

Using the multiplicative rules of **i**, **j**, **k** stated earlier, we can expand this equation into

$$\begin{aligned}\mathbf{p}' = &\mathbf{i}\{X[q_0^2 + q_1^2 - q_2^2 - q_3^2] + Y[2q_3q_0 + 2q_1q_2] + Z[2q_1q_3 - 2q_0q_2]\} \\ &+ \mathbf{j}\{X[2q_1q_2 - 2q_3q_0] + Y[q_0^2 - q_1^2 + q_2^2 - q_3^2] + Z[2q_1q_0 + 2q_3q_2]\} \\ &+ \mathbf{k}\{X[2q_0q_2 + 2q_1q_3] + Y[2q_2q_3 - 2q_0q_1] + Z[q_0^2 - q_1^2 - q_2^2 + q_3^2]\}\end{aligned}$$

2.5. Quaternion Representation in Coordinate Transformations

In matrix form this equation becomes

$$\mathbf{p'} = \begin{pmatrix} X' \\ Y' \\ Z' \end{pmatrix}$$

$$= \begin{pmatrix} q_0^2+q_1^2-q_2^2-q_3^2 & 2(q_3q_0+q_1q_2) & 2(q_1q_3-q_0q_2) \\ 2(q_1q_2-q_3q_0) & q_0^2-q_1^2+q_2^2-q_3^2 & 2(q_1q_0+q_3q_2) \\ 2(q_0q_2+q_1q_3) & 2(q_2q_3-q_0q_1) & q_0^2-q_1^2-q_2^2+q_3^2 \end{pmatrix} \begin{pmatrix} X \\ Y \\ Z \end{pmatrix} \quad (2.63a)$$

$$\begin{bmatrix} X \\ Y \\ Z \end{bmatrix} = \begin{bmatrix} q_0^2+q_1^2-q_2^2-q_3^2 & 2(q_1q_2-q_3q_0) & 2(q_1q_3+q_0q_2) \\ 2(q_1q_2+q_0q_3) & q_0^2-q_1^2+q_2^2-q_3^2 & 2(q_2q_3+q_0q_1) \\ 2(q_1q_3-q_0q_2) & 2(q_2q_3+q_0q_1) & q_0^2-q_1^2-q_2^2+q_3^2 \end{bmatrix} \begin{bmatrix} X' \\ Y' \\ Z' \end{bmatrix} \quad (2.63b)$$

From the preceding discussion, it is clear that a rotating matrix using quaternion elements can be defined that is similar to the one using direction cosines. Toward this end, use will be made of Fig. 2-9 and Eq. (2.35) to show the equivalence. From Eq. (2.35) we can write

$$[C] = \begin{bmatrix} C_{11} & C_{12} & C_{13} \\ C_{21} & C_{22} & C_{23} \\ C_{31} & C_{32} & C_{33} \end{bmatrix}$$

Also, the quaternion can be written in the form

$$[Q] = (q_0 + q_1\mathbf{i} + q_2\mathbf{j} + q_3\mathbf{k})$$

The quaternion can also be expressed as a 4×4 matrix. Thus

$$[Q] = \begin{bmatrix} q_0 & q_1 & q_2 & q_3 \\ -q_1 & q_0 & -q_3 & q_2 \\ -q_2 & q_3 & q_0 & -q_1 \\ -q_3 & -q_2 & q_1 & q_0 \end{bmatrix}$$

where, as before, q_0, q_1, q_2, q_3 are the real components. From the above discussion, we see that there exists an equivalence between the elements of the quaternion (q_0, q_1, q_2, q_3) and the direction cosine elements. Therefore, by equating the corresponding elements of the matrix Eq. (2.63b) with the corresponding direction cosine matrix $[C]$ of Eq. (2.35), we have the following nine equations:

$$\begin{aligned}
C_{11} &= \cos\theta \cos\psi & &= q_0^2+q_1^2-q_2^2-q_3^2 \\
C_{12} &= \cos\theta \sin\psi & &= 2(q_1q_2-q_0q_3) \\
C_{13} &= -\sin\theta & &= 2(q_1q_3+q_0q_2) \\
C_{21} &= \sin\phi \sin\theta \cos\psi - \cos\phi \sin\psi &&= 2(q_1q_2+q_0q_3) \\
C_{22} &= \sin\phi \sin\theta \sin\psi + \cos\phi \cos\psi &&= q_0^2-q_1^2+q_2^2-q_3^2
\end{aligned}$$

$$C_{23} = \sin \phi \cos \theta \qquad = 2(q_2 q_3 - q_1 q_0)$$
$$C_{31} = \cos \phi \sin \theta \cos \psi + \sin \phi \sin \psi \qquad = 2(q_1 q_3 - q_0 q_2)$$
$$C_{32} = \cos \phi \sin \theta \sin \psi - \sin \phi \cos \psi \qquad = 2(q_2 q_3 + q_0 q_1)$$
$$C_{33} = \cos \phi \cos \theta \qquad = q_0^2 - q_1^2 - q_2^2 + q_3^2$$

A primary advantage in using quaternions lies in the fact that only four differential equations are necessary for finding the new transformation matrix, whereas the direction cosine method requires nine. The process of finding the quaternion components as functions of the Euler angles is quite straightforward. Consider the two coordinate systems illustrated in Fig. 2-16. Furthermore, it will be assumed that the coordinate system (X, Y, Z) is fixed in space, while the coordinate system (X', Y', Z') is moving in some arbitrary manner; however, both coordinate systems have the same origin. Using Euler's theorem, the (X', Y', Z') coordinate system is rotated through the angle μ about some fixed axis, which makes angles α, β, and γ with the (X, Y, Z) axes, respectively. Note that this axis of rotation makes the same angles α, β, γ with the (X', Y', Z') axes, also. From the four-parameter system $(\mu, \alpha, \beta, \gamma)$, another transformation matrix will be derived that will aid in the development of the differential equations of the four quaternion parameters.

Now let the quaternion that describes the rotation of system a into system b be denoted by Q_a^b. Consequently, since the quaternion is composed of four elements, we can denote it by

$$Q_a^b = (q_0, q_1, q_2, q_3) = q_0 1 + q_1 \mathbf{i} + q_2 \mathbf{j} + q_3 \mathbf{k}$$
$$= (q_0, \mathbf{q})$$

Conceptually, q_1, q_2, q_3 define a vector in space and q_0 is the amount of rotation about that vector. Referring to Fig. 2-16, it is clear that the axis of rotation or vector $\mathbf{n}$ is the axis about which system a is rotated in order to coincide with system b. If, as indicated above, μ is the amount of rotation and α, β, and γ are the angles between the axis of rotation and the coordinate system (X, Y, Z), then the elements of the quaternion which describe this rotation are by definition

$$q_0 = \cos \frac{\mu}{2} \qquad (2.64a)$$

$$q_1 = \sin \frac{\mu}{2} \cos \alpha \qquad (2.64b)$$

$$q_2 = \sin \frac{\mu}{2} \cos \beta \qquad (2.64c)$$

$$q_3 = \sin \frac{\mu}{2} \cos \gamma \qquad (2.64d)$$

2.5. Quaternion Representation in Coordinate Transformations

or in vector form

$$[Q] = \begin{bmatrix} \cos\dfrac{\mu}{2} \\ \sin\dfrac{\mu}{2}\cos\alpha \\ \sin\dfrac{\mu}{2}\cos\beta \\ \sin\dfrac{\mu}{2}\cos\gamma \end{bmatrix}$$

From these equations and the definition of the quaternion, we have

$$[Q] = q_0 + q_1\mathbf{i} + q_2\mathbf{j} + q_3\mathbf{k}$$
$$= N\left[\mathbf{n}\sin\left(\dfrac{\mu}{2}\right) + \cos\left(\dfrac{\mu}{2}\right)\right] \qquad 0 \le \mu \le \pi$$

where

$$N = q_0^2 + q_1^2 + q_2^2 + q_3^2$$
$$\sin\left(\dfrac{\mu}{2}\right) = \dfrac{\pm\sqrt{q_1^2 + q_2^2 + q_3^2}}{N}$$
$$\cos\left(\dfrac{\mu}{2}\right) = \dfrac{q_0}{N}$$

Specifically, we note that the vector $\mathbf{n}$ is the invariant Euler axis of rotation and μ is the rotation about that axis that takes system a into system b in the least angle sense. The vector $\mathbf{n}$ is given by

$$\mathbf{n} = \dfrac{q_1\mathbf{i} + q_2\mathbf{j} + q_3\mathbf{k}}{\sqrt{q_1^2 + q_2^2 + q_3^2}}$$

It is apparent that the rotation μ is smaller than the algebraic sum of the Euler angles ϕ, θ, ψ and that the angle μ is the shortest angular path between the two coordinate systems.

As before, let $[C]$ be a general rotation matrix and $[Q]$ the equivalent quaternion. It can be shown that the correspondence between $[C]$ and $[Q]$ is

$$[C] = \begin{bmatrix} 1 - 2(q_2^2 + q_3^2) & 2(q_1 q_2 - q_3 q_0) & 2(q_3 q_1 + q_2 q_0) \\ 2(q_3 q_0 + q_1 q_2) & 1 - 2(q_3^2 + q_1^2) & 2(q_2 q_3 - q_0 q_1) \\ 2(q_3 q_1 - q_2 q_0) & 2(q_2 q_3 + q_0 q_1) & 1 - 2(q_1^2 + q_2^2) \end{bmatrix} \qquad (2.65)$$

Therefore, if (x, y, z) is a body-fixed coordinate system and (X, Y, Z) is an earth-fixed coordinate system, then

$$\begin{bmatrix} X \\ Y \\ Z \end{bmatrix} = [C] \begin{bmatrix} x \\ y \\ z \end{bmatrix}$$

and the inverse transformation from earth-fixed to body-fixed coordinates is given by

$$\begin{bmatrix} x \\ y \\ z \end{bmatrix} = [C]^{-1} \begin{bmatrix} X \\ Y \\ Z \end{bmatrix}$$

where $[C]^{-1} = [C]^T$. That is, the orthogonality of $[C]$ allows us to use the transpose of $[C]$ instead of the inverse.

Using Eq. (2.63a), and after some algebraic manipulation, we can obtain the equations for the quaternion elements in terms of the direction cosine elements. First, we note that the trace of Eq. (2.63a) is

$$\text{tr}(C) = 3q_0^2 - q_1^2 - q_2^2 - q_3^2$$

Then, from Eq. (2.64a), we have

$$q_0 = \cos\left(\frac{\mu}{2}\right)$$

Starting with this equation, we can write

$$q_0 = \cos\left(\frac{\mu}{2}\right)$$
$$= \frac{1}{2}\sqrt{4\cos^2\left(\frac{\mu}{2}\right)}$$
$$= \frac{1}{2}\sqrt{1 + 3\cos^2\left(\frac{\mu}{2}\right) - 1 + \cos^2\left(\frac{\mu}{2}\right)}$$
$$= \frac{1}{2}\sqrt{1 + 3\cos^2\left(\frac{\mu}{2}\right) - \sin^2\left(\frac{\mu}{2}\right)[\cos^2\alpha + \cos^2\beta + \cos^2\gamma]}$$
$$= \frac{1}{2}\sqrt{1 + 3\cos^2\left(\frac{\mu}{2}\right) - \sin^2\left(\frac{\mu}{2}\right)\cos^2\alpha - \sin^2\left(\frac{\mu}{2}\right)\cos^2\beta - \sin^2\left(\frac{\mu}{2}\right)\cos^2\gamma}$$
$$= \frac{1}{2}\sqrt{1 + 3q_0^2 - q_1^2 - q_2^2 - q_3^2}$$
$$= \frac{1}{2}\sqrt{1 + \underbrace{q_0^2 + q_1^2 - q_2^2 - q_3^2}_{C_{11}} + \underbrace{q_0^2 - q_1^2 + q_2^2 - q_3^2}_{C_{22}} + \underbrace{q_0^2 - q_1^2 - q_2^2 + q_3^2}_{C_{33}}}$$
$$= \frac{1}{2}\sqrt{1 + \text{tr}(C)} \qquad (2.66)$$

2.5. Quaternion Representation in Coordinate Transformations

where tr is the trace of the matrix $[C]$, that is, $\text{tr}(C) = C_{11} + C_{22} + C_{23}$. Equation (2.66) can also be written in the form

$$4q_0^2 = 1 + \text{tr}(C)$$

The other quaternion elements are obtained from the relations

$$q_1 = \frac{C_{32} - C_{23}}{4q_0} \tag{2.67}$$

$$q_2 = \frac{C_{13} - C_{31}}{4q_0} \tag{2.68}$$

$$q_3 = \frac{C_{21} - C_{12}}{4q_0} \tag{2.69}$$

Similarly, in terms of the diagonal terms, we have

$$q_1 = \frac{C_{12} + C_{23}}{4q_2} \tag{2.70}$$

$$q_2 = \frac{C_{32} + C_{23}}{4q_3} \tag{2.71}$$

$$q_3 = \frac{C_{13} + C_{31}}{4q_1} \tag{2.72}$$

Consequently, the above equations readily yield the solution for the quaternion elements. In order to start the procedure, when the elements are programmed on a computer, select either q_0 or one of the q_i ($i = 1, 2, 3$) to start the computation and use the positive square root in Eq. (2.66).

Since there is a single redundancy in a four-parameter description of coordinate rotations, as opposed to six for the direction cosines, the quaternion parameters satisfy a single constraint equation (orthogonality of quaternions) [15]:

$$q_0^2 + q_1^2 + q_2^2 + q_3^2 = 1 \tag{2.73}$$

In terms of the Euler angles ψ, θ, and ϕ, it can be shown that q_0, q_1, q_2, q_3 are given by

$$q_0 = \cos\frac{\psi}{2}\cos\frac{\theta}{2}\cos\frac{\phi}{2} - \sin\frac{\psi}{2}\sin\frac{\theta}{2}\sin\frac{\phi}{2} \tag{2.74a}$$

$$q_1 = \sin\frac{\theta}{2}\sin\frac{\phi}{2}\cos\frac{\psi}{2} + \sin\frac{\psi}{2}\cos\frac{\theta}{2}\cos\frac{\phi}{2} \tag{2.74b}$$

$$q_2 = \sin\frac{\theta}{2}\cos\frac{\psi}{2}\cos\frac{\phi}{2} - \sin\frac{\psi}{2}\sin\frac{\phi}{2}\cos\frac{\theta}{2} \tag{2.74c}$$

$$q_3 = \sin\frac{\phi}{2}\cos\frac{\psi}{2}\cos\frac{\theta}{2} + \sin\frac{\psi}{2}\sin\frac{\theta}{2}\cos\frac{\phi}{2} \tag{2.74d}$$

Equations (2.74a)–(2.74d) determine uniquely the quaternion components for any given set of Euler angles. The inverse relationship (the Euler angles as functions of the quaternion components) can be obtained directly from these equations. It should be pointed out, however, that although these relations define the quaternion parameters, they are not convenient for initialization. In addition to requiring sine and cosine evaluations, the triple products of those functions suffer from potential numerical inaccuracies on a short-wordlength computer. Finally, the rotation that yielded Eq. (2.63b) can also be written in the form

$$[C] = \begin{bmatrix} 1 - 2\sin^2\frac{\mu}{2}\sin^2\alpha & 2\left(\sin^2\frac{\mu}{2}\cos\alpha\cos\beta - \sin\frac{\mu}{2}\cos\frac{\mu}{2}\cos\gamma\right) & 2\left(\cos\alpha\cos\gamma\sin^2\frac{\mu}{2} + \sin\frac{\mu}{2}\cos\frac{\mu}{2}\cos\beta\right) \\ 2\left(\sin^2\frac{\mu}{2}\cos\alpha\cos\beta - \sin\frac{\mu}{2}\cos\frac{\mu}{2}\cos\gamma\right) & 1 - 2\sin^2\frac{\mu}{2}\sin^2\beta & 2\left(\sin^2\frac{\mu}{2}\cos\beta\cos\gamma - \sin\frac{\mu}{2}\cos\frac{\mu}{2}\cos\alpha\right) \\ 2\left(\cos\alpha\cos\gamma\sin^2\frac{\mu}{2} - \sin\frac{\mu}{2}\cos\frac{\mu}{2}\cos\beta\right) & 2\left(\sin^2\frac{\mu}{2}\cos\beta\cos\gamma - \sin\frac{\mu}{2}\cos\frac{\mu}{2}\cos\alpha\right) & 1 - 2\sin^2\frac{\mu}{2}\sin^2\gamma \end{bmatrix}$$

The associated differential equations for the quaternion elements are given in the form

$$\dot{q}_0 = -\tfrac{1}{2}(\mathbf{q}^T\boldsymbol{\omega}) \qquad (2.75)$$

with $\mathbf{q} = q_1\mathbf{i} + q_2\mathbf{j} + q_3\mathbf{k}$ and

$$\dot{\mathbf{q}} = \tfrac{1}{2}[q_0\boldsymbol{\omega} - \boldsymbol{\omega}\times\mathbf{q}] = \tfrac{1}{2}[q_0\boldsymbol{\omega} + \mathbf{q}\times\boldsymbol{\omega}] \qquad (2.76)$$

where $\boldsymbol{\omega} = \omega_x\mathbf{i} + \omega_y\mathbf{j} + \omega_z\mathbf{k}$ is the angular velocity (or Darboux) vector, the components of which are direct outputs of the gyroscopes and are measured in the moving axes. From the above, we can now write a more compact

2.5. Quaternion Representation in Coordinate Transformations

differential equation in the form

$$d[Q]/dt = \tfrac{1}{2}\{B[Q]\} \tag{2.77}$$

where

$$B = \begin{bmatrix} 0 & -\omega_x & -\omega_y & -\omega_z \\ \omega_x & 0 & \omega_z & -\omega_y \\ \omega_y & -\omega_z & 0 & \omega_x \\ \omega_z & \omega_y & -\omega_x & 0 \end{bmatrix} = \begin{bmatrix} 0 & -\omega^T \\ \omega & -\Omega \end{bmatrix}$$

In terms of the scalar components, we have

$$\dot{q}_0 = -\tfrac{1}{2}[q_1\omega_x + q_2\omega_y + q_3\omega_z] \tag{2.78a}$$

$$\dot{q}_1 = \tfrac{1}{2}[q_0\omega_x + q_2\omega_z - q_3\omega_y] \tag{2.78b}$$

$$\dot{q}_2 = \tfrac{1}{2}[q_0\omega_y - q_1\omega_z + q_3\omega_x] \tag{2.78c}$$

$$\dot{q}_3 = \tfrac{1}{2}[q_0\omega_z + q_1\omega_y - q_2\omega_x] \tag{2.78d}$$

These are linear differential equations that can be integrated easily, given the initial conditions $Q(t_0)$ and the forcing function $\omega(t)$, which is the relative angular velocity between the two coordinate frames in question. In strapdown (analytic) navigation systems, once Q is obtained, the attitude matrix is immediately given by the Euler four-parameter matrix Eq. (2.63a). It should be pointed out here, however, that since the outputs of the gyroscopes are physical quantities, they are never exact. Therefore, in order to solve Eq. (2.77), one must use numerical methods.

Quaternions follow the rules of matrix algebra. That is, they obey the associative and commutative laws of addition and the associative and distributive laws of multiplication, but, as we have seen earlier, they do not obey the commutative law of multiplication. Let three quaternions be designated as Q_1, Q_2, and Q_3. Then, the above properties may be indicated as follows:

Associative addition: $(Q_1 + Q_2) + Q_3 = Q_1 + (Q_2 + Q_3)$
Commutative addition: $Q_1 + Q_2 = Q_2 + Q_1$
Associative multiplication: $(Q_1 Q_2) Q_3 = Q_1 (Q_2 Q_3)$
Distributive multiplication: $Q_1 (Q_2 + Q_3) = Q_1 Q_2 + Q_1 Q_3$

(*Note*: For the commutative multiplication, $Q_1 Q_2 \neq Q_2 Q_1$.)

An example will be used to illustrate the relationship between the direction cosines and quaternions. For this example, we will write a computer program to compute the coordinate transformation matrix between, say, the a-frame and b-frame, given ordered rotations (Euler angles) of yaw (ψ), pitch (θ), and roll (ϕ). That is, the input data is ψ, θ, and ϕ, and the output data is C_a^b. In particular, we will write a program that takes C_a^b as the input and gives the quaternion Q_a^b as the output. The computer program will be interactive, allowing an input of ψ, θ, and ϕ to yield C_a^b and Q_a^b as output. By inputing values for ψ, θ, and ϕ, the program will compute the corresponding quaternion q_1, q_2, q_3, q_4 ($[Q] = (q_1 + q_2\mathbf{i} + q_3\mathbf{j} + q_4\mathbf{k})$) as output. The direction cosine matrix C_a^b used is the same as that given by Eq. (2.35), and for the transformation matrix for the quaternion Q_a^b we will use Eq. (2.63b). Note that the signs of the off-diagonal terms change; consequently, the signs in Eqs. (2.67)–(2.69) will also change. Following is a program listing for the example written in PASCAL.

```
PROGRAM QUATERNION (INPUT,OUTPUT);

(************************************************)
(*   Given ordered rotation (Euler angles) of   *)
(*   yaw (xi), pitch (theta), and roll (phi),   *)
(*   this program will compute the coordinate   *)
(*   transformation matrix between the a-frame  *)
(*   and b-frame. Then it will take the         *)
(*   transformation matrix as input and gives   *)
(*   the quaternion Q1,Q2,Q3,Q4 as output.      *)

const pi = 3.1415927; n = 3;
type
     DCM = array[1..n,1..n] of real;

var  C : DCM;
     theta,phi,xi,a,b,,c,d,e,f,x,y,z : real;
     q1,q2,q3,q4,tr,deltatr : real;
     done,q1equalzero : boolean;
     ch : char;
                              (Note: xi here is psi ( ψ ))
(******************************************************************)

procedure readindata;   (* read in all the Euler angles *)

begin
     writeln;writeln;
     write('enter xi = ');
     readln(xi); writeln;
     write('enter theta = ');
     readln(theta); writeln;
     write('enter phi = ');
     readln(phi); writeln
end;

procedure formdcm;   (* form direction cosine matrix *)

begin
     x := xi*pi/180;
     y := theta*pi/180;
     z := phi*pi/180;
```

2.5. Quaternion Representation in Coordinate Transformations 65

```
      a := cos(x); b := sin(x);
      c := cos(y); d := sin(y);
      e := cos(z); f := sin(z);
      C[1,1] := a*c; C[1,2] := c*b; C[1,3] := -d;
      C[2,1] := -e*b + f*d*a; C[2,2] := e*a + f*d*b; C[2,3] := f*c;
      C[3,1] := f*b + e*d*a; C[3,2] := -f*a + e*d*b; C[3,3] := e*c
end;

procedure computeq;   (* compute the quaternion *)

begin
   tr := C[1,1] + C[2,2] + C[3,3]; q1equalzero := false;
   deltatr := tr + 1;
   if (deltatr < 0.00001) and (deltatr > -0.0001) then (* tr = -1 *)
      q1equalzero := true
   else (* <> 0, we're o.k *)
      begin
         q1 := sqrt(1 + tr)/2;
         q2 := (C[2,3] - C[3,2])/(4*q1);
         q3 := (C[3,1] - C[1,3])/(4*q1);
         q4 := (C[1,2] - C[2,1])/(4*q1)
      end
end;

procedure print; (*write out direction cosine matrix and quaternion*)

var i,j : integer;

begin
   writeln;
   writeln('yaw(xi) = ',xi:4:6);writeln;
   writeln('pitch(theta) = ',theta:4:6);writeln;
   writeln('roll(phi) = ',phi:4:6);
   writeln;
   writeln('Direction Cosine Matrix...');writeln;
   for i := 1 to n do
      begin
         for j := 1 to n do
            write(C[i,j]:3:6,'       '); writeln
      end;
   if q1equalzero then   divider = 0 *)
      begin
         writeln;
         writeln('We can not compute quaternion for you,because');
         writeln('this method won't work with tr = -1 or Q1 = 0')
      end
   else (* we're o.k. *)
      begin
         writeln;
         writeln('Quaternion from input DCM ');
         writeln;
         writeln('Q1 = ',q1:5:6);
         writeln('Q2 = ',q2:5:6);
         writeln('Q3 = ',q3:5:6);
         writeln('Q4 = ',q4:5:6);
      end;
```

```
      writeln;
      (* redo the problem *)
      write('are you done ? [y/n] ');
      readln(ch);writeln;
      if ch <> 'n' then done := true
      else
          done := false
 end;

(*************************************************************)

(* main program *)
begin
   while not done do
      begin
         readindata;
         formdcm;
         computeq;
         print
      end
end.
% pix EE534.p
Execution begins...

enter xi = 45
enter theta = 75
enter phi = 25

yaw(xi) = 45.000000
pitch(theta) = 75.000000
roll(phi) = 25.000000

Direction Cosine Matrix...

  0.183013        0.183013       -0.965926
 -0.352203        0.929510        0.109382
  0.917856        0.320183        0.234570

Quaternion from input DCM

Q1 =   0.766011
Q2 =  -0.068799
Q3 =   0.614802
Q4 =   0.174676

are you done ? [y/n] n

enter xi = 55
enter theta = 90
enter phi = 15

yaw(xi) = 55.000000
pitch(theta) = 90.000000
roll(phi) = 15.000000
```

2.5. Quaternion Representation in Coordinate Transformations

```
-0.000000    -0.000000    -1.000000
-0.642788     0.766044    -0.000000
 0.766044     0.642788    -0.000000

Quaternion from input DCM

Q1 =  0.664463
Q2 = -0.241845
Q3 =  0.664463
Q4 =  0.241845

are you done   [y/n] y

Execution terminated.

188 statements executed in 0.22 seconds cpu time.
%
```

This section can be summarized by listing the advantages and disadvantages of each of the three transformation methods (Euler angles, direction cosines, and quaternions) as shown in Table 2-3.

TABLE 2-3

Advantages and Disadvantages of the Three Transformation Methods

	Advantages	Disadvantages
Euler angles	Only three differential equations are needed	Differential equations are nonlinear
	Direct initialization from ϕ_0, θ_0, and ψ_0	Singularity occurs as the angles approach $\pm 90°$
		Transformation matrix is not directly available
		Order of rotation is important
Direction cosines	Differential equations are linear	Nine linear differential equations (reducible to six)
	No singularities	Euler angles are not directly available, which are required for initial calculations
	Transformation matrix can be calculated directly	Computational burden
Quaternions	Only four linear coupled differential equations are needed	Initial calculations using Euler angles required if coordinate systems do not coincide at $t=0$
	No singularities; avoids the gimbal-lock problem associated with the Euler angles	Euler angles are not readily available
	Computationally simple	Transformation matrix not directly available

TABLE 2-4
Basic Algebraic Properties of Quaternions

Quaternion equation	$[Q] = q_0 + \mathbf{q} = q_0 + q_1\mathbf{i} + q_2\mathbf{j} + q_3\mathbf{k}$ $= (q_0, \mathbf{q})$
Addition	$[Q] + [S] = (q_0 + s_0) + (q_1 + s_1)\mathbf{i} + (q_2 + s_2)\mathbf{j} + (q_3 + s_3)\mathbf{k}$
Sutraction	$[Q] - [S] = (q_0 - s_0, q_1 - s_1, q_2 - s_2, q_3 - s_3)$
Multiplication	$[Q][S] = (q_0 s_0 - q_1 s_1 - q_2 s_2 - q_3 s_3)$ $+ (q_0 s_1 + q_1 s_0 + q_2 s_3 - q_3 s_2)\mathbf{i}$ $+ (q_0 s_2 - q_1 s_3 + q_2 s_0 + q_3 s_1)\mathbf{j}$ $+ (q_0 s_3 + q_1 s_2 - q_2 s_1 + q_3 s_0)\mathbf{k}$
Multiplication by a scalar	$\lambda[Q] = \lambda q_0 + \lambda q_1\mathbf{i} + \lambda q_2\mathbf{j} + \lambda q_3\mathbf{k}$
Unit vectors (bases)	$\mathbf{i}^2 = \mathbf{j}^2 = \mathbf{k}^2 = -1$ $\mathbf{ij} = -\mathbf{ji} = \mathbf{k}$ $\mathbf{jk} = -\mathbf{kj} = \mathbf{i}$ $\mathbf{ki} = -\mathbf{ik} = \mathbf{j}$
Conjugate	$[Q]^* = q_0 - q_1\mathbf{i} - q_2\mathbf{j} - q_3\mathbf{k}$ $[Q][Q]^* = q_0^2 + q_1^2 + q_2^2 + q_3^2$
Norm	$N(Q) = [Q][Q]^* = [Q]^*[Q]$
Inverse	$[Q][Q]^{-1} = [Q]^{-1}[Q] = 1$ $[Q]^{-1} = [Q]^*/N(Q)$
Equality	$[Q] = [S]$ if and only if $q_0 = s_0, q_1 = s_1, q_2 = s_2, q_3 = s_3$
Unit quaternion	$[Q][Q]^* = 1 = q_0^2 + q_1^2 + q_2^2 + q_3^2$

Finally, Table 2-4 summarizes the various algebraic properties of the quaternions.

2.6. THE UNIVERSAL TRANSVERSE MERCATOR GRID REFERENCE SYSTEM

The Universal Transverse Mercator (UTM) grid reference system is a coordinate system mainly used by the military as the standard position reference between the latitude of 84° north and 80° south. More importantly, this system virtually eliminates the convergence between meridians and large-scale variations of conventional latitude–longitude coordinates. The UTM system is also suitable for great circle navigation. In essence, the UTM projection is a mercator projection, whereby the earth is modeled by a sphere. The cylinder used in the UTM projection is tangent to the spherical earth at a great circle passing through the poles [1]. Furthermore, the intersection of the great circle with the equator forms the origin of the "northing" and "easting" coordinates.

Since the UTM is based on the transverse mercator projection applied to maps on the earth's surface, the earth is divided in quadrilaterals or "grid zones" 6° east–west by 8° north–south between the latitudes of 84°N and 80°S. However, between 72°N and 84°N these quadrilaterals do not always measure 6° wide. The Air Force, on the other hand, uses the UTM system for the latitudes between 80°S and 84°N, and the Universal Polar Stereographic (UPS) system from 84°N and 80°S to the respective poles. Another position reference system is the Military Grid Reference System (MGRS), which was developed for use with the UTM and UPS grids.

Historically, the use of military grid systems can be traced to World War I, when the French superimposed a military grid on maps of small areas in order to control artillery fire [1]. After the end of that war, a number of nations adapted military grid systems for their own military forces. Primarily, a military grid is composed of two series of equally spaced parallel lines perpendicular to each other. The north–south grid lines are known as northings and the east–west grid lines as eastings. Thus, a position reference system is any coordinate system that permits the designation of a point or an area on the earth's surface, commonly by an alphanumeric (letters and numbers) representation. On military maps of scale 1:75,000 and larger, the distance between successive grid lines represents 1000 meters (m) at the scale of the map. Present-day inertial navigation systems are capable of operating completely and continuously in the UTM coordinate system by solving a minimal number of algorithms in the navigation computer. Conversion for latitude–longitude to UTM presents no difficulties. The navigation computer continuously computes and displays present position via the CDU (Control and Display Unit) in geodetic latitude–longitude or alphanumeric representation. Some systems provide common grids directly, which can be converted by a computer operation to a standard military grid. For example, certain military applications of Loran C/D require that the receiver locations be given in UTM coordinates as well as geodetic latitude and longitude.

2.6.1. UTM Grid Zone Description

As stated above, the globe is divided into quadrilaterals or grid zones 6° east–west by 8° north–south between 84°N and 80°S latitudes, each of which is given a unique identification called the grid zone designation (GZD). The columns, which are 6° wide, are identified by the UTM zone numbers. That is, starting with the 180° meridian and proceeding easterly, the columns are numbered 1 through 60, consecutively. The rows, 8° high (except for the last, which is 12° high), are identified by letters. Starting from 80°S and proceeding northerly to 84°N, the rows are lettered alphabetically C through X, with the letters I and O omitted. Figure 2-17 illustrates the UTM grid designation [1].

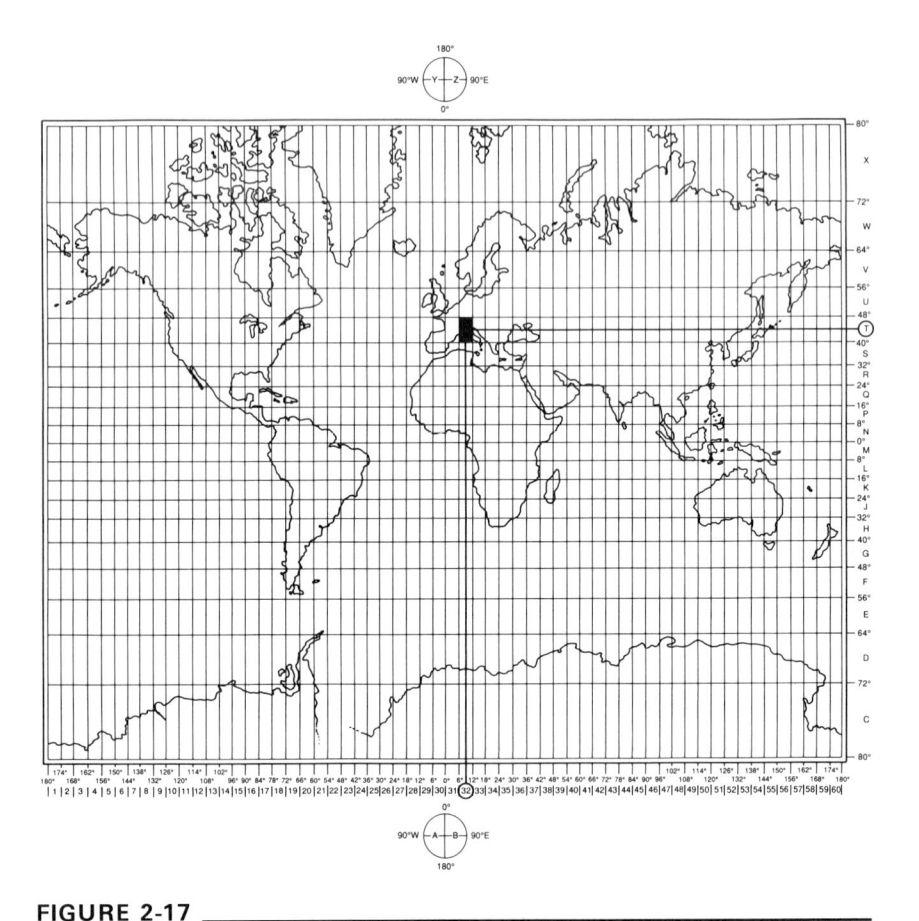

FIGURE 2-17
Designation of the UTM grid zones [1].

Next, the grid zone designation of any quadrilateral is determined by reading first to the right (column designation), and then up (row designation). Furthermore, each 6° × 8° area between 84°N and 80°S is subdivided into squares based on the UTM grid for that zone. Each column of squares, 100,000 m on a side is identified by a letter, as is each row of squares (see Fig. 2-18). By further subdivision, an object on the UTM map can be located down to 1-m square (that is, 1-m resolution). Reference 1 gives a complete breakdown of this identification system.

From Fig. 2-18, it can be seen that the numerical coordinates of a point are based on two measurements: (1) the distance from the equator and (2) the distance from the central meridian in each grid column. The central meridian is arbitrarily established as 500,000 m east, and the equator is "zero

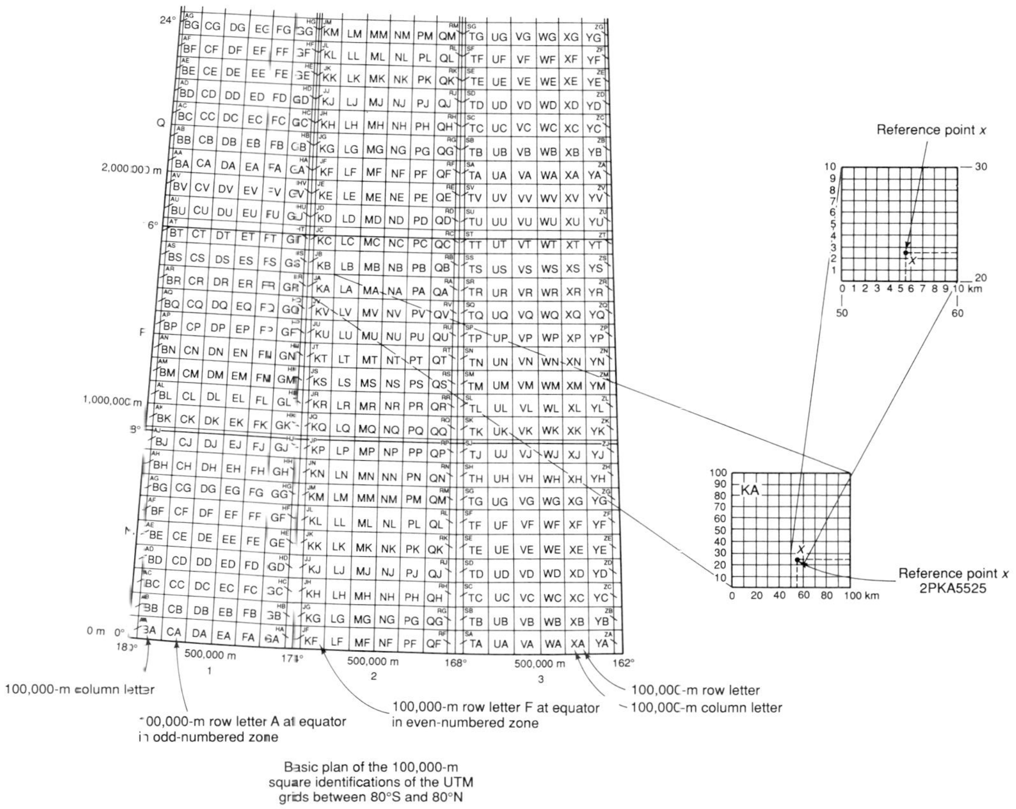

FIGURE 2-18
Designation of 100,000-m squares of the UTM grid between 80°S and 84°N [1].

northing" for the northern hemisphere. With reference to Fig. 2-18, it is a simple procedure for locating a point in the UTM grid system. For example, let us locate the point whose standard reference is 2PKA5525. This reference point can be explained as follows:

2P: Identifies the $6° \times 8°$ grid zone in which the point is located.

KA: Identifies the 100,000-m (100-km) square, in which point X is located.

55: The numbers 5 and 5 are both easting figures and are both read to the right. The first (5) identifies the interval easting beyond which point X falls. The second digit (5) is estimated; it represents the value of the imaginary 1000-m (1-km) easting column, which includes the point.

25: The numbers 2 and 5 are both northing figures and are both read up. The first (2) identifies the interval beyond which the point falls. The second digit (5) is estimated; it defines the 1000-m estimated northing required to complete the area dimensions of the reference point in the example.

In diagram form, these definitions are illustrated in Fig. 2-19. (Note that the last two digits or 100 m and 10 m of both easting and northing resolution are indicated as zero, but this need not be the case.)

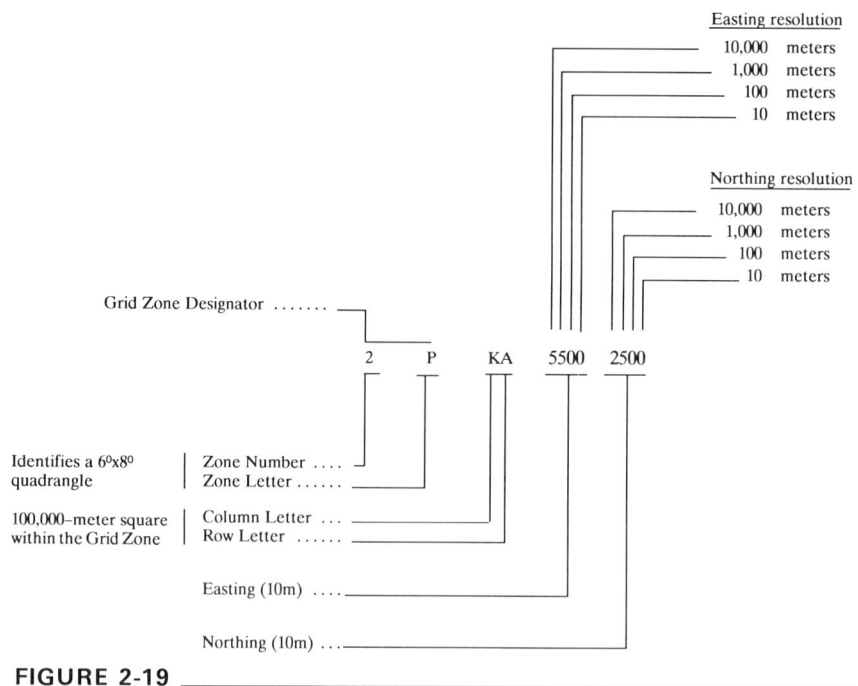

FIGURE 2-19
UTM coordinate definition.

2.6. The Universal Transverse Mercator Grid Reference System

A point of interest within a zone is defined by two numbers: (1) a northing and (2) an easting coordinate. These are distances on a grid given in meters (m) or kilometers (km) from the origin. The origin is established as being the intersection of the equator with the central meridian, modified by a "false easting" of 500 km and a "false northing" of 10,000 km. Consequently, since any point that lies on a grid boundary is equidistant to the central meridian in both zones, the easting values at a boundary can be expressed as $500 + X$ for the eastern boundary and $500 - X$ for the western boundary, where X equals the distance to the central meridian. These concepts are illustrated in Fig. 2-20.

Both northing and easting are always positive numbers and increase in their respective directions. Moreover, the hemisphere must also be indicated, since the same northing coordinates appear in either the northern or southern hemisphere. In order to see where the numbers in Fig. 2.20 originate, the following calculation will make it clear. Let the equatorial radius R_{eq} take the value $R_{eq} = 6382$ km (a spherical earth model has been assumed).

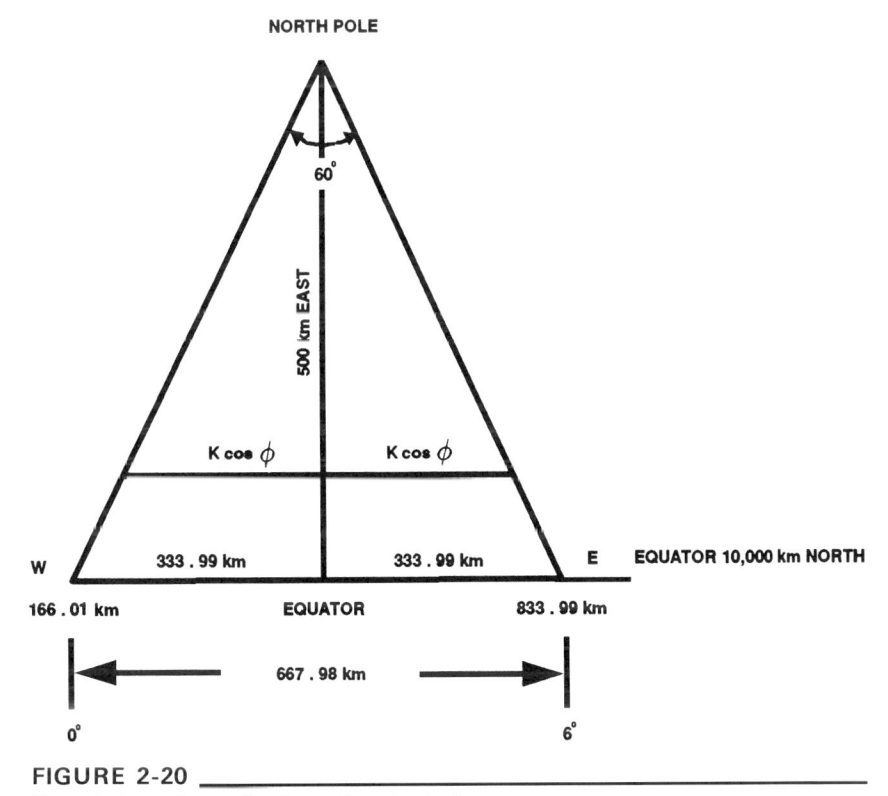

FIGURE 2-20
Numerical coordinates for the northern hemisphere.

Then
$$C_{eq} = 2\pi R_{eq} = 6.28(6382) = 40{,}078.96 \text{ km}$$
Therefore
$$\text{Width of a } 6° \text{ grid column at the equator} = C_{eq}/60$$
$$= 667.983 \text{ km}$$

Using the definitions of the east boundary and west boundary of a grid column at the equator described above, we have

$$\text{East boundary} = 500 + \tfrac{1}{2}(667.98) = 833.99 \text{ km}$$
$$\text{West boundary} = 500 - \tfrac{1}{2}(667.98) = 166.01 \text{ km}$$

From Fig. 2-20 it is obvious that the width of a grid column decreases as a function of the cosine of latitude in the east–west column. Mathematically, this decrease is $K \cos \phi$, where ϕ is the latitude and $K = 333.99$ km. Therefore, the east and west boundaries can be expressed as follows:

$$\text{East boundary} = 500 + K \cos \phi$$
$$\text{West boundary} = 500 - K \cos \phi$$

An aircraft inertial navigation system can operate in this coordinate system since there is no restriction in operation from zone to zone.

2.6.2. The UTM Equations

The UTM grid reference system may be derived from several ellipsoidal models of the earth, while a Mercator projection must be based on latitude–longitude coordinates or an equivalent set of spherical coordinates, which are defined in terms of the particular spheroid used. In summary, the UTM grid has the following specifications [14]:

1. Projection is Transverse Mercator (Gauss–Krueger type) in zones 6° wide by 8° high.
2. The spheroid model varies depending on geographic location. (Constants such as ellipticity and polar–equatorial radii are normally stored in the navigation computer for various earth models.)
3. The length unit is the meter.
4. Latitude of origin is the equator (0°).
5. Longitude of origin is the central meridian of each zone.
6. False easting is 500,000 m.
7. False northing is 0 m for the northern hemisphere and 10,000,000 m for the southern hemisphere.
8. Latitude limits of the system are 80°S to 84°N.

2.6. The Universal Transverse Mercator Grid Reference System

9. Zones are numbered 1 through 60 eastward around the globe, starting at a longitude of 180°, as shown in Fig. 2-17.
10. Scale factor at the central meridian is 0.9996.

The equations used in the computation of the UTM grid coordinates, northing N and easting E', from geographic coordinates (latitude–longitude) are given in Ref. 14 as follows:

$$N = (I) + (II)p^2 + (III)p^4 + A_6 \qquad (2.79)$$

$$E' = (IV)p + (V)p^3 + B_5 \qquad (2.80)$$

where
ϕ = latitude
λ = longitude, in units of seconds
λ_0 = longitude of the origin (the central meridian) of the projection, in units of seconds
$\Delta\lambda$ = difference of longitude from the central meridian
  = $\lambda - \lambda_0$ in the eastern hemisphere
  = $\lambda_0 - \lambda$ in the western hemisphere
a = semimajor axis of the spheroid
b = semiminor axis of the spheroid
f = flattening (or ellipticity) = $(a-b)/a$
e^2 = eccentricity squared = $(a^2 - b^2)/a^2$
$e'^2 = (a^2 - b^2)/b^2 = e^2/(1-e^2)$
ρ = radius of curvature in the meridian or constant longitude (north–south) = $a(1-e^2)/(1 - e^2 \sin^2 \phi)^{3/2}$
ν = radius of curvature in the prime vertical or constant latitude (east–west); also defined as the normal to the spheroid terminating at the minor axis = $a/(1 - e^2 \sin^2 \phi)^{1/2} = \rho(1 + e'^2 \cos^2 \phi)$
S = true meridional distance on the spheroid from the equator
k_0 = central scale factor; an arbitrary reduction applied to all geodetic lengths to reduce the maximum scale distortion of the projection; for the UTM, $k_0 = 0.9996$
FN = false northing
FE = false easting
E' = grid distance from the central meridian (always positive)
E = grid easting = E' + 500,000 when a point is east of the meridian, 500,000 − E' when the point is west of the central meridian
N = grid northing
$p = 0.0001 \, \Delta\lambda$
$q = 0.000001 E'$

The equation for the Roman numerals appearing in Eqs. (2.79) and (2.80) are as follows:

$$(I) = S k_0 \qquad (2.81)$$

$$(\text{II}) = \frac{\nu \sin\phi \cos\phi \sin^2 1''}{2} \cdot k_0 \cdot 10^8 \qquad (2.82)$$

$$(\text{III}) = \frac{\sin^4 1'' \nu \sin\phi \cos^3 \phi}{24} (5 - \tan^2\phi + 9e'^2 \cos^2\phi$$

$$+ 4e'^4 \cos^4\phi) \cdot k_0 \cdot 10^{16} \qquad (2.83)$$

$$(\text{IV}) = \nu \cos\phi \sin 1'' \cdot k_0 \cdot 10^4 \qquad (2.84)$$

$$(\text{V}) = \frac{\sin^3 1'' \nu \cos^3 \phi}{6} (1 - \tan^2\phi + e'^2 \cos^2\phi) \cdot k_0 \cdot 10^{12} \qquad (2.85)$$

$$A_6 = p^6 \frac{\sin^6 1'' \nu \sin\phi \cos^5\phi}{720} (61 - 58 \tan^2\phi + \tan^4\phi$$

$$+ 270 e'^2 \cos^2\phi - 330 e'^2 \sin^2\phi) \times k_0 \times 10^{24} \qquad (2.86)$$

$$B_5 = p^5 \frac{\sin^5 1'' \nu \cos^5\phi}{120} (5 - 18 \tan^2\phi + \tan^4\phi$$

$$+ 14 e'^2 \cos^2\phi - 58 e'^2 \sin^2\phi) \cdot k_0 \cdot 10^{20} \qquad (2.87)$$

South of the equator, the northing N becomes

$$N = 10^7 - [(\text{I}) + (\text{II})p^2 + (\text{III})p^4 + A_6] \qquad (2.88)$$

Computation (or conversion) of geographic coordinates from UTM grid coordinates is similar to the above procedure and can be realized using the following equations [14]:

$$\phi = \phi' - (\text{VII})q^2 + (\text{VIII})q^4 - D_6 \qquad (2.89)$$

$$\Delta\lambda = (\text{IX})q - (\text{X})q^3 + E_5 \qquad (2.90)$$

where ϕ' = latitude of the foot of the perpendicular from the point to the central meridian
$q = 0.00000 E'$

Roman numerals (VII) through (X) are given by

$$(\text{VII}) = \frac{\tan\phi'}{2\nu^2 \sin 1''} (1 + e'^2 \cos^2\phi) \cdot \frac{1}{k_0^2} \cdot 10^{12} \qquad (2.91)$$

$$(\text{VIII}) = \frac{\tan\phi'}{24\nu^4 \sin 1''} (5 + 3\tan^2\phi + 6e'^2 \cos^2\phi - 6e'^2 \sin^2\phi$$

$$- 3e'^4 \cos^4\phi - 9e'^4 \cos^2\phi \sin^2\phi) \times \frac{1}{k_0^4} \times 10^{24} \qquad (2.92)$$

2.6. The Universal Transverse Mercator Grid Reference System

$$(IX) = \frac{\sec \phi'}{\nu \sin 1''} \cdot \frac{1}{k_0} \cdot 10^6 \qquad (2.\;)$$

$$(X) = \frac{\sec \phi'}{6\nu^3 \sin 1''}(1 + 2\tan^2 \phi + e'^2 \cos^2 \phi) \cdot \frac{1}{k_0^3} \cdot 10^{18} \qquad (2.94)$$

The UTM to latitude–longitude conversion can accept, but is not limited to, the following six model spheroids: (1) Clarke 1866, (2) Clarke 1880, (3) Bessel, (4) International, (5) Everest, and (6) Australian National. Spheroid constants used in the models are given in Appendix B, Table B-1. Simplifcations to Eqs. (2.79) and (2.80) are possible. Choosing parameters from the International spheroid model, Eqs. (2.79) and (2.80) can be written, after some algebra, in a simpler form:

$$N = 6365.10\phi + [1596.50\Delta\lambda^2 - 16.24]\sin 2\phi \qquad (2.95)$$

$$E' = 6386.10\Delta\lambda \cos \phi \text{ km} \qquad (2.96)$$

where ϕ and $\Delta\lambda$ are in radians. The UTM grid system can be updated using a straightforward routine. The update routine first calculates the gradients of the UTM conversion as a function of changes in latitude and longitude and then uses these gradients along with the continuous values of $\Delta\phi$ and $\Delta\lambda$ to provide values of ΔN and ΔE for use between each of the full solutions. The updated values of northing and easting are as follows:

$$N_n = N_{n-1} + \Delta N$$

$$E_n = E_{n-1} + \Delta E$$

where

$$\Delta N = \frac{\partial N}{\partial \lambda}\Delta\lambda + \frac{\partial N}{\partial \phi}\Delta\phi$$

$$\Delta E = \frac{\partial E}{\partial \lambda}\Delta\lambda + \frac{\partial E}{\partial \phi}\Delta\phi$$

2.6.3. Computation of Convergence

Convergence is defined as the difference between grid north (GN) and true north (TN) for any given position. The central meridian in each grid strip is established in the grid system as grid north. Lines parallel to this meridian throughout a grid column represent grid north. Since the meridians converge at the poles, grid north will coincide with true north only at the central meridian. When a grid boundary is crossed, the grid convergence angle must be changed to indicate convergence relative to a new central meridian. The meridian convergence must be calculated for the initial starting point and

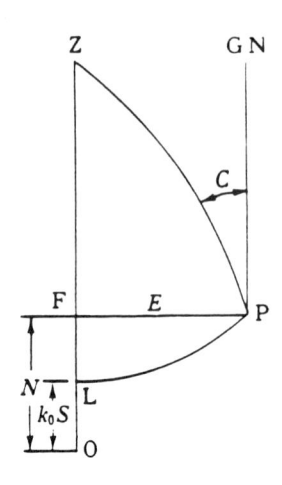

P = the point under consideration
F = the foot of the perpendicular from P to the central meridian
O = the origin
O Z = the central meridian
L P = the parallel of the latitude of P
Z P = the meridian of P
O L = $k_0 S$, the meridional arc from the equator
L F = the ordinate of curvature
O F = N, the grid northing
F P = E', the grid distance from the central meridian
G N = grid north
C = the convergence of the meridians; i.e., the angle at P between true north and grid north

FIGURE 2-21
Meridian convergence [14].

zone crossings, respectively. This convergence must be in each case referred to the mean meridian of the initial starting point. Figure 2-21 illustrates the concept of convergence.

The equation for the computation of the convergence from geographic coordinates is [14]

$$C = (\text{XII})p + (\text{XIII})p^3 + C_5 \qquad (2.97)$$

where

$$(\text{XII}) = \sin \phi \cdot 10^4$$

$$(\text{XIII}) = \frac{\sin^2 1'' \sin \phi \cos^2 \phi}{3} (1 + 3e'^2 \cos^2 \phi + 2e'^4 \cos^4 \phi) \cdot 10^{12}$$

$$C_5 = p^5 \frac{\sin^4 1'' \sin \phi \cos^4 \phi}{15} (2 - \tan^2 \phi) \cdot 10^{20}$$

and p has been defined in Section 2.6.2. The equation for convergence from UTM coordinates is given in the form [14]

$$C = (\text{XV})q - (\text{XVI})q^3 + F_5 \qquad (2.98)$$

where

$$(\text{XV}) = \frac{\tan \phi'}{\nu \sin 1''} \cdot \frac{1}{k_0} \cdot 10^6$$

2.7. Rotation Vector

$$(XVI) = \frac{\tan \phi'}{3v^3 \sin 1''} (1 + \tan^2 \phi - e'^2 \cos^2 \phi - 2e'^4 \cos^4 \phi) \cdot \frac{1}{k_0^3} \cdot 10^{18}$$

$$F_5 = q^5 \frac{\tan \phi'}{15 v^5 \sin 1''} (2 + 5 \tan^2 \phi + 3 \tan^4 \phi) \cdot \frac{1}{k_0^5} \cdot 10^{30}$$

and q has been defined previously.

2.7. ROTATION VECTOR

As discussed earlier, a coordinate frame can be carried into another frame, arbitrarily oriented with respect to it, by a rotation around a single, fixed axis. In the literature, this vector is called the *rotation vector* ϕ. (It should be noted at the outset that the rotation vector is not related to the Euler angle ϕ, the roll angle of an aircraft.) Therefore, no matter what was the motion history that carried, say, the b-frame, to its present orientation, there exists a single-axis rotation that would carry another frame, the n-frame, into coincidence with the b-frame. Usually, it is convenient to express the aforementioned rotation as a vector ϕ, which is directed along the axis of rotation and having a magnitude equal to the rotation angle in radians. One must be careful, however, not to assign all vector properties to the rotation vector (e.g., the result of two successive rotations about noncollinear axes is not represented by the sum of the two rotation vectors). The kinetic equation of the rotation vector can be defined as

$$\phi = \phi \mathbf{1}_\phi$$

where ϕ is the angle of rotation and $\mathbf{1}_\phi$ is the column vector representing the axis of rotation. For three general rotation angles (α, β, γ), the rotation vector assumes the form [9]

$$\phi = \phi \mathbf{1}_\phi - \phi \begin{bmatrix} \cos \alpha \\ \cos \beta \\ \cos \gamma \end{bmatrix}$$

The rotation vector ϕ has the property such that it is an eigenvector of the direction cosine matrix (DCM), associated with the unity eigenvalue. We note that there are three eigenvalues of the DCM; the other two eigenvalues are purely imaginary and represent the rotation magnitude or phase rotation. Consider now two Cartesian reference frames, n and b, and define the orthogonal basis sets for the two reference frame as (x^n, y^n, z^n) and (x^b, y^b, z^b). From Section 2.2, it is noted that the transformation of any vector $\mathbf{v}$ from the b-frame to the n-frame may be thought of as describing how $\mathbf{v}$ would look to an observer sitting on a coordinate frame originally coincident with b, which rotates through $-\boldsymbol{\phi}$ and ends at n. However, the fixed **v** appears to that observer to have rotated through $+\boldsymbol{\phi}$. The rotation vector can be used either directly in the mathematical operations to transform $\mathbf{v}^b$ into $\mathbf{v}^n$, or by forming the DCM C_b^n and performing the matrix times vector multiplication. That is, $C_b^n \boldsymbol{\phi}^b = \boldsymbol{\phi}^n$. In general, the DCM can be written in the form

$$C(\phi, \boldsymbol{\phi}) = I + \frac{\sin \phi}{\phi} [[\boldsymbol{\phi}]] + \frac{1-\cos \phi}{\phi^2} [[\boldsymbol{\phi}]]^2$$

where

$$[[\boldsymbol{\phi}]] \equiv \begin{bmatrix} 0 & \phi_3 & -\phi_2 \\ -\phi_3 & 0 & \phi_1 \\ \phi_2 & -\phi_1 & 0 \end{bmatrix}$$

and the related kinematic equation is

$$\frac{d}{dt} C = [[\omega]] C$$

A direct transformation using the rotation vector is [9]

$$\mathbf{v}^n = \mathbf{v}^b + \frac{\sin \phi}{\phi} \boldsymbol{\phi} \times \mathbf{v}^b + \frac{1-\cos \phi}{\phi^2} [\boldsymbol{\phi} \times (\boldsymbol{\phi} \times \mathbf{v}^b)] \qquad (2.99)$$

Furthermore, the direction cosine matrix transformation of $\mathbf{v}^b$ into $\mathbf{v}^n$ is

$$\mathbf{v}^n = \left[\cos \phi I + \frac{1-\cos \phi}{\phi^2} \boldsymbol{\phi} \boldsymbol{\phi}^T + \frac{\sin \phi}{\phi} \Phi^k \right] \mathbf{v}^b \qquad (2.100)$$

where ϕ is the magnitude of $\boldsymbol{\phi}$, and Φ^k is the usual skew–symmetric matrix of the form

$$\Phi^k = \begin{bmatrix} 0 & \phi_3 & -\phi_2 \\ -\phi_3 & 0 & \phi_1 \\ \phi_2 & -\phi_1 & 0 \end{bmatrix}$$

The rotation vector transformation operates on the $\mathbf{v}^b$ vector in the same manner as the DCM. In fact, having $\boldsymbol{\phi}$, the DCM C_b^n is obtained in the form

$$C_b^n = \cos \phi I + \frac{\sin \phi}{\phi} \Phi^k + \frac{1-\cos \phi}{\phi^2} \boldsymbol{\phi} \boldsymbol{\phi}^T \qquad (2.101)$$

The coefficients

$$\frac{\sin \phi}{\phi} \approx 1 - \frac{\phi^2}{6}$$

$$\frac{1-\cos \phi}{\phi^2} \approx \frac{1}{2} - \frac{\phi^2}{24}$$

2.7. Rotation Vector

as $\phi \to 0$, are well-defined for small ϕ. An alternate form of this transformation is

$$C_b^n = I + \frac{\sin \phi}{\phi} \Phi^k + \frac{1-\cos \phi}{\phi^2} (\Phi^k)^2 \qquad (2.102)$$

Another direct transformation form of the vector $\mathbf{v}^b$ into the vector $\mathbf{v}^n$ is

$$\mathbf{v}^n = \cos \phi \, \mathbf{v}^b + \frac{\sin \phi}{\phi} \boldsymbol{\phi} \times \mathbf{v}^b + \frac{1-\cos \phi}{\phi^2} (\boldsymbol{\phi}^T \mathbf{v}^b) \boldsymbol{\phi} \qquad (2.103)$$

These relationships are useful in the analysis of strapdown inertial navigation system algorithms. In general, the rotation vector differential equation can be written in the form [8, 9]

$$\dot{\boldsymbol{\phi}} = \boldsymbol{\omega} + \tfrac{1}{2} \boldsymbol{\phi} \times \boldsymbol{\omega} + \frac{1}{\phi^2} \left(1 - \frac{\phi \sin \phi}{2(1-\cos \phi)} \right) \boldsymbol{\phi} \times (\boldsymbol{\phi} \times \boldsymbol{\omega}) \qquad (2.104\text{a})$$

or

$$\dot{\boldsymbol{\phi}} = \boldsymbol{\omega} + \tfrac{1}{2} \boldsymbol{\phi} \times \boldsymbol{\omega} + \frac{1}{\phi^2} [1 - (\phi/2) \cot(\phi/2)] \boldsymbol{\phi} \times (\boldsymbol{\phi} \times \boldsymbol{\omega}) \qquad (2.104\text{b})$$

where $\boldsymbol{\phi}$ represents the rotation vector with magnitude $\phi = (\boldsymbol{\phi}^T \boldsymbol{\phi})^{1/2}$ and $\boldsymbol{\omega}$ represents the angular velocity vector. (Note that the $\boldsymbol{\omega}$ can be identified as a gyroscope output vector.) The last two terms in Eq. (2.104a) are referred to as noncommutative rate vectors. Since the noncommutativity (i.e., coning) rate is obviously contained in the rotation vector differential equation, it can be effectively used to improve the accuracy of strapdown attitude algorithms. From the preceding discussion, in terms of the two coordinate frames n and b, the rotation vector can be obtained from a solution to the differential equation given that the relative angular velocity between the two frames is a forcing function $\boldsymbol{\omega}_{nb}^b$. This differential equation is nonlinear in $\boldsymbol{\phi}$. Therefore

$$\dot{\boldsymbol{\phi}} = \boldsymbol{\omega}_{nb}^b \times \tfrac{1}{2} \boldsymbol{\phi} \times \boldsymbol{\omega}_{nb}^b + \frac{1}{\phi^2} \left\{ 1 - \frac{\phi \sin \phi}{2(1-\cos \phi)} \right\} [\boldsymbol{\phi} \times (\boldsymbol{\phi} \times \boldsymbol{\omega}_{nb}^b)]$$

$$= \left[I + \tfrac{1}{2} \Phi^k + \frac{1}{\phi^2} \left\{ 1 - \frac{\phi \sin \phi}{2(1-\cos \phi)} \right\} (\Phi^k)^2 \right] \boldsymbol{\omega}_{nb}^b \qquad (2.105)$$

which is similar to Eq. (2.104a). Finally, since the rotation vector is a three-parameter transformation operator, it contains a singularity at $\phi = 2n\pi$. As discussed in Section 2.5, a related transformation operator that does not contain singularities is the quaternion. In designing and/or selecting strapdown algorithms, it is customary to let ϕ be small. Thus, Eq. (2.104a) can be approximated by the equation [8]

$$\dot{\boldsymbol{\phi}} = \boldsymbol{\omega} + \tfrac{1}{2} \boldsymbol{\phi} \times \boldsymbol{\omega} + \tfrac{1}{12} \boldsymbol{\phi} \times (\boldsymbol{\phi} \times \boldsymbol{\omega}) \qquad (2.106)$$

REFERENCES

1. "Air Navigation," AFM 51-40 (NAVAIR 00-80V-49), Departments of the Air Force and the Navy, March 15, 1983.
2. Brand, L.: *Vector and Tensor Analysis*, Wiley, New York, 1948.
3. Britting, K. R.: *Inertial Navigation Systems Analysis*, Wiley-Interscience, New York, 1971.
4. Broxmeyer, C.: *Inertial Navigation Systems*, McGraw-Hill, New York, 1964.
5. Carlson, N. A.: "Fast Geodetic Coordinate Transformations," AIAA Guidance and Control Conference, August 11–13, 1980, Paper AIAA-80-1771-CP, pp. 328–337.
6. Goldstein, H.: *Classical Mechanics*, Addison-Wesley, Reading, Mass., 1965.
7. Halfman, R. L.: *Dynamics: Particles, Rigid Bodies, and Systems*, Vol. I, Addison-Wesley, Reading, Mass., 1962.
8. Jiang, Y. F., and Lin, Y. P.: "On the Rotation Vector Differential Equation," *IEEE Transactions on Aerospace and Electronic Systems*, Vol. AES-27(1), January 1991, pp. 181–183.
9. Lewantowicz, Z. H.: "Fundamentals of Aerospace Instruments and Navigation Systems," Course Notes for EENG 534, Air Force Institute of Technology, October, 1989.
10. Lupash, L. O.: "A New Algorithm for the Computation of the Geodetic Coordinates as a Function of Earth Centered Earth-Fixed Coordinates," *AIAA Journal of Guidance, Control, and Dynamics*, **8**(6), 787–789, November–December 1985.
11. Pitman, G. R., Jr. (ed.): *Intertial Guidance*, Wiley, New York, 1962.
12. Roy, A. E.: *Orbital Motion*, Adam Hilger Ltd., Bristol, U.K., 1982.
13. Thomson, W. T.: *Introduction to Space Dynamics*, Wiley, New York, 1963.
14. "Universal Transverse Mercator Grid," Department of the Army Technical Manual, TM 5-241-8, July 7, 1958.
15. VanBronkhorst, A.: "Strapdown Systems Analysis," in *Strap-Down Inertial Systems*, AGARD Lecture Series No. 95, 1978, pp. 3-1–3-22.
16. Whittaker, E. T.: *A Treatise on the Analytical Dynamics of Particles and Rigid Bodies*, 4th ed., Cambridge University Press, 1964.

3
Inertial Sensors

3.1. INTRODUCTION

This chapter presents a discussion of the ring laser and fiber-optic gyros. As stated in Chapter 1, the literature is replete with the treatment of the conventional spinning-mass gyroscope and linear accelerometers; therefore, these inertial sensors will not be treated here. Instead, this chapter will be devoted exclusively to the theory and applications of the ring laser and fiber-optic gyros since they represent the latest state of the art in inertial sensor technology. Conventional spinning-mass gyroscopes perform best when operating on gimbaled platforms; however, they suffer performance degradation when employed in the strapdown mode. This characteristic precludes taking advantage of the improvements in size, weight, power, maintainability, and reliability that strapdown systems offer. Also, spinning-mass gyroscopes bear a high cost of ownership and have inherent error sources that limit their operational capabilities.

The ring laser gyro (RLG) operates on the principles of general relativity, whereas the conventional spinning-mass gyroscope operation is based on the storage of mechanical energy. The essential feature of the laser gyro is a resonant optical cavity containing two oppositely directed traveling light waves generated by stimulated emission of radiation. Detection of rotation with light was demonstrated by Sagnac in 1913. Specifically, Sagnac demonstrated that two light waves acquired a phase difference by propagating in opposite directions around a loop interferometer. The laser approach avoids many of the error sources that limit the performance of the conventional gyroscope. It has an inherent capability to operate in the strapdown mode

83

because of its scale factor linearity over the full dynamic range and is easily interfaced with digital systems. The RLG has made the transition from research and development into production, where it is now routinely used in many aircraft–missile inertial navigation and/or inertial reference systems. Its ruggedness and functional simplicity make it especially attractive for strapdown applications. In recent years, an alternate and convenient method for increasing the sensitivity of passive Sagnac systems was demonstrated; this is the fiber-optic gyro (FOG). As its name implies, the FOG uses an optical fiber, which may be wrapped many times around a small cylinder to increase the phase difference produced by rotation. In the sections that follow, these two sensors will be discussed in detail.

3.2. THE RING LASER GYRO

3.2.1. Introduction

The initial effort that gave impetus to the development of the RLG was performed in 1911 by the French physicist G. Sagnac [10]. Sagnac, in his original experiment, used the "Sagnac interferometer," in which a beam-splitter divided an incident beam so that one component beam traversed the perimeter of a rectangle in a clockwise direction and the other in a counterclockwise direction prior to recombination. That is, optical rotation sensors function by detecting a differential shift in optical path length between two beams propagating in opposite directions around a closed path. As a result, the rotation induces a fringe shift in the interference pattern proportional to the rotation rate. However, in practice the Sagnac device was limited in sensitivity. Consequently, in order to detect the 15-deg/h rotation of the earth, Michelson and Gale in 1925 [7] had to extend the interferometer dimensions to about 0.5 km. Therefore, it can be said that the Michelson–Gale experiment marked the beginning of this activity.

In 1958, A. L. Schawlow and C. H. Townes (discussed in Ref. 4) proposed a method of constructing a maser (microwave amplification by stimulated emission of radiation) for optical wavelengths by using a resonant cavity whose dimensions were millions of times the wavelength of light. In the optical maser, they used a reflecting box, with two small plane mirrors facing each other and separated by a length L, into which ammonia atoms were energized. Specifically, a photon traveling within the mirrored device would interact with other energized atoms to emit other photons. Thus, photons traveling perpendicular to the plane of the mirrors would strike the mirror and be reflected back toward the other mirror. Consequently, with each succeeding passage the wave would grow in intensity until it was strong enough to burst through one of the mirrors as a flash of coherent light. Such a system is called a *Fabry–Pérot resonator* or *Fabry–Pérot interferometer*. A photon emitted by stimulation of another photon is in phase with the first, since the frequency of both is the same, and because they travel in the same direction; the emitted beam has space, time, and directional coherence. In July 1960, T. H. Maiman of the Hughes Aircraft Company succeeded in developing an optical laser using a ruby crystal, and before the end of 1960, five materials had been successfully tested in different laboratories. In Maiman's laser, the ruby contained about 0.05% chromium, giving the crystal a light-pink hue. Subsequently, in February 1961, scientists at Bell Telephone Laboratories announced the achievement of a continuously operating gaseous optical laser. Although structurally much different from the solid-state laser, the basic principles are the same. The device, 1 m long, used a mixture of gases as an active medium. The first laser used neon and helium gases, in proportions of 90% and 10%, respectively, at a pressure of 1–2 mm Hg. This mixture produced five coherent infrared emissions, the strongest at 11,530 Å* units (1.153 µm). Next in the evolution of the gas laser was the development of a helium–neon device that could continuously emit a beam of visible coherent light at a wavelength of 6328 Å (0.6328 µm). Rosenthal [9] in 1962 proposed a configuration whose high sensitivity would be derived from the extraordinary coherence properties of laser radiation. This was accomplished by combining the optical-generation and rotation-sensing functions in a laser oscillator with a ring-shaped cavity, typically a square or a triangle. Rotation then induces a difference in the generation frequencies for the two traveling waves that propagate in opposite directions around the ring (closed path).

After its initial demonstration by Macek and Davis [6] in 1963, and more than 15 years of research, development, and testing, the use of the RLG as an inertial angular rate sensor has gone into operational use in both civil and military applications. A number of airlines are now flying RLGs in an inertial reference system (IRS) configuration. Each aircraft carries three inertial platforms with three laser gyros and three accelerometers. A RLG inertial navigation system (INS), as we shall see later, provides the same data as the conventional gimbaled systems, specifically: acceleration, velocity, present position, attitude, heading, flight guidance, and aircraft steering commands. Furthermore, the INS combines laser rate sensors, accelerometers, processors, computers, and power supplies with extensive input/output functions and operations. The onboard navigation computer calculates coordinate rotations, attitude references, navigation, and flight guidance. The Boeing Company has selected the Honeywell RLG for use in an IRS configuration for its new 737-300, 757, and 767 commercial

---

*1 angstrom (Å) unit = 1×10^{-8} cm = 1×10^{-10} m = 1×10^{-1} nm; 1 micrometer (µm) = 1×10^{-6} m; 1 nanometer = 1 nm = 1×10^{-9} m.

transports, while the manufacturer of the A310 and A320 Airbus selected the Litton RLG system. The armed services, after many years of testing and evaluation, are beginning to use RLG inertial navigation systems for missiles, helicopters, aircraft, and shipboard fire-control system applications. Intrinsically, RLGs are ideal strapdown instruments. Low-ownership-cost strapdown navigation, guidance, and attitude and heading reference systems (AHRS) are required for many applications. Therefore, RLGs are highly desirable angular rate sensors for use in such systems. Also, because of their inherent mechanical simplicity, very high dynamic inputs can be tolerated, especially in military applications. Ring laser gyros have a digital output that simplifies their interface with microprocessors and navigation computers; that is, they require less interface hardware. The RLG, in various configurations, has been actively pursued as a viable inertial sensor by other industrial firms, most notably Litton Guidance–Control Systems, Northrop-Electronics Division, Honeywell, and the Autonetics Group-Rockwell International.

3.2.2. Laser Gyro Fundamentals

The RLG is a unique rate integrating inertial sensor, whereby two distinct laser beams of the same frequency are emitted from a gas discharge tube, one clockwise and the other anticlockwise, and travel the same path in opposite directions around a closed ring. Moreover, since the RLG is basically a solid-state sensor, with no moving parts to achieve inertial sensing capability, it has great potential for widespread application. Specifically, the common RLG uses a mixture of He–Ne gas to generate coherent monochromatic radiation in two directions. Therefore, instead of using a spinning mass, as in the conventional gyroscopes, it detects and measures differential angular rotations by measuring the frequency difference between the two contrarotating beams. Basically, the ring can be any geometric shape; the most common is triangular or square in shape with reflecting optics at the corners. The optical energy is confined by the utilization of ultra-high-quality mirrors. Reflectivity and scattering of these mirrors must be highly controlled in order to achieve desired performance. Mirror technology, therefore, is considered to be a key element of all RLG designs. A thorough understanding of how a laser gyro works requires an understanding of quantum mechanics and optical physics. In this section, an attempt will be made to explain in basic terms how a laser gyro works and the differences between the various configurations. The word "laser" is an acronym for "light amplification by stimulated emission of radiation." Originally, it was called an "optical maser," because, as was noted earlier, the original research was an outgrowth of research conducted on microwave frequency devices. Lasers may be made of a solid material such as a synthetic ruby or a gas such as a mixture of helium and neon.

3.2.3. Theoretical Background

3.2.3.1. The Passive Sagnac Interferometer

Sagnac's experiment consisted of rotating a square interferometer that included mirrors, a source of monochromatic light, and a detector about an axis perpendicular to the plane of the optical path. Now consider an ideal circular interferometer, where the light is constrained to travel along the circumference, as shown in Fig. 3-1.

Light is introduced in the interferometer at point A and is split by the beamsplitter, thereby forming two beams, one traveling clockwise and the other counterclockwise. Both beams are recombined at the original beamsplitter, after passing once around the circular ring. In the absence of rotation, the transit time for the light to travel the complete path is the same for both beams and is given by [2] $t = 2\pi R/c$, where c is the speed of light and R is the radius of the circular path. Now, if the interferometer is rotated at a constant angular velocity Ω, the transit times of the two beams to return to the beamsplitter will differ for each beam, since the beamsplitter is also rotating, moving to a new location B during the interval of time it takes the light to complete its circuit. As a result, and with respect to inertial space, the light moving in the direction of rotation must travel a longer distance than the light traveling in the opposite direction. Therefore, the two transit times will differ from the case where there is no rotation.

Let X be the inertial space distance between points A and B. Furthermore, let the positive sign (+) refer to the beam traveling in the direction of rotation and the negative sign (−) refer to the beam traveling opposite to the direction of rotation. Here we will assume that the speed of light remains

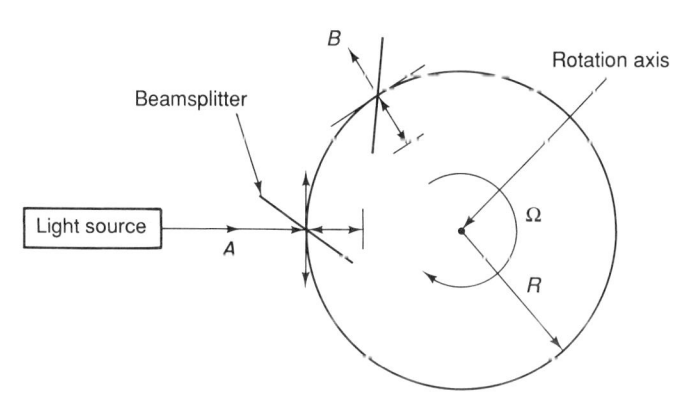

FIGURE 3-1
Ideal rotating Sagnac interferometer

invariant. Thus, the total closed-path transit for the light is given by [2, 5]

$$ct_\pm = 2\pi R \pm X_\pm \tag{3.1}$$

where $X_\pm = R\Omega t_\pm$. The net transit time can be written as

$$\Delta t = t_+ - t_- = t_\pm = \frac{2\pi R}{c} \pm \frac{R\Omega t_\pm}{c} \tag{3.2}$$

or solving Eq. (3.2) to $t_\pm$ gives

$$t_\pm = \frac{2\pi R}{c \pm R\Omega} \tag{3.3}$$

Next, in order to find the optical path difference ΔL between the two beams, it is necessary to find the transit time Δt. Thus

$$\begin{aligned}\Delta t = t_+ - t_- &= 2\pi R \left[\frac{1}{c - R\Omega} - \frac{1}{c + R\Omega}\right] \\ &= 2\pi R \left[\frac{2R\Omega}{c^2 - R^2\Omega^2}\right] \\ &= \frac{4\pi R^2 \Omega}{c^2 - R^2 \Omega^2}\end{aligned} \tag{3.4}$$

Dividing the numerator and denominator by c^2 we have

$$\Delta t = \frac{\frac{4\pi R}{c}\left(\frac{R\Omega}{c}\right)}{1 - \left(\frac{R\Omega}{c}\right)^2}$$

Using the binomial series for $(R\Omega/c)$ results in

$$\Delta t = (4\pi R/c)(R\Omega/c)[1 + (R\Omega/c) + (R\Omega/c)^2 + \cdots]$$

where to a first approximation in $(R\Omega/c)$

$$\Delta t = \frac{4\pi \Omega R^2}{c^2} \tag{3.5}$$

Equation (3.5), called the "Sagnac effect," is considered as the basis of the RLG [11]. The optical path difference ΔL is equal to $c\,\Delta t$; therefore, the basic equation for the optical path difference of the rotating interferometer is

$$\Delta L = c\, \Delta t = c \left[\frac{4\pi \Omega R^2}{c^2} \right] = \frac{4\pi \Omega R^2}{c}$$

$$= \frac{4A\Omega}{c} \qquad (3.6)$$

where A is the area enclosed by the circular optical path ($A = \pi R^2$). Note that Eq. (3.6) is general and applies to any geometric closed path. One problem arising here is that the path difference is small, even with a large area.

3.2.3.2. The Active Ring Laser Interferometer

The passive Sagnac interferometer discussed in the previous section is not practical for measuring very low angular input rates, since the ratio of the total area enclosed to the wavelength must be very large. That is, it lacks sensitivity because the path difference for light traveling in the two directions is much less than a wavelength. Sensitivity can be improved by replacing the beamsplitter with a mirror to form a resonant circuital optical cavity supporting traveling-wave modes for the counterrotating beams. Specifically, the improvement in sensitivity arises from the fact that the laser frequency is dependent on the cavity length. Rosenthal suggested that these modes could be made self-sustaining by placing the lasing medium in the cavity [9]. The two oppositely traveling waves form a standing wave within the cavity, and the amplitudes and frequencies are constrained to be equal. Ideally, the two oppositely directed traveling waves oscillate independently, each with its own frequency and amplitude. The fractional difference between these two frequencies corresponds to the fractional difference in optical path lengths traveled by each wave and, therefore, is proportional to the angular velocity. Therefore, in order to sustain oscillation, there must be enough gain in the medium to overcome losses in the cavity, and each beam must have an integral number of wavelengths around the ring. That is, the condition for oscillation is

$$N\lambda_{\pm} = L_{\pm} \qquad (3.7)$$

where $L_{\pm}$ is the optical path length of each beam, N is a large integer (typically 10^5 to 10^6), and $\lambda_{\pm}$ is the wavelength. It should be pointed out that the cavity geometry determines the wavelengths of a given mode. Now,

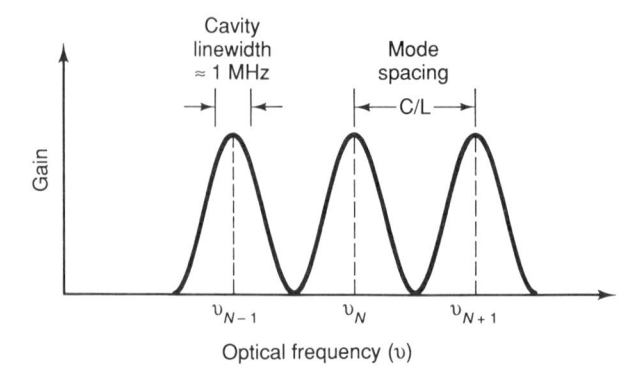

FIGURE 3-2
Empty-cavity modes.

the fractional frequency shift $\Delta v/v$ equals the fractional path length $\Delta L/L$ [5]. Therefore, the relation between $\Delta v = v_+ - v_-$ and $\Delta L = L_+ - L_-$ is

$$\Delta v / v = \Delta L / L \tag{3.8}$$

Equation (3.8) states that small path-length changes lead to small frequency changes. Adjacent longitudinal resonant modes of the empty cavity are separated in frequency by c/L, as illustrated in Fig. 3-2. The line width of the empty cavity is due to cavity losses. Since $\lambda = c/v$, solving Eq. (3.8) for Δv, and substituting Eq. (3.6) for ΔL, gives the resultant beat frequency between the two waves as

$$\Delta v = v_+ - v_- = (4A/L\lambda)\Omega \tag{3.9}$$

where Δv = beat frequency (Hz)
 A = area enclosed by the cavity (cm^2)
 λ = wavelength of the laser light (m)
 L = total optical path length (or simply cavity length) (cm)
 Ω = inertial rotation rate (rad/s)

This beat frequency is sensed as a rotation rate of the inteference fringe patterns past a photodetector. Equation (3.9) is the ideal RLG equation. The quantity $(4A/L\lambda)$ is the geometric or ideal scale factor S. The laser gyro is a rate integrating gyro* that gives N counts when turned through an angle

---

*Laser gyros can also be rate gyros if one observes the frequency of the phase shift.

3.2. The Ring Laser Gyro

θ. Therefore, integrating both sides of Eq. (3.9) over the interval time t yields

$$\int_{t_1}^{t_2} \Delta v \, dt = S \int_{t_1}^{t_2} \Omega \, dt$$

$$N = S\theta \tag{3.10}$$

where N = total phase shift or beats (pulses) counted during the measurement time
θ = total inertial angle of rotation

In order to get an idea of the magnitude of the measurable beat frequency Δv, as given by Eq. (3.9), consider an equilateral triangular RLG with the following characteristics:

One side of the triangle: 7.239 cm
Height of the triangle: 6.2687 cm
Total optical path length: $L = 21.717$ cm
Area of the triangle: $A = 2 \times \frac{1}{2}bh = (3.6195)(6.2687) = 22.6895$ cm^2
Transition: 0.6328 μm
Frequency: $v_0 = \dfrac{c}{\lambda} = \dfrac{2.997925 \times 10^8 \text{ m/s}}{0.6328 \times 10^{-6} \text{ m}} = 4.7375$ Hz
Input angular velocity: $\Omega = 1$ deg/h $= 4.85 \times 10^{-6}$ rad/s

Therefore, the measurable beat frequency is

$$\Delta v = (4A/L\lambda)\Omega = \frac{4(2.27 \times 10^{-3} \text{ m}^2)(4.85 \times 10^{-6} \text{ rad/s})}{(2.172 \times 10^{-1} \text{ m})(0.6328 \times 10^{-6} \text{ m})}$$

$$= 0.32 \text{ Hz}$$

At this point, it should be noted that a "rate integrating" gyro provides an incremental angle as an output. This represents the angle change over a given interval of time. As a result, attitude is created by appropriately accumulating or summing the angle increments so obtained. The "rate" gyro, on the other hand, provides an instantaneous angular rate at the specific instant at which it is sampled. However, the angular displacement over a certain interval must be determined by assuming that the observed instantaneous rate holds over the entire interval. Furthermore, when a rate gyro is subject to wideband random vibration, the approximate integration of the instantaneous rate leads to angular random walk, causing attitude drift and other navigation errors. Moreover, the instrument has a digital output quantized in incremental input axis rotation. Scale factor sensitivity can be increased by increasing the enclosed area A while at the same time decreasing the wavelength λ or the optical path length L. The typical geometric scale factors for some common geometric forms are as follows:

Equilateral triangle Let each side be of length a. Then, $a = L/3$. The area of the equilateral triangle is $A = (1/2)a^2 \sin 60° = (a^2/4)\sqrt{3} = L^2/20.785$.

Equilateral Triangle:

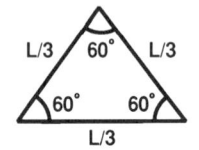

Therefore

$$S = \frac{4A}{L\lambda} = \frac{4}{\lambda L}\left[\frac{\sqrt{3}}{4}\left(\frac{L}{3}\right)^2\right] = \frac{1}{\lambda}\left(\frac{L}{3\sqrt{3}}\right)$$

Square For a square, each side is $L/4$. The area is $A = L^2/16$. Therefore

$$S = \frac{4A}{L\lambda} = \frac{4}{\lambda L}\left[\frac{L}{4} \times \frac{L}{4}\right] = \frac{1}{\lambda}\left[\frac{L}{4}\right]$$

Square:

General form $S = (1/\lambda)$(diameter of inscribed circle).

The accuracy of RLG is proportional to the area enclosed by the path and inversely proportional to the path length. For example, a square laser gyro encloses a greater area for a given path length than a triangular one, thus having greater potential accuracy. Also, a square configuration can be packaged into a smaller-sized inertial navigation unit. Sensitivity can also be improved in a fiber-optic ring circuit by providing a long optical path. The scale factor given above relates counts/s output to radians/s input, where 1 count is equivalent to 1 pulse. Therefore, the units of S will commonly be given by counts per radian. Sometimes, the scale factor is expressed in counts

3.2. The Ring Laser Gyro

per arcsecond or arcsecond per count, a more useful form in system applications. For example, the Sperry SLG-15 laser gyro has a scale factor of 3.5 arcsec/pulse, while the Honeywell GG-1328 laser gyro has a scale factor of 3.147 arcsec/pulse (or 65,000 counts/radian). In the ideal scale factor case, the wavelength λ can take discrete values only in a resonant ring. Consequently, only one value is associated at any time with the sustained oscillation that is within the laser bandwidth. Therefore, if λ changes by a discrete value, the scale factor will also change. Figure 3-3 depicts three RLG geometric configurations. In particular, this figure illustrates the concept of the equivalent circular ring laser. For example, the equivalent circular ring laser is the one that gives the same ideal scale factor as the actual polygonal ring laser. In the case of an equilateral triangle, the equivalent circular ring laser can be shown to be the inscribed circle. Moreover, the triangular path represents the actual optical path. If the triangular ring is stationary, and if there is no bias mechanism present, a resonant optical standing wave is formed by the two counterrotating traveling waves. Each of these two waves must satisfy the condition of an integer number of wavelengths N around the ring. Since each node of the standing wave occurs with a separation of $\lambda/2$ along the beam, the total number of nodes N_t is given by $N_t = 2N$.

3.2.3.3. The "Lasing" Action

The operation of a laser is based on quantum physics, and the phenomenon that makes the laser possible is that of "stimulated emission of radiation." From physics, we are all familiar with the process of "spontaneous emission," in which an atom in an excited state E_i can emit a quantum of radiation of frequency v_{ij}, thereby dropping into a lower energy state E_j, according to the relation [13]

$$E_i - E_j = h v_{ij} \tag{3.11}$$

where h is Planck's constant ($=6.62 \times 10^{-27}$ erg s) and hv is one photon (or "light quantum"). These jumps occur at a rate A_{ij} with a resultant spatially isotropic rate of emission of power $N_i A_{ij} h v_{ij}$, where N_i is the population of atoms in the excited state. The rate of these stimulated jumps is proportional to the energy density $u(v_{ij})$ of the radiation and to the population difference, $N_i - N_j$ between the upper and lower energy states. In 1917, Einstein pointed out that an excited atom can revert to a lower state through photon emission via two distinctive mechanisms: (1) in one instance the atom emits energy spontaneously, or (2) it is triggered into emission by the presence of electromagnetic radiation of the proper frequency. The latter process, as we stated above, is known as *stimulated emission*, which is the key to the operation of the laser. A remarkable feature of this process is that the emitted photon is in phase with, has the polarization of, and propagates in the same direction

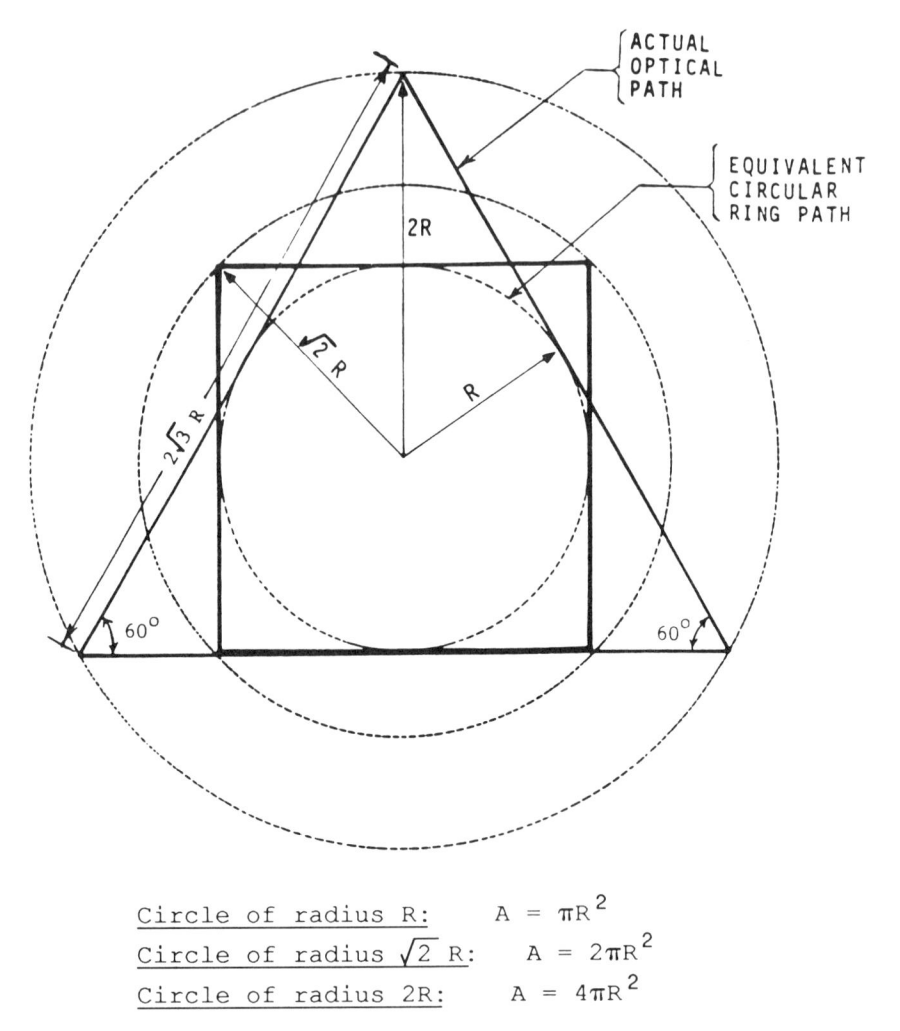

Circle of radius R: $A = \pi R^2$
Circle of radius $\sqrt{2}$ R: $A = 2\pi R^2$
Circle of radius 2R: $A = 4\pi R^2$

FIGURE 3-3
Equilateral triangle ring and equivalent circular ring for determination of scale factor relationships.

as the stimulating wave [8]. This then is the process that gives rise to the amplification and directional properties of lasers.

Let us now see what happens in the lasing medium. In order to obtain lasing action, two conditions must be fulfilled [5]. First, a gain (or amplification) medium to overcome the losses must be present. The gain medium used in the modern RLG consists of a tube filled with a mixture of helium and neon gases at a very low pressure of about 3–7 torr* and operating at the visible region of 6328 Å. However, other mixtures of gases are being investigated. As we shall see below, in order for the gain medium to supply a gain, a population inversion must be generated. When a voltage is applied across a metallic anode (source) and a cathode (sink), the voltage ionizes the gas, producing an orange glow discharge. Now, excited helium atoms collide with neon atoms, thereby transferring energy to the neon atoms and raising them temporarily to higher energy levels. Each orbit is associated with a specific energy level that increases as a function of distance from the nucleus. A few neon atoms spontaneously emit a photon as they fall back to a normal energy level. These photons strike other excited neon atoms, stimulating the release of new photons that, in turn, do the same. A cascade of photons is thus produced in all directions. The wavelength of the photon depends on the amount of energy released, which in turn depends on the change in orbit of the electron. For two energy levels E_1 and E_2 we obtain [13]

$$E_1 - E_2 = h\nu \qquad (3.12)$$

From the cascade of photons metioned above and in Eq. (3.12) we can write

$$\text{Atom in } E_1 \text{ level} + \text{photon } (h\nu) \Leftrightarrow \text{excited atom}$$

Figure 3-4 shows the physical interpretations of these concepts. If N is the number of atoms, the equation for the curve in Fig. 3-4 takes the form

$$N = N_0 \exp[-(E_2 - E_1)/kT] = N_0 \exp[-(\Delta E)/kT] \qquad (3.13)$$

where k is Boltzmann's constant ($=1.3804 \times 10^{-16}$ erg/deg) and T is the absolute temperature. Equation (3.13) is known as *Boltzmann's equation*, expressing the thermal equilibrium of the population.

In order to obtain light amplification, we need a larger population of atoms in the E_2 energy level, that is, $N_2 > N_1$. However, in order to obtain population inversion, the atom must have at least three energy levels. Population inversion requires that, in addition to keeping the upper levels populated, the lower levels be kept depopulated. This is effected primarily through collisions of the neon atoms with the cavity walls, which leads to an inverse radial dependence of gain. Figure 3-5 shows this effect. Level E_3 is metastable; that is, it is the level in which most of the population of neon atoms must be inverted before laser action is possible. A more detailed account of what is happening in the various energy levels can be obtained by referring to Fig. 3-6 [4].

From Fig. 3-6, we note that many helium atoms, after dropping down from several upper levels, accumulate in the long-lived 2^1S and 2^3S states.

---

*The sea-level pressure is by definition 1.0132×10^5 newtons/m^2. Also, 1 atm = 760 torr = 1013.793 mb (millibars).

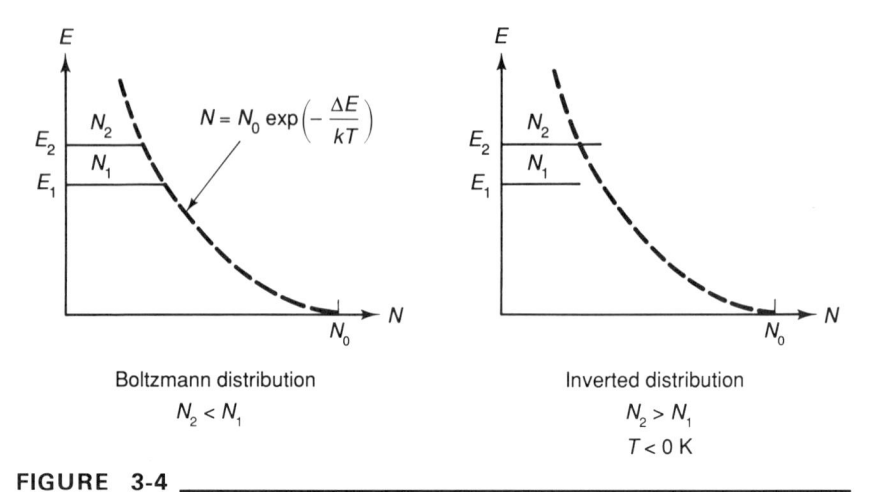

FIGURE 3-4
Energy distribution and population inversion.

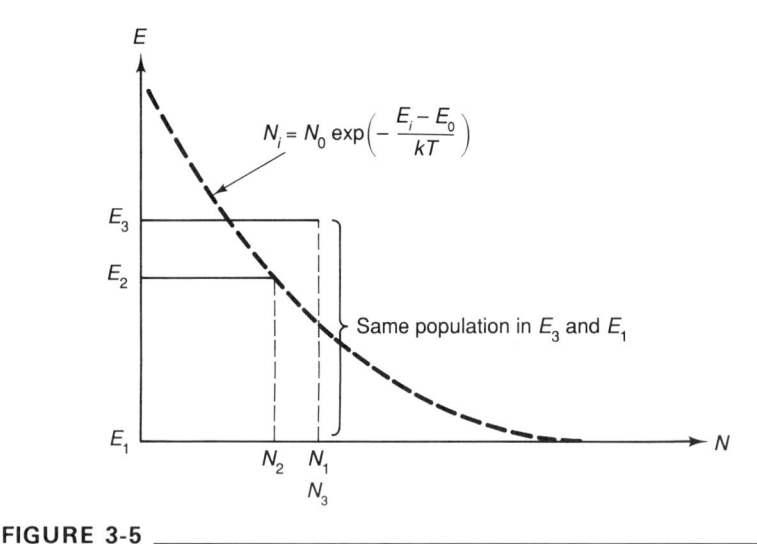

FIGURE 3-5
Energy distribution after intense pumping.

These, as we have seen, are metastable states from which there are no allowed radiative transitions. The excited He atoms inelastically collide with the transfer energy to the ground-state Ne atoms, raising them in turn to the $3s_2$ and $2s_2$ states. These are the upper laser levels and there then exists a population inversion with respect to the lower $3s_2$ and $2p_4$ states. Spontaneous photons initiate stimulated emission, and a chain reaction begins. The

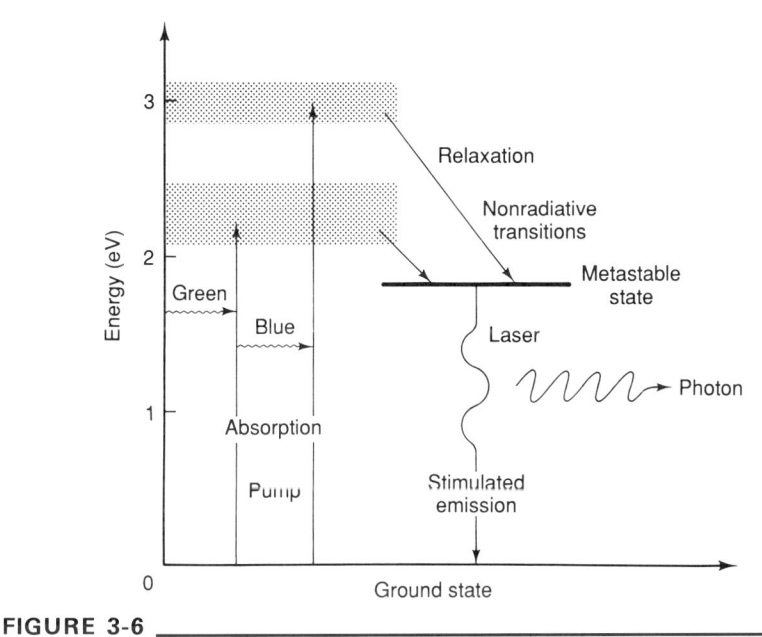

FIGURE 3-6
Helium–neon energy levels [4].

dominant laser transitions correspond to 1152.3 and 3391.2 nm in the infrared, and 632.8 nm in the visible spectrum. The p states drain off into the $1s$ state, thus remaining uncrowded themselves and thereby continuously sustaining the inversion. The $1s$ level is metastable, so that $1s$ atoms return to the ground state after losing energy to the walls of the cavity. For this reason, the plasma tube diameter inversely affects the gain and is, accordingly, an important design parameter.

In addition to the longitudinal or axial modes of oscillation, which correspond to standing waves set up along the cavity z axis, transverse modes can be sustained as well. These cavity resonances are approximately transverse electric and magnetic or TEM_{mn} modes. The m and n subscripts represent the integer number of transverse nodal lines in the x and y directions. A complete specification of each node has, therefore, the form TEM_{mnq}, where q is the longitudinal mode number. Thus, for each transverse mode (m, n), there can be many longitudinal q modes. The lowest-order modes that will oscillate are TEM_{00q}, TEM_{00q+1}, and TEM_{00q+2}. Also, the lowest-order transverse mode that will oscillate is the TEM_{00} mode, which is perhaps the one most widely used. Above all, the flux density of the TEM_{00} transverse mode is ideally Gaussian over the beam's cross section and has the following characteristics: (1) there are no phase shifts in the electric field, and so it is spatially coherent; (2) the beam's angular divergence is the smallest; and (3) it can be focused to the smallest-sized spot. Figure 3-7 illustrates these ideas. Population inversion can be achieved by the following methods: (1) spatial separation of excited molecules (ammonia maser), (2) optical pumping (dielectric solids, liquids, gases, semiconductors), (3) electric discharges (gases), (4) chemical reactions (gases), (5) electron–hole pair production (doped semiconductor), and (6) rapid expansion of hot-gas mixtures.

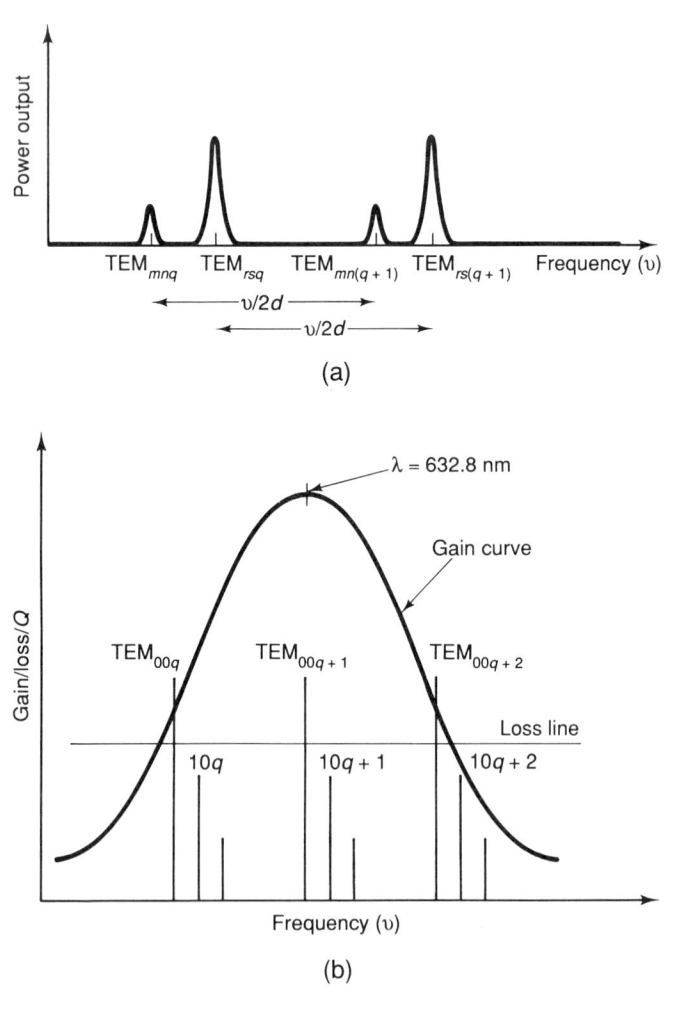

FIGURE 3-7
Characteristic frequencies and field patterns: (a) laser mode nomenclature; (b) gain medium properties indicating the oscillating modes.

3.2. The Ring Laser Gyro

A resonant cavity is required to fulfill the second condition. Mirrors are placed at each end of a triangular cavity. Two of the mirrors possess 100% reflectivity, while the third mirror allows partial transmission of less than 0.1% for signal detection. Also, two of the mirrors will be flat while the third is normally concave spherical. Photons traveling parallel to the cavity will be reflected back and forth many times. The mirrors, which are usually of the hardcoated multilayer-dielectric (MLD) type, become the end of a resonant cavity, producing a positive feedback necessary to sustain oscillation. It must be pointed out, however, that the laser principle works for any closed path. The only reason for choosing a triangular configuration for the closed path is because of the ease of alignment. To understand the mechanics of what is happening in the cavity, let us look at the resonant modes. As mentioned earlier, in order to sustain laser oscillation (as contrasted to single-pass amplification), optical feedback must be used to obtain an optical resonator. This is easily explained by considering the Fabry–Pérot etalon, which, as we have seen, is the simplest resonant cavity. This interferometer provides a practical means of obtaining mode separation necessary to produce oscillation in one or a few modes at optical frequencies, while providing the optical feedback necessary for low gain transitions. For example, the axial mode spacing in a laser with 1-m separation ($\Delta v = c/2L = 150$ MHz) is in excess of the natural linewidth and enables five or six dominant modes across the full linewidth in a cavity without conducting walls. The structure of the electromagnetic field E within the space defined by the mirrors is determined by the boundary condition $\mathbf{E} = 0$ at $x = 0, L$. Conditions for resonance then exist according to the following equations:

$$L = \left(\frac{\lambda}{2}\right) n = \left(\frac{\lambda}{2}\right) \frac{c}{v}$$

where n is a large integer (approximately 10^5). Then

$$v_n = \frac{n}{2}\left(\frac{c}{L}\right), \quad v_{n+1} = (n+1)\frac{c}{2L}$$

$$\Delta v = |v_{n+1} - v_n| = \frac{c}{2L} \tag{3.14}$$

Finally, the quality factor, or Q, of a cavity is a figure of merit of an energy storing system and is defined by

$$Q = \frac{2\pi(\text{average energy stored})}{\text{energy dissipated/cycle}} = \frac{2\pi\varepsilon}{PT} = \frac{2\pi\varepsilon}{\left(-\dfrac{d\varepsilon}{dt}\right)T}$$

where P is the dissipated power, ε is the average energy stored, and T is the period of oscillation; Q takes the final form

$$Q = \frac{v_n}{\Delta v_c} \qquad (3.15)$$

where

$$\Delta v_c = \frac{1}{2\pi(\text{photon lifetime})}$$

A high Q means low dissipation and narrow linewidth. Typical values for Q are in the order of 10^8–10^9. The length of the cavity is many thousands of times the optical wavelength. Consequently, the cavity is resonant at a large number of frequencies, but most of these frequencies do not receive sufficient gain to oscillate. When any mode (or resonant frequency) receives a gain greater than its losses, the radiation builds up to a steady-state value and is emitted through the mirror, which allows partial transmission. The emission is the laser radiation, which is essentially a single frequency. While propagating in a resonant cavity, however, an electromagnetic wave is subject to a number of cavity losses, among which the following are typical: (1) transmission and absorption, (2) scattering by optical inhomogeneities and surface imperfection, (3) diffraction at the extremes of the laser cavity, (4) absorption in the amplifying medium, and (5) losses and mode conversion due to imperfect imaging by the mirrors. Table 3-1 gives some of the properties of stimulated emission of the laser beam.

3.2.3.4. Definition of Coherence

If a light source is coherent, its wavefield is characterized by an instantaneous wave amplitude that can be expressed by a unique harmonic function of time and position. Stated another way, the ratio of the complex wave amplitude at two points in the wavefield, no matter how far apart, does not change with time. Mathematically, an ideal coherent source and its wavefield can be expressed by

$$\tilde{E}(r, t) = \frac{A}{r} \exp[i(kr - \omega t + \phi)] \qquad (3.16)$$

TABLE 3-1
Equivalence between Stimulated Emission and Laser Beams

Properties of	
Stimulated emission	**Laser beam**
In phase	Coherence
Same frequency	Monochromatic
Same direction	Directional
Same polarization	

where $k = 2\pi/\lambda$, $\omega = 2\pi f$, A is the amplitude, and r is the radius from the source to the wavefront.

3.3. THE RING LASER GYRO PRINCIPLE

Commonly, the RLG is a triangular or square cavity filled with gas, in which two oppositely traveling light waves are generated by stimulated emission of radiation. In this case, one speaks of a two-mode, continuous wave (cw), active laser gyro. If, on the other hand, the lasing medium is external to the cavity, as in the fiber-optic gyro, then one speaks of a passive laser gyro. The laser gyro combines the properties of an optical oscillator and general relativity to produce the function of the conventional mechanical gyroscopes. Specifically, the simplest ideal RLG consists of a He–Ne plasma cavity or medium with gain, three or four mirrors used to form a closed optical circuit (ring resonator), and associated electronics for detection of the two counterrotating beams. Such a combination can oscillate a traveling wave going around the ring in a clockwise (CW) direction and a traveling wave going around in a counterclockwise (CCW) direction. The beams, formed by the lasing action described in Section 3.2.3.3, travel around the enclosed area in both directions simultaneously. Both clockwise and counterclockwise beams occupy the same space at the same time and oscillate at the same frequency when no input rate is present. Since the light beams have the same frequency, they form a standing-wave pattern. This standing-wave pattern is characterized by the equation

$$E(x, t) = 2E_0 \sin kx \cos \Omega t$$

where E_0 is the amplitude, Ω is the rotation rate (or angular frequency in rad/s), and k is the propagation constant (kx is in units of radians). The profile of this wave does not move through space. The laser beam takes a finite amount of time to make one trip around the ring. Rotation of the base on which the laser cavity and the mirrors are mounted causes the frequency of the beam traveling in the direction of rotation to decrease and the frequency of the oppositely directed beam to increase. Thus, the two beams will have a frequency difference $\Delta v = v_1 - v_2$, which is proportional to the input rate. This difference is detected by heterodyning the two beams on a photosensitive device. The output is the integral of the input rate. This is kept on track by counting pulses. The number of these pulses is proportional to the integrated angle. That is, in practice the laser gyro is operated in an integrating mode, in which each cycle of the difference frequency is counted as a unit of angular displacement. The signal thus detected has a high signal-to-noise ratio. Today's laser gyros are capable of sensing rotation rates down to 0.001 deg/h (or 0.0001 earth rate). Frequency differences detected by the laser gyro are in the order of less than 0.1 Hz. More specifically, the precise frequency of each beam depends on both the laser gain medium and the length of the closed path. A typical triangular ring is shown in Fig. 3-8. The cavity path-length difference ΔL between the clockwise and counterclockwise beams will vary as

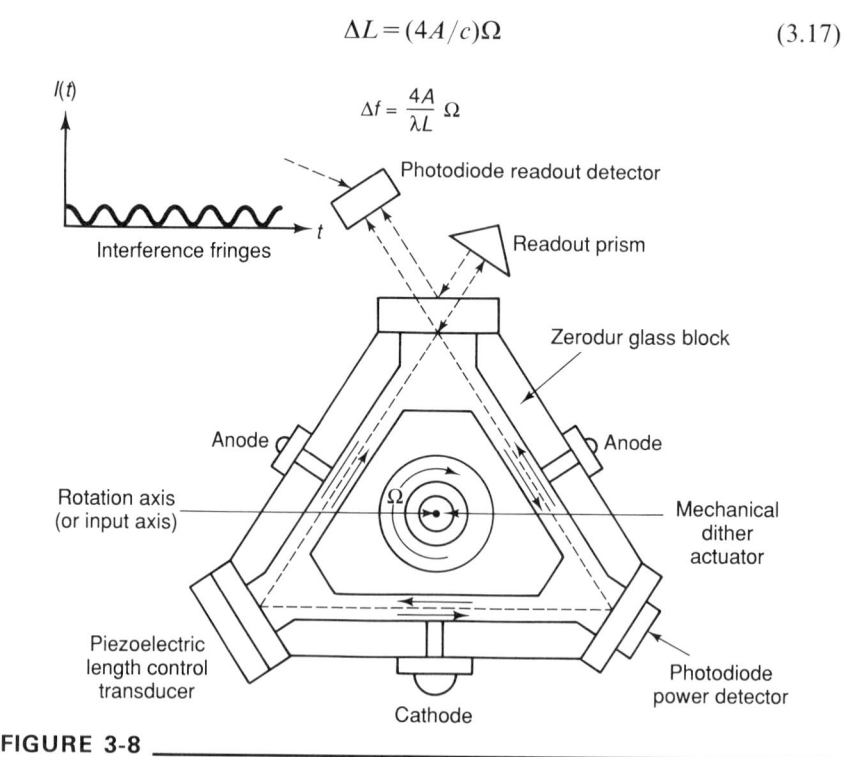

$$\Delta L = (4A/c)\Omega \qquad (3.17)$$

FIGURE 3-8
Schematic representation of a triangular ring laser gyro.

where A is the projected area of the optical ring orthogonal to the angular velocity Ω and c is the speed of light in vacuum. In other words, optical rotation sensors function by detecting a differential shift in optical path length between two beams propagating in opposite directions around a closed path. Since the fractional frequency shift ($\Delta v/v$) of the resonant modes of the laser equals the fractional path-length change ($\Delta L/L$), as given by Eq. (3.8), the frequency difference of the counterrotating beams is given by Eq. (3.9) as

$$\Delta v = (4A/L\lambda)\Omega \qquad (3.18)$$

where L is the cavity length and λ is the wavelength. Simply, by measuring Δv, one obtains the angular velocity Ω. This is the ideal RLG equation. That is, in an ideal RLG, the output rate (or beat frequency) is proportional to the input rate. The slope of the line relating the two quantities is found from Eq. (3.9).

The helium–neon RLG is capable of operating at two wavelengths: 0.6328 and 1.15 μm. The specific frequency is governed by the mirrors and the path length. The shorter wavelength has greater resolution and, therefore, greater accuracy. The longer wavelength provides greater gain for a given length of gain medium. Thus, the shorter wavelength is more desirable for larger, more accurate laser gyros and vice versa.

In the design and fabrication of a RLG, the major components are (1) the block material, (2) mirrors, (3) the gain medium, (4) the readout mechanism, and (5) associated electronics. These components will now be described in more detail.

Laser Gyro Block Material In this section, we will consider the triangular RLG. Most RLG designs have monolithic bodies, machined blocks of ultra-low-temperature coefficient-of-expansion glass. Two common block materials used are the Cer-Vit (ceramic vitreous) manufactured by Owens-Illinois, and Zerodur manufactured by the Schott Optical Company. The material used for the laser gyro body must have special properties. One is a very low coefficient of expansion to reduce the amount of movement required of the piezoelectric path-length control device, and to reduce temperature sensitivity. Another property of monolithic laser gyros is imperviousness to helium. Ease of fabrication, low cost, and other factors are also desirable. Only the two materials, Cer-Vit and Zerodur, both forms of glass, possess the first two attributes. Other forms of ultra-low-expansion glass are available, but they are not impervious to helium. Regardless of the block material used, one solid block of material is drilled to provide a path for the laser beams. Flat surfaces are machined at the correct angle for mounting the mirrors. Electrodes are then installed and the entire cavity is filled with the laser gain medium. In an alternate RLG design, a separate lasing tube versus a totally machined block is utilized. That is, the gain medium is contained in a separate gain tube that fits between two mirrors in the laser body. An advantage of the modular design is ease of fabrication and repair. The mirrors and other optical components must often be replaced or adjusted during assembly and test. In a monolithic block, the laser gain medium will be lost. Another advantage is that the entire body need not be sealed to prevent the loss of helium. A disadvantage of the modular design is the short gain region. In a monolithic design, the gain region extends around approximately one-half of the path. Since the monolithic RLG is constructed in the form of a solid block, it offers greater stability and ruggedness. In the modular design, it is limited to a distance less than that between two adjacent mirrors. This means that modular RLGs are generally larger than monolithic laser gyros of comparable accuracy.

Mirrors The mirrors are the most important components of a RLG. Depending on the particular manufacturer design, RLGs use three or four mirrors. For example, multioscillator or four-mode laser gyros must use an even number of mirrors, usually four. The common triangular two-mode (or two-frequency) laser gyros use three mirrors. The number of mirrors used is determined by design tradeoffs. Regardless of the design, the mirrors are commonly placed in optical contact with the block, forming a stress-free, stable hermetic seal. Each mirror causes the loss of a portion of the laser energy through absorption or transmission. Also, a portion of the energy is backscattered, resulting in lock-in, which will be the topic of discussion in Section 3.3.1. Backscattering results from surface roughness and optical index variations in the coatings. Specifically, the reflectance and the angular orientation of the mirrors affect the laser gain, and the backscattering of light at the mirror surfaces adversely affects the lock-band error. Mirrors used by most manufacturers are of the MLD type. The high reflectance (>99%) required for laser gyro operation is commonly achieved with MLD coatings such as silicon dioxide and titanium dioxide. The advantage of these materials are durability and resistance to contamination and deterioration by the He–Ne plasma. Multilayer dielectric mirrors are made by depositing layers of alternately high and low index of refraction on a highly polished substrate. The thickness of each layer is such that light entering the top surface at a specific angle of incidence travels one-quarter wavelength before arriving at the bottom surface. A portion of the light is reflected from each surface of the multilayer stack. The rest passes through to the next layer where a portion is again reflected. Moreover, the thickness of each layer must be tailored to both the wavelength of the laser beam and to the angle of incidence. The mirrors must be extremely smooth and free of defects in order to reduce backscatter as much as possible, thereby reducing the lock-in band. Fabrication of laser gyro mirrors is evidently an art.

Backscatter is maximum in the direction normal to the plane of the mirror and decreases at smaller angles. For example, a laser beam striking a mirror at an angle of incidence of 60°, as in a triangular ring laser gyro, will experience greater backscatter than one with an angle of incidence of 45°. Consequently, a four-mirror laser gyro will have more scattering surfaces, but less backscatter from each surface. Ring laser gyro performance can be strongly affected by backscattered light, resulting in mode coupling. Use of only three mirrors limits the magnitude of these problems. Therefore, as mentioned above, a tradeoff design at the system level will be required. If mirrors are expensive and of low quality, it is better to use only three. As mirror cost decreases and quality increases, four-mirror laser gyros become more desirable. A variant of the multilayer dielectric process is to use magnetic mirrors. Magnetic mirrors are, generally, identical to other multilayer dielectric mirrors, except that they have a layer of magnetic material between the substrate and the dielectric stack. Magnetic mirrors used in the two-frequency laser gyros must have an alternating magnetic field applied. They use the transverse magneto-Kerr effect required for linearly polarized light. The magnetic layer is composed of a garnet material similar to that used for bubble memories. Specifically, the magneto-Kerr effect is the reflective analog of the Faraday effect. (A Faraday cell is a piece of crystal that changes the phase of a light wave traveling through it as a function of an applied magnetic field; the effect is nonreciprocal.) A mirror containing a layer of magnetic material under the reflective surface advances or retards the phase of a reflected light beam, depending on the direction of the incident and reflected beams. Since a magnetic mirror reduces the backscatter and temperature sensitivity problems associated with a Faraday cell, it has nearly supplanted the Faraday cell in current usage. Use of Faraday cells has been noted to exhibit strain birefringence, which generates path-length differences and, hence, false rotation information. Magnetic mirrors used in multioscillator laser gyros have a constant magnetic field applied. They use the polar magneto-Kerr effect required for use with circularly polarized light. The magnetic layer is composed of a mixture of manganese and bismuth.

Gain Medium The gain medium is capable of producing laser action over a finite band of frequencies due principally to Doppler broadening, the effect of atoms within the gain medium. The precise frequency within the band is determined by the path length around the ring. Consequently, since the ring is a resonant cavity, a whole number of half waves form around the ring. Normally, the gain is kept sufficiently low that only one mode, a specified number of half wavelengths, is propagated in each direction. The path length is usually controlled so that the frequency remains in the center of the curve of gain versus frequency. As mentioned above, the medium is capable of supplying gain at the desired cavity resonance. Thus, if the gain supplied by the active medium to a resonant mode is greater than the cavity losses, oscillation will occur. However, the resonant modes are considerably narrower in frequency than the bandwidth of the normal spontaneous atomic transition. These modes will be the ones that are sustained in the cavity; as a result the emerging beam is restricted to a region close to those frequencies.

The gain medium used in most RLGs is a high-purity trinary mixture of helium (^{3}He) and two isotopes of neon (^{20}Ne and ^{22}Ne) at a ratio of He/Ne approximately 13:1 and at the proper pressure. In one design, this mixture is made up of 93% ^{3}He with the balance consisting of the two neon isotopes with the percentages of 52% ^{20}Ne and 48% ^{22}Ne, respectively. In an alternate design, a 1:1 (i.e., 50% each) mixture is used for the two neon isotopes. The helium is required to absorb a portion of the energy in the neon atoms to bring them to the proper energy state to provide an output at the proper frequency. The two neon isotopes have peak gains at slightly different frequencies. Each counterrotating beam favors one isotope. Without the two isotopes, the two beams will compete for the same population of atoms. Thus, only one beam will result. On the other hand, if the pressure is increased, beyond the proper value, mode competition will appear even in a laser that has a 1:1 mixture of ^{20}Ne and ^{22}Ne. This is likely to happen because the holes burned in the gain curve will become wider with increasing pressure.

Readout Mechanism In order to sense the difference in frequencies between the two counterrotating laser beams, one of the mirrors is made partially transmissive, typically less the 0.1%. The two laser beams, after emerging from the mirror, are combined by use of a corner prism so that they are nearly parallel, resulting in a fringe pattern that is detected by a set of photodetectors (or photodiodes). As the laser gyro rotates, the fringes appear to move across the photodetectors. The photodetectors then sense the beat frequency by heterodyning the two optical frequencies, that is, by counting the fringes. The fringe pattern will shift in a direction determined by the relative magnitude of the frequencies of the two laser beams. When the ring is rotated clockwise the fringe pattern moves in one direction, while the pattern will reverse when the direction of the ring is reversed. The number of fringes is a measure of angular displacement, and the fringe repetition rate corresponds to the rotation rate of the ring. That is, each fringe pattern moves at a rate that is directly proportional to the frequency difference between the two counterrotating beams. This is then converted into a digital output. As a result, the output pulse rate is proportional to the input rate, and the cumulative pulse count is a measure of the change in the gyro orientation with respect to some reference point.

Laser Gyro Electronics The RLG electronics consist of the following circuits: discharge current control, path-length control, dither drive (if dither is used), and readout amplifiers and direction logic. The triangular RLG is designed with two anodes and a cathode, so that the discharge is initiated by applying two balanced high dc voltages between the anodes and a common cathode. The value of this voltage will vary from one laser gyro design to another. Typically, this voltage is twice the operating potential required to sustain the discharge or continuous operation. After discharge is initiated, it is immediately regulated by the current control circuit. Path-length control is accomplished by a control circuit, which operates on the intensity of the laser beam. A piezoelectric transducer is attached to one of the mirrors in order to keep the cavity length of the ring constant during environmental changes. Any changes in path length manifest themselves as changes in scale factor and null stability, which reduce the sensor's overall accuracy. In closed-loop operation, the path-length control circuit operates as a null-seeking system to stabilize the cavity length at the maximum intensity. Ring laser gyros using dithering techniques can improve the performance by introducing a rotational bias so that the gyro operates outside the lock-in band or region. In essence, the bias consists of sinusoidally oscillating the gyro block at a dither frequency between 100 and 500 Hz, depending on the design. The bias averages to zero over each cycle, and as a result, operation occurs mostly in the linear region. The frequency and amplitude of the dither are such that the laser gyro now becomes responsive to very low angular input rates. The bias thus introduced is then subtracted from the output signal. The dither must be symmetric, or errors will be introduced. The two counterrotating laser beams are optically combined outside the cavity into a fringe pattern. The motion of the fringe pattern, as we have seen, is determined by a dual photodetector, amplified and converted into a pulse rate. In addition, a logic circuit operates the output pulses that are proportional to clockwise rotation and those proportional to the counterclockwise rotation.

3.3.1. Laser Gyro Error Sources

In the previous sections we described how the relativistic properties of electromagnetic radiation, in this case monochromatic laser light, can be used to construct an angular rate motion sensor. As we have seen, the laser gyro operating principle is based on the property that two counterrotating light beams will interact to produce an interference pattern that can be detected by a photodetector. Theoretically, without instrument imperfections, the laser gyro should be an ideal single-degree-of-freedom incremental rate-integrating inertial sensor. In practice, however, inherent electromagnetic properties, optical instrument imperfections, properties of materials used to construct the instrument, and detector limitations become sources of error. Any effect that causes the input/output relationship to deviate from the ideal case is an error source. Ideally, the RLG displays a frequency difference as a linear function of input rate passing through the origin. In the design of a RLG, the errors that are crucial to its operation may be grouped into three types: (1) null shift, (2) lock-in, and (3) mode pulling. References 2 and 4 give a detailed account of these errors. These error sources are illustrated in Fig. 3-9. Such errors cause the laser gyro to deviate from the ideal straight line. Early investigations of RLGs revealed that the linear relationship of Eq. (3.9) was not obeyed at low rotation rates, where the frequency difference remained locked to zero. These error sources will now be discussed.

Null Shift A laser gyro can exhibit a null shift, i.e., a nonzero beatnote for a zero-input rotation. Null shift arises when the optical path is anisotropic for the oppositely directed waves. Causes may be attributed to a nonreciprocal index of refraction for the two beams, noise in the bias used to eliminate lock-in, drift, nonsymmetry of an applied bias (to avoid lock-in), active medium flow (e.g., Langmuir flow), tube temperature gradients,

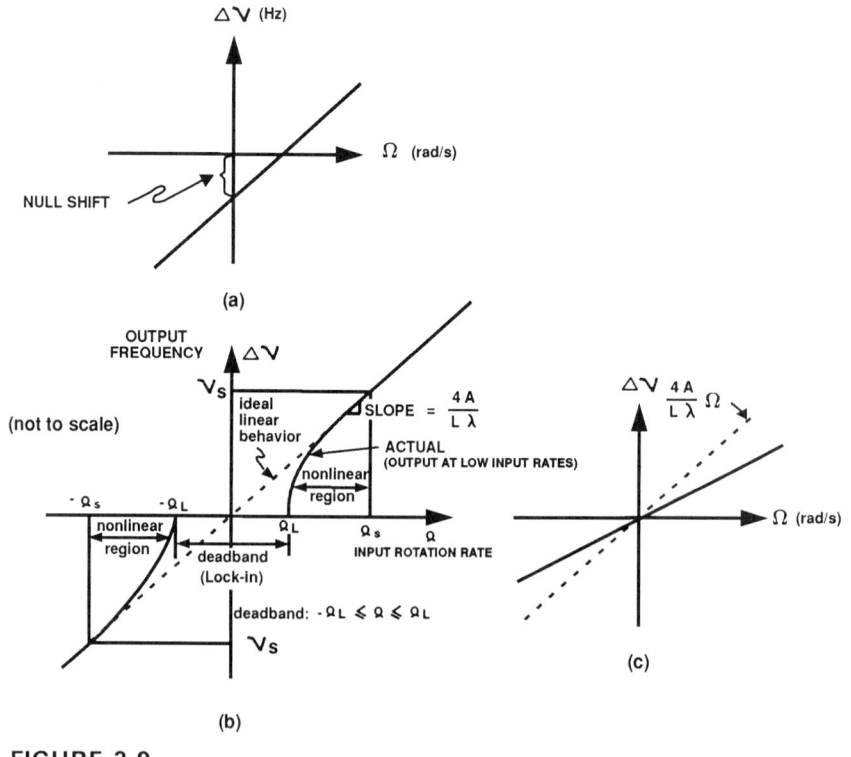

FIGURE 3-9

Unbiased two-mode ring laser gyro errors affecting the output: (a) null shift; (b) lock-in; (c) mode pulling.

and current differentials. The Langmuir flow bias is associated with gas-flow effects in a plasma discharge, which produces a frequency difference not related to rotational motion. Also, the Langmuir flow of neon atoms is caused by a negatively charged cavity wall. A symmetric discharge utilizing two anodes and one cathode allows balancing the oppositely directed anode currents, which reduce the Langmuir flow effects. Null shifts can occur due to anisotropic anomalous dispersion effects and atomic transition. Anisotropy is caused by the well-known "Fresnel–Fizeau effect." Further, null shifts caused by nonreciprocal saturation effects in the active gain medium can arise from any element in the cavity causing nonreciprocal loss for the two waves: anisotropic scattering effects and magnetic interactions.

Lock-in The error that receives the most attention is the lock-in phenomenon. Lock-in is a fundamental physical phenomenon associated with all oscillating devices. Laser gyros exhibit the same phenomenon. The most important result of the lock-in phenomenon is that the scale factor S becomes a function of the rotation rate Ω. In laser gyros, the coupling mechanism is the reflection of light from one laser beam back along the path where it interacts with the oppositely traveling beam. This, as we have seen, is called *backscatter* and is caused primarily by imperfect mirrors. More specifically, backscattering, localized losses, and polarization anisotropies cause the mode frequencies to lock at low rotation rates. This results in a zero beatnote for a nonzero rotation [1, 4]. That is, the lock-in is the deadzone, in which there is zero output for input angular rates of less than Ω_L. Mathematically, the difference in frequency $\Delta \nu$ can be expressed as

$$\Delta \nu = \begin{cases} 0 & \text{for } \Omega^2 \leq \Omega_L^2 \\ (4A/L\lambda)\sqrt{\Omega^2 - \Omega_L^2} & \text{for } \Omega^2 > \Omega_L^2 \end{cases} \quad (3.19)^*$$

Equation (3.19) approaches the ideal linear behavior of Eq. (4.56) only in the limit of high angular rates (see Fig. 3-9*b*). Therefore, the laser gyro beat frequency is unresponsive until an angular input rate greater than the lock-in threshold ($\pm \Omega_L$) is sensed. A bias must be applied such that with a zero or low input rate there is enough of a frequency difference between the two beams to avoid the lock-in region. Going one step further, Eq. (3.19) can also be written in the form

$$\Delta \nu = S(\Omega_{in} - \Omega_B)\sqrt{1 - \left(\frac{\Omega_L}{\Omega_{in} - \Omega_B}\right)^2}, \quad \Omega_L < (\Omega_{in} - \Omega_B) \quad (3.20)$$

*Equation (3.19) says that the gyro sees an output beat frequency if $\Omega > \Omega_L$. Consequently, there are small nonlinear effects, and if Ω_L varies, so will the output. If $\Omega < \Omega_L$, the two beams are frequency locked.

where Ω_B is the bias for zero input rate. We note that from Fig. 3-9b, the bias Ω_B is identical to the dither rate Ω_D. Lock-in typically occurs at rotation rates of approximately 0.1 deg/s. There are many different schemes for avoiding the previously discussed lock-in problem. In the two-frequency laser gyro, biasing methods are commonly used to keep the gyro out of the lock-in region: (1) the magnetooptic (nonreciprocal phase shift introduced between the counterrotating beams) bias or Kerr effect, which includes the Faraday effect and magnetic mirror, and (2) mechanical dither. Here we will briefly consider only the mechanical dither. The mechanical dither is a sinusoidal, symmetric signal, alternating the bias about the zero-input rate. In this method, the entire laser gyro block is rotated sinusoidally at peak rates of 50–250 deg/s. Typical values of dither frequency are on the order of 100–500 Hz and amplitudes are between ±100 and 500 arcsec, depending on the laser gyro design. Some authors (e.g., [3]) give the value of the dither frequency as $\omega_D = 400$ Hz and the amplitude $\Omega_D = 100$ kHz, respectively. As we have noted earlier, the RLG is inherently an integrating rate gyro, with a digital output. A simplified expression for the output, considering the lock-in effect, is given by [1, 2]

$$\dot{\psi} = \Omega - \Omega_L \sin(\psi + \beta)$$

where ψ is the instantaneous phase difference between the two counterrotating beams (ψ includes input and null shifts), Ω is the input rotation rate, Ω_L is the lock-in rate, and β is the effect of backscattering phase shifts. For a sinusoidal oscillating bias, the mathematical model of the laser gyro, which includes the dithering effects, can be written in the form [2]

$$\dot{\psi} \cong \Omega - \Omega_L \sin(\psi + \beta) + \Omega_D \sin \omega_D t$$

where $\dot{\psi}$ is the beat frequency, Ω_D is the bias magnitude, and ω_D is the dither rate. It should be pointed out that in any alternating bias method, the bias must be perfectly symmetric.

One of the easiest methods to implement, as mentioned above, is to rotate the laser gyro about its axis at a constant rate above the lock-in threshold and subtract this rate from the gyro output. The problem is to accurately control this constant bias. To be used, for example, in a fighter aircraft, the laser gyro must be able to measure rotation rates of 400 deg/s with an accuracy of 0.01 deg/h. No constant rotation rate can be controlled to this degree of accuracy. A refinement of this technique is to periodically reverse the direction of rotation. If the magnitude and duration of rotation in both directions are the same, they cancel each other out. This periodic reversal is normally accomplished by means of a torsional spring and a piezoelectric driver which produces a sinusoidal rotation rate called, as we have seen, *mechanical dither*. Figure 3-10 depicts the mechanical dither.

3.3. The Ring Laser Gyro Principle

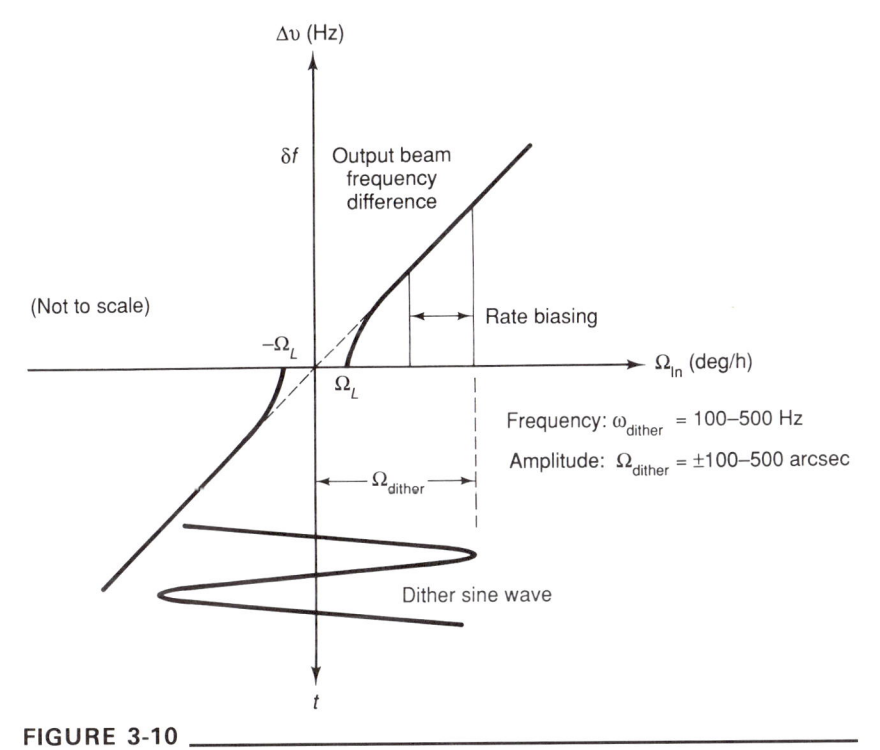

FIGURE 3-10
Lock-in avoidance scheme region of operation.

The bias averages to zero over each cycle and operation occurs mostly in the linear region. The dither eliminates, so to speak, the deadband and nonlinearity associated with lock-in and replaces it with a measurement uncertainty about the ideal input/output curve, as illustrated in Fig. 3-11. This uncertainty is commonly called the *random-walk** error. The *S* in this

---

*Random walk (also known as "Brownian motion" and the "Wiener process") is one of the most fundamental of stochastic processes and is obtained by passing white noise through an integrator. Mathematically, random walk is expressed by the differential equation

$$\frac{dx(t)}{dt} = w(t)$$

$$x(0) = 0$$

where $w(t)$ is a zero-mean white noise process. Random walk can also be expressed by the "difference equation"

$$x_{n+1} = x_n + w_n \qquad n = 0, 1, \ldots$$

$$x_0 = 0$$

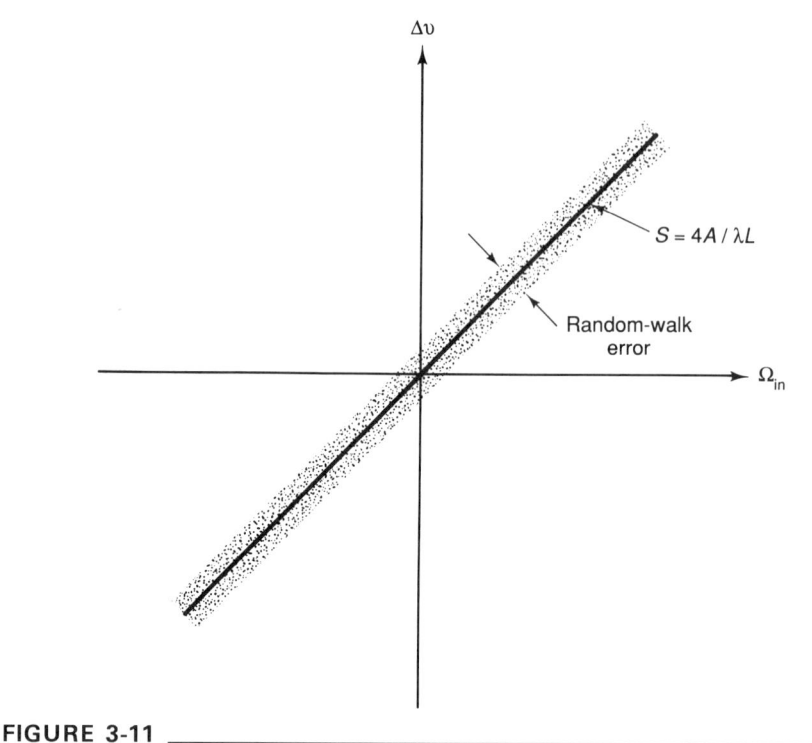

FIGURE 3-11
Input/output relationship of a dithered laser gyro.

figure represents the ideal scale factor. Dithering is currently the most common means of lock-in compensation. There are two advantages to this scheme. One is that the rotation rate periodically passes through zero where lock-in occurs. This results in a random-walk angular position error. The other is that the vibration of the gyro body may affect other components or the gyro may be affected by other vibration sources. The laser gyro, as we shall see in the next section, need not be mechanically dithered. The same result may be achieved by the four-wave technique.

Mode Pulling Since the laser radiation is generated in a medium with dispersive properties, the scale factor S will vary as the frequency of the mode of oscillation varies with position on the gain curve. Variations in the scale factor can be caused by mode pulling and pushing. In considering the effects of saturation on the dispersion properties of the active medium, there is a mode pushing correction to the oscillation frequencies due to the hole burning correction in the dispersion curve. That is, one beam "pushes" on the other. If there is a loss difference between the two beams, caused, for

TABLE 3-2

Ring Laser Gyro Error Budget Parameters

Parameter	Value
Bias error	
Bias repeatability	0.001 deg/h (1σ)
Bias stability	0.005 deg/h (1σ) [0.05 deg/h z axis]
Random walk	0.004 deg/$\sqrt{h}$ (1σ) [0.04 deg/h z axis]
Scale factor	
Stability	<5 ppm (1σ)
Asymmetry	<1 ppm (1σ)
Nonlinearity	<5 ppm (1σ)
Input axis alignment	
Error value	±1 milliradian maximum
Stability	10 microradians maximum
Axis-to-axis orthogonality	<2 arcsec
Rate range	0 to ±400 deg/s each axis
Temperature range	−54°C to +71°C
Shock (10 ms, $\frac{1}{2}$ sine)	20g
Vibration	5g
Life	
Operating	50,000 h
Storage	>5 years
Pulse weight (resolution)	≤2 arcsec
Input power	5 W (watts)

example, by anisotropic or other effects, a differential loss null shift will result. As the anisotropic effects change with temperature, so will the null shift. Thus, for each beam there is a linear pulling, that is, the detuning of the oscillation frequency from the cavity frequency. This depends on the plasma dispersion and the amount by which the oscillation frequency differs from the atomic resonance frequency [2]. Furthermore, this pulling also depends on the beam loss. Normally, this type of pulling can be related to a shift in the scale factor. Table 3-2 lists the desired RLG performance parameters that will meet a 1.852-km/h navigation system CEP rate or better.

3.3.2. Multioscillator Ring Laser Gyros

As we have seen in the previous section, two-frequency RLGs tend to lock in at low input rates, causing drift errors. Currently, these laser gyros are mechanically dithered or rate biased in order to avoid the lock-in region. However, it can be shown that the lock-in problem can also be circumvented by producing four lasing modes instead of two (called *multioscillating*), thus eliminating the need for moving parts.

The multioscillator RLG consists of two independent, two-frequency laser gyros sharing the same cavity and having a common optical path, but biased in opposite senses (i.e., having different polarizations). These two modes must oscillate independently in order to accurately measure rotation. Consequently, in the differential output of these two laser gyros the bias cancels out, so that any signals generated as a result of an input rotation would add. Furthermore, the gyro output never passes through the lock-in region since the bias need not be dithered. Thus, multioscillator laser gyros produce four laser beams at four separate frequencies rather than the two laser beams of the two-mode laser gyros to avoid lock-in. Two of the beams, one traveling in each direction, are left circularly polarized, while the other two are right circularly polarized. The laser gyro is configured first by creating a four-mirror (one of the mirrors is the output or transmissive mirror) square cavity path in a two-frequency gyro as shown in Fig. 3-12a, and as in the two-mode laser gyro, nearly equal amounts of ^{20}Ne and ^{22}Ne are used to reduce mode competition. A crystalline quartz element is then placed in

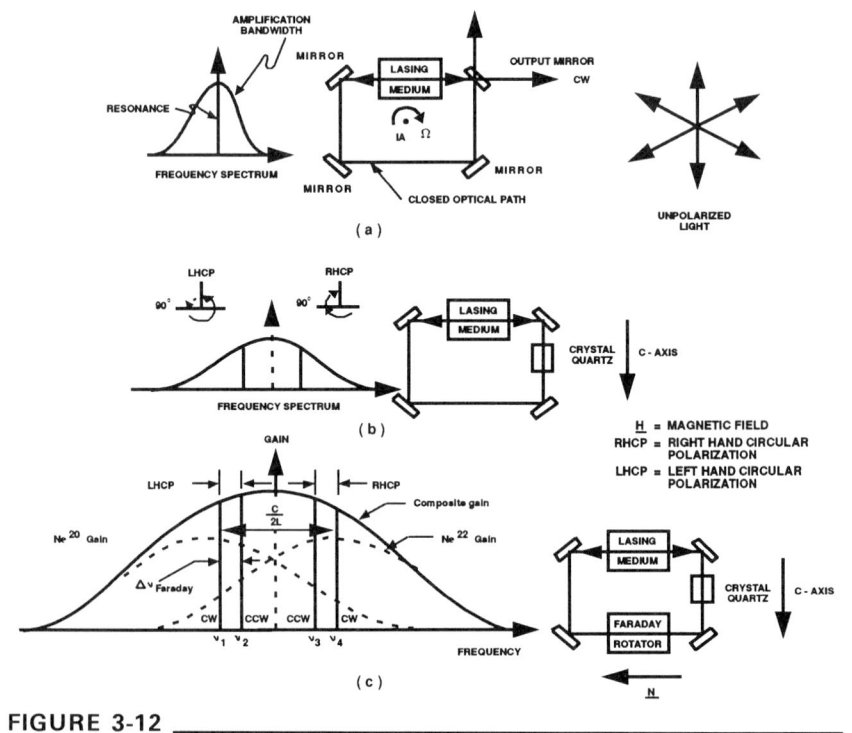

FIGURE 3-12
Multioscillator ring laser gyro principle: (a) four-mirror laser gyro; (b) laser gyro showing the two polarizations; (c) laser gyro showing addition of a Faraday rotator.

3.3. The Ring Laser Gyro Principle

one of the paths as illustrated in Fig. 3-12b. This device has what is called *rotary birefringence*, which polarizes the light passing through it into only two distinct polarizations, both circular (left-hand and right-hand). Specifically, the quartz crystal is a reciprocal polarization rotator. This device imparts rotation to the polarization of the light beams and acts equally on both the clockwise and counterclockwise beams; hence the description "reciprocal." This differs from the two-frequency laser gyro, in which no polarization constraints are made. But more important, due to the orientation of the quartz, the beam of the left-hand circular polarization (LHCP) and the beam of the right-hand circular polarization (RHCP) see nominally different path lengths and consequently different optical lasing frequencies. This frequency difference can be made large enough to avoid mode coupling. Figure 3-12c shows the addition of a magnetic field device, of magnetic field intensity **H**, which further splits the beams (that is, the frequency degeneracy between the CW and CCW traveling beams is eliminated) and a Faraday bias is added. More specifically, the addition of a Faraday polarization rotator (or cell) introduces a reciprocal (direction-independent) polarization anisotropy, such that orthogonally polarized modes propagating in the same direction experience slightly different optical path lengths during traversal of the same optical path. The polarization rotator splits the frequencies of the resulting RHCP and LHCP modes. As a result, with the crystal and a fixed magnetic field applied, four resonant traveling-wave frequencies are generated. By assigning frequencies as indicated in Fig. 3-12c, the individual laser gyro equations are [4]

$$\Delta v_L = v_2 - v_1 = -S\Omega + \text{Faraday bias}$$

$$\Delta v_R = v_4 - v_3 = S\Omega + \text{Faraday bias}$$

where S is the ideal scale factor. The difference of these two quantities is therefore $2S\Omega$ with no bias contribution. In practice, the bias is normally chosen to be positive and at the same time much greater than $|S\Omega|$, so that the gyro operates far from the lock-in region (see Fig. 3-13). Combining these frequency shifts differentially, we note that the measured beat frequency is

$$\begin{aligned}
\Delta v &= \Delta v_R - \Delta v_L \\
&= (v_4 - v_3) - (v_2 - v_1) \\
&= (2S)\Omega \\
&= \left(\frac{8A}{L\lambda}\right)\Omega
\end{aligned} \quad (3.21)$$

From the preceding discussion, it will be noted that a number of advantages are apparent in the multioscillator approach. First, the mechanical

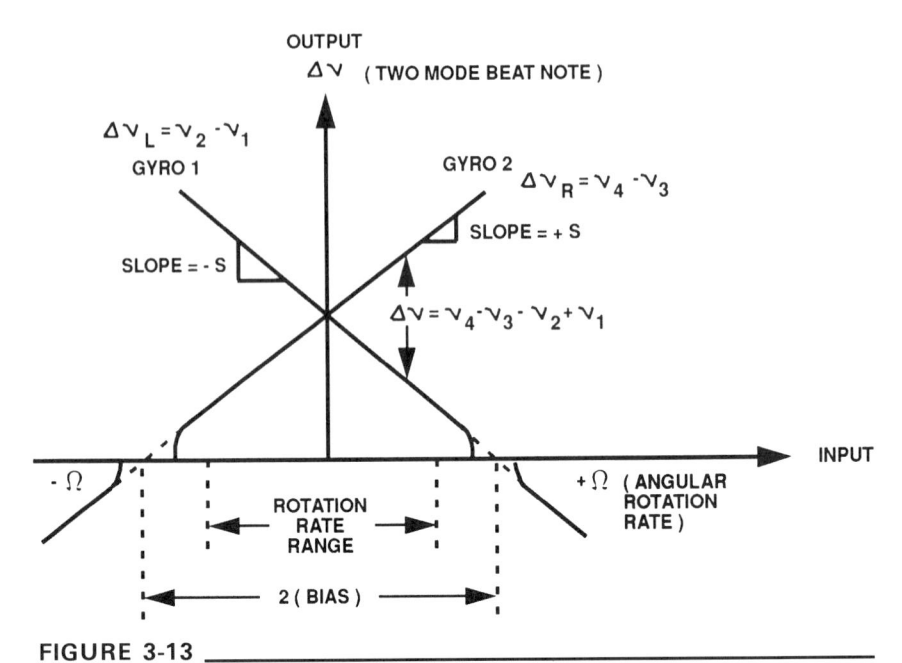

FIGURE 3-13
Multioscillator response to rotation in inertial space.

dither is eliminated, thus reducing the noise and errors because no lock-in region is entered as seen in Fig. 3-13. Also, by using passive bias techniques, the output is insensitive to bias drifts (through subtraction of the beat frequencies), and the sensitivity is twice as great as in the two-frequency RLG ($\Delta v = 2S\Omega$). Moreover, because of their intrinsic low noise, multioscillator RLGs are well suited for applications requiring rapid position update, high resolution, and high accuracy. A change in the frequency separation within one pair of like polarization is the same as that of the other. Likewise, a change in the frequency separation between oppositely polarized pairs will not directly affect the output. Therefore dithering is not required and there is no periodic passing through the lock-in region. Thus, the lock-in problem is eliminated. All of these factors contribute to the advantages of a strapdown inertial navigation system, particularly reduced size and power, and increased accuracy and reliability through the elimination of mechanical moving parts.

Rather than using a crystal to separate the frequencies of opposite polarization, the same effect may be achieved by use of out-of-plane geometry, or nonplanar cavity. A nonplanar ring resonator is a closed polygonal ring, whose segments do not lie all in a single plane. Nonplanar rings belong to the broader class of nonorthogonal optical systems, that is, systems that do not possess meridional planes of symmetry. Consequently, such systems are difficult to analyze because the principal axes for the phase fronts and intensities are seldom orthogonal. However, independent investigations of nonplanar rings have been demonstrated to be inherently advantageous for multioscillator laser gyros. It should be pointed out that the sensing properties of such rings are simple extensions of the conventional planar rings, and that a nonplanar ring is a single-axis rotation sensor, even though it is a multiplane instrument. The Faraday rotator results in frequency splitting directly attributable to a rotation or phase shift of the plane of polarization, which is positive for CW left-hand circularly polarized (LHCP) and CCW (RHCP) light and negative for the other two modes. Using the geometry of nonplanar cavity instead of a Faraday rotator, the phase shift phenomenon can be re-created. This approach provides a greatly improved means of achieving the polarization frequency splitting without the use of the quartz polarization rotator. However, this approach is still in the research stage, showing some promise.

Historically, the original research on the multioscillator RLG principle can be traced to the mid-1960s, when active research and development was started by the United Aircraft Corporation's Research Division. In 1968, a patent was filed by United Aircraft on this technology in what came to be known as the "differential laser gyro" (DILAG). The DILAG consisted essentially of two laser gyros sharing the same cavity and taking advantage of the properties of light polarization. Toward the end of the 1970s, United Aircraft decided to discontinue further research and development of the DILAG. In the early 1970s, the Raytheon Company started research and development of the multioscillator laser gyro, filing for a patent in 1972. Two years later, in 1974, Raytheon demonstrated a 55-cm path-length multioscillator with a bias error of 0.001 deg/h. A multioscillator research program was also initiated in 1974 by Litton-Guidance and Control Systems. The multioscillator's potential for high accuracy, rapid alignment, and accurate pointing and tracking applications was immediately recognized with contracts for further development awarded by the U.S. Air Force and NASA. In 1982, Raytheon decided to discontinue work on the multioscillator laser gyro, and in 1985, Litton bought the rights to Raytheon's multioscillator technology. Over the years, Litton further refined the instrument, achieving, among other things, excellent bias stability, excellent scale factor linearity, and stability, and demonstrated its applicability in high-accuracy strapdown inertial navigation systems.

3.3.3. Alternate RLG Designs: Passive Laser Gyros

The use of the "Sagnac Effect" as a means of rotation measurements has been treated extensively in the literature. A passive laser gyro is one that does not have the gain medium within the resonant cavity. Instead, a linear laser produces a beam that is split into two beams that are introduced into the resonant cavity where they travel in opposite directions. Here the laser output frequency is held constant. The path length of the resonant cavity is adjusted so that it is an integral number of half wavelengths for one of the two beams. As the passive laser gyro rotates, the path length is adjusted to compensate for the apparent path-length changes discussed earlier in this section (3.3). The frequency of the other beam is changed by means of an acoustooptic modulator so that the beam is resonant at the altered path length. The amount by which the frequency is changed is a function of the rotation rate. Like the active RLG, the relative shift of resonant frequencies associated with the counterpropagating waves is given by the Sagnac equation for laser gyro operation [Eq. (3.9)]. The precision with which the frequency difference Δv can be measured depends on the cavity linewidth. The passive laser gyro concept does not have a lock-in problem since both laser beams are derived from one constant-frequency source. However, the passive RLG is very sensitive to environmental fluctuations because alignment of the cavity mirrors must be precisely maintained during the time of operation. The greatest disadvantages of the passive RLG appear to be larger size and a greater number of optical components, which translate into higher cost.

3.4. FIBER-OPTIC GYROS

The development of the fiber-optic gyro (FOG) dates to the mid-1970s. The fiber-optic gyro, like the conventional RLG, makes use of the "Sagnac effect" as a means of measuring rotation rate. Fiber-optic gyros use optical fiber as the light path, in contrast to RLGs in which light is beamed around a cavity. In essence, therefore, the FOG is a rotation rate or angular "rate" sensor that uses optical fiber as the propagation medium for light waves. Rotation is measured by analyzing the phase shift of light caused by the Sagnac effect. Ideal behavior of the FOG assumes perfect reciprocity, that is, the paths of the clockwise- and counterclockwise-traveling beams are identical. Since the beams traveled the same path (although in opposite directions) and assuming the fiber-optic is isotropic, the only possible source of the phase difference is inertial rotation of the fiber. Any nonreciprocity that may arise from nonlinear index changes caused by unequal intensities in the counterpropagating directions is effectively handled by using a broadband source, such as a superluminescent diode. Nonreciprocity caused by external magnetic fields due to residual Faraday effect in the fiber may be reduced substantially by magnetic shielding. As stated in the previous sections, the Sagnac effect is the differential phase shift induced by rotating an

3.4. Fiber-Optic Gyros

optical system in which light travels clockwise and counterclockwise around the system. Therefore, the FOG measures rotation rates about an axis perpendicular to the plane of the fiber-optic coil by means of ring interferometry, whereby the necessary closed light path is provided by a coil of optical fiber. [For this reason, these types of gyros are commonly referred to as *interferometric fiber-optic gyros* (IFOG).] Depending on the application, the length of the optical fiber can be anywhere from about 50 m long to 1 km long; thus it can be wound into a coil of low volume.

The basic fiber ring interferometer design (see Fig. 3-14) consists of four components in the following order: (1) a semiconductor laser diode that acts as the light source, (2) a beamsplitter, (3) the coil of optical fiber to transmit the laser beams, and (4) a photodetector.

Light from the laser diode is divided or split into two beams of almost equal intensity (i.e., phase and frequency). One of these beams is passed clockwise through the fiber coil, and the other beam counterclockwise. The two beams are recombined or superimposed when they exit the fiber, and the resulting interference is monitored. The rigid mount provides a defined

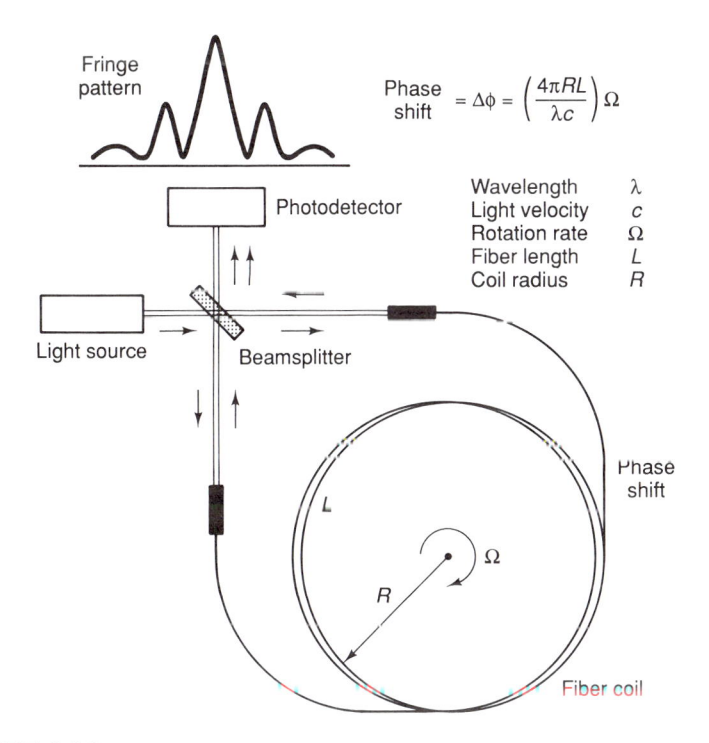

Phase shift $= \Delta\phi = \left(\dfrac{4\pi RL}{\lambda c}\right)\Omega$

Wavelength λ
Light velocity c
Rotation rate Ω
Fiber length L
Coil radius R

FIGURE 3-14
Principle of the fiber-optic gyro.

reference plane for the gyro axis. When there is no fiber ring rotation, transit times of the light in both directions around the fiber ring are identical. With fiber ring rotation, the transit times are different, resulting in reduced intensity at the detector. That is, as mentioned above, as the coil is rotated about an axis perpendicular to the plane of the fiber-optic coil, the light traveling in the direction of rotation has a longer propagation time than the other beam, due to the Sagnac effect. In other words, rotation results in a difference of propagation time between the clockwise and counterclockwise beams, which is manifested as a relative phase shift between the beams, and hence reduced intensity at the detector. More specifically, the phase difference causes the interference of the two beams to change in intensity. However, it should be noted that in real systems the change in intensity is very small for useful rotation rates, so that special techniques must be introduced in order to enhance the resolution. The amplitude of the phase shift is proportional to the coil's angular velocity, and the direction of the phase shift is indicative of the direction of rotation. The applied input rotation rate is transformed by the fiber coil into a phase difference between the counterrotating beams of light. Consequently, the phase shift results in a change in the intensity of the interference signal, which is converted into an electrical signal by the photodetector. Digital electronics process this analog signal into a digital signal proportional to the rotation rate. Digital mechanization is possible by analog-to-digital conversion of the photodetector intensity every t seconds. The resulting output is a digital word equal to the angular rotation rate, Ω, every t seconds. Specifically, t is chosen sufficiently small so as to preclude any loss of data in computing angular displacement from angular rate samples. A simple-chip microprocessor is used for scaling and output handling of the gyro signal and for scale factor stabilization.

Mathematically, the relative phase shift is expressed as

$$\Delta\phi = \phi_{CW} - \phi_{CCW} = \left(\frac{4\pi RL}{\lambda c}\right)\Omega \qquad (3.22)$$

where L is the fiber length, λ is the diode laser wavelength, R is the coil radius, Ω is the inertial rotation rate, and c is the free-space velocity of light. The phase difference will produce a change in the intensity of the fringe pattern and is, thus, the basic observable for sensing rotation. This equation states that by measuring the output of the photodetector, one can determine the Sagnac phase shift, $\Delta\phi$, proportional to the inertial rate Ω. In certain designs of FOGs, the accuracy/sensitivity ratio increases as a function of the number of turns of fiber in the sensing coil, in addition to being a function of the area enclosed by the coil. Stated in other words, the phase shift is directly proportional to the area A, so that the sensitivity can be improved by increasing the number of turns of the fiber coil. Consequently,

3.4. Fiber-Optic Gyros

an extension of the above discussion can account for a multiple turn solenoid of the single loop. For a solenoid of N turns, the length of the fiber L, is given by $L = 2\pi R N$, and the area $A = \pi R^2$. Therefore, Eq. (3.22) takes the form

$$\Delta\phi = \left(\frac{8\pi NA}{\lambda c}\right)\Omega \quad (3.23)$$

where A is the area enclosed by the fiber loop. The quantity in the parentheses can be combined into a constant K, which is the sensor's scale factor. From the preceding discussion, it is apparent that the Sagnac effect can be viewed in terms of either a time delay or as a phase shift. Although mathematically equivalent, these concepts form a natural breakdown for consideration of processing techniques. Since the phase of the interference pattern is directly dependent on the frequency of the interfering beams, the Sagnac phase shift can be compensated for by means of a change in frequency of one of the beams. For a fiber of length L, the frequency shift or difference can be obtained from the expression

$$\Delta f = \left(\frac{2R}{n\lambda}\right)\Omega \quad (3.24)$$

where n is the index of refraction of the fiber.

The fiber-optic gyro offers a possible solution to two of the major design limitations of the RLG: the lock-in and the scale factor, which is limited by the size of the instrument. Because in IFOGs the light source is normally placed outside the sensing cavity, the gain medium does not amplify any effects from backscatter which would couple the CW and CWW beams. Thus, it does not suffer from the lock-in phenomenon. The scale factor can be raised by simply adding more turns of fiber and increasing the mean area enclosed by each. However, it should be pointed out that, signal losses in the fiber impose a practical limit to the fiber coil's length. The major difference between RLGs and FOGs is that RLGs depend on the necessary conditions and characteristics of a laser. Fiber-optic gyros, on the other hand, achieve optimum performance using a broadband light source dominated by spontaneous rather than stimulated emission. In particular, the ring laser gyro is a resonator that supports a standing wave for an indefinite period of time, whereas the IFOG stores the light waves only for one transit time. Therefore, the FOG can use cheaper, more rugged solid-state light sources such as a superradiant diode. Also, the IFOG is a rate gyroscope in contrast to the RLG, which is a rate-integrating gyro. Note that the phase shift in an IFOG is analogous to the frequency shift that is used as an indication of motion in an RLG. The IFOG also has the potential to achieve a large advantage in terms of size, weight, power, packaging flexibility, assembly, and cost. There are some problems that need to be overcome. It was hoped that because the scale factor could be increased simply by adding more coils to the fiber ring, smaller diameters could be used without sacrificing sensitivity. The development of fiber optics has been driven by the requirements for long-distance communication links and has resulted in single-mode optical waveguides with losses that are approximately at the theoretical limit. Relatively poor fibers have losses that are determined by impurity absorption and scattering by surface and volume irregularities. Recently, however, fiber technology has advanced to the point that losses are now determined by the fundamental thermodynamic limits set by the optical interaction with elementary excitations of the fiber materials. The advantages of the FOG are (1) no moving parts, (2) instant turn-on, (3) high reliability, (4) no g-sensitivity, (5) resistance to high shock and vibration, (6) wide dynamic range, (7) no preventive maintenance, (8) long shelf life, and (9) low cost. Some typical FOG characteristics are listed below.

Power dissipation: 15 W
Bias (drift) uncertainty: 10–0.01 deg/h
Scale factor error: 1000–10 ppm
Bandwidth: 500 Hz
Range: ±400 to ±600 deg/s
Operating temperature: −40 to +70°C
Characteristic path length: 50 m–5 km

Recently, it has been determined that the IFOG can also be configured as a rate integrating gyroscope, thereby exhibiting the desired output characteristics suitable for navigation applications. This is a departure from the historical point of view that the IFOG is strictly a rate gyro. Since the IFOG output is generally noisy, there is often an advantage to filtering or oversampling and averaging its output. However, the rate integrating characteristic holds only if the gyro is sampled and accumulated every transit time. In essence then, the IFOG bears a great resemblance to the ring laser gyro.

Modern advanced technology makes it possible to design an FOG to include all the components, including fiber coil, into an integrated optical chip. This large-scale integration of optic components translates into low-cost, accurate, and reliable gyroscopes the size of a conventional computer chip. In fact, the solid-state gyroscope, like the RLG, has no moving parts and consists primarily of a spool of fiber optics, an integrated optics chip, and a single solid-state laser source. All indications are that the present state of the art in gyroscope technology is evolving toward interferometric fiber-optic gyros. While RLG technology for high-accuracy inertial navigation

TABLE 3-3
Typical RLG Characteristics

Path length	6–50 cm
Bore size	0.2–0.3 cm
Beam diameter	0.05 cm
Geometrical shape of the block	Triangle, square, rectangle
Mode of operation	2-frequency or 4-frequency
Gas (total pressure)	~5 torr
Gas mixture	He/Ne 13:1
	^{20}Ne/^{22}Ne 1:1
Operating wavelength (transition)	0.6328 μm
	1.1523 μm
Excitation voltage	1500–3000 V dc
Power	3–8 W
Discharge current	1–2 mA
Laser block material	Cer-Vit or Zerodur
Block construction	Monolithic, modular
Readout device	Silicon photodetector
Mirror curvature	4–5 m
Lock-in threshold	100–1000 deg/h
Resolution (pulse weight)	2–3.14 arcsec/pulse

systems is quite mature, the IFOG is improving and will achieve high accuracy levels in the not-too-distant future. Medium-accuracy (e.g., 0.1–1.0 deg/h) IFOGs for tactical applications are available at the present time. Typical fiber-optic gyro applications include the following: (1) commercial/military aircraft navigation and flight-control systems, (2) guided missiles, (3) attitude and heading reference systems, (4) antenna and gun stabilization, and (5) civil applications such as robotics and traffic management control. Table 3-3 presents some typical RLG characteristics, while Table 3-4 compares the performance characteristics of the two-mode RLG and the fiber-optic gyro.

3.5. THE RING LASER GYRO ERROR MODEL

The general analytical model for the RLG parallels that for a single-degree-of-freedom spinning-mass gyroscope, where the errors associated with mass properties have been removed. As a result, this error model can be expressed as [12]

$$\Omega_{out} = \Omega_B + (1 + \varepsilon)[\Omega_{in} + \gamma_z \Omega_y - \gamma_y \Omega_z]$$
$$\varepsilon = \varepsilon_0 + f(|\Omega_{in}|) + g(\Omega_{in}) \tag{3.25}$$
$$\Omega_B = B_0 + n_1 + n_2$$

TABLE 3-4

Performance Comparison

	Ring laser gyro	**Analog fiber-optic gyro**
Output	$\Delta v = \dfrac{4A}{L\lambda} \Omega$ Frequency proportional to rate or counts per turning angle	$\Delta\phi = \dfrac{8\pi N A \Omega}{\lambda c}$ Voltage proportional to rate
Characteristic path length	6–50 cm	50 m–5 km
Scale factor	<1 ppm	10^3–10 ppm
Dynamic range	0–500 deg/s	0–800 deg/s
Critical issue	Lock-in compensation	Reduction in scattering polarization control
Thermal errors	Packaging, electrode placement, path-length control	Packaging, thermal control and compensation

where
  Ω_{out} = laser gyro output signal
  Ω_B = gyro bias error
  ε = gyro scale factor error
  Ω_{in} = gyro input rate
  Ω_y, Ω_z = angular rotation rates of the gyro case normal to the input axis
  γ_y, γ_z = misalignments of the gyro lasing plane relative to the nominal gyro input axis
  ε_0 = "fixed" scale factor error
  $f(|\Omega_{in}|)$ = symmetrical (relative to positive and negative input rates) linearity error
  $g(\Omega_{in})$ = generalized linearity error (containing symmetric and asymmetric components)
  B_0 = "fixed" bias error
  n_1 = random bias error with unbounded integral value*
  n_2 = random bias error with bounded integral value

---

*The "random bias" (or "random constant") model is described by the differential equation

$$\frac{dx(t)}{dt} = 0$$

or as a difference equation

$$x_{k+1} = x_k.$$

The random bias is generated from the output of an integrator with no input.

Bias and scale factor are two key parameters that define the performance of a ring laser gyro. *Bias* is defined as the average output angular pulse rate after the inertial-rate inputs have been removed and is expressed in degrees per hour. Or more simply stated, the laser gyro bias is defined as the gyro output for zero input angular rate. It is the change in bias that produces system error, and it is of primary importance in evaluating laser gyro performance. In order to determine the bias stability, the laser gyro is commonly mounted to a fixed, stable mount. The number of gyro output pulses that are accumulated during the sampling interval ΔT are recorded. The average bias over the sampling interval is then calculated after the component of the earth rate along the gyro input axis has been removed. Finally, the standard deviation of a set of n ($n \geq 20$) consecutive measurements is calculated. Short-term bias stability is measured with sampling intervals between 1 and 100 s, while for long-term bias stability the sampling interval would normally range between 100 and 10,000 s. Statistically, white noise* describes the best bias runs for sampling intervals of up to about 400 s; however, random walk or some other process is predominant for longer intervals.

Scale factor is defined as the ratio that relates the inertial-frame rotation about the gyro's input axis to the gyro output expressed in arcseconds per pulse. Stated more simply, the laser gyro scale factor relates the number of output pulses to a corresponding gyro rotation angle about its input axis. In the laboratory, it is convenient to record scale factor data in terms of the number of gyro output pulses for one input revolution. The scale factor data can be obtained by mounting the laser gyro on a precision rate table with its input axis parallel to the table's axis of rotation. Consequently, the number of gyro output pulses are recorded for each revolution. This is repeated several times (e.g., 10 times or more) at each rotation rate for both CW and CCW directions. Finally, the mean and standard deviation errors are computed. The standard deviation about the scale factor mean is used to describe the scale factor nonlinearity. With regard to Eq. (3.25), let us examine bias and scale factor errors in some detail.

Gain-Medium-Dependent Bias Errors The B_0 "fixed" bias term is caused mainly by circulation flow phenomena in the lasing cavity that cause differential optical path-length variations between the clockwise and counterclockwise laser beams, by forward-scattering effects caused by laser cavity interference with the laser (i.e., beam interactions with imperfect mirror surfaces that produce differential phase shifts between the laser beams), and by residual errors introduced by the lock-in compensation device. The

---

*A white noise is a noise which has a constant amount of power content across all frequencies. White noise is not correlated in time. That is, the magnitude of noise signal at one instant of time does not indicate the magnitude of noise signal present at any other instant of time.

circulation flow phenomena (i.e., the velocity of the plasma within the cavity), caused by the high-voltage electrical field (Langmuir flow) and temperature gradients, will result in a constant frequency shift between the two counterrotating light beams, which will be detected as an angular input rate. This is due to the "Fizeau effect." The n_1 error is a white- or colored-noise effect generated within the lasing cavity. A classical cause of the mechanically dithered laser gyros is the random angle error introduced each time the gyro input rate is cycled through the lock-in region (twice each dither cycle). In general, n_1 is caused by random instabilities in the bias producing mechanisms in the lasing cavity. The n_1 error is typically measured in terms of the rms (root mean square) value of its integral over a specified time period, which is long compared to the n_1 noise process correlation time. As is the case with classical zero-mean random-noise processes, the average magnitude of the square of the integral of n_1 builds linearly with time. The n_2 bounded noise term is caused by scale factor errors in the mechanism used to eliminate the lock-in compensation bias from the output of laser gyros employing alternating bias. For laser gyros employing alternating electrooptical bias for lock-in compensation, n_2 is caused by scale factor uncertainties in the applied electrooptical bias. The errors due to input axis misalignments are caused by the difference between the inferred and the actual input axis orientation. Mechanical, thermal, or other phenomena that contribute to bending or distortion of the gyro will cause misalignments in the input axis position. In general, the B_0, ε_0, γ_y, and γ_z terms in the analytical model are measurable and predictable to a large extent for purposes of compensation. The remaining errors are generally unpredictable, and are controllable only through gyro design and manufacturing practices established to satisfy application requirements.

Scale Factor Errors The presence of a gain medium inside the laser cavity causes the value of the ideal scale factor S to deviate by the factor $(1+\varepsilon)$. The "fixed" scale factor error coefficient ε_0 is caused mainly by fluctuations in the gain medium parameters and variations in the cavity perimeter. This source of error is due to the dependence of the medium refractive index on the frequency v of the transmitted light. This causes the prism spectra of white light and is known as "dispersion effect." For the RLG, this means that the optical cavity length L and thus its scale factor are functions of frequency. The symmetrical scale factor error term, $f(|\Omega_{in}|)$ is the residual effect of lock-in for laser gyros employing mechanical dither compensation. The $g(\Omega_{in})$ term is the residual effect of lock-in for laser gyros incorporating nonmechanical lock-in compensation. In general, the magnitude of the scale factor linearity error for a given input rate is proportional to the degree to which the biased gyro input is removed from the lock-in region divided by the input rate being sensed. Since the width of the lock-in region is proportional to the lock-in rate, a low scale factor linearity error

3.5. The Ring Laser Gyro Error Model

is achieved with a high ratio of applied bias to lock-in rate. In its simplest form, a model ring laser gyro output can be written as

$$\Omega_{out} = S\Omega_{in} + \Omega_{bias} \tag{3.26a}$$

where S is the ideal scale factor. Another commonly used RLG error model consists of three basic types of errors: bias, scale factor, and input axis misalignment errors. Mathematically, this error model takes the form

$$\Omega_{out} = B + \varepsilon\Omega_{in} + \gamma\Omega_a \tag{3.26b}$$

where Ω_{out} = total laser gyro output error
 B = bias error
 ε = scale factor error
 Ω_{in} = input rotation rate
 Ω_a = rotation rate about an input axis perpendicular to the input axis of the gyro
 γ = input axis misalignment

In Eq. (3.26b), the bias, scale factor, and input axis misalignment can be further decomposed into components that include temperature gradients and magnetic field effects. We will now examine these error terms individually. The bias error term may be written in the form

$$B = B_0 + B_t\Delta T + B_{\nabla t}\nabla T + B_m M + B_a A + N_{RW}(t)\sqrt{t} \tag{3.27}$$

where B_0 = fixed bias error term in the gyro
 B_t = bias error due to temperature difference from nominal
 ΔT = temperature change from nominal
 $B_{\nabla t}$ = bias error due to temperature gradients
 B_m = bias error due to magnetic fields
 M = magnetic field strength
 B_a = bias error due to an acceleration along the input axis
 A = acceleration along the input axis
 $N_{RW}(t)$ = random component consisting of angular random walk and bias random walk
 ∇T = temperature gradient

Theoretically, the random-walk contribution to the bias consists of two terms as follows:

$$N_{RW}(t) = \sqrt{N_{RW,DN}^2 + N_{RW,QN}^2} \tag{3.28}$$

where $N_{RW,DN}$ = random-walk contribution due to the dither noise (this term will be zero for an undithered laser gyro)
 $N_{RW,QN}$ = quantum noise contribution to random walk due to spontaneous emission in the laser gyro

The fixed bias or bias repeatability term is a function of the plasma's stability, gas-flow conditions, and cavity alignment. Temperature changes cause a bias error by changing the pressure within the cavity. This pressure change causes gain changes in the cavity path, and consequently in the laser intensity. Furthermore, temperature gradients cause an asymmetric flow of the gases. A magnetic field will change the properties of light that is not linearly polarized, while accelerations can cause misalignment of the cavity by bending the laser gyro block. The random term is caused mainly by backscatter in the cavity and is due to mirror imperfections. The error generated when a RLG is dithered by a sinusoidal signal, which is several orders of magnitude greater than the lock-in threshold, can be characterized by a one-dimensional, random-walk (integrated white noise) process. In the ideal case, the sinusoidal dither signal remains the same from cycle to cycle. However, even if a small amount of random noise is present, it will be sufficient to affect the laser gyro's input/output relationship. The random walk process is given by the equation

$$N_{RW,DN} = \sqrt{\frac{S}{2\pi\Omega_D}} \Omega_L \quad (\deg/\sqrt{h}) \quad (3.29)$$

where S = ideal scale factor (arcsec/count) (count ≡ pulse)
Ω_D = peak dither rate (deg/h = arcsec/s) (1 Hz = deg/h)
Ω_L = lock-in rate (deg/h)

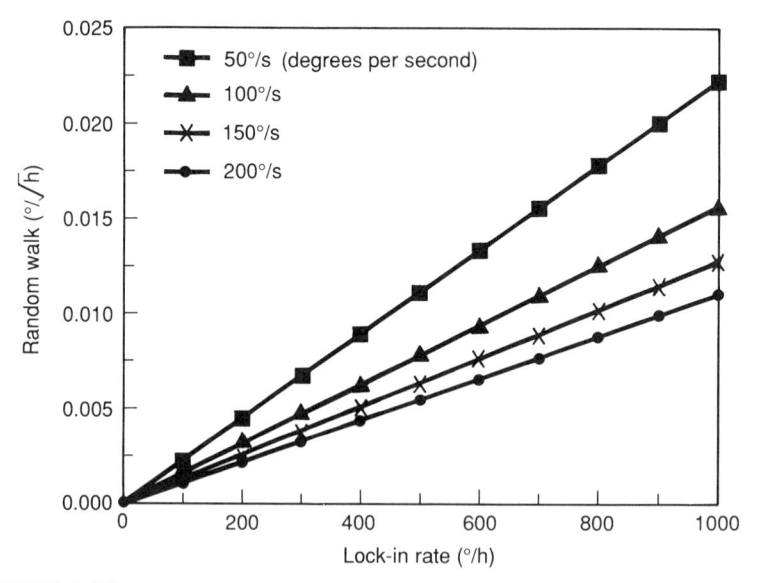

FIGURE 3-15
Random walk versus lock-in and dither amplitude.

3.5. The Ring Laser Gyro Error Model

For such a random process, the mean-value error is zero, while the rms error is expressed as

$$\varepsilon_{\text{rms}}(t) = \text{rms error}(t) = N_{\text{RW,Dn}}\sqrt{t} \quad (\text{arcsec}) \quad (3.30)$$

where t is the time in seconds. With current technology, a random walk of $0.001 - 0.005 \text{ deg}/\sqrt{\text{h}}$ or better can be achieved.

It is evident from Eq. (3.29) that the magnitude of the random-walk error is a function of the dither amplitude and the lock-in rate. This relationship is depicted in Fig. 3-15. The figure was obtained by programming Eq. (3.29) using **FORTRAN** 77. A program listing for the random walk is presented here:

```
C Calculate Random Walk (RWx) vs Lock-in (OL) for set values of
C Dither Amplitude (OD)
      PROGRAM RW
      PI = ACOS(-1.0)
      DO I = 0,10
         OL = FLOAT(I)*100.0             ! Lock-in [deg/h]
         OD = 50.0*3600.0**2             ! Dither = 50 deg/s
         RW1= SQRT(1.0/(PI*OD))*OL
         OD = 100.0*3600.0**2            ! Dither = 100 deg/s
         RW2= SQRT(1.0/(PI*OD))*OL
         OD = 150.0*3600.0**2            ! Dither = 150 deg/s
         RW3= SQRT(1.0/(PI*OD))*OL
         OD = 200.0*3600.0**2            ! Dither = 200 deg/s
         RW4= SQRT(1.0/(PI*OD))*OL
         WRITE (*,'(5F10.4)') OL,RW1,RW2,RW3,RW4
      ENDDO
      END
```

The scale factor term can be expressed mathematically in the form

$$\varepsilon = \varepsilon_0 + \varepsilon_t \Delta T + \varepsilon_{\nabla t} \nabla T + \varepsilon_m M + \varepsilon_a A + \varepsilon_w W + N_{\text{RW}}(t) \quad (3.31)$$

where
- ε_0 = fixed (or nominal) scale factor error
- ε_t = scale factor error due to the temperature difference from the nominal
- ΔT = temperature change from nominal
- $\varepsilon_{\nabla t}$ = scale factor error due to temperature gradients
- ∇T = temperature gradient
- ε_m = scale factor error due to magnetic fields
- M = magnetic field strength
- ε_a = scale factor error due to acceleration
- A = acceleration
- ε_w = scale factor error due to the input rate change
- W = input rate change
- $N_{\text{RW}}(t)$ = random component modeled as a random-walk process

The scale factor fixed term is a function of long-term losses in the cavity. These losses are caused by current changes, plasma bombardment, and laser misalignment. The scale factor linearity is, as we have seen, a function of the lock-in phenomenon. Lastly, an asymmetric scale factor can be caused by a loss in one of the counterrotating beams or asymmetric lock-in.

In a similar manner, we can express the input axis misalignment equation as follows:

$$\gamma = \gamma_0 + \gamma_t \Delta T + \gamma_{\nabla t} \nabla T + \gamma_m M + \gamma_a A + N_{RW}(t) \qquad (3.32)$$

where
γ_0 = fixed (or nominal) input axis misalignment
γ_t = input axis misalignment due to the temperature difference from the nominal
ΔT = temperature change from the nominal
$\gamma_{\nabla t}$ = input axis misalignment due to temperature gradient
∇T = temperature gradient
γ_m = input axis misalignment due to magnetic fields
M = magnetic field strength
γ_a = input axis misalignment due to acceleration
A = acceleration
$R_{RW}(t)$ = random component modeled as input axis random walk

The input axis misalignment term is due to the difference between where the input axis is mathematically modeled and where the input axis really is. Anything that bends or distorts the laser gyro will cause a misalignment in the input axis position. All remaining terms are due to same error sources as the bias and scale factor errors.

3.6. CONCLUSION

More than 20 years of flight testing, including numerous studies, have confirmed that the dominant RLG inertial navigation system position error propagation is that due to white-noise random drift, which builds up as $\sqrt{t}$. (In a conventional gimbaled INS the dominant long-term position error propagation is due to the z- or vertical-axis ramp drift, which builds up as a function of t^2.) It must be pointed out, however, that other error sources, such as Markov random drift, will increase the conventional navigation error. Thus, in a RLG navigation system the random drift is nearly a pure white noise. For the aforementioned reasons, extensive flight tests by the military services demonstrated that the RLG is an ideal inertial rate sensor suitable for strapdown inertial navigation systems, flight control, and weapon delivery capability under high dynamic conditions. The ring laser gyro presents a technological opportunity to overcome cost-of-ownership,

3.6. Conclusion

TABLE 3-5

Typical Ring Laser Gyro INS Performance Requirements

Position accuracy	0.1852–1.852 km/h CEP rate
Velocity accuracy	<0.800 m/s (1σ) per axis
Attitude accuracy	2.5 arcmin minimum (rms)
Heading accuracy	3.0 arcmin minimum (rms)
Rate capability	100 deg/s each axis (for commercial transports), 400 deg/s each axis (military applications)
Acceleration capability	10g each axis
Reaction time (−54 to +71°C)	<5 min
Reliability	>5000 h MTBF[a]
Test provisions	BIT (including sensors)

[a] Mean time between failures.

reliability, and performance deficiencies encountered with conventional inertial navigation systems. As shown earlier (p. 36), strapdown inertial navigation systems provide all the information and outputs normally provided by a gimbaled INS. In addition, RLGs provide body rate information for aircraft flight control systems, thus obviating the need for separate angular rate sensors, which are not part of the navigation system. (Note that flight control uses for the RLG presume adequate redundancy and proper positioning.) Table 3-5 summarizes a typical RLG inertial navigation system's performance.

The inherent digital output of the laser gyro has made possible extensive use of the built-in test (BIT) for performance monitoring, failure detection, and malfunction isolation. For example, each laser gyro outputs seven "health" signals that can provide an indication of performance degradation long before a failure occurs. Furthermore, BIT signals are brought out for the accelerometers, central processor, timing and control, power supplies, and interface functions. Specifically, each laser gyro has its own programmable read-only memory (PROM) containing known error coefficients (from the error models given above) that are read into the computer memory at system turn on. As a result, the laser gyros can be replaced readily without the need to perform system-level recalibration, obviating the need for costly and sophisticated ground test equipment in the field and/or the high cost of depot maintenance.

The RLG attributes that make it an ideal angular rate sensor for strapdown inertial navigation systems, aircraft flight control, and other applications are summarized below:

- Digital output
- High angular rate capability
- High accuracy; arcsec resolution over the full range
- Wide dynamic range

- Fast reaction time
- No moving parts
- No inherent g-sensitivity
- No frictional motion or inherent vibration sensitivity
- Low thermal sensitivity
- Adaptability to BIT and fault isolation
- Low cost of ownership
- Reduction of the overall INS cost, power, size, and weight
- High reliability, long operating life, and long shelf life (or storage)
- No need for recalibration or field maintenance

In addition, two other important system advantages could be listed:(1) the scale factor is fixed by the geometry selected by the designer so that it remains constant; and (2) the input axis alignment is stable and completely independent of the input rate, since there is no gimbal to rotate further away from null with increasing input rate. The RLG has certain disadvantages, also. For example, in the dithered version, the gyro imposes mechanical dither reaction torques on the inertial platform, and its housekeeping electronics are more complex than those of the conventional mechanical gyroscopes. Also, while the conventional gyroscope requires relatively low voltage for the spin motor, the laser gyro requires a high starting (lasing) voltage. However, when compared against cost of ownership, maintenance, reliability, and lifespan, the laser gyro's advantages outweigh its disadvantages.

A disadvantage or limitation of the RLG is encountered in ground gyrocompassing, that is, as we shall see later, prior to entering the navigation mode or flight for an unaided system. The problem in gyrocompassing is that for low rotation rates, the two counterrotating beams have no frequency difference in a RLG. This is because backscattered radiation causes these two oscillators to become synchronous and they are pulled to a common frequency of oscillation. Consequently, this is not a surprising result and the effect will always be present since backscattering cannot be completely eliminated from a structure that confines the beams to some geometric pattern. As we have seen earlier, several biasing techniques (e.g., dither, four-mode, rate biasing) have been developed to minimize this effect. These techniques attempt to overcome the deviations from the ideal RLG associated with the phase locking, and improve the overall system performance.

REFERENCES

1. Anderson, D., Chow, W. W., Sanders, V., and Scully, M. O.: "Novel Multioscillator Approach to the Problem of Locking in Two-Mode Ring Laser Gyros: II," *Applied Optics*, **18**, 941–942, 1979.

2. Aronowitz, F.: "The Laser Gyro," in *Laser Applications*, Vol. I, M. Ross (ed.), Academic Press, New York, 1971.
3. Chow, W. W., Hambenne, J. B., Hutchings, T. J., Sanders, V. E., Sargent, M., and Scully, M. O.: "Multioscillator Laser Gyros," *IEEE Journal of Quantum Electronics*, **QE-16**(9), 919–935, September 1980.
4. Hecht, E., and Zajac, A.: *Optics*, 3rd Printing, Addison-Wesley, Reading, Mass., 1976.
5. Killpatrick, J.: "The Laser Gyro," *IEEE Spectrum*, **4**(10), 44–55, October 1967.
6. Macek, W. M., and Davis, D. T. M., Jr.: "Rotation Rate Sensing with Traveling-Wave Ring Lasers," *Applied Physics Letters*, **2**, 67–68, February 1963.
7. Michelson, A. A., and Gale, H. G.: "The Effect of the Earth's Rotation on the Velocity of Light," *Astrophysics Journal*, **61**, 137–140, April 1925.
8. Post, E. J.: "Sagnac Effect," *Reviews in Modern Physics*, **39**(2), 475–493, April 1967.
9. Rosenthal, A. H.: "Regenerative Circulatory Multiple Beam Interferometry for the Study of Light Propagation Effects," *Journal of Optical Society of America*, **52**, 1143–1148, 1962.
10. Sagnac, G.: *Comptes Rendus*, **157**, 1913.
11. Sagnac, G.: *Journal de Physique*, **4**, Ser. 5, p. 177, 1914.
12. Savage, P.: "Strapdown Sensors," AGARD Lecture Series No. 95, Strap-Down Inertial Systems, pp. 2-1–2-46, 1978.
13. Siouris, G. M.: "A Survey of New Inertial Sensor Technology," *Zeitschrift für Flugwissenschaften und Weltraumforschung*, **1**(5) (Cologne), 346–353, September–October 1977.

4 Kinematic Compensation Equations

4.1. INTRODUCTION

Navigation is the science of directing a vehicle to a destination by determining its position from observation of landmarks, celestial bodies, or radio beams. Inertial navigation, on the other hand, accomplishes this task without the aforementioned observations. The self-contained determination of the instantaneous position and other parameters of motion of a vehicle is accomplished in inertial navigation systems by measuring specific force, angular velocity, and time in a previously selected coordinate system. Specifically, the fundamental concept of inertial navigation is that vehicle velocity and position are determined through real-time integration of the governing differential equations, with measured specific force as an input. The analysis of inertial navigation errors is conceptually fairly straightforward. As we shall see in the next section, the mathematical relations of inertial navigation are a little more than the standard kinematic equations for the acceleration of a rigid body, whereby the constraints of the problem tend to force the form of the equations of motion. The two most common constraints are (1) the position of the body is calculated from acceleration computations made in a coordinate frame in which Newton's law of motion applies and that is rotating at a rate that, in general, differs from the angular rotation rate of the body; and (2) the accelerometers are insensitive to the gravitational (mass attraction) acceleration, which acts on both the case and proof mass of the accelerometers; gravitational and nongravitational acceleration are separated in the mathematical formulation of the problem. The former is obtained from calculations using an a priori gravity model, while the latter is obtained from corrected measurements. Inertial navigation is accomplished by double integration of accelerations that are compensated for by the earth's gravitational attraction. The accelerations or specific forces are sensed by the accelerometers, and the knowledge or control of the coordinate reference frame is maintained by the gyroscopes. Errors in the inertial navigation system are a function of (1) initial condition errors, (2) gravitational mass attraction compensation errors, (3) coordinate frame transformation errors, and (4) sensor errors such as accelerometers, gyroscopes, and external navigation aids when used. However, the error equations become somewhat complicated because of the different coordinate frames involved and the many error sources inherent in the instruments. Generally, bias errors in the accelerometers induce the classical Schuler sinusoids with an increasing envelope, while bias errors in the gyroscopes integrate into ramping errors in navigated position. Furthermore, many sensor errors are excited by the dynamic environment of the vehicle. Consequently, error excitation is a function of mechanization, and, as we noted previously, the gimbals provide a dynamically benign environment for the inertial sensors, as compared to the strapdown mechanization.

Typically, a simple inertial navigation system will possess the following characteristics:

1. A platform with three orthogonally placed accelerometers for measuring specific force along the horizontal and vertical axes, mounted on a three-axis gyrostabilized platform, which maintains constant orientation relative to the fixed stars.
2. The platform is maintained in a geographic coordinate system by servomechanisms, which include onboard digital computers, a clock, and other avionics.
3. Coriolis and centripetal accelerations may be subtracted from the output signals of the accelerometers, with the corrected accelerometer output signals going into the corresponding integrators.

The sensitive elements of the gyrostabilized platform are three single-degree-of-freedom gyroscopes or two 2-degree-of-freedom gyroscopes whose angular momenta are oriented perpendicular to the axes of stabilization. As already indicated in the previous chapter, the gyrostabilized platform of the inertial system is intended to maintain a "reference" coordinate system, which is fixed relative to the stars. If the gyroscope has a low drift rate, the angular deflection of the platform from the inertial coordinate system will remain small after a long time interval. That is, in the presence of disturbance torques about the gyroscope precession axes, the gyrostabilized platform will exhibit a small drift rate.

Today, aircraft and/or ship inertial navigation systems with global capability and mission time of 10 h or more have become commonplace for both commercial and military applications. With the proliferation of inertial navigation systems in commercial as well as military applications, standardization becomes an important factor in the procurement of these systems. Standardization of inertial navigation systems is therefore the driving factor in all military navigation and guidance systems. In order to standardize the interfaces used in avionics systems, most systems today are being designed with a MIL-STD-1553A,B multiplex bus (MUX BUS) for communication between subsystems. Also, testing of hardware and flight software are required to interface with other elements of the avionics system via the MIL-STD-1553 MUX BUS. The accuracy and stability of an inertial navigation system is generally improved by supplementing the inertial system with independent or external reference information. Commonly, continuous reference velocity has long been used to damp the Schuler frequencies. For example, position information obtained from Loran or Omega may be used to damp the inherent 24-h oscillations. Finally, an integrated approach to avionics subsystems is desirable. The motivation for the integrated system approach is to take advantage of size, weight, and power savings offered by this approach.

4.2. ROTATING COORDINATES; GENERAL RELATIVE MOTION EQUATIONS

Before developing the general navigation equations, we will discuss in this section the relative motion equations of a rigid body in Cartesian coordinates. With reference to Fig. 4-1, assume that the rigid body is fixed at point O of a fixed Cartesian coordinate system (X, Y, Z), the I-frame, with origin at O.

Furthermore, let a typical point in the body have a position $\mathbf{r}$ and velocity $\dot{\mathbf{r}}$ relative to the origin O. Then

$$\mathbf{r} = X\mathbf{I} + Y\mathbf{J} + Z\mathbf{K} \qquad (4.1a)$$

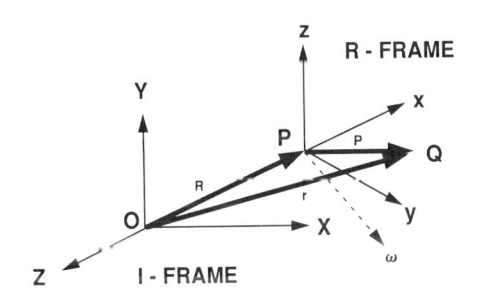

FIGURE 4-1
Definition of reference axes.

$$\dot{\mathbf{r}} = \frac{d\mathbf{r}}{dt} = \dot{X}\mathbf{I} + \dot{Y}\mathbf{J} + \dot{Z}\mathbf{K} \tag{4.1b}$$

where $(\mathbf{I}, \mathbf{J}, \mathbf{K})$ are unit vectors along the X, Y, Z axes, respectively. For a rigid body with a fixed point, we also know that

$$\dot{\mathbf{r}} = \boldsymbol{\omega} \times \mathbf{r} \tag{4.2}$$

where $\boldsymbol{\omega}$ is the angular velocity vector given by

$$\boldsymbol{\omega} = \Omega_X \mathbf{I} + \Omega_Y \mathbf{J} + \Omega_Z \mathbf{K}$$

in the fixed coordinate system. Equation (4.2) represents the simplest motion, or pure rotation about a fixed axis in space. Since small angles can be represented as vectors, an angular displacement $\Delta\boldsymbol{\phi}$ can be approximated by the expression [6]

$$\Delta\boldsymbol{\phi} = \int_0^{\Delta t} \boldsymbol{\omega} \, dt$$

or

$$\Delta\boldsymbol{\phi} = \boldsymbol{\omega}\Delta t$$

if Δt is a small time interval. The limiting position of the axis of rotation through the fixed point is called the "instantaneous axis of rotation," and $\boldsymbol{\omega}$ has the direction of the instantaneous axis [15]. Thus, in Eq. (4.2) we have

$$\boldsymbol{\omega} = \lim_{\Delta t \to 0} \Delta\boldsymbol{\phi}/\Delta t \tag{4.3}$$

Carrying out the vector operation indicated in Eq. (4.2), we obtain

$$\boldsymbol{\omega} \times \mathbf{r} = [\Omega_Y Z - \Omega_Z Y]\mathbf{I} + [\Omega_Z X - \Omega_X Z]\mathbf{J} + [\Omega_X Y - \Omega_Y X]\mathbf{K} \tag{4.4}$$

where in component form

$$v_X = \dot{X} = \Omega_Y Z - \Omega_Z Y$$
$$v_Y = \dot{Y} = \Omega_Z X - \Omega_X Z$$
$$v_Z = \dot{Z} = \Omega_X Y - \Omega_Y X$$

The fixed-point restriction assumed above can now be removed. Introduce now a second Cartesian system (x, y, z), the R-frame, with unit vectors $(\mathbf{i}, \mathbf{j}, \mathbf{k})$ along the x, y, z axes, respectively. For the general rigid body, let P be any point in the body and O the origin of the space axes. Next, we note that the velocity of any other point Q in the body with respect to the space axes at O is given by [6]

$$\mathbf{v}_Q = \mathbf{v}_P + \boldsymbol{\omega} \times \mathbf{r}_{Q/P} \tag{4.5}$$

where $\mathbf{r}_{Q/P}$ is the relative distance of Q with respect to P. Consequently, the relative velocity of Q with respect to P is, by definition

$$\mathbf{v}_{Q/P} = \mathbf{v}_Q - \mathbf{v}_P$$

4.2. Rotating Coordinates; General Relative Motion Equations

The space axes may be used to compute $\mathbf{v}_P$, while the body axes based at point P are used to compute $\boldsymbol{\omega} \times \mathbf{r}_{Q/P}$. From the foregoing discussion, a fundamental result relating the frame derivatives of any vector function $\mathbf{A}(t)$ described in two coordinate systems that are in relative rotation may be obtained. The vector function $A(t)$ can be given in the two different coordinate systems (x, y, z) and (X, Y, Z):

$$\mathbf{A} = A_X \mathbf{I} + A_Y \mathbf{J} + A_Z \mathbf{K} = A_x \mathbf{i} + A_y \mathbf{j} + A_z \mathbf{k}$$

and the derivatives as

$$\dot{\mathbf{A}} = \frac{d\mathbf{A}}{dt} = \dot{A}_X \mathbf{I} + \dot{A}_Y \mathbf{J} + \dot{A}_Z \mathbf{K}$$

$$\dot{\mathbf{A}} = \frac{\delta \mathbf{A}}{\delta t} + A_x \dot{\mathbf{i}} + A_y \dot{\mathbf{j}} + A_z \dot{\mathbf{k}}$$

A rotating rigid body has an angular velocity $\boldsymbol{\omega}$. The derivatives of the unit vectors $(\mathbf{i}, \mathbf{j}, \mathbf{k})$ in the fixed system are given by

$$\frac{d\mathbf{i}}{dt} = \boldsymbol{\omega} \times \mathbf{i}, \qquad \frac{d\mathbf{j}}{dt} = \boldsymbol{\omega} \times \mathbf{j}, \qquad \frac{d\mathbf{k}}{dt} = \boldsymbol{\omega} \times \mathbf{k}$$

where $\boldsymbol{\omega}$ is now the angular velocity of the coordinate axes. The frame derivatives of a vector in two coordinate systems are equal, except for a term proportional to the vector and to the rate of rotation of one system with respect to the other. Thus

$$\left(\frac{d\mathbf{A}}{dt} \right)_{XYZ} \equiv \frac{d\mathbf{A}}{dt} = \frac{\delta \mathbf{A}}{\delta t} + \boldsymbol{\omega} \times \mathbf{A} \equiv \left(\frac{d\mathbf{A}}{dt} \right)_{xyz} + \boldsymbol{\omega} \times \mathbf{A} \qquad (4.6)$$

where $\boldsymbol{\omega}$ is the angular velocity of the (x, y, z) coordinate system relative to the (X, Y, Z) coordinate system. This is a very important equation for the discussion of the next section. Notice that, from a kinematic point of view, it makes no difference which system is considered as fixed and which as rotating. That is,

$$\frac{\delta \mathbf{A}}{\delta t} = \frac{d\mathbf{A}}{dt} - \boldsymbol{\omega} \times \mathbf{A} = \frac{d\mathbf{A}}{dt} + (-\boldsymbol{\omega}) \times \mathbf{A} \qquad (4.7)$$

where $-\boldsymbol{\omega}$ is the angular velocity of the (X, Y, Z) system relative to the (x, y, z) system. Now suppose that the two moving points, P and Q, have positions $\mathbf{R}$ and $\mathbf{r}$, respectively, with respect to the fixed point O. Let the relative position vector of point Q with respect to point P be denoted to $\boldsymbol{\rho}$. Consequently, the relative motion equations are $\mathbf{r} = \mathbf{R} + \boldsymbol{\rho}$, $\dot{\mathbf{r}} = \dot{\mathbf{R}} + \dot{\boldsymbol{\rho}}$, $\ddot{\mathbf{r}} = \ddot{\mathbf{R}} + \ddot{\boldsymbol{\rho}}$ (see Fig. 4.1). Furthermore, let the coordinate system (X, Y, Z) with origin at O be fixed, while the coordinate system (x, y, z) with origin at P is moving with an angular velocity $\boldsymbol{\omega}$. Using Eq. (4.7), the derivative of the relative position vector $\mathbf{p}$ in the moving coordinate system is given by

$$\dot{\mathbf{p}} = \frac{d\mathbf{p}}{dt} = \frac{\delta \mathbf{p}}{\delta t} + \boldsymbol{\omega} \times \mathbf{p} \qquad (4.8)$$

Having a representation of $\mathbf{p}$ in moving coordinates, we now can compute its relative acceleration as follows [11,12]:

$$\frac{d^2\mathbf{p}}{dt^2} = \frac{d\dot{\mathbf{p}}}{dt} = \frac{\delta \dot{\mathbf{p}}}{\delta t} + \boldsymbol{\omega} \times \dot{\mathbf{p}}$$

$$= \frac{\delta^2 \mathbf{p}}{\delta t^2} + \frac{\delta}{\delta t}(\boldsymbol{\omega} \times \mathbf{p}) + \boldsymbol{\omega} \times \left(\frac{\delta \mathbf{p}}{\delta t} + \boldsymbol{\omega} \times \mathbf{p}\right)$$

$$= \frac{\delta^2 \mathbf{p}}{\delta t^2} + \frac{\delta \boldsymbol{\omega}}{\delta t} \times \mathbf{p} + 2\boldsymbol{\omega} \times \frac{\delta \mathbf{p}}{\delta t} + \boldsymbol{\omega} \times (\boldsymbol{\omega} \times \mathbf{p}) \qquad (4.9)$$

Since $\delta\boldsymbol{\omega}/\delta t = d\boldsymbol{\omega}/dt = \dot{\boldsymbol{\omega}}$, we can write Eq. (4.9) in the form

$$\frac{d^2\mathbf{p}}{dt^2} = \frac{\delta^2 \mathbf{p}}{\delta t^2} + 2\boldsymbol{\omega} \times \frac{\delta \mathbf{p}}{\delta t} + \dot{\boldsymbol{\omega}} \times \mathbf{p} + \boldsymbol{\omega} \times (\boldsymbol{\omega} \times \mathbf{p}) \qquad (4.10)$$

The complete relative motion equations are summarized below:

$$\mathbf{r} = \mathbf{R} + \mathbf{p} \qquad (4.11a)$$

$$\frac{d\mathbf{r}}{dt} = \frac{d\mathbf{R}}{dt} + \frac{\delta \mathbf{p}}{\delta t} + \boldsymbol{\omega} \times \mathbf{p} \qquad (4.11b)$$

$$\frac{d^2\mathbf{r}}{dt^2} = \frac{d^2\mathbf{R}}{dt^2} + \frac{\delta^2 \mathbf{p}}{\delta t^2} + \dot{\boldsymbol{\omega}} \times \mathbf{p} + \boldsymbol{\omega} \times (\boldsymbol{\omega} \times \mathbf{p}) + 2\boldsymbol{\omega} \times \frac{\delta \mathbf{p}}{\delta t} \qquad (4.11c)$$

The significance of the various terms in Eq. (4.11) is as follows:

$\frac{d^2\mathbf{r}}{dt^2}, \frac{d^2\mathbf{R}}{dt^2}, \frac{\delta^2 \mathbf{p}}{\delta t^2}$: Linear acceleration terms

$\boldsymbol{\omega} \times (\boldsymbol{\omega} \times \mathbf{p})$: Centripetal acceleration

$\dot{\boldsymbol{\omega}} \times \mathbf{p}$: Represents the tangential component of acceleration due to the angular acceleration vector $d\boldsymbol{\omega}/dt$

$2\boldsymbol{\omega} \times \frac{\delta \mathbf{p}}{\delta t}$: Coriolis acceleration, named after its discoverer

4.3. GENERAL NAVIGATION EQUATIONS

In this section we will discuss and develop several undamped navigation mechanizations. *Navigation mechanization* refers to the equations and procedures used with a particular inertial navigation system in order to generate

4.3. General Navigation Equations

position and velocity information. We will derive the necessary mathematical relationships, which allow a completely self-contained inertial navigation system implementation. For convenience, a spherical earth model will be assumed. However, the results can be easily extended to an ellipsoidal earth model. All terrestrial inertial navigation systems, regardless of the nominal platform orientation (e.g., local-level, space-stable, strapdown), contain undamped oscillatory dynamics at both the Schuler (84.4-min period) and earth (24-h period) rates. This is mostly the case with slow moving vehicles. Damping, as we shall see later, is provided by using externally measured altitude reference, such as from a barometric or radar altimeter, to the altitude channel of an unaided inertial navigation system.

We begin our discussion by noting that the differential equation of motion of inertial navigation of a vehicle relative to an inertial frame* can be written in vector form as [12,13]

$$\dot{\mathbf{R}} = \mathbf{V} \qquad (4.12a)$$

$$\left[\frac{d\mathbf{V}}{dt}\right]_I = \mathbf{A} + \mathbf{g}_m(\mathbf{R}) \qquad (4.12b)$$

where
$\mathbf{R}$ = geocentric position vector
$\mathbf{V}$ = velocity of the vehicle relative to the inertial frame
$= [V_x \quad V_y \quad V_z]^T$
$\mathbf{A}$ = nongravitational specific force (acceleration due to all nongravitational causes) (this is the acceleration sensed or measured by an accelerometer; it is the nonfield contact force per unit mass exerted on the instrument set)
$\mathbf{g}_m(\mathbf{R})$ = gravitational acceleration due to mass attraction, considered positive toward the center of the earth

In Eq. (4.12), the gravitational acceleration due to the moon, the sun, and the stars has been neglected, since this acceleration is very small. Writing Eq. (4.12) in the form

$$\mathbf{A} = [d^2\mathbf{R}/dt^2]_I - \mathbf{g}_m(\mathbf{R})$$

this equation states that the specific force (accelerometer output) $\mathbf{A}$ is proportional to the inertial acceleration of the system due to all forces except gravity. The term $[d^2\mathbf{R}/dt^2]_I$ represents the inertial acceleration with respect to the center of the nonrotating earth, that is, an inertial reference frame. The fact that the earth is rotating, however, is of no consequence to navigation in inertial coordinates. An inertial frame must be considered since this is the

---

*As we have seen in Chapter 2, an inertial frame is nonrotating and nonaccelerating relative to inertial space. For inertial orientation, we use the distant stars.

only frame in which Newton's laws of motion are valid. Equation (4.12) therefore represents the motion of the vehicle in inertial coordinates, or more specifically, in an earth-centered inertial (ECI) frame. The subscript I in Eq. (4.12b) denotes that the rate of change is relative to the inertial frame. Moreover, since the earth is rotating and moving with respect to inertial space, a transformation is necessary in order to relate measurements taken in inertial space to observations of position, velocity, and acceleration in a moving vehicle. This is necessary if inertial instruments are used as sensors in an environment such as that of a large rotating central attracting body like the earth. For this reason, consider the vector output of an ideal set of accelerometers mounted on a platform (P). As we noted earlier, an ideal accelerometer measures the specific force, that is, the difference between the inertial acceleration and gravitational acceleration. Therefore, for the earth-centered inertial system, Eq. (4.12b) can be written in the form

$$\mathbf{A}_I^P = C_I^P[\ddot{\mathbf{R}}^I - \mathbf{g}_m^I(\mathbf{R})] \tag{4.13}$$

where $\ddot{\mathbf{R}}^I$ is the inertially referenced acceleration vector and C_I^P is the transformation matrix from inertial to platform coordinates. Now we wish to express the earth-centered inertial acceleration in terms of the specific force and the gravitational acceleration. This can be done by rearranging Eq. (4.13) as follows:

$$\ddot{\mathbf{R}}^I = C_P^I \mathbf{A}^P + \mathbf{g}_m^I(\mathbf{R}) \tag{4.14}$$

Consequently, inertial navigation is based on solving this equation for velocity and position by the onboard navigation computer. Figure 4-2 illustrates Eq. (4.14) in block diagram form.

For navigation at or near the surface of the earth, we need to refer the position and velocity of the vehicle to an earth-fixed coordinate system,

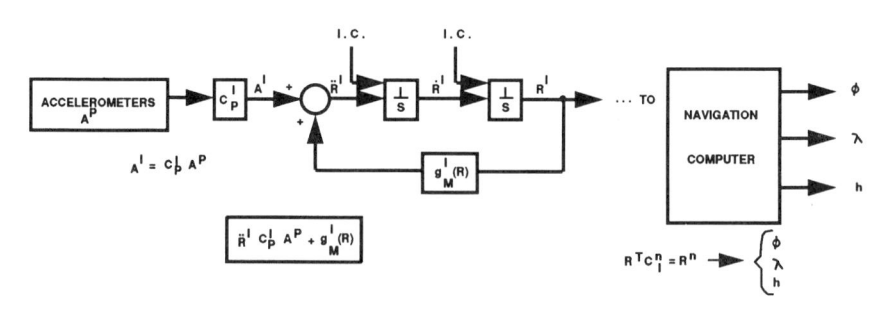

FIGURE 4-2

An analytic system for computing velocity and position in inertial coordinates.

4.3. General Navigation Equations

which rotates with the earth. From the Law of Coriolis, the expression relating inertial (viz., ECI) and earth-fixed velocities is given by

$$\left[\frac{d\mathbf{R}}{dt}\right]_I = \left[\frac{d\mathbf{R}}{dt}\right]_E + \mathbf{\Omega} \times \mathbf{R} = \mathbf{V} + \mathbf{\Omega} \times \mathbf{R}$$
$$\mathbf{V} = V_x\mathbf{i} + V_y\mathbf{j} + V_z\mathbf{k}$$
(4.15)

where $\mathbf{\Omega}$ is the angular rate of the earth relative to the inertial frame (or sidereal rate vector) and $\mathbf{V}$ is the true velocity of the vehicle with respect to the earth. It should be noted that Coriolis acceleration is present when a vehicle is moving with some velocity $[d\mathbf{R}/dt]$, with respect to the moving coordinate frame (such as the earth's surface). Differentiating Eq. (4.15) with respect to the inertial coordinates, and since $d\mathbf{\Omega}/dt = 0$, we have

$$\left[\frac{d^2\mathbf{R}}{dt^2}\right]_I = \left[\frac{d\mathbf{V}}{dt}\right]_I + \mathbf{\Omega} \times \left[\frac{d\mathbf{R}}{dt}\right]_I \quad (4.16)$$

Substituting $[d\mathbf{R}/dt]_I$ from Eq. (4.15) into Eq. (4.16) yields

$$\left[\frac{d^2\mathbf{R}}{dt^2}\right]_I = \left[\frac{d\mathbf{V}}{dt}\right]_I + \mathbf{\Omega} \times \mathbf{V} + \mathbf{\Omega} \times (\mathbf{\Omega} \times \mathbf{R}) \quad (4.17)$$

As we noted earlier, the output of the accelerometer gives quantities that are measured along the platform (or system axes). Differentiation and/or integration of these components is, therefore, carried out with respect to the platform axes. Consequently, the derivative of the velocity $\mathbf{V}$ with respect to the platform axes is an essential quantity and can be related to the derivative with respect to inertial space by the expression

$$\left[\frac{d\mathbf{V}}{dt}\right]_I = \left[\frac{d\mathbf{V}}{dt}\right]_P + \mathbf{\omega} \times \mathbf{V} \quad (4.18)$$

where $\mathbf{\omega}$ is the angular rate of the platform with respect to inertial space, which is also referred to as the *spatial rate*. Substituting Eq. (4.18) into Eq. (4.17) results in

$$\left[\frac{d^2\mathbf{R}}{dt^2}\right]_I = \left[\frac{d\mathbf{V}}{dt}\right]_P + (\mathbf{\omega} + \mathbf{\Omega}) \times \mathbf{V} + \mathbf{\Omega} \times (\mathbf{\Omega} \times \mathbf{R}) \quad (4.19)$$

Finally, substituting Eq. (4.19) into Eq. (4.12b) gives the expression [16]

$$\mathbf{A} = \left[\frac{d\mathbf{V}}{dt}\right]_P + (\mathbf{\omega} + \mathbf{\Omega}) \times \mathbf{V} + \mathbf{\Omega} \times (\mathbf{\Omega} \times \mathbf{R}) - \mathbf{g}_m(\mathbf{R}) \quad (4.20)$$

Since the centripetal acceleration of the earth term $\mathbf{\Omega} \times (\mathbf{\Omega} \times \mathbf{R})$ is a function of position on the earth only, it can be combined with the mass attraction gravity term to give the apparent (or local plumb-bob) gravity vector as

$$g(\mathbf{R}) = g_m(\mathbf{R}) - \mathbf{\Omega} \times (\mathbf{\Omega} \times \mathbf{R}) \approx \omega_s^2 \mathbf{R} \tag{4.21}$$

[ω_s = Schuler angular frequency = $\sqrt{g(R)/R}$], where $\mathbf{R}$ is the distance from the center of the earth to the vehicle. The term $g(\mathbf{R})$ is perhaps the dominant feedback term in inertial navigation systems and is responsible for the principal modes of behavior of such systems (i.e., Schuler oscillations in the horizontal channels and strong instability in the vertical channel). Again, substituting Eq. (4.21) into Eq. (4.20) gives the expression

$$\mathbf{A} = \left[\frac{d\mathbf{V}}{dt}\right]_P + (\mathbf{\omega} + \mathbf{\Omega}) \times \mathbf{V} - g(\mathbf{R}) \tag{4.22}$$

This is a generalized mechanization equation; that is, it does not refer to any particular type of system coordinate frame. The locally level platform coordinate frame used in this system results in the spatial rate being the sum of the earth rate and vehicle (or platform) angular rate $\mathbf{\rho}$ with respect to the earth-fixed frame. The parameter $\mathbf{\rho}$ is also known as the *transport rate*. Mathematically, this can be stated as

$$\mathbf{\omega} = \mathbf{\rho} + \mathbf{\Omega} \tag{4.23}$$

Substituting Eq. (4.23) into Eq. (4.22), and rearranging terms, results in [7]

$$\left[\frac{d\mathbf{V}}{dt}\right]_P = \mathbf{A} - (\mathbf{\rho} + 2\mathbf{\Omega}) \times \mathbf{V} + g(\mathbf{R}) \tag{4.24}$$

This is the generalized navigation equation for a vehicle, expressed in the platform (P), or computational frame, which is referenced to the earth. Performing the vector operation $(\mathbf{\rho} + 2\mathbf{\Omega}) \times \mathbf{V}$ in Eq. (4.24), we have [4]

$$(\mathbf{\rho} + 2\mathbf{\Omega}) \times \mathbf{V} = [(\rho_y + 2\Omega_y)V_z - (\rho_z + 2\Omega_z)V_y]\mathbf{i}$$
$$+ [(\rho_z + 2\Omega_z)V_x - (\rho_x + 2\Omega_x)V_z]\mathbf{j}$$
$$+ [(\rho_x + 2\Omega_x)V_y - (\rho_y + 2\Omega_y)V_x]\mathbf{k}$$

where $(\mathbf{i}, \mathbf{j}, \mathbf{k})$ represent orthogonal unit vectors along the moving axes (x, y, z), respectively. Therefore, the generalized equation can be expressed along the platform x, y, z axes in the following form:

$$\dot{V}_x = A_x - (\rho_y + 2\Omega_y)V_z + (\rho_z + 2\Omega_z)V_y + g_x \tag{4.25a}$$

$$\dot{V}_y = A_y - (\rho_z + 2\Omega_z)V_x + (\rho_x + 2\Omega_x)V_z + g_y \tag{4.25b}$$

$$\dot{V}_z = A_z - (\rho_x + 2\Omega_x)V_y + (\rho_y + 2\Omega_y)V_x + g_z \tag{4.25c}$$

4.3. General Navigation Equations

In general, the gravitational model is based on a spherical harmonic expansion of the gravitational potential $U(R, \phi)$. There are two commonly used expansions of the gravitational potential. These are

1. The spherical or zonal harmonics that depend on the geocentric latitude only.
2. Tesseral and sectoral harmonics that depend on both latitude and longitude.

For certain inertial navigation system applications, the contributions of the Tesseral and sectoral harmonics, which indicate deviations from rotational symmetry, can be neglected without compromising system performance or accuracy. The derivation of the gravitational potential is based on the reference ellipsoid. Moreover, this derivation assumes that the earth's mass distribution is symmetric about the polar axis (z_e axis), and that the external gravitational potential due to mass attraction, or "normal" potential $U(R, \phi)$, is at a distance R from the center of the earth (assumed to be independent of longitude). Therefore, the gravitational potential $U(R, \phi)$ in an earth-fixed, earth-centered, right-handed orthogonal coordinate system with the positive x_e axis located at the intersection of the Greenwich meridional plane and the equatorial plane, the positive z_e axis along the earth's polar axis, and the y_e axis positive 90° east of the x_e axis, can be expressed in terms of spherical harmonics in the following form:

$$U(R, \phi) = -\frac{\mu}{R}\left[1 - \sum_{n=2}^{\infty} J_n\left(\frac{a}{R}\right)^n P_n(\sin \phi)\right]$$

where
  μ = the earth's gravitational constant
  a = mean equatorial radius of the earth (or semimajor axis)
  $R = [x_e^2 + y_e^2 + z_e^2]^{1/2}$ is the magnitude of the geocentric position vector
  ϕ = geocentric latitude ($\sin \phi = z_e/R$)
  J_n = coefficients of zonal harmonics of the earth potential function (constants determined from observations of orbit perturbations of artificial satellites)
  $P_n(\sin \phi)$ = associated Legendre polynomials of the first kind as functions of ϕ and degree n

The first term on the right-hand side of this equation, μ/R, which is the mean value, is the simplified gravitational potential of the earth and is due to a spherically mass symmetric body. The remaining terms of the infinite series account for the fact that the earth is not a spherically symmetric body, but is bulged at the equator, flattened at the poles, and is generally asymmetric. Specifically, the second harmonic term in the infinite series J_2 results from the earth's flattening, the meridional cross section being an ellipse rather than a circle. The third harmonic coefficient J_3 corresponds to a tendency toward a triangular shape. The fourth harmonic coefficient J_4 approximates a square. If symmetry with respect to the equatorial plane is assumed, then all the odd harmonics in the infinite series vanish. That is, $J_1 = J_3 = J_5 = \cdots = 0$. Finally, note that since R is very large, all terms within the brackets are small compared with unity, so that $U(R, \phi) = \mu/R$. (Some selected values for the constants J_n are as follows; $J_2 = 1082.63 + 0.01 \times 10^{-6}$, $J_3 = -2.51 + 0.01 \times 10^{-6}$, $J_4 = -1.60 + 0.01 \times 10^{-6}$, and $J_5 = -0.13 + 0.13 + 0.01 \times 10^{-6}$.)

In the analysis that follows, we will assume that the direction of the gravity vector **g** coincides with the "normal" gravity vector, and is defined to be perpendicular to the reference ellipsoid, positive in the downward direction. Mathematically, this can be stated as

$$\mathbf{g}(\mathbf{R}) = -g_z \mathbf{1}_z.$$

For a spherical earth model, $\mathbf{g}(\mathbf{R}) = -\mu \mathbf{R}/R^3$, and $\mathbf{R} = [x \ y \ z]^T$. More specifically, since $\mathbf{g}(\mathbf{R})$ is acting entirely along $-\mathbf{R}$ in a spherical earth model, the components g_x and g_y vanish; that is, $\mathbf{g}(\mathbf{R}) = [0 \ 0 \ g_z]^T$ and $\mathbf{R} = [0 \ 0 \ z]^T$. Stated another way, the components of the apparent gravity vector along the platform x, y axes can be neglected because their magnitude is $<10^{-5}g$ at the altitude considered. However, $\mathbf{g}(\mathbf{R})$ is not acting quite along $-\mathbf{R}$, that is, not quite toward the center of the earth. Moreover, a more accurate model for $\mathbf{g}(\mathbf{R})$ will be required for precise navigation, as the dominant characteristic of the gravitational field is that of an oblate spheroid, which is symmetric in latitude and independent of longitude.

The gravitation vector is given as the gradient of the potential as

$$\mathbf{g}(\mathbf{R}) = [g_x \ g_y \ g_z]^T$$

where

$$g_x = \frac{\partial U}{\partial x}, \qquad g_y = \frac{\partial U}{\partial y}, \qquad g_z = \frac{\partial U}{\partial z}$$

The above equations can be expressed in a rather simple form if a particular choice is made of the ECI coordinate frame. In this frame, the z axis is assumed to be directed along the earth's North Polar axis, the x, y axes in the equatorial plane with the x axis defined by the intersection of the reference meridian and the plane of the mean astronomic equator, and the

4.3. General Navigation Equations

y axis 90°-east of the x axis, completing a right-handed ECI orthogonal system. Retaining only the first three terms of the expansion of $U(R, \phi)$, we have for a spheroidal earth model [3]

$$g_x = -\frac{\mu}{R^2}\left[1 - \tfrac{3}{2}J_2\left(\frac{a}{R}\right)^2\left(5\frac{z^2}{R^2} - 1\right)\right.$$
$$\left. - \tfrac{5}{8}J_4\left(\frac{a}{R}\right)^4\left(63\frac{z^4}{R^4} - 42\frac{z^2}{R^2} + 3\right)\right]\frac{x}{R} \qquad (4.26a)$$

$$g_y = -\frac{\mu}{R^2}\left[1 - \tfrac{3}{2}J_2\left(\frac{a}{R}\right)^2\left(5\frac{z^2}{R^2} - 1\right)\right.$$
$$\left. - \tfrac{5}{8}J_4\left(\frac{a}{R}\right)^4\left(63\frac{z^4}{R^4} - 42\frac{z^2}{R^2} + 3\right)\right]\frac{y}{R} \qquad (4.26b)$$

$$g_z = -\frac{\mu}{R^2}\left[1 - \tfrac{3}{2}J_2\left(\frac{a}{R}\right)^2\left(5\frac{z^2}{R^2} - 3\right)\right.$$
$$\left. - \tfrac{5}{8}J_4\left(\frac{a}{R}\right)^4\left(63\frac{z^4}{R^4} - 70\frac{z^2}{R^2} + 15\right)\right]\frac{z}{R} \qquad (4.26c)$$

$$\frac{x}{R} = \cos \lambda \cos \phi \qquad R = [x^2 + y^2 + z^2]^{1/2}$$

$$\frac{y}{R} = \sin \lambda \cos \phi$$

$$\frac{z}{R} = \sin \phi$$

where ϕ is the geographic latitude. In a spheroidal earth model, the local plumb-bob (or geodetic) vertical does not go through the earth's center, leading to the use of geodetic (or geographic) latitude concept. Thus, the inertial platform's vertical axis and the geodetic vertical do not coincide.

From the above discussion we can see that navigation in ECI coordinates involves integration of a simple set of differential equations driven by the measured specific force Λ. As mentioned earlier, for most terrestrial applications, it is more convenient to refer the position and velocity of the vehicle to an earth-fixed Cartesian coordinate frame, which rotates with the earth. In this coordinate frame, the equations of motion must account for the rotation of the coordinate frame. This is accomplished, as we have seen, by using the Law of Coriolis. Assume now a north–east–up (NEU) coordinate

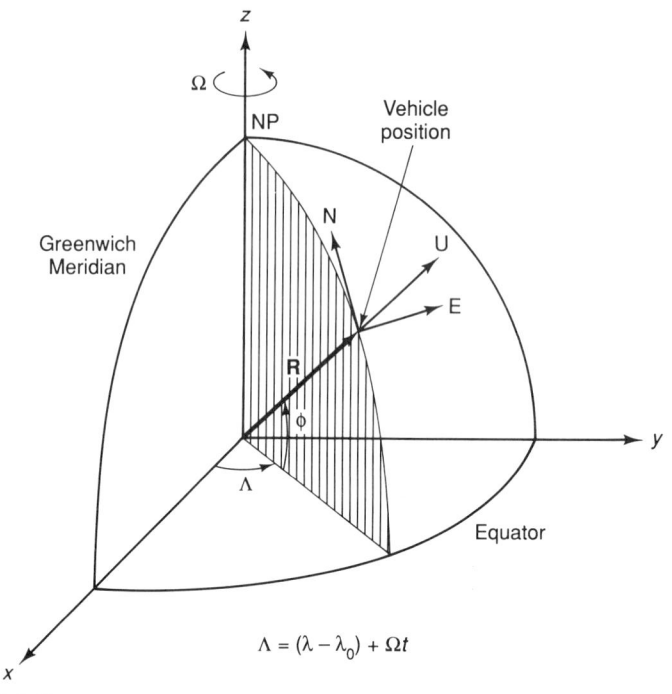

FIGURE 4-3
Definition of the coordinate axes.

frame as shown in Fig. 4-3. In this system, the components of the earth rate are given by

$$\mathbf{\Omega} = 0 \cdot \mathbf{i} + \Omega \cos \phi \cdot \mathbf{j} + \Omega \sin \phi \cdot \mathbf{k}$$
$$\Omega_x = \Omega_E = 0$$
$$\Omega_y = \Omega_N = \Omega \cos \phi$$
$$\Omega_z = \Omega_U = \Omega \sin \phi$$

where $\mathbf{i}$, $\mathbf{j}$, $\mathbf{k}$ are unit vectors along E, N, U, respectively. Then Eq. (4.25) takes the simpler form

$$\dot{V}_x = A_x - (\rho_y + 2\Omega_y) V_z + (\rho_z + 2\Omega_z) V_y \tag{4.27a}$$
$$\dot{V}_y = A_y - (\rho_z + 2\Omega_z) V_x + (\rho_x + 2\Omega_x) V_x \tag{4.27b}$$
$$\dot{V}_z = A_z - \rho_x V_y + (\rho_y + 2\Omega_y) V_x - g_z \tag{4.27b}$$

Consider an aircraft flying at an altitude h above the earth. The radii of curvature of a constant longitude (north–south) line and of a constant

4.3. General Navigation Equations

latitude (east–west) line on the surface of an ellipsoidal earth are functions of geographic latitude and are given by [5]

$$R_\phi = \frac{R_0(1-\varepsilon^2)}{(1-\varepsilon^2 \sin^2 \phi)^{3/2}} + h$$

$$R_\lambda = \frac{R_0}{(1-\varepsilon^2 \sin^2 \phi)^{1/2}} + h$$

respectively, where R_0 is the equatorial radius of the earth, ϕ is the geographic latitude, h is the altitude above the oblate spheroid, and ε is the eccentricity. In other words, R_ϕ is the meridional radius of curvature for north–south motion, and R_λ is the normal radius of curvature for east–west motion. Note that both R_ϕ and R_λ are derived at the surface of the earth ($h=0$) for an ellipsoidal earth. From Ref. 13, the radii of curvature* of the reference ellipsoid can be approximated with sufficient accuracy as

$$R_{N-S} = R_\phi = R_0[1 - 2f + 3f \sin^2 \phi] = R_0[1 + f(3 \sin^2 \phi - 2)]$$

$$R_{E-W} = R_\lambda = R_0[1 + f \sin^2 \phi]$$

where f is the flattening ($f = 1 - b/a$, a is the semimajor axis and b is the semiminor axis of the reference ellipsoid). Thus, the east, north, and up components of the angular rate of a locally level platform ρ with respect to the earth-fixed frame, are given in the form [11]

$$\rho_E = -\frac{V_N}{R_\phi + h}$$

$$\rho_N = \frac{V_E}{R_\lambda + h}$$

$$\rho_U = \rho_N \tan \phi = \left(\frac{V_E}{R_\lambda + h}\right) \tan \phi$$

or, in terms of the earth's flattering (or ellipticity), f

$$\rho_E = -\frac{V_N}{R_0\left[1 - 2f + 3f \sin^2 \phi + \dfrac{h}{R_0}\right]}$$

---

*Both radii of curvature can be simplified using the binomial expansion, i.e.,

$$(1-x)^{-n} = 1 + nx + \frac{n(n+1)}{2!} x^2 + \cdots$$

by letting n be $\frac{3}{2}$ and $\frac{1}{2}$, respectively.

$$\rho_N = \frac{V_E}{R_0\left[1 + f\sin^2\phi + \dfrac{h}{R_0}\right]}$$

$$\rho_U = \frac{V_E \tan\phi}{R_0\left[1 + f\sin^2\phi + \dfrac{h}{R_0}\right]}$$

Since $\omega_x = \rho_x + \Omega_x$, $\omega_y = \rho_y + \Omega_y$, and $\omega_z = \rho_z + \Omega_z$, then

$$\omega_x = \rho_E + \Omega_x = -\frac{V_N}{R_0\left[1 - 2f + 3f\sin^2\phi + \dfrac{h}{R_0}\right]}$$

$$\omega_y = \rho_N + \Omega_y = \frac{V_E}{R_0\left[1 + f\sin^2\phi + \dfrac{h}{R_0}\right]} + \Omega\cos\phi$$

$$\omega_z = \rho_U + \Omega_z = \frac{V_E \tan\phi}{R_0\left[1 + f\sin^2\phi + \dfrac{h}{R_0}\right]} + \Omega\sin\phi$$

Note that these are the craft (transport) rate components of the gyroscope torquing rate ($\omega = \rho + \Omega$). For this aircraft, the current coordinates in an earth-centered earth-fixed (ECEF) system can be computed by a transformation from geodetic to earth-fixed coordinates. Thus [5]

$$x_e = (R_\lambda + h)\cos\phi\cos\Lambda$$

$$y_e = (R_\lambda + h)\cos\phi\sin\Lambda$$

$$z_e = [R_\lambda(1 - \varepsilon^2) + h]\sin\phi$$

For the aircraft under consideration, (x_e, y_e, z_e) are the coordinates at the current time at the midpoint of the integration interval. The inverse transformation, that is, ECEF coordinates to geodetic latitude ϕ, longitude λ, and altitude h, can be accomplished as follows:

$$\lambda = \tan^{-1}\left(\frac{y_e}{x_e}\right)$$

$$\phi = \tan^{-1}\left[\frac{R_\lambda + h}{\left(\dfrac{b}{a}\right)^2 R_\lambda + h}\right] \frac{z_e}{\sqrt{x_e^2 + y_e^2}}$$

4.3. General Navigation Equations

$$h = \frac{z_e}{\sin \phi} - (1 - \varepsilon^2) R_\lambda; \quad 1 - \varepsilon^2 = \frac{b^2}{a^2}, \quad \varepsilon^2 = 2f$$

where a is the semimajor axis and b is the semiminor axis of the ellipsoid. The geodetic latitude and altitude cannot be solved in closed form. An iterative process must be used in order to solve for these parameters. Referring to Fig. 4-3, the transformation that converts the ECI(x, y, z) coordinates to the local level (L-L) UEN frame is obtained from

$$\begin{bmatrix} U \\ E \\ N \end{bmatrix} = \begin{bmatrix} \cos \phi & 0 & \sin \phi \\ 0 & 1 & 0 \\ -\sin \phi & 0 & \cos \phi \end{bmatrix} \begin{bmatrix} \cos \Lambda & \sin \Lambda & 0 \\ -\sin \Lambda & \cos \Lambda & 0 \\ 0 & 0 & 1 \end{bmatrix} \begin{bmatrix} x \\ y \\ z \end{bmatrix}$$

$$= \begin{bmatrix} \cos \phi \cos \Lambda & \cos \phi \sin \Lambda & \sin \phi \\ -\sin \Lambda & \cos \Lambda & 0 \\ -\sin \phi \cos \Lambda & -\sin \phi \sin \Lambda & \cos \phi \end{bmatrix} \begin{bmatrix} x \\ y \\ z \end{bmatrix}$$

$$= C_{ECI}^{L-LUEN} \begin{bmatrix} x \\ y \\ z \end{bmatrix}$$

where the longitude $\Lambda = (\lambda - \lambda_0) + \Omega t$ and λ_0 is the longitude at $t = 0$. The longitude λ_0 may be assumed to be zero. Finally, if the longitude λ is in units of degrees/180, then the sidereal hour angle Ωt must be divided by π. Referring once again to Eq. (4.24), the part $(\boldsymbol{\rho} + 2\boldsymbol{\Omega})\mathbf{V}$ can be written in matrix form as

$$\mathbf{M} = (\boldsymbol{\rho} + 2\boldsymbol{\Omega}) \times \mathbf{V} = \begin{bmatrix} \mathbf{1}_U & \mathbf{1}_E & \mathbf{1}_N \\ (\Omega_U + 2\Omega_U) & (\Omega_E + 2\Omega_E) & (\Omega_N + 2\Omega_N) \\ V_U & V_E & V_N \end{bmatrix}$$

and since $\Omega_E = 0$, $\Omega_N = \Omega \cos \phi$, $\Omega_U = \Omega \sin \phi$, then in component form we have

$$M_U = -\frac{V_N^2}{R_\lambda + h} - \left(\frac{V_E^2}{R_\phi + h}\right) - 2\Omega \cos \phi \cdot V_E$$

$$M_E = \frac{V_U V_E}{R_\psi + h} + 2\Omega \cos \phi \cdot V_U$$

$$- \left(\frac{V_N V_E}{R_\phi + h}\right) \tan \phi - 2\Omega \sin \phi \cdot V_N$$

$$M_N = \frac{V_U V_N}{R_\lambda + h} + 2\Omega \sin \phi \cdot V_E$$

$$+ \left(\frac{V_E^2}{R_\phi + h}\right) \tan \phi$$

In terms of these equations, the total acceleration can be expressed in the form of Eq. (4.19)

$$\frac{d^2 \mathbf{R}}{dt^2} = \left(\frac{\mathbf{V} - \mathbf{V}_0}{\Delta t}\right) + M_U \mathbf{1}_U + M_E \mathbf{1}_E + M_N \mathbf{1}_N + \mathbf{\Omega} \times (\mathbf{\Omega} \times \mathbf{R})$$

in which case we have substituted $(d\mathbf{V}/dt)_P = (\mathbf{V} - \mathbf{V}_0)/\Delta t$, where $\mathbf{V}$ is the velocity vector at time t and $\mathbf{V}_0$ is the velocity vector at $(t - \Delta t)$. In certain applications, the centripetal acceleration term $\mathbf{\Omega} \times (\mathbf{\Omega} \times \mathbf{R})$ can be neglected since the error contribution is small. Then, the east(x), north(y), and vertical (z) components of the rates of change of velocity measured with respect to the earth-fixed coordinates are as follows:

$$\dot{V}_x = A_x - \left(\frac{V_E V_U}{R_\phi + h}\right) - 2\Omega V_U \cos \phi$$

$$+ \left(\frac{V_N V_E}{R_\phi + h}\right) \tan \phi + 2\Omega V_N \sin \phi$$

$$\dot{V}_y = A_y - \left(\frac{V_E^2}{R_\phi + h}\right) \tan \phi - 2\Omega V_E \sin \phi$$

$$- \left(\frac{V_U V_N}{R_\lambda + h}\right) - R\Omega^2 \sin \phi \cos \phi$$

$$\dot{V}_z = A_z + \left(\frac{V_N^2}{R_\lambda + h}\right) + \left(\frac{V_E^2}{R_\phi + h}\right)$$

$$+ 2\Omega V_E \cos \phi - g_{m0}\left(1 - \frac{2h}{R_0}\right) + R\Omega^2 \cos^2 \phi$$

Therefore

$$V_x = \int_0^t \dot{V}_x \, dt$$

$$V_y = \int_0^t \dot{V}_y \, dt$$

$$V_z = \int_0^t \dot{V}_z \, dt$$

4.3. General Navigation Equations

Assuming $t=0$, $\Lambda = \lambda$, so that the latitude and longitude rates are

$$\dot{\phi} = \frac{\int_0^t \dot{V}_y \, dt}{R_\phi + h}$$

$$\dot{\lambda} = \frac{\int_0^t \dot{V}_x \, dt}{(R_\lambda + h \cos \phi)}$$

respectively, so that the present position of the platform in terms of (ϕ, λ, h) may therefore be computed as follows:

$$\phi - \phi_0 = \int_0^t \dot{\phi} \, dt = \int_0^t \left[\frac{\int_0^t \dot{V}_y \, dt}{R_\phi + h} \right] dt$$

$$\lambda - \lambda_0 = \int_0^t \dot{\lambda} \, dt = \int_0^t \left[\frac{\int_0^t \dot{V}_x \, dt}{(R_\lambda + h) \cos \phi} \right] dt$$

$$h - h_0 = \int_0^t \dot{h} \, dt = \int_0^t \int_0^t \dot{V}_z \, dt$$

where (ϕ_0, λ_0, h_0) represent the initial position of the platform in geographic coordinates at time zero.

Equations (4.27a)–(4.27c) can be integrated by the inertial navigation system's onboard navigation computer to calculate continuously the y(north), x(east), and z(vertical or up) velocities as follows:

$$V_N = V_N(0) + \int_0^t \dot{V}_N \, dt$$

$$V_E = V_E(0) + \int_0^t \dot{V}_E \, dt$$

$$V_U = V_U(0) + \int_0^t \dot{V}_U \, dt$$

From the preceding discussion, we note that if the plumb-bob gravity vector defined in Eq. (4.21) is assumed (for error modeling purposes) to act entirely in the radial direction, then the mass attraction component of $\mathbf{g}(\mathbf{R})$ is governed by the inverse square law as follows:

$$\mathbf{g}_m(\mathbf{R}) = g_{m0}(R_0/R)^2 \mathbf{1}_R \qquad (4.28)$$

where $g_{m0}(\mathbf{R})$ = mass attraction gravity at the surface of the earth
            R_0 = equatorial radius of the earth
            R = radial distance from the center of the earth to the vehicle
            $\mathbf{1}_R$ = unit vector positive in the $-\mathbf{R}$ direction = $\mathbf{R}/R$

In the present discussion, $\mathbf{g}_m(\mathbf{R})$ will be considered positive toward the earth's center at the point $\mathbf{R}$. The mass attraction gravity component at moderate altitudes h can be accurately approximated by expanding Eq. (4.28) in a Taylor series. Ignoring higher-order terms, the expansion yields

$$\mathbf{g}_m(\mathbf{R}) = g_{m0}(1 - 2h/R_0)\mathbf{1}_R \tag{4.29}$$

Carrying these results one step further, we note that the vertical component of the centripetal acceleration can be found from the triple vector product $\mathbf{\Omega} \times (\mathbf{\Omega} \times \mathbf{R})$ in scalar form. Thus, the vertical component of the centripetal acceleration is $\Omega^2 R \cos^2 \phi$, where ϕ is the geographic latitude and Ω is the earth's rotational rate. The centripetal acceleration has a maximum value of approximately $0.003g$. Next, we note that the angular rate vector $\boldsymbol{\rho}$ defines the rate at which the local-level reference frame is precessing relative to an earth-fixed set of axes, which results from vehicle translational motion. It should be pointed out here that a "local-level" coordinate frame is one which two axes of the frame are always normal to $\mathbf{g}(\mathbf{R})$. From basic kinematic considerations, the vector $\boldsymbol{\rho}$ may be expressed as

$$\boldsymbol{\rho} = \left(\frac{\mathbf{V}}{R}\right)\mathbf{1}_R + \rho_R \mathbf{1}_R \approx \mathbf{V} \times \frac{\mathbf{R}}{R^2} \tag{4.30}$$

where $\mathbf{1}_R =$ a unit vector in the radial direction $= \mathbf{R}/R$ and $\rho_R =$ precessional rate applied about the vertical axis.

The precessional rate of the local-level frame, due to the earth's angular velocity Ω, is explicitly defined by the transformation matrix relating the local-level frame and an earth-referenced frame. Specifically, the local-level frame is defined as having one axis along the radial direction, with the other two axes perpendicular to the first and defining the level plane. The precessional rate of the local-level frame about the two level axes is completely defined by the requirement that these axes remain at all times horizontal. Therefore, the precessional rate about the vertical axis is dependent on the particular navigation equation implementation. For example, as we shall see later, in a conventional north–east–down (NED) mechanization, the vertical axis is precessed at a rate that keeps the two level axis pointing north and east at all times.

Before the start of a flight, for instance, the inertial coordinates of vehicle position relative to the earth must be entered into the onboard navigation computer. Furthermore, an alignment procedure must be carried out so that the accelerometer triad is coincident with the computational reference frame. This is commonly accomplished by torquing the platform gyroscopes until the platform axes coincide with the computational reference axes. Subsequent alignment can be obtained by continuous torquing of the gyroscopes. These torquing rates consist of the earth's rotational rate and the rotational rate of the computational frame resulting from vehicle motion with respect to the earth. The signals that are applied to the platform gyroscopes are $\rho_x + \Omega_x$, $\rho_y + \Omega_y$, and $\rho_z + \Omega_z$.

4.4 STANDARD MECHANIZATION EQUATIONS

In this section we will discuss the various navigation system implementation equations that are available to the inertial navigation system designer. In particular, we will investigate the inertial navigation system position and velocity equations, which are commonly expressed and calculated in a locally level navigation coordinate frame. As demonstrated in Ref. 3, however, the basic error differential equations for any inertial navigation system may be written in standard coordinates, regardless of the physical mechanization or internal navigation variables. Furthermore, under certain assumptions, the homogeneous or unforced part of these differential equations is identical for any arbitrarily configured terrestrial inertial navigation system.

Three of the most commonly used navigation coordinate frame implementations are (1) geographic (or latitude–longitude), (2) wander–azimuth, and (3) space-stable. The first two methods differ only in the azimuth or heading orientation of the navigation frame's horizontal axes relative to north. As we shall see in the next section, the geographic implementation of the inertial navigation position rate equations calculates latitude and longitude rate directly as functions of computed vehicle north and east velocity components. These signals are then continuously integrated to calculate current latitude and longitude. Actually, as we saw in Section 2.1, there are several common choices available to the designer for the "P" or platform frame. (This is the coordinate frame in which the gyroscopes and accelerometers are fixed.) Listed below are some of the coordinate frames which are used in connection with the system mechanizations:

Inertial A Cartesian coordinate system, nonrotating with respect to the distant, fixed stars. The inertial coordinates are also designated as space-stable or space-stable with respect to inertial space. This system is attractive for simulation purposes. However, this system requires a considerable amount of computation related to the inertial coordinates.

Earth-fixed This is a Cartesian coordinate system that is closely related to the inertial coordinates. The earth-fixed coordinate system also requires a considerable amount of computations.

Local level The two horizontal coordinate axes are always normal to the gravity vector $\mathbf{g}(\mathbf{R})$. That is, the horizontal plane formed by the two axes is tangent to the reference ellipsoid. Also, the third or vertical axis is normal to the reference ellipsoid. The locally level system includes the north-slaved (or north-pointing) and wander–azimuth systems.

Platform The platform frame's axes are parallel to the nominal accelerometer input axes.

True frame This frame corresponds to the ideal or error-free orientation of the inertial platform at the vehicle's actual position. This frame is mechanization-dependent.

Computer The computer frame is defined as the frame in which the navigation equation mechanization actually occurs. Moreover, this reference frame is specified by the navigation system outputs of position and velocity. It should be pointed out, however, that because of system errors, this frame will not be the same as the true frame in which the equations are nominally mechanized.

Tangent plane This is a local-level system which is fixed at one point on the earth only.

Strapdown The coordinate axes are fixed to the vehicle.

We note that the true frame, platform frame, and computer frame are usually coincident. However, because of inertial navigation system errors, the system designer must account for small-angle misalignments between these frames. Over the years, there has been considerable discussion concerning what constitutes the "best" choice for system mechanization. There is no single best choice, but an individual choice can be made by the inertial navigation system designer, on the basis of factors such as mission requirements, instruments available, system complexity, system environmental constraints, onboard navigation computer computation speed and storage capacity, and ease of implementation of the navigation system equations and algorithms. Future navigation and guidance systems will be required to perform increasingly complex tasks. It is safe to say that the burden will be on the flight algorithms to help meet these requirements. Moreover, these algorithms will be designed to provide automatic adaptation to changing mission demands and environments. The onboard algorithms must maximize system performance as measured by autonomy, accuracy, in-flight adaptability, mission flexibility, and reliability. In addition, these algorithms must be computationally efficient, robust, self-starting, and capable of functioning autonomously. Improvements in computer memory and speed enhance the feasibility of techniques that implement optimal adaptive guidance algorithms onboard the vehicle. For vehicles operating in a near-earth environment, strapdown systems appear to be gaining in acceptance by navigation system designers, mainly because these systems have the potential advantages of lower cost, simplicity, and reliability. In this book, we will use different arrangements of the local-level axes, such as north–east–down, east–north–up, and west–

north–up, so that the navigation system designer can select the one that best fulfills the objective, while at the same time provide the student with some expertise in dealing with different systems.

In inertial mechanizations, the gyroscopes sense rotations with respect to inertial space, thus maintaining the platform fixed in inertial space. The transformation to any other desired frame, such as an ECEF frame is accomplished by constructing the desired transformation from known dependence of the desired coordinate frame (e.g., a clock would be necessary to transform to an earth-fixed frame). Moreover, in an inertial frame, the gyroscopes will "tumble" in the gravity field, at a rate determined by the earth rate and the vehicle velocity. As already mentioned, a locally level system allows the two horizontal gyroscopes to maintain their output axes vertical and their input axes in the horizontal plane. In this arrangement, the g-sensitive drift is due mainly to the mass unbalance about the output axis of the gyroscope, and as a result is virtually absent for the horizontal gyroscopes in this mechanization. In addition, gyroscope drift-induced position errors for this mechanization are caused predominantly by the horizontal gyroscopes. Note that the vertical gyroscope is also subject to g-sensitive drift. A local-level system will have a bounded oscillatory latitude error and a growing longitude error in response to a constant gyroscope drift. An inertial system, on the other hand, will have unbounded latitude and longitude errors in response to the same drift.

4.4.1. Latitude–Longitude Mechanization

Let the vehicle position be defined by a coordinate system such that the y axis points north, the x axis east, and the z axis up as shown in Fig. 4-4. In this arrangement, since the y accelerometer points north, it senses north–south acceleration. In addition, an ellipsoidal earth model will be assumed.

The gyroscope torquing or precession rates ω_x, ω_y, and ω_z with respect to inertial space are [3]

$$\omega_x = \omega_E = \rho_E \tag{4.31a}$$

$$\omega_y = \omega_N = \rho_N + \Omega \cos \phi \tag{4.31b}$$

$$\omega_z = \omega_z = \rho_z + \Omega \sin \phi \tag{4.31c}$$

The two rates ω_x and ω_y represent the level angular rates of the platform required to maintain the platform level, while ω_z represents the platform azimuth rate required to maintain the desired platform orientation to north; that is, ω_z defines a north-pointing system mechanization. Next, we need the craft rate equations, as follows:

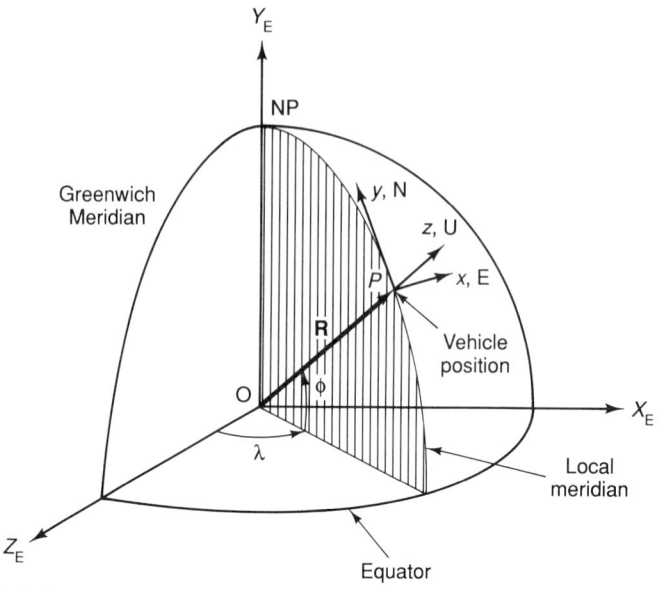

FIGURE 4-4
Coordinate axes for latitude–longitude mechanization.

$$\rho_E = -\frac{V_N}{R_e}\left[1 - \frac{h}{R_e} - f(1 - 3\cos^2\phi)\right] \quad (4.32a)$$

$$\rho_N = \frac{V_E}{R_e}\left[1 - \frac{h}{R_e} - f(1 - \cos^2\phi)\right] \quad (4.32b)$$

$$\rho_z = \rho_N \tan\phi \quad (4.32c)$$

Consequently, in order to maintain the platform level, the latitude and longitude gimbal axes must be driven with the following rates:

$$\dot{\phi} = -\rho_E \quad (4.33)$$

$$\dot{\lambda} = \frac{\rho_N}{\cos\phi} \quad (4.34)$$

From Eq. (4.22), the level and vertical velocity equations take the form

$$\dot{V}_E = A_E - (\omega_N + \Omega\cos\phi)V_z + (\omega_z + \Omega\sin\phi)V_N \quad (4.35)$$

$$\dot{V}_N = A_N - (\omega_z + \Omega\sin\phi)V_E + \omega_E V_z \quad (4.36)$$

$$\dot{V}_z = A_z - \omega_E V_N + (\omega_N + \Omega\cos\phi)V_E - g_z + K_2(h_B - h) \quad (4.37)$$

4.4. Standard Mechanization Equations

with

$$\dot{h} = V_z + K_1(h_B - h) \tag{4.38}$$

where h_B is the barometric altitude.

All the preceding equations must be initialized before the start of the program. That is, the initialization is accomplished by placing $V_E(0)$, $V_N(0)$, etc. Also, we note that use was made of $\mathbf{g}(\mathbf{R}) = -g_z \mathbf{1}_z$. The various terms and or coefficients appearing in the vertical velocity channel will be treated in more detail in the next subsection.

Now consider a spherical earth model. The gyroscope torquing rates ω_x, ω_y, and ω_z with respect to inertial space are

$$\omega_x = \omega_E = -\dot{\phi} \tag{4.39a}$$

$$\omega_y = \omega_N = \dot{\lambda} \cos \phi + \Omega \cos \phi = \frac{V_x}{R} + \Omega \cos \phi \tag{4.39b}$$

$$\omega_z = \omega_z = \dot{\lambda} \sin \phi + \Omega \sin \phi = \frac{V_x}{R} \tan \phi + \Omega \sin \phi \tag{4.39c}$$

As before, in order to maintain the platform level, the latitude and longitude gimbal axes must be driven with the rates

$$\dot{\phi} = \frac{V_y}{R} = \frac{V_N}{R} \tag{4.40}$$

$$\dot{\lambda} = \frac{V_x}{R \cos \phi} = \left(\frac{V_E}{R}\right) \sec \phi \tag{4.41}$$

The latitude and longitude are obtained by integrating Eqs. (4.40) and (4.41). Now, integrating Eq. (4.40), one obtains

$$\phi = \phi(0) + \left(\frac{V_y}{R}\right) t \tag{4.42}$$

For the longitude we can write

$$\lambda = \lambda(0) + \int_0^t \dot{\lambda} \, dt \tag{4.43}$$

Therefore, Eqs. (4.31a)–(4.31c) can be written as

$$\omega_x = -\frac{V_y}{R} \tag{4.44a}$$

$$\omega_y = \frac{V_x}{R} + \Omega \cos\left[\phi(0) + \left(\frac{V_y}{R}\right) t\right] \tag{4.44b}$$

$$\omega_z = \frac{V_x}{R} \tan\left[\phi(0) + \left(\frac{V_y}{R}\right)t\right] + \Omega \sin\left[\phi(0) + \left(\frac{V_y}{R}\right)t\right] \quad (4.44c)$$

Next, we note that in terms of Eq. (4.42), the components of the earth rate are given by

$$\Omega_x = 0 \quad (4.45a)$$

$$\Omega_y = \Omega \cos\left[\phi(0) + \left(\frac{V_y}{R}\right)t\right] \quad (4.45b)$$

$$\Omega_z = \Omega \sin\left[\phi(0) + \left(\frac{V_y}{R}\right)t\right] \quad (4.46c)$$

so that the platform (or craft) rates relative to the earth are

$$\rho_x = \omega_x - \Omega_x = -\frac{V_y}{R} \quad (4.46a)$$

$$\rho_y = \omega_y - \Omega_y = \frac{V_x}{R} \quad (4.46b)$$

$$\rho_z = \omega_z - \Omega_z = \frac{V_x}{R} \tan\left[\phi(0) + \left(\frac{V_y}{R}\right)t\right] \quad (4.46c)$$

The gravity components are taken to be

$$g_x = 0 \quad (4.47a)$$

$$g_y = 0 \quad (4.47b)$$

$$g_z = -\mathbf{g}(\mathbf{R}) = -g_0(R_e/R)^2 \quad (4.47c)$$

where g_0 is the acceleration of gravity at the earth's surface.

Substituting Eqs. (4.47a)–(4.47c) into Eq. (4.24), one obtains

$$A_x = \dot{V}_x - (\rho_z + 2\Omega_z)V_y + (\rho_y + 2\Omega_y)V_z \quad (4.48a)$$

$$A_y = \dot{V}_y - \rho_x V_z \quad (4.48b)$$

$$A_z = \dot{V}_z - (\rho_y + 2\Omega_y)V_x + \rho_x V_y + g_z \quad (4.48c)$$

Finally, the magnitude of the vector ω can be obtained from Eqs. (4.46a)–(4.46c) as

$$|\omega| = \omega = \sqrt{\left(\frac{V_y}{R}\right)^2 + \left[\Omega + \left(\frac{V_x}{R}\right) \cdot \frac{1}{\cos\left[\phi(0) + \left(\frac{V_y}{R}\right)t\right]}\right]^2} \quad (4.49)$$

Equation (4.49) cannot be solved in closed form since most of the coefficients are time-dependent. However, one can obtain an idea of the magnitude of this equation by assuming that the absolute value of Eq. (4.42) is, say, equal to or less than $60°$. Mathematically, $|\phi(0)+(V_y/R)t| \leq 60°$.

Latitude and longitude are weakly coupled to the vertical channel. Thus, as we shall see later, vertical channel navigation is often decoupled and performed in conjunction with the combined outputs of the vertical accelerometer and an altitude-indicating sensor, such as a barometric altimeter. A disadvantage of the latitude–longitude mechanization arises when navigating at or near the polar regions where the vehicle latitude approaches $\pm 90°$. The required rotational rate of the computation frame becomes infinite in the azimuth as indicated by Eq. (4.46c). Also, the computation of longitude rate as given by Eq. (4.41) results in a singularity, so that the system becomes discontinuous, and ω_z becomes unbounded. Consequently, a navigation system mechanized in latitude–longitude (north-pointing) is limited to latitudes at which the maximum allowable precessional rate of its vertical axis will not be exceeded and, therefore, does not have a worldwide navigation capability. That is, a local-level north-pointing navigation system is limited to latitudes of, say, $70°$ or less. On the other hand, there are a number of advantages in using this mechanization: (1) it requires less computation time than other mechanizations; (2) if the platform axes are kept aligned with the axes of the navigation frame, the angles between the various adjacent gimbals are equal to the conventionally defined roll, pitch, and heading angles of the vehicle; and (3) it turns out that two accelerometers suffice for this mechanization.

4.4.2. Wander–Azimuth Mechanization

A number of all-earth navigation methods have been proposed that circumvent the problem of singularities at the poles described in the latitude–longitude, local-level, north-pointing mechanization of the previous section. A completely general solution of the all-earth navigation problem can be achieved, among other mechanizations, by the so-called wander–azimuth mechanization. This choice is motivated by the fact that a local-level, wander–azimuth implementation allows a true worldwide capability and is utilized in many major inertial navigation systems. In a wander–azimuth, local-level mechanization, the platform is aligned so that it is perpendicular to the local geodetic vertical. Consequently, the gyroscope that senses rotation about the vertical is left untorqued and will not maintain a particular terrestrial heading reference. However, in practice torques are applied to the vertical (or altitude) channel in order to cancel the gyroscope bias error. Orientation with respect to north will, therefore, vary with time and vehicle position by the wander angle α. Local-level, wander–azimuth navigation systems are therefore used when navigating at latitudes above 70°. In navigation systems where a higher accuracy is required, a "double precision" (or two-word) wander angle value is commonly used in order to minimize the "wander-angle drift." The two-word wander angle is taken as the sum of the values α (most significant part) and $\Delta\alpha$ (a least significant part).

The analysis of the wander–azimuth mechanization will be carried out for a local-vertical, geodetic frame (see Fig. 4-5). Unless otherwise specified, the shape of the earth will be assumed to be that of an oblate spheroid. Moreover, in the derivation of the equations that follow, the wander angle α will be defined to positive counterclockwise from north. Also, we will take the (x, y, z)-frame to coincide with the computational and platform frames. The positive direction of the z axis is up, directed along the geodetic latitude, and the positive direction of the y axis for $\alpha=0$ is north, while the x, y axes form a plane that is locally level, forming the gyroscope x and y axes, respectively. The Y_E axis is in the direction of the earth's angular velocity vector (North Pole), the Z_E axis is in the equatorial plane through the Greenwich Meridian, and the X_E axis is in the equatorial plane completing the triad ($\mathbf{i}_{X_E} = \mathbf{j}_{Y_E} \times \mathbf{k}_{Z_E}$). The navigation computations make use of the direction cosine matrix C, in the mechanization of transformation from the platform frame to the earth-fixed frame. In this method, the local-level axes are precessed at a generalized angular rate $\mathbf{\rho}$, relative to the earth-reference axes, and the transformation between the two sets of axes is numerically established on a continuous basis by solving the direction cosine differential equation [18]

$$\dot{C} = C\{\mathbf{\rho}\} \tag{4.50}$$

where $\{\mathbf{\rho}\}$ is a skew–symmetric matrix representing the angular velocity of the platform frame with respect to the earth-fixed frame expressed in the platform frame coordinates, and is given by

$$\{\mathbf{\rho}\} = \begin{bmatrix} 0 & -\rho_z & \rho_y \\ \rho_z & 0 & -\rho_x \\ -\rho_y & \rho_x & 0 \end{bmatrix} \tag{4.51}$$

That is, the direction cosine matrix C is updated using a first-order approximate solution of Eq. (4.50). The level components of $\mathbf{\rho}$ maintain the level axes perpendicular to the gravity vector, while the precessional rate about the vertical axis is completely arbitrary in this implementation concept. (In strapdown navigation systems, the vertical component of $\mathbf{\rho}$ is commonly chosen to be zero). In platform systems, it is common to have the vertical component of $\mathbf{\rho}$ equal and opposite to the vertical component of $\mathbf{\Omega}$. This is commonly referred to as "free azimuth," which has the advantage that the vertical axis of the platform need not be physically precessed, resulting in less hardware needed to mechanize the system (see Section 2.3.3). The velocity

4.4. Standard Mechanization Equations

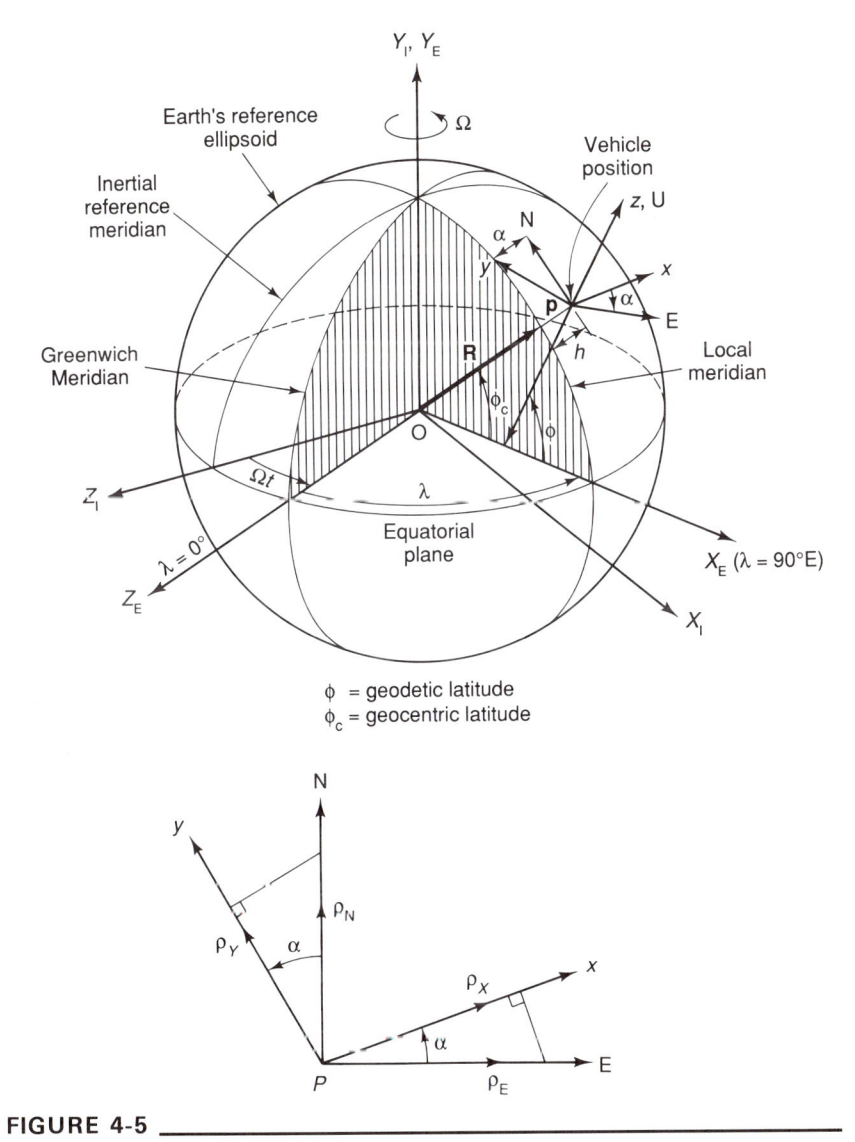

FIGURE 4-5
Coordinate axes for wander–azimuth mechanization.

increments $\Delta \mathbf{V}$ sensed by the accelerometers and resolved along the platform frame axes are combined with the V/R terms and with the Coriolis correction terms. The computed horizontal velocity is resolved through the wander angle Ω into east and north components. Computation of the vertical velocity component V_z requires the magnitude of the local gravity. The computed horizontal velocity is used in the computation of the angular rate $\boldsymbol{\rho}$ of the platform frame with respect to the earth-fixed frame. The $\{\mathbf{\rho}\}$ matrix is used to update the direction cosine matrix transformation between the platform and earth-fixed frames. As stated above, the updating is based on the solution of the differential equation $\dot{C} = C\{\mathbf{\rho}\}$. Double precision is normally used in solving this equation. The latitude, longitude, and wander angle are computed from the C-matrix terms.

In order to uniquely define the orientation of the local-level frame in a wander azimuth implementation, a set of direction cosines relating the platform local-level axes to the earth-fixed axes is utilized. Figure 4-5 illustrates the definition of the wander–azimuth axes for an oblate spheroid earth. Symmetry about the Y_E, Y_I semiminor axis is assumed.

In general, the local-level frame will have an azimuth rotation α relative to north. This angle, which we defined as the wander angle, varies as the vehicle moves over the earth from its initial position. (Note that for a north-pointing system, the wander angle $\alpha = 0$). Following the procedures described in Chapter 2, the transformation matrix C_E^P that transforms the earth-fixed coordinates into the platform coordinates is obtained by a series of rotations. The order and direction of rotations is as follows: (1) a positive rotation of λ about Y_E, (2) a negative rotation of ϕ about the rotated X_E axis, and (3) a positive rotation of α about the rotated Z_E axis. However, before we proceed with the earth-centered earth-fixed (ECEF) coordinate transformation, it is necessary to transform the earth-centered inertial frame (ECI) into the ECEF frame. These two coordinate frames are related by the simple transformation matrix

$$C_I^E = \begin{bmatrix} \cos \Omega t & 0 & -\sin \Omega t \\ 0 & 1 & 0 \\ \sin \Omega t & 0 & \cos \Omega t \end{bmatrix}$$

Therefore, the transformation matrix takes the form

$$C_E^P = \begin{bmatrix} \cos \alpha & \sin \alpha & 0 \\ -\sin \alpha & \cos \alpha & 0 \\ 0 & 0 & 1 \end{bmatrix} \begin{bmatrix} 1 & 0 & 0 \\ 0 & \cos \phi & -\sin \phi \\ 0 & \sin \phi & \cos \phi \end{bmatrix} \begin{bmatrix} \cos \lambda & 0 & -\sin \lambda \\ 0 & 1 & 0 \\ \sin \lambda & 0 & \cos \lambda \end{bmatrix}$$

Wander Angle ← Latitude ← Longitude

$$= \begin{bmatrix} \cos \alpha \cos \lambda - \sin \alpha \sin \phi \sin \lambda & \sin \alpha \cos \phi & -\cos \alpha \sin \lambda - \sin \alpha \sin \phi \cos \lambda \\ -\sin \alpha \cos \lambda - \cos \alpha \sin \phi \sin \lambda & \cos \alpha \cos \phi & \sin \alpha \sin \lambda - \cos \alpha \sin \phi \cos \lambda \\ \cos \phi \sin \lambda & \sin \phi & \cos \phi \cos \lambda \end{bmatrix} \quad (4.52)$$

4.4. Standard Mechanization Equations

The elements of the transformation matrix [Eq. (4.52)] can also be written in the form

$$C_E^P = \begin{bmatrix} C_{Xx} & C_{Yx} & C_{Zx} \\ C_{Xy} & C_{Yy} & C_{Zy} \\ C_{Xz} & C_{Yz} & C_{Zz} \end{bmatrix}$$

where $C_{Xx} = \cos\alpha \cos\lambda - \sin\alpha \sin\phi \sin\lambda$
$C_{Xy} = -\sin\alpha \cos\lambda - \cos\alpha \sin\phi \sin\lambda$
$C_{Xz} = \cos\phi \sin\lambda$
$C_{Yx} = \sin\alpha \cos\phi$
$C_{Yy} = \cos\alpha \cos\phi$
$C_{Yz} = \sin\phi$
$C_{Zx} = -\cos\alpha \sin\lambda - \sin\alpha \sin\phi \cos\lambda$
$C_{Zy} = \sin\alpha \sin\lambda - \cos\alpha \sin\phi \cos\lambda$
$C_{Zz} = \cos\phi \cos\lambda$

In terms of the direction cosine elements, the latitude ϕ, longitude λ, and wander angle α can be calculated from the relations

$$\phi = \sin^{-1} C_{Yz} \tag{4.53}$$

$$\lambda = \tan^{-1} [C_{Xz}/(C_{Xx}C_{Yy} - C_{Yx}C_{Xy})] \tag{4.54}$$

$$\alpha = \tan^{-1} [C_{Yx}/C_{Yy}] \tag{4.55}$$

The vehicle north and east velocity components are obtained by a transformation of the x and y velocity components as follows:

$$V_E = V_x \cos\alpha - V_y \sin\alpha \tag{4.56}$$

$$V_N = V_x \sin\alpha + V_y \cos\alpha \tag{4.57}$$

and

$$V_x = V_E \cos\alpha + V_N \sin\alpha$$

$$V_y = -V_E \sin\alpha + V_N \cos\alpha$$

The aircraft ground velocity is given by

$$V_g = \sqrt{V_x^2 + V_y^2}$$

From Section 2.3.4 Eq. (2.38), the wander angle is governed by

$$\dot\alpha = \omega_{zc} - \omega_{zp}$$

Therefore, in terms of the above velocity components, the wander angle can also be written as

$$\dot{\alpha} = -(V_E/R)\tan\phi \qquad (4.58)$$

Likewise, the level craft rate components ρ_x and ρ_y with respect to the earth can be obtained using Fig. 4-5.

$$\rho_x = \rho_E \cos\alpha + \rho_N \sin\alpha \qquad (4.59)$$

$$\rho_y = -\rho_E \sin\alpha + \rho_N \cos\alpha \qquad (4.60)$$

Substituting the values for ρ_E and ρ_N defined previously, we obtain

$$\rho_x = -\left(\frac{V_N}{R_\phi + h}\right)\cos\alpha + \left(\frac{V_E}{R_\lambda + h}\right)\sin\alpha \qquad (4.61)$$

$$\rho_y = \left(\frac{V_N}{R_\phi + h}\right)\sin\alpha + \left(\frac{V_E}{R_\lambda + h}\right)\cos\alpha \qquad (4.62)$$

Note that for the wander–azimuth mechanization $\rho_z = 0$, but the gyroscope torquing rate is $\omega_z = \Omega \sin\phi$. Carrying the above results one step further, we substitute Eqs. (4.56) and (4.57) into Eqs. (4.61) and (4.56), resulting in [4]

$$\rho_x = -V_y\left(\frac{1}{R_y}\right) - V_x\left(\frac{1}{Q}\right) \qquad (4.63)$$

$$\rho_y = V_x\left(\frac{1}{R_x}\right) + V_y\left(\frac{1}{Q}\right) \qquad (4.64)$$

where

$$\frac{1}{R_x} = \frac{\cos^2\alpha}{R_\lambda + h} + \frac{\sin^2\alpha}{R_\phi + h}$$

$$\frac{1}{R_y} = \frac{\sin^2\alpha}{R_\lambda + h} + \frac{\cos^2\alpha}{R_\phi + h}$$

$$\frac{1}{Q} = \left[\frac{1}{R_\phi + h} - \frac{1}{R_\lambda + h}\right]\cos\alpha\sin\alpha$$

Equations (4.63) and (4.64) are the exact level craft rate equations for a vehicle flying at an altitude h above the reference ellipsoid. Approximations to Eqs. (4.63) and (4.64) can be obtained by substituting the respective values for R_λ and R_ϕ and expanding the results into a first-order series.

4.4. Standard Mechanization Equations

Performing these operations, we obtain the approximate level craft rates as follows [4]:

$$\rho_x = -\frac{V_y}{R_e}\left[1 - \frac{h}{R_e} - f(1 - 3\cos^2\alpha\cos^2\phi - \sin^2\alpha\cos^2\alpha)\right]$$

$$-\frac{V_x}{R_e}[2f\sin\alpha\cos\alpha\cos^2\phi] \quad (4.65)$$

$$\rho_y = \frac{V_x}{R_e}\left[1 - \frac{h}{R_e} - f(1 - 3\sin^2\alpha\cos^2\phi - \cos^2\alpha\cos^2\phi)\right]$$

$$+\frac{V_y}{R_e}[2f\sin\alpha\cos\alpha\cos^2\phi] \quad (4.66)$$

where R_e is the equatorial radius of the earth and f is the flattening. In general, the craft rate can be written in vector notation from Fig. 4-5 as

$$\boldsymbol{\rho} = \rho_x\mathbf{1}_x + \rho_y\mathbf{1}_y + \rho_z\mathbf{1}_z$$
$$= -\dot{\phi}\mathbf{1}_E + \dot{\lambda}\mathbf{1}_Y + \dot{\alpha}\mathbf{1}_z \quad (4.67)$$

where $\mathbf{1}_x$, $\mathbf{1}_y$, $\mathbf{1}_z$ are the unit vectors along the respective platform axes. Therefore, for the wander–azimuth mechanization

$$\boldsymbol{\rho} = \begin{bmatrix} \dfrac{-V_x(2f\sin\alpha\cos\alpha\cos^2\phi) - V_y\left[1 - \dfrac{h}{R_e} - f(1 - 3\cos^2\alpha\cos^2\phi - \sin^2\alpha\cos^2\phi)\right]}{R_e} \\ \dfrac{V_x\left[1 - \dfrac{h}{R_e} - f(1 - 3\cos^2\phi\sin^2\alpha - \cos^2\phi\cos^2\alpha)\right] + V_y(2f\sin\alpha\cos\alpha\cos^2\phi)}{R_e} \\ 0 \end{bmatrix} \quad (4.68)$$

However, it should be pointed out that, in the general case, the vertical component of the craft rate ρ_z is not zero, but takes the form

$$\rho_z = \left[\frac{\rho_y\cos\alpha\cos\phi + \rho_x\sin\alpha\cos\phi}{1 - \sin^2\phi}\right]\sin\phi + \dot{\alpha}$$
$$= \dot{\lambda}\sin\phi + \dot{\alpha} \quad (4.69)$$

Thus, the value of ρ_z will depend on the type of mechanization selected. As we noted in Section 2.3, the vertical gyroscope torquing rate is

$$\omega_z = \rho_z + \Omega\sin\phi = (\dot{\lambda} + \Omega)\sin\phi + \dot{\alpha} \quad (4.70)$$

where

$$\dot{\lambda} = \frac{\rho_y \cos\alpha \cos\phi + \rho_x \sin\alpha \cos\phi}{1 - \sin^2\phi}$$

Thus, for the general mechanization case, and depending what value of α is chosen, Eq. (4.68) will contain a ρ_z component:

$$\boldsymbol{\rho} = \begin{bmatrix} \dfrac{-V_x(2f\sin\alpha\cos\alpha\cos^2\phi) - V_y\left[1 - \dfrac{h}{R_e} - f(1 - 3\cos^2\alpha\cos^2\phi - \cos^2\phi\sin^2\alpha)\right]}{R_e} \\ \dfrac{V_x\left[1 - \dfrac{h}{R_e} - f(1 - 3\sin^2\alpha\cos^2\phi - \cos^2\alpha\cos^2\phi)\right] + V_y(2f\sin\alpha\cos\alpha\cos^2\phi)}{R_e} \\ \dot{\alpha} + (\dot{\lambda} + \Omega)\sin\phi \end{bmatrix} \quad (4.71)$$

The value of ρ_z will, therefore, depend on the type of mechanization used. Once it is known, the general direction cosine (or position) differential equations can be obtained. Let $i = X, Y, Z$. Then the direction cosine rates are obtained as follows:

$$\dot{C}_{ix} = \lim_{\Delta t \to 0} \frac{C_{ix}(t + \Delta t) - C_{ix}(t)}{\Delta t} = C_{iy}\rho_z - C_{iz}\rho_y$$

$$\dot{C}_{iy} = \lim_{\Delta t \to 0} \frac{C_{iy}(t + \Delta t) - C_{iy}(t)}{\Delta t} = C_{iz}\rho_x - C_{ix}\rho_z$$

$$\dot{C}_{iz} = \lim_{\Delta t \to 0} \frac{C_{iz}(t + \Delta t) - C_{iz}(t)}{\Delta t} = C_{ix}\rho_y - C_{iy}\rho_x$$

These results can be repeated in a cyclical order by letting $i = X, Y, Z$, resulting in the following equations (note that the nine direction cosine rate equations result from this process, but only six are needed for navigation or flight control):

$$\begin{aligned} \dot{C}_{Xx} &= C_{Xy}\rho_z - C_{Xz}\rho_y \\ \dot{C}_{Xy} &= C_{Xz}\rho_x - C_{Xx}\rho_z \\ \dot{C}_{Xz} &= C_{Xx}\rho_y - C_{Xy}\rho_x \\ \dot{C}_{Yx} &= C_{Yy}\rho_z - C_{Yz}\rho_y \\ \dot{C}_{Yy} &= C_{Yz}\rho_x - C_{Yx}\rho_z \\ \dot{C}_{Yz} &= C_{Yx}\rho_y - C_{Yy}\rho_x \end{aligned} \quad (4.72)$$

4.4. Standard Mechanization Equations

For the wander-azimuth mechanization under consideration, $\rho_z = 0$; therefore, the direction cosine rate differential equations simplify to

$$\begin{aligned}
\dot{C}_{Xx} &= -C_{Xz}\rho_y \\
\dot{C}_{Xy} &= C_{Xz}\rho_x \\
\dot{C}_{Xz} &= C_{Xx}\rho_y - C_{Xy}\rho_x \\
\dot{C}_{Yx} &= -C_{Yz}\rho_y \\
\dot{C}_{Yy} &= C_{Yz}\rho_x \\
\dot{C}_{Yz} &= C_{yx}\rho_y - C_{Xy}\rho_x
\end{aligned} \qquad (4.73a)$$

The processing of the direction cosine rates and direction cosines depends on the particular application. For example, some systems use processing rates of about 50 ms. The direction cosine differential equations can now be updated as follows:

$$\begin{aligned}
C_{Xx_i} &= C_{Xx_{i-1}} + \dot{C}_{Xx_i}\Delta t \\
C_{Xy_i} &= C_{Xy_{i-1}} + \dot{C}_{Xy_i}\Delta t \\
C_{Xz_i} &= C_{Xz_{i-1}} + \dot{C}_{Xz_i}\Delta t \\
C_{Yx_i} &= C_{Yx_{i-1}} + \dot{C}_{Yx_i}\Delta t \\
C_{Yy_i} &= C_{Yy_{i-1}} + \dot{C}_{Yy_i}\Delta t \\
C_{Yz_i} &= C_{Yz_{i-1}} + \dot{C}_{Yz_i}\Delta t
\end{aligned} \qquad (4.73b)$$

The level velocity differential equations can now be given in terms of the direction cosine as

$$\dot{V}_x = A_x - (\rho_y + 2\Omega C_{Yy})V_z + (2\Omega C_{Yz})V_y \qquad (4.74)$$

$$\dot{V}_y = A_y + (\rho_x + 2\Omega C_{Yx})V_z - (2\Omega C_{Yz})V_x \qquad (4.75)$$

For the purposes of processing the level velocities, they can be written in the form

$$V_{xi} = V_{x_{i-1}} + \dot{V}_x \Delta t + \delta V_x$$

$$V_{yi} = V_{y_{i-1}} + \dot{V}_y \Delta t + \delta V_y$$

where $\delta V_{x,y}$ is the corrected incremental velocity change in the x, y axes and Δt is the time increment. Specifically, the processing is performed each time the inertial platform alignment function computes the level velocity corrections

$$V_{x_i} = V_{x_i} + \Delta V_x$$

$$V_{y_i} = V_{y_i} + \Delta V_y$$

where ΔV_x and ΔV_y are the velocity corrections. As in the case of the direction cosines, the level velocities are processed every 50 ms. So far, very little has been said about the vertical channel. Commonly, a barometric altimeter is used as an external altitude reference to stabilize the vertical (altitude) channel. From Eq. (4.25c), the complete vertical velocity equation can be obtained from [9]

$$\dot{V}_z = A_z + (\rho_y + 2\Omega C_{Yy})V_x - (\rho_x + 2\Omega C_{Yx})V_y - g_z - K_2 \Delta h - \delta\hat{a} \quad (4.76)$$

with
$$\delta\hat{a} = K_3 \Delta h$$
$$\Delta h = h - h_B$$
$$\dot{h} = V_z - K_1 \Delta h = V_z + K_1(h_B - h)$$
$$g_z = -g_0[1 - (2h/R_0) + K_4 \sin^2 \phi] \approx -g_0[1 - (2h/R_0)]$$

where $K_1, K_2, K_3 =$ coefficients used in the altitude channel baroinertial loop ($K_1 = 3.0 \times 10^{-2}$ s^{-1}, $K_2 = 3.0 \times 10^{-4}$ s^{-2}, $K_3 = 1.0 \times 10^{-6}$ s^{-3}) (these coefficients are determined from specifications on the loop damping ratio and time constant)
$K_4 =$ gravity model coefficient $= 5.2888 \times 10^{-3}$ (the value of this coefficient is for gravity given in ft/s^2)
$g_0 =$ magnitude of gravity at the equator and at sea level
$\delta\hat{a} =$ estimated vertical acceleration
$h =$ computed altitude above the reference ellipsoid in the vertical channel
$h_B =$ barometric altitude
$R_0 =$ earth's equatorial radius

The system altitude can also be processed every 50 ms according to the expression

$$V_{z_i} = V_z + \dot{V}_z \Delta t + \delta V_z$$

where δV_z is the incremental z-axis velocity corrected for the vehicle nonorthogonality coordinates. From Eq. (4.76), the simplest relationship between the vertical velocity and altitude can be expressed as $V_z = \dot{h}$.

Note that in certain applications, the vertical velocity V_z may be zero, except for a few times during the mission. In practice, however, the vertical velocity V_z cannot be obtained from Eq. (4.25c) because this equation is basically unstable. Moreover, if the acceleration of gravity up component g_z is reduced to a first-order approximation, we have

$$g_z \approx -g_0 + (2g_0/R_0)h$$

The second term on the right-hand side of the equation is the potential positive feedback term, which will result in an exponentially diverging altitude error. A detailed description of the vertical channel will be deferred for a later section.

4.4. Standard Mechanization Equations

As mentioned earlier, there are various coordinate frames available to the inertial system designer. For example, let us now select another arrangement for the platform axes as shown in Fig. 4-6.

In this frame, we will take the x axis in the north direction, the y axis east, and the z axis down. Also, we will take the wander angle α to be a negative (i.e., to the right of north). The order of rotation for this design is as follows: (1) a rotation about the Y axis through the angle λ, (2) a negative rotation about the E axis through the angle $-\phi$, and (3) a negative rotation about the new Z axis which is vertical and opposite to the z axis through the angle $-\alpha$. Performing these rotations, we have

$$\begin{bmatrix} x \\ y \\ z \end{bmatrix} = \begin{bmatrix} \sin\alpha & \cos\alpha & 0 \\ \cos\alpha & -\sin\alpha & 0 \\ 0 & 0 & -1 \end{bmatrix} \begin{bmatrix} 1 & 0 & 0 \\ 0 & \cos\phi & -\sin\phi \\ 0 & \sin\phi & \cos\phi \end{bmatrix} \begin{bmatrix} \cos\lambda & 0 & -\sin\lambda \\ 0 & 1 & 0 \\ \sin\lambda & 0 & \cos\lambda \end{bmatrix} \begin{bmatrix} X \\ Y \\ Z \end{bmatrix}$$

Wander Angle ← Latitude ← Longitude

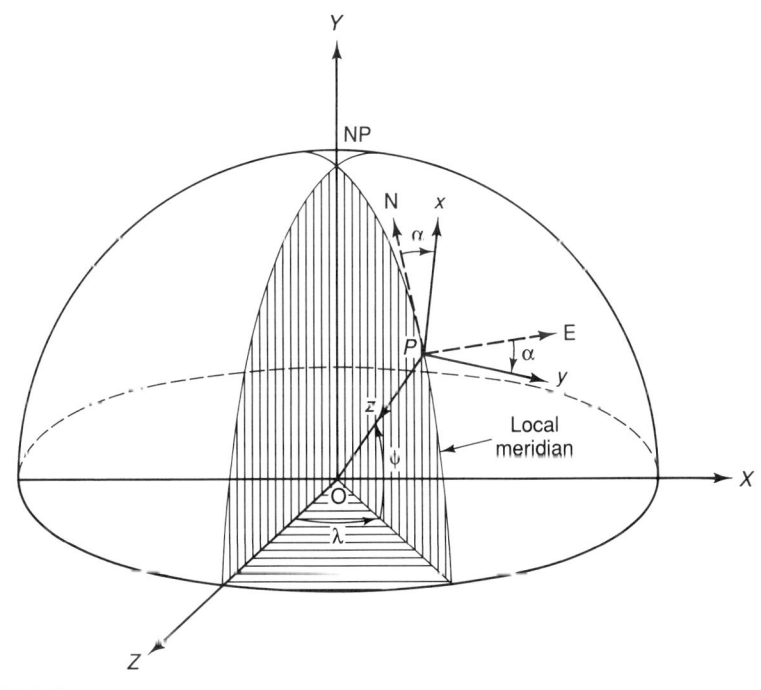

FIGURE 4-6
Wander–azimuth coordinate definition for the example.

Therefore, the direction cosines are

$$C_{Xx} = \cos \lambda \sin \alpha - \sin \phi \sin \lambda \cos \alpha$$
$$C_{Xy} = \cos \lambda \cos \alpha + \sin \phi \sin \lambda \sin \alpha$$
$$C_{Xz} = -\cos \phi \sin \lambda$$
$$C_{Yx} = \cos \phi \cos \alpha$$
$$C_{Yy} = -\cos \phi \sin \alpha$$
$$C_{Yz} = -\sin \alpha$$
$$C_{Zx} = -\sin \lambda \sin \alpha - \sin \phi \cos \lambda \cos \alpha$$
$$C_{Zy} = -\sin \lambda \cos \alpha + \sin \phi \cos \lambda \sin \alpha$$
$$C_{Zz} = -\cos \phi \cos \lambda$$

The latitude, longitude, and wander angle can be expressed as functions of the direction cosines. Thus

$$\phi = -\sin^{-1} C_{Yz}$$
$$\lambda = \tan^{-1}(C_{Xz}/C_{Zz})$$
$$\alpha = -\tan^{-1}(C_{Yy}/C_{Yz})$$

Using the previously derived results and concepts, we have the following, in summary form:

1. Angular rate of platform with respect to the inertial space ω (spatial or gyroscope torquing rate):

$$\omega = \rho + \Omega$$

   where ρ is the vector of the angular rate of the platform with respect to the earth-fixed frame (craft rate component of gyroscope torquing rate) and Ω is the angular rate of the earth-fixed frame with respect to the inertial frame.

2. Total gravity vector:

$$\mathbf{g} = [g_x \ g_y \ g_z]^T = [0 \ 0 \ g_z]^T$$

3. Angular rate of the earth-fixed frame with respect to the inertial frame:

$$\Omega = \begin{bmatrix} \Omega_x \\ \Omega_y \\ \Omega_z \end{bmatrix} = \begin{bmatrix} C_{Yx} \\ C_{Yy} \\ C_{Yz} \end{bmatrix} \Omega = \begin{bmatrix} \sin \alpha \cos \phi \\ \cos \alpha \cos \phi \\ \sin \phi \end{bmatrix} \Omega$$

4.4. Standard Mechanization Equations

4. Angular rate of the platform with respect to the earth-fixed frame:

$$\boldsymbol{\rho} = -\dot{\phi}\mathbf{E} + \dot{\lambda}\mathbf{Y} + \dot{\alpha}\mathbf{z}$$
$$= -\dot{\phi}(\sin\alpha \cdot \mathbf{x} + \cos\alpha \cdot \mathbf{y}) + \dot{\lambda}(\cos\phi \cdot \mathbf{N} - \sin\phi \cdot \mathbf{z}) + \dot{\alpha} \cdot \mathbf{z}$$
$$= -\dot{\phi}\sin\alpha \cdot \mathbf{x} - \dot{\phi}\cos\alpha \cdot \mathbf{y} - \dot{\lambda}\sin\phi \cdot \mathbf{z} + \dot{\alpha}\mathbf{z} + \dot{\lambda}$$
$$\times \cos\phi(-\sin\alpha \cdot \mathbf{y} + \cos\alpha \cdot \mathbf{x})$$
$$= \begin{bmatrix} -\dot{\phi}\sin\alpha + \dot{\lambda}\cos\phi\cos\alpha \\ -\dot{\phi}\cos\alpha - \dot{\lambda}\cos\phi\sin\alpha \\ \dot{\alpha} - \dot{\lambda}\sin\phi \end{bmatrix}$$

5. Latitude and longitude angular velocities in terms of the vehicle velocity:

$$\dot{\phi} = \frac{V_N}{R_\phi + h}$$

$$\dot{\lambda} = \frac{V_g}{(R_\lambda + h)\cos\phi}$$

$$V_E = V_x \sin\alpha + V_y \cos\alpha$$
$$V_N = V_x \cos\alpha - V_y \sin\alpha$$
$$V_x = V_N \cos\alpha + V_E \sin\alpha$$
$$V_y = -V_N \sin\alpha + V_E \cos\alpha$$

The platform angular rate $\boldsymbol{\rho}$ can now be put into vector form as in Eq. (4.71):

$$\boldsymbol{\rho} = \begin{bmatrix} \dfrac{2fC_{Yx}C_{Yy}V_x + \left[1 - \dfrac{h}{R_e} + f(2 - 3C_{Yz}^2 - 2C_{Yx}^2)\right]V_y}{R_e} \\ \dfrac{-\left[1 - \dfrac{h}{R_e} + f(2 - 3C_{Yz}^2 - 2C_{Yy}^2)\right]V_x - 2fC_{Yx}C_{Yy}V_y}{R_e} \\ \omega_z - \Omega C_{Yz} \end{bmatrix}$$

Finally, the vehicle velocity in component form is

$$\dot{V}_x = A_x + (\omega_z + \Omega_z)V_y - (\omega_y + \Omega_y)V_z$$
$$\dot{V}_y = A_y + (\omega_x + \Omega_x)V_z - (\omega_z + \Omega_z)V_x$$
$$\dot{V}_z = A_z + (\omega_y + \Omega_y)V_x - (\omega_x + \Omega_x)V_y + g_z$$

Integration of these equations yields the velocities as follows:

$$V_x = V_x(0) + \int_{t_0}^{t} \dot{V}_x \, dt$$

$$V_y = V_y(0) + \int_{t_0}^{t} \dot{V}_y \, dt$$

$$V_z = V_z(0) + \int_{t_0}^{t} \dot{V}_z \, dt$$

As an example, let us consider a software implementation of the unipolar nominal mechanization equations to compute the velocity, latitude, longi-

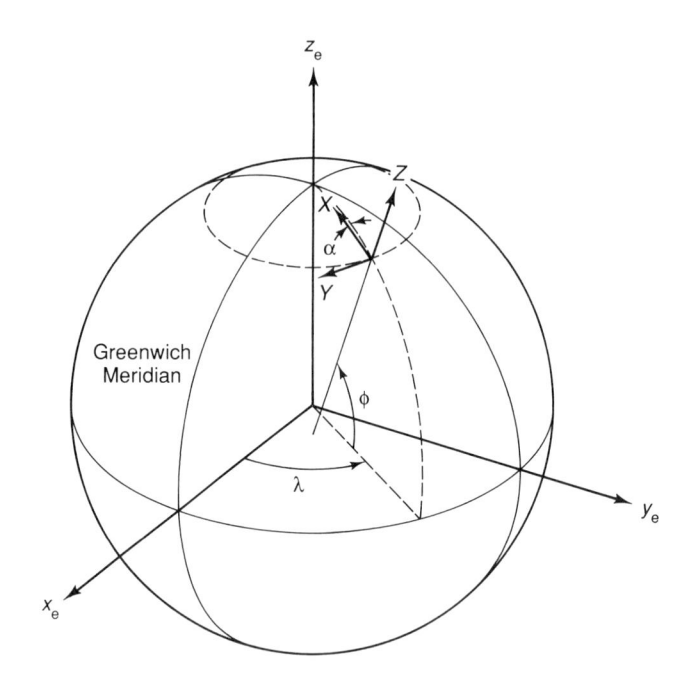

ϕ = geodetic latitude
λ = longitude
α = heading angle
$\lambda = \lambda_0 \pm (\alpha_0 - \alpha)$ = longitude for unipolar navigation

X, Y, Z: Navigation Cartesian coordinate frame
x_e, y_e, z_e: Earth-fixed Cartesian coordinates

4.4. Standard Mechanization Equations

tude, and heading angle as a function of time. The program* written in FORTRAN uses the fourth-order Runge–Kutta integration method, with the integration step size fixed at the start (see Chapter 2, Section 2.3 for the unipolar equations).

Direction Cosines

Let the navigation frame unipolar position direction cosines be denoted by p_X, p_Y, p_Z. Then

$$p_X = \cos \phi \cos \alpha$$
$$p_Y = -\cos \phi \sin \alpha$$
$$p_Z = \sin \phi$$

(Note: $p_X^2 + p_Y^2 + p_Z^2 = 1$ is an additional redundant parameter.)

Direction Cosine Propagation

$$\dot{p}_X = p_Y \rho_Z - p_Z \rho_Y$$
$$\dot{p}_Y = p_Z \rho_X - p_X \rho_Z$$
$$\dot{p}_Z = p_X \rho_Y - p_Y \rho_X$$

Platform Angular Rates (with Respect to the Earth)

$$\rho_X = -\frac{v_Y}{(r_{ew}+h)} - k p_Y (p_X v_X + p_Y v_Y)$$

$$\rho_Y = \frac{v_X}{(r_{ns}+h)} + k p_Y (p_X v_Y - p_Y v_X)$$

$$\rho_Z = \frac{p_X v_Y - p_Y v_X}{(r_{ew}+h)(p_Z \pm 1)}$$

$$k = \frac{e^2 (1-e^2)^{-1} r_{ns}}{(r_{ew}+h)(r_{ns}+h)}$$

Platform Torquing

$$w_X = \rho_X + w_e p_X$$
$$w_Y = \rho_Y + w_e p_Y$$
$$w_Z = \rho_Z + w_e p_Z$$

---

*The program was written by Dr. Gerald G. Cano, Lt. Col. U.S. Air Force Reserves.

Velocity Propagation

$$\dot{v}_X = a_X + g_X + (\rho_Z + 2w_e p_Z)v_Y - (\rho_Y + 2w_e p_Y)v_Z$$
$$\dot{v}_Y = a_Y + g_Y + (\rho_X + 2w_e p_X)v_Z - (\rho_Z + 2w_e p_Z)v_X$$
$$\dot{v}_Z = a_Z + g_Z + (\rho_Y + 2w_e p_Y)v_X - (\rho_X + 2w_e p_X)v_Y$$

Gravity Mass Attraction Components

$$g_X = \frac{GMX}{R^3}\left\{1 + \frac{J_2}{R^2}\left(\frac{5Z^2}{R^2} - 1\right) + \frac{J_4}{R^4}\left(\frac{63Z^4}{R^4} - \frac{42Z^2}{R^2} + 3\right)\right\}$$

$$g_Y = \frac{GMY}{R^3}\left\{1 + \frac{J_2}{R^2}\left(\frac{5Z^2}{R^2} - 1\right) + \frac{J_4}{R^4}\left(\frac{63Z^4}{R^4} - \frac{42Z^2}{R^2} + 3\right)\right\}$$

$$g_Z = \frac{GMZ}{R^3}\left\{1 + \frac{J_2}{R^2}\left(\frac{5Z^2}{R^2} - 3\right) + \frac{J_4}{R^4}\left(\frac{63Z^4}{R^4} - \frac{70Z^2}{R^2} + 15\right)\right\}$$

Implementation of Equations

The following lists define symbols for variables and constants in the equations describing the system.

Variables

E:	eccentricity or flattening of earth; about 1/300 depending on where used
rhoj	angular rate in 3 directions, where x, y, or z can be substituted for j
P(3,j)	direction cosines in 3 coordinates; $j=[1,n]$
dPdt(3,4)	derivative of P(i,j) with respect to time; 3 coordinates; 4 estimates per iteration
V(3,j):	velocity in 3 coordinates; $j=[1,n]$
dVdt(3,4)	derivative of V(i,j) with respect to time; 3 coordinates; 4 estimates per iteration
Pos(3,j):	position in 3 coordinates; $j=[1,n]$
dPosdt(3,4)	derivative of Pos(i,j) with respect to time; 3 coordinates; 4 estimates per iteration
G(3)	gravitational acceleration in 3 coordinates
A(3)	inertial acceleration in 3 coordinates
altitude	height above surface of earth

4.4. Standard Mechanization Equations

Phi(j)   latitude in radians; $j=[1,n]$
Lambda(j)   longitude in radians; $j=[1,n]$
Alpha(j)   navigation heading angle; $j=[1,n]$
delta   interval used in Runge–Kutta (R-K)
timint   total integration interval

Note convention for first subscript of a double subscript variable: $1 =>X$ coordinate; $2 =>Y$ coordinate; $3 =>Z$ coordinate.

Constants

RADeq   equatorial radius of earth
GM   gravitational constant
H2   constant used to compute gravitational acceleration (zonal harmonic)
H4   constant used to compute gravitational acceleration (zonal harmonic)
We   earth's rotational rate in rad/s
conK   constant used in computing ρ values; function

Components of gravitational acceleration are first computed, followed by R*ew*, R*ns*, and K. These are constant for the duration of integration period, an implementation acceptable to the sponsor.

Next, derivatives with respect to time for three direction cosines, three velocity components, and three position components are computed for use in the fourth-order R-K method. This entails first approximating four values for each computer of direction cosines, velocity, and angular rate. Note that the time rate of change of position is just velocity. The equations are shown below. Nslope goes from one to four: DELTeq is a function of the value of Nslope.

Approximation of Direction Cosines

Px = Phld(1) + DELTeq*dPdt(1,Nslope)
Py = Phld(2) + DELTeq*dPdt(2,Nslope)
Pz = Phld(3) + DELTeq*dPdt(3,Nslope)

Approximation of Velocity

Vx = Vhld(1) + DELTeq*dVdt(1,Nslope)
Vy = Vhld(2) + DELTeq*dVdt(2,Nslope)
Vz = Vhld(3) + DELTeq*dVdt(3,Nslope)

Approximations of Angular Rate

rhox = −conK*Py*(Px*Vx + Py*Vy) − Vy/(Rew + altitude)
rhoy = conK*Py*(Px*Vy − Py*Vx) + Vx/(Rns + altitude)
rhoz = (Px*Vy − Py*Vx)/(Rew + altitude)*(Pz + 1)

Approximations for Derivatives of Direction Cosines

dPdt(1,Nslope) = Py*rhoz − Pz*rhoy
dPdt(2,Nslope) = Pz*rhox − Px*rhoz
dPdt(3,Nslope) = Px*rhoy − Py*rhox

Approximations for Derivatives of Velocity

dVdt(1,Nslope) =
   A(1) + G(1) + (rhoz + 2*We*Pz)*Vy − (rhoy + 2*We*Py)*Vz
dVdt(2,Nslope) =
   A(2) + G(2) + (rhox + 2*We*Px)*Vz − (rhoz + 2*We*Pz)*Vx
dVdt(3,Nslope) =
   A(3) + G(3) + (rhoy + 2*We*Py)*Vx − (rhox + 2*We*Px)*Vy

Approximations for Derivatives of Position

dPosdt(1,Nslope) = Vx
dPosdt(2,Nslope) = Vy
dPosdt(3,Nslope) = Vz

Phi is latitude, *Lambda* is longitude and *Alpha* is navigation heading angle.

Once these are computed, the fourth order R-K method is applied to direction cosines, velocity, and position. This is done in the following *do loop* from the program:
For I = 1,3:

   Phld(I) = Phld(I) + (.1*delta/6.)*(dPdt(I,1) + 2.*dPdt(I,2)
         + 2.*dPdt(I,3) + dPdt(I,4))
   Vhld(I) = Vhld(I) + (.1*delta/6.)*(dVdt(I,1) + 2.*dVdt(I,2)
         + 2.*dVdt(I,3) + dVdt(I,4))
   Poshld(I) = Poshld(I) + (.1*delta/6.)*(dPosdt(I,1) + 2.*dPosdt(I,2)
         + 2.*dPosdt(I,3) + dPosdt(I,4))

Finally, latitude, longitude, and navigation heading angle are computed from these using the appropriate identities.

   Phi(Itimes) = ASIN(Pos(3,Itimes)/RADeq)
Lambda(Itimes) = ATAN(Pos(2,Itimes)/Pos(1,Itimes))
   Alpha(Itimes) = ACOS(P(1,Itimes)/COS(Phi(Itimes)))

Program Use

Source code for the program (INS.FOR) follows this section. The program is interactive, allowing the user to supply initial conditions. The user is also requested to supply the period for the simulation in seconds (*Enter integration time interval in second*) and the integration step size (*Enter output step size*). The period for simulation divided by the step size should be less than or equal to 1024. The program actually takes the step size and divides it by ten for a finer integration; however, only output values for whole-number multiples of the specified step size are returned. For instance, if the integration step size is 10 s, the integration is done in steps of one second, but only multiples of 10 s are returned.

Output information is returned to the screen in the form of a table: three velocity components (meters/second), latitude (phi), longitude (lambda), and navigation heading angle (all in degrees) are returned as a function of time into the problem, beginning at time zero. At the discretion of the user, a solution may also be directed to a file that also retains initial conditions (altitude and three components of acceleration); other initial conditions are displayed at time zero seconds into the problem. At the end of the run, the program queries the user to determine if another run is desired. The user is then given the option to enter new initial conditions or use the last computed values as the initial conditions. To use the last computed values (or any conditions previously entered), merely strike "/" and a carriage return when queried. If, at the beginning of the simulation session, the option to output to file was selected, additional runs are appended to the same file.

```
************************************************************
*       PROGRAM: INS.FOR                                    *
*       VERSION: I                                          *
*       CREATED: 14 NOVEMBER 1989                           *
************************************************************

CCCC    PURPOSE: Compute velocity and position for an INS from UNIPOLAR
C                NOMINAL MECHANIZATION equations. Integration is done
C                with a 4th order Runge-Kutta method and fixed steps.

CCCC    INPUTS:  DURATION of integration interval in seconds
C                Integration STEP size in seconds
C                Initial conditions: POSITION COSINES, VELOCITY,
C                                    POSITION, NAVIGATION HEADING ANGLE
```

```
CCCC  DEFINITION OF VARIABLES:
C          E          = eccentricity or flattening of earth;
C                       1/298.30
C          rho(3)     = angular rate in 3 directions
C          P(3,j)     = direction cosines in 3 coordinates; j=[1,n]
C          dPdt(3,4)= derivative of P(i,j) with respect to time;
C                       3 coordinates; 4 estimates per iteration
C          V(3,j)     = velocity in 3 coordinates; j=[1,n]
C          dVdt(3,4)= derivative of V(i,j) with respect to time;
C                       3 coordinates; 4 estimates per iteration
C          Pos(3,j)   = position in 3 coordinates; j=[1,n]
C          dPosdt(3,4)=derivative of Pos(i,j) with respect to
C                       time; 3 coordinates; 4 estimates per
C                       iteration
C          G(3)       = gravitational acceleration in 3 coordinates
C          A(3)       = inertial acceleration in 3 coordinates
C          Altitude = height above the surface of the earth
C          Phi(j)     = latitude in radians; j=[1,n]
C          Lambda(j)= longitude in radians; j=[1,n]
C          Alpha(j)   = navigation heading angle; j=[1,n]
C          delta      = interval used in Runge-Kutta method
C          timint     = total integration interval

CCCC  Note convention for 1st subscript: 1 => X-coordinate;
C     2 => Y-coordinate; 3 => Z-coordinate

CCCC  DEFINITION OF CONSTANTS
C            RADeq    = equatorial radius of the earth
C            GM       = gravitational constant
C            H2       = constant used to compute the gravitational
C                       acceleration (zonal harmonic)
C            H4       = constant used to compute the gravitational
C                       acceleration (zonal harmonic)
C            We       = earth's rotational rate in rad/s
C            conK     = constant used in computing rho values;
C                       function of RADeq

      IMPLICIT REAL (L)

      DIMENSION P(3,1024), dPdt(3,4), V(3,1024), Pos(3,1024),
     1 dPosdt(3,4), dVdt(3,4), Phi(1024), Lambda(1024), G(3), A(3),
     1 Alpha(1024), Phld(3), Vhld(3), Poshld(3)

      CHARACTER*1 TOFILE,AGAIN

      CHARACTER*64 FNAME

      DATA RADeq /6378137/, GM /3986005 E+8/, H2
     1 /-6.606185379 E+10/, H4 /2.452177656 E+21/,
     1 We /7.292115 E-5/, Pi /3.14168/
```

4.4. Standard Mechanization Equations

```
              WRITE(*,1997)
              WRITE(*,1998)
              READ(*,6003) TOFILE
              IF(TOFILE.EQ.'N'.OR.TOFILE.EQ.'n') GOTO 1
              WRITE(*6000)
              READ(*,6001) FNAME
              OPEN(1,FILE=FNAME,STATUS='NEW')

CCCC    Query for user supplied inputs:
C               1) Latitude (Phi) and longitude (Lambda)
C               2) Initial velocity
C               3) Acceleration
C               4) Altitude
C               5) Delta (fixed) for integration
C               6) Time period of integration

              INTVAL = 1

1             WRITE(*,1999)
              WRITE(*,2000)

              WRITE(*,4001)
              READ(*,*) E

              Phi(1) = Phi(INTVAL+1)
              WRITE(*,2001)
              READ(*,*) Phi(1)

              Lambda(1) = Lambda(INTVAL+1)
              WRITE(*,2002)
              READ(*,*) Lambda(1)

              Alpha(1) = Alpha(INTVAL+1)
              WRITE(*,2003)
              READ(*,*) Alpha(1)

              WRITE(*,2010)
              V(1,1) = V(1,INTVAL+1)
              WRITE(*,2011)
              READ(*,*) V(1,1)
              V(2,1) = V(2,INTVAL)
              WRITE(*,2012)
              READ(*,*) V(2,1)
              V(3,1) = V(3,INTVAL)
              WRITE(*,2013)
              READ(*,*) V(3,1)

              WRITE(*,2020)
              WRITE(*,2021)
              READ(*,*) A(1)
              WRITE(*,2022)
              READ(*,*) A(2)
              WRITE(*,2023)
              READ(*,*) A(3)
```

```
          WRITE(*,2030)
          WRITE(*,2031)
          READ(*,*) altitude

          WRITE(*,2039)
          WRITE(*,2040)
          WRITE(*,2041)
          READ(*,*) delta

          WRITE(*,2050)
          WRITE(*,2051)
          READ(*,*) timint

CCCC  Convert input longitude and latitude from degrees to radians

          Phi(1) = Phi(1)*Pi/180.
          Lambda(1) = Lambda(1)*Pi/180.
          Alpha(1) = Alpha(1)*Pi/180.

CCCC  Compute components for gravitational mass attraction
CCCC  Also compute initial positions in three coordinates
C         Z/RADeq = SIN(Phi(1))
C         X/RADeq = COS(Phi)*COS(Lambda)

          Z = RADeq * SIN(Phi(1))

          G(1) = (GM/RADeq**2)*(1. + (H2/RADeq**2)*(((5.*Z**2)/
        1 RADeq**2) - 1.) + (H4/RADeq**4)*(((63.*Z**4)/RADeq**4)
        1 -((42.*Z**2) + 3.))

          G(1) = G(1)*COS(Phi(1))*COS(Lambda(1))
          Pos(1,1) = RADeq*COS(Phi(1))*COS(Lambda(1))

C         Y/RADeq = COS(Phi(1))*SIN(Lambda(1))

          G(2) = (GM/RADeq**2)*(1. + (H2/RADeq**2)*(((5.*Z**2)/
        1 RADeq**2) - 1.) + (H4/RADeq**4)*(((63.*Z**4)/
        1 - ((42.*Z**2)/RADeq**2) + 3.))

          G(2) = G(2)*COS(Phi(1))SIN(Lambda(1))
          Pos(2,1) = RADeq*COS(Phi(1))*SIN(Lambda(1))

C         Z/RADeq = SIN(Phi(1))

          G(3) = (GM/RADeq**2)*(1. + (H2/RADeq**2)*(((5.*Z**2)/
        1 RADeq**2) - 3.) + (H4/RADeq**4)*(((63.*Z**4)/RADeq**4)
        1 -((70.*Z**2)/RADeq**2) + 15.))

          G(3) = G(3)*SIN(Phi(1))
          Pos(3,1) = Z
```

4.4. Standard Mechanization Equations

```
CCCC  Compute Rew, Rns, and conK; Phi is the aircraft's latitude

      Rns = RADeq*(1. - E**2)/SQRT((1. - (E**2)*((SIN(Phi(1)))
     1 **2))**3 )

      Rew = RADeq/SQRT(1. -(E**2)*(SIN(Phi(1)))**2)
      conK= (E**2)*Rns/((Rew + altitude)*(Rns + altitude)*
     1 (1. - E**2))

C     Note: Rns and Rew will remain constant in the rho equations
C     though they involve Phi.

CCCC  Direction Cosines

      P(1,1) =  COS(Phi(1))*COS(Alpha(1))
      P(2,1) = -COS(Phi(1))*SIN(Alpha(1))
      P(3,1) =  SIN(Phi(1))

CCCC  Compute velocity and position for the required interval

      intval = IFIX(timint/delta)
      IF(intval.GT.1024) intval = 1024

CCCC  Initialize hold values

      DO 288 I = 1,3
      Poshld(I) = Pos(I,1)
      Phld(I) = P(I,1)
      Vhld(I) = V(I,1)

288   CONTINUE

      DO 1000 Itimes = 2, INTVAL+1

CCCC  Compute all necessary derivatives for fourth order
C     Runge-Kutta method

      DO 301 J = 1,10
      DO 300 Nslope = 1,4

         DELTeq= .5*delta/10.
         IF(Nslope.EQ.1) DELTeq = 0.
         IF(Nslope.EQ.4) DELTeq = delta/10.

         Px = Phld(1) + DELTeq*dPdt(1,Nslope)
         Py = Phld(2) + DELTeq*dPdt(2,Nslope)
         Pz = Phld(3) + DELTeq*dPdt(3,Nslope)

         Vx = Vhld(1) + DELTeq*dVdt(1,Nslope)
         Vy = Vhld(2) + DELTeq*dVdt(2,Nslope)
         Vz = Vhld(3) + DELTeq*dVdt(3,Nslope)
```

```
              rhox = -conK*Py*(Px*Vx + Py*Vy) - Vy/(Rew + altitude)
              rhoy =  conK*Py*(Px*Vy - Py*Vx) + Vx/(Rns + altitude)
              rhoz =  (Px*Vy - Py*Vx)/(Rew + altitude)*(Pz + 1)

              dPdt(1,Nslope) = Py*rhoz - Pz*rhoy
              dPdt(2,Nslope) = Pz*rhox - Px*rhoz
              dPdt(3,Nslope) = Px*rhoy - Py*rhox

              dVdt(1,Nslope) = A(1) + G(1) + (rhoz + 2*We*Pz)*Vy -
             1 (rhoy + 2*We*Py)*Vz
              dVdt(2,Nslope) = A(2) + G(2) + (rhox + 2*We*Px)*Vz -
             1 (rhoz + 2*We*Pz)*Vx
              dVdt(3,Nslope) = A(3) + G(3) + (rhoy + 2*We*Py)*Vx -
             1 (rhox + 2*We*Px)*Vy

              dPosdt(1,Nslope) = Vx
              dPosdt(2,Nslope) = Vy
              dPosdt(3,Nslope) = Vz

300   CONTINUE

CCCC  At this point, all necessary derivative approximations have
C     been computed. Direction cosines, velocity, and position are
C     now computed with a fourth order Runge-Kutta method.
              DO 600 I = 1,3

              Phld(I) = Phld(I) + (.1*delta/6.)*(dPdt(I,1)
             1 + 2.*dPdt(I,2) + 2.*dPdt(I,3) + dPdt(I,4))
              Vhld(I) = Vhld(I) + (.1*delta/6.)*(dVdt(I,1)
             1 + 2.*dVdt(I,2) + 2.*dVdt(I,3) + dVdt(I,4))
              Poshld(I) = Poshld(I) + (.1*delta/6.)*
             1 (dPosdt(I,1) + 2*dPosdt(I,2) + 2*dPosdt(I,3) + dPosdt(I,4))

600   CONTINUE

              DO 601 I = 1,3

              P(I,Itimes)   = Phld(I)
              V(I,Itimes)   = Vhld(I)
              Pos(I,Itimes) = Poshld(I)

601   CONTINUE

301   CONTINUE

CCCC  Latitude (Phi) and longitude (Lambda) are now computed using
C     identities commented in gravitational component computations.
C     Pos(1,Itimes), Pos(2,Itimes), and Pos(3,Itimes) contain X,Y,Z.
C     The navigational heading angle is computed using a direction
C     cosine and the computed value for Phi.

              IF(Pos(3,Itimes).LE.RADeq) GOTO 998
```

4.4. Standard Mechanization Equations

```
            WRITE(*,4002)
            INTVAL = Itimes
            GOTO 1001

998         Phi(Itimes)    = ASIN(Pos(2,Itimes)/RADeq)
            Lambda(Itimes) = ATAN(Pos(2,Itimes)/Pos(1,Itimes))
            Alpha(Itimes)  = ACOS(P(1,Itimes)/COS(Phi(Itimes)))

CCCC    Computations are complete for current point; if less than 1024
C       continue in computing loop.

1000    CONTINUE

CCCC    Output velocity and position data to the screen.

1001        WRITE(*,2060)
            WRITE(*,2061)
            DO 1200 I = 1,INTVAL+1

CCCC    Convert latitude and longitude from radians to degrees.

            Phi(I)   = Phi(I)*180./Pi
            Lambda(I)= Lambda(I)*180./Pi
            Alpha(I) = Alpha(I)*180./Pi

            Idel = IFIX((I-1)*delta)
            WRITE(*,2062) Idel, V(1,I), V(2,I), V(3,I), Phi(I),
           1 Lambda((I), Alpha(I)

1200    CONTINUE

            IF(TOFILE.EQ.'N'.ORTOFILE.EQ.'n') GOTO 1202

CCCC    WRITE DATA TO FILE

            WRITE(1,2063) ALTITUDE
            WRITE(1,2064) A(1), A(2), A(3)
            WRITE(1,2060)
            WRITE(1,2061)
            DO 1201 I = 1,INTVAL+1
            Idel = IFIX((I-1)*delta)

            WRITE(1,2062) Idel, V(1,I), V(2,I), V(3,I), Phi(I),
           1 Lambda(I), Alpha(I)

1201    CONTINUE

            IF(Pos(3,INTVAL).LE.RADeq) GOTO 1202
            WRITE(1,4002)

CCCC    OPTION GIVEN TO CONTINUE OR END
```

```
1202    WRITE(*,5000)
        WRITE(*,5001)
        WRITE(*,5002)
        READ(*,6003) AGAIN
        IF(AGAIN.EQ.'Y'.OR.AGAIN.EQ.'y') GOTO 1
        IF(TOFILE.EQ.'N'.OR.TOFILE.EQ.'n') GOTO 9999
        CLOSE(1)
9999    STOP

1997    FORMAT (' Default output to screen.', )
1998    FORMAT (' Do you want output to file (Y/N)?, )

1999    FORMAT (' Enter new value or "/<CR>" for next prompt.', )

2000    FORMAT (' Initialize coordinates in degrees:', )
2001    FORMAT (' Latitude > ')
2002    FORMAT (' Longitude > ')
2003    FORMAT (' Navigation heading angle > ')

2010    FORMAT (' Initialize velocities (3 coordinates) in m/sec:', )
2011    FORMAT (' X-component of velocity > ')
2012    FORMAT (' Y-component of velocity > ')
2013    FORMAT (' Z-component of velocity > ')

2020    FORMAT (' Initialize accel (3 coordinates) in m/sec**2:', )
2021    FORMAT (' X-component of acceleration > ')
2022    FORMAT (' Y-component of acceleration > ')
2023    FORMAT (' Z-component of acceleration > ')

2030    FORMAT (' Enter altitude in meters: ', )
2031    FORMAT (' Altitude > ')

2039    FORMAT (' In one run, integration interval <=1024.', )
2040    FORMAT (' Enter output step size: ', )
2041    FORMAT (' Step size > ')

2050    FORMAT (' Enter integration time interval in seconds: ', )
2051    FORMAT (' Interval > ')

2060    FORMAT (' TIME',4X,'Vx',7X,'Vy',7X,'Vz',7X,'Phi',7X,'Lambda',
       1 7X,'Alpha')
2061    FORMAT (' sec',4X,'m/sec',5X,'m/sec',5X,'m/sec',5X,'degrees',
       1 5X,'degrees')
2062    FORMAT (1X,I4,3X,F9.4,2X,F9.4,2X,F9.4,2X,F9.5,2X,F9.5,2X,F9.5)
2063    FORMAT ('Initial altitude:'2X,F9.4,1X,'meters')
2064    FORMAT ('Initial acceleration in m/sec/sec:',2X,'Ax = ',F5.1,
       1 2X,'Ay = ',F5.1,2X, 'Az = ',F5.1)

4001    FORMAT (' Enter eccentricity (1/298.30) as decimal > ', )
4002    FORMAT (' ERROR! Cannot compute Phi (ASIN > 1).', )

5000    FORMAT (' You have the option to continue .', )
5001    FORMAT (' You may reset the initial conditions or continue with
       1 the latest values.', )
5002    FORMAT (' Do you wish to continue (Y/N)? ')
```

```
6000    FORMAT (' Enter file name to store output: ')
6001    FORMAT (A64)
6003    FORMAT (A)

        END
```

4.4.3. Space-Stabilized Mechanization

The essential difference between a local-vertical and space-stabilized platform configuration is the absence of platform torquing on the space-stabilized case. As a result, the error equations for a space-stabilized configuration are simpler than those of the local-vertical system. Moreover, for a space-stabilized platform mounted in a vehicle, the accelerometers attached to the platform will measure accelerations in a coordinate system fixed in inertial space. Specifically, in the space-stabilized mechanization, the accelerometer triad is held in an earth-centered nonrotating inertial frame by a three-axis gyroscopic stabilized platform. In the space-stable mechanization, an inertial rather than a geographic reference frame is used for the navigation computations. As in every "all attitude" gimbaled inertial navigation system, four gimbals are normally required to isolate the inertial platform from vehicle angular motion. In the navigate mode, the gyroscopes are untorqued, or at most they are torqued at a very low level in order to compensate for the known gyroscope drift rates. That is, the inertial platform is uncommanded, so within the limits imposed by the gyroscope drift, the platform will remain inertially nonrotating. Stated in another way, the commanded platform inertial angular velocity is equal to the desired platform angular velocity, which is equal to zero ($\omega_I^P = 0$). Furthermore, we have seen in Chapter 2 that the gyroscope and platform axes are nonorthogonal and are related by a small-angle transformation. Therefore, since the platform rotation is very small, it can be said that the angular velocity of the gyroscope frame ω_I^G is equal to the angular velocity of the platform, ω_I^P. Thus, $\omega_I^G = \omega_I^P$. Theoretically, the platform axes could be aligned with the axes of any inertially nonrotating frame, it will be assumed here that the I-frame is being instrumented [10].

Space-stable navigation systems have been used since the early 1950s. Space stabilized systems are mostly used in spacecraft and missile platform mechanizations, since in these applications no geographic navigation information is needed. Also, space-stable navigation systems are used in some terrestrial aircraft and marine navigation applications. Honeywell's Standard Precision Navigator/Gimbaled Electrically Suspended Gyro Airborne Navigation System (SPN/GEANS) is one example of an all-attitude terrestrial space stabilized aircraft INS mechanization. SPN/GEANS employs four primary coordinate frames for the navigation and attitude computations: (1) the inertial reference frame, which is the computational frame; (2) the local (geodetic) vertical north frame; (3) the body or vehicle frame; and (4) the platform space-stabilized frame (the platform reference axes have a fixed orientation with respect to the computational frame axis).

When the platform axes are aligned with the axes of the I-frame, the outputs of the three platform mounted accelerometers with their sensitive axes aligned with the platform axes would be the three components of

$$\mathbf{A}^A = C_I^A \ddot{\mathbf{R}}^I - \mathbf{g}^A$$

where C_I^A is a coordinate transformation matrix that relates the inertial axes to the accelerometer axes A, $\ddot{\mathbf{R}}^I$ is the inertially referenced acceleration, and $\mathbf{g}^A$ is the gravitational field acceleration, and $\mathbf{A}^A$ is the specific force. The position vector components are sent to the gravitation computer (see Fig. 4-2), which computes $\mathbf{g}^I$ as a function of $\mathbf{R}^I$ using, as we have noted, a complex expression for the gravitational potential. Moreover, the position vector is also sent to the navigation computer, which uses $\mathbf{R}^I$ to compute geographic latitude and longitude. The computation done by the computer can be easily understood if it is noted that

$$\mathbf{R}^I = C_I^n \mathbf{R}^n = C_I^E C_E^n \mathbf{R}^n$$

where C_E^n is the earth-to-navigational (or computational) frame transformation. For a spherical earth model, with coordinate axes arrangement as indicated in Fig. 4-3, the earth's geocentric position vector is expressed by

$$\mathbf{R}^I = \begin{bmatrix} R_{Ix} \\ R_{Iy} \\ R_{Iz} \end{bmatrix} = \begin{bmatrix} R \cos \phi \cos (\lambda + \Omega t) \\ R \cos \phi \sin (\lambda + \Omega t) \\ R \sin \phi \end{bmatrix}$$

This equation states that for a spherical earth model, the I-frame and the E-frame coincide at the time $t = 0$. Therefore

$$\Lambda = \tan^{-1} \left(\frac{R_{Iy}}{R_{Ix}} \right) \qquad (4.77)$$

where $\Lambda = (\lambda - \lambda_0) + \Omega t$. Consequently, the geographic longitude can be obtained from

$$\lambda = \lambda_0 + \tan^{-1} \left(\frac{R_{Iy}}{R_{Ix}} \right) - \Omega t \qquad (4.78)$$

where λ_0 is the longitude east of the Greenwich Meridian at $t = 0$. From Eq. (4.78) we note that since it contains the sidereal hour angle term Ωt, it can be seen that the space-stabilized system's computer must contain a time

4.4. Standard Mechanization Equations

reference. The geocentric latitude can be obtained from

$$\sin \phi = \frac{R_{Iz}}{|\mathbf{R}^I|} \tag{4.79}$$

From Eq. (4.79), we can write the geographic latitude in terms of the deviation of the normal δ as

$$\phi = \sin^{-1}\left(\frac{R_{Iz}}{|\mathbf{R}^I|}\right) + \delta \tag{4.80}$$

where

$$\delta = f \sin 2\phi_g$$

The primary input to the space navigation is velocity increment data in inertial reference (space-stable) coordinates. The velocity increment data when divided by the computation interval can be viewed as an equivalent average sensed acceleration. Standard position outputs, that is, geodetic latitude and longitude, are derived by performing inverse trigonometric functions on the inertial Cartesian position components. Moreover, these angles are also used for a transformation relating local-level north coordinates to inertial reference coordinates. Consequently, this transformation is used in deriving earth-relative velocity in local-level north coordinates. However, this velocity is subject to singularities at the earth's poles, but recovers after the vehicle moves away from the pole because the basic Schuler loop is mechanized in inertial Cartesian coordinates. Stated another way, performing the computations in inertial Cartesian coordinates, singularities (internal to the Schuler loop) at the earth's poles are avoided. A simplified block diagram for space-stable navigation is shown in Fig. 4-7.

It may be noted that position and velocity data that are furnished to the Kalman filter module are compared with the reference position and velocity data to generate an observation residual during airborne alignment and aided navigation. Position and velocity corrections are accepted from the filter upon processing of the observation residuals. The function of the Kalman filter will be explained more fully in a later chapter. For the space-stable navigation, consider now the coordinate frame illustrated in Fig. 4-8. The Z_I axis is parallel to the earth's North Polar axis, while the Y_I axis is defined as being parallel to east at the start of the navigation mode. This coordinate frame is used as the reference for the inertial integrations. Consequently, the resulting values of the X_I, Y_I, Z_I are the coordinates of vehicle position relative to the center of the earth.

Assuming that the wander angle $\alpha = 0$, the relationship between the local-level-north coordinates and the inertial coordinates can be obtained as follows: (1) a rotation about the Z_I axes through the sidereal hour angle

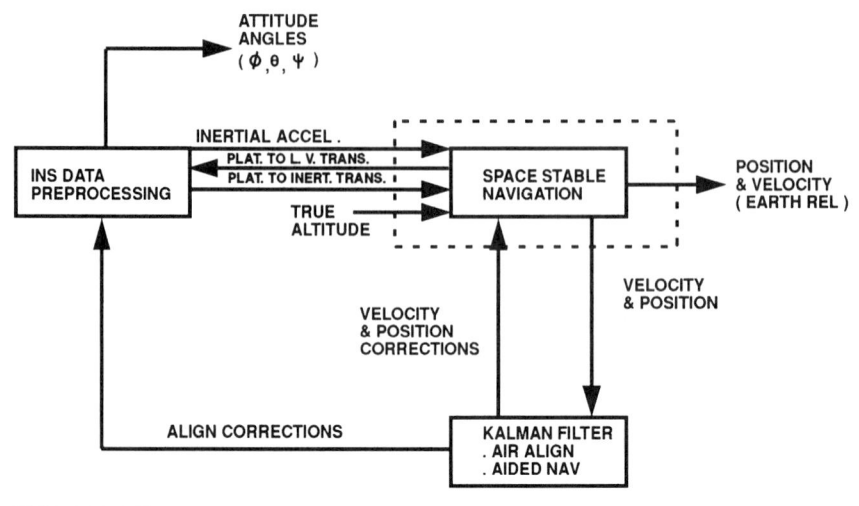

FIGURE 4-7
Simplified space-stable navigation block diagram.

rotation of the earth (Ωt) plus the change in longitude ($\lambda - \lambda_0$) from the start of navigation and (2) a negative rotation about the Y_I axis through the geodetic latitude angle ($-\phi$) to obtain a vertical, east, north coordinate set. Thus

$$\begin{bmatrix} U \\ E \\ N \end{bmatrix} = [Y(-\phi)][Z(\lambda - \lambda_0 + \Omega t)] \begin{bmatrix} X_I \\ Y_I \\ Z_I \end{bmatrix}$$

$$= \begin{bmatrix} \cos\phi \cos\Lambda & \cos\phi \sin\Lambda & \sin\phi \\ -\sin\Lambda & \cos\Lambda & 0 \\ -\sin\phi \cos\Lambda & -\sin\phi \sin\lambda & \cos\phi \end{bmatrix} \begin{bmatrix} X_I \\ Y_I \\ Z_I \end{bmatrix}$$

where Λ is the change in celestial longitude given by $\Lambda = (\lambda - \lambda_0) + \Omega t$. At $t = 0$, $\Lambda = \lambda + \Omega t$. In terms of the equatorial components of inertial position X_I and Y_I, Λ is given by

$$\Lambda = \tan^{-1}\left(\frac{Y_I}{X_I}\right) \tag{4.81}$$

which is similar to Eq. (4.77). The geodetic longitude can therefore be computed from

$$\lambda = \lambda_0 + \Lambda - \Omega t$$

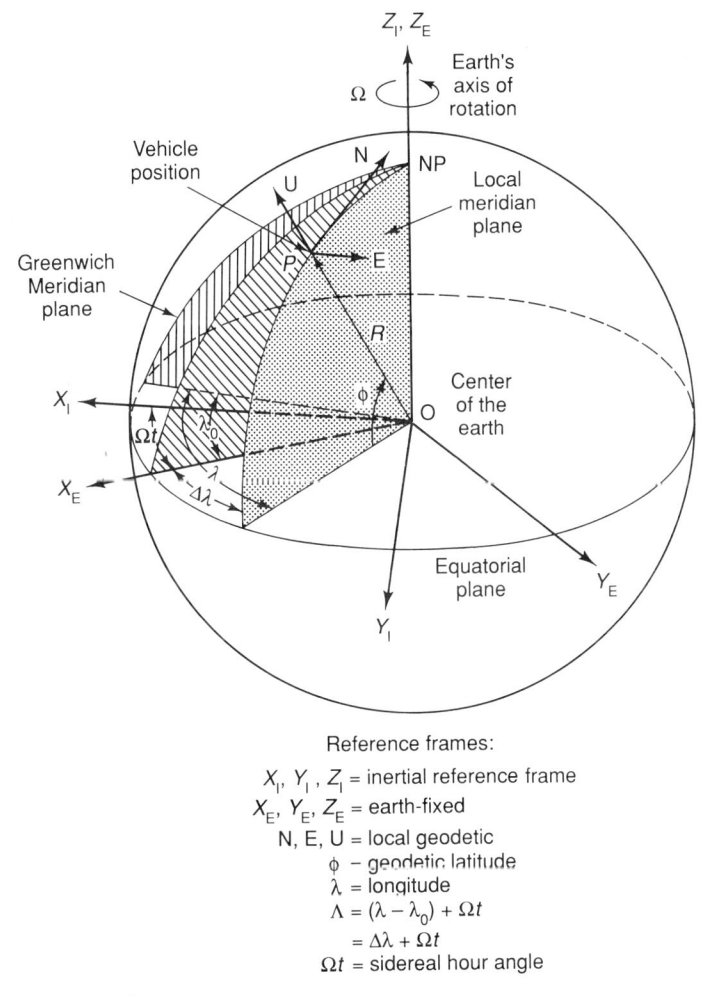

4.4. Standard Mechanization Equations

FIGURE 4-8
Coordinates for space-stable navigation.

Reference frames:
X_I, Y_I, Z_I = inertial reference frame
X_E, Y_E, Z_E = earth-fixed
N, E, U = local geodetic
ϕ = geodetic latitude
λ = longitude
$\Lambda = (\lambda - \lambda_0) + \Omega t$
$\quad = \Delta\lambda + \Omega t$
Ωt = sidereal hour angle

The geodetic latitude ϕ is computed from

$$\phi = \tan^{-1}\left[\left(\frac{1}{1-\varepsilon^2}\right)\frac{Z_I}{\sqrt{X_I^2 + Y_I^2}}\right] \qquad (4.82)$$

where ε is the eccentricity of the meridional ellipse. Next, we note that for a space-stable inertial navigation system using a local vertical wander–azimuth computational (or navigational) frame, the platform frame is related to the computational frame by a transformation that is expressed as the product of three rotation matrices. Mathematically, this can be written in the form [10]

$$C_P^C = C_E^C C_I^E C_P^I$$

An explanation of these matrices is in order. First, the platform to inertial frame transformation C_P^I is constant since we have considered the platform to be nondrifting. Also C_P^I is constant in the navigate mode. Its initial value depends on the alignment method that is mechanized. Second, the transformation from the inertial to the earth-fixed frame C_I^E is realized by a single-axis rotation $Z(\Omega t)$. This accounts, as we have seen earlier, for the rotation of the earth from the start of navigation until the start of the present navigation computation cycle. Therefore, its initial value is I. Third, the transformation from the earth-fixed frame to the computational frame C_E^C is the transpose of the direction cosine matrix computed in the local vertical wander–azimuth algorithm. In order to minimize the computations required, the incremental velocities $\Delta \mathbf{V}^P$ accumulated over one sample interval in the space-stable platform frame are premultiplied successively by each of the three individual rotation matrices C_P^I, C_I^E, and C_E^C to obtain $\Delta \mathbf{V}^C$; for example, $\Delta \mathbf{V}^C = C_P^C \Delta \mathbf{V}^P$. In space-stable navigation, the basic computation cycle for the navigation equations is usually 8 Hz. However, lower iteration rates on the order of 4 Hz are also used, particularly in the presence of mild vehicle dynamics. On the other hand, higher rates, say, 16 Hz, present no problem if numeric precision is sufficient. The space-stable navigation module of Fig. 4-7 will now be expanded as shown in Fig. 4-9, explaining in more detail the various computations that take place in space-stable navigation.

Compensated sensed acceleration data A_X, A_Y, A_Z in inertial reference coordinates enters the inertial integrations block. The gravity vector (i.e., the earth's mass attraction) is computed by the gravity computer, which also includes the effect of known or estimated gravity deflections. As a result, and from the basic equation of motion [Eq. (4.13)], the effect of the mass attraction of the earth is subtracted from the sensed acceleration. The remaining acceleration, due to vehicle motion with respect to inertial space, is effectively integrated twice in inertial reference coordinates. The output of the first integrator is inertial relative velocity, while the output of the second integration is position relative to the center of the earth in Cartesian coordinates. The position thus obtained is used in calculating the mass attraction for the next computation cycle. In effect, this forms the basic Schuler loop commonly associated with all inertial navigation systems [8]. Next, the vertical position damping module furnishes compensation terms in order to damp the unstable vertical channel. Conventional vertical loop damping, that is, fixed gains with feedback inputs to acceleration and velocity, is performed using a baroaltimeter as a reference. The position and velocity

4.4. Standard Mechanization Equations

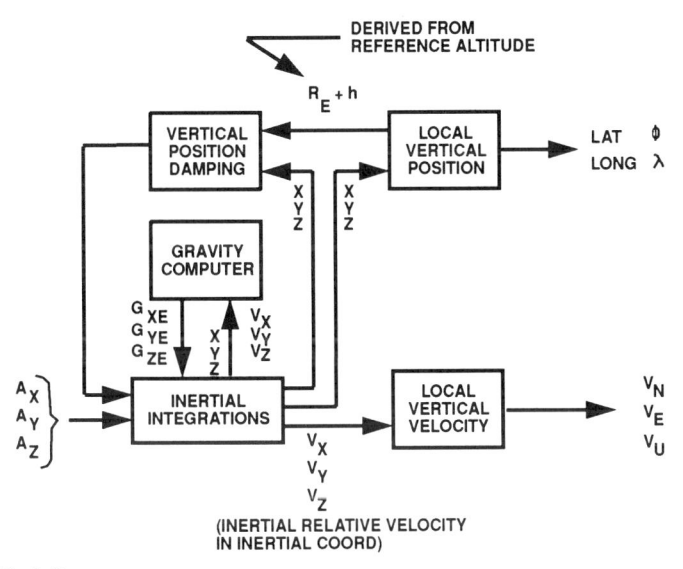

FIGURE 4-9

Basic elements of space-stable navigation.

components from the inertial integration module are used by the local vertical position and velocity modules to compute the output navigation parameters. The former module computes the geodetic latitude and longitude from the position vector, while the latter module uses the inertial position and velocity in local coordinates. Inertial-to-local vertical north transformation is used in this computation. Note that a transformation from the platform inner element to the local vertical north, which is required in the computation of vehicle attitude by the INS data processing module, is also computed in the local vertical velocity module. Finally, the INS data preprocessing module provides the input A_X, A_Y, A_Z (average acceleration) over the computation interval with respect to the inertial reference coordinates. Also, transformation from inertial reference space to local vertical north reference space will be an output from this module. The discussion presented above can also be followed by referring to Fig. 4-10.

The space-stabilized system has the following advantages:

1. Errors resulting from the gyroscopes torque generator are avoided, since no platform torquing is required.
2. Since the computations are carried out in an inertially nonrotating frame, no Coriolis or other accelerometer computation signals are required.

The space-stabilized system has the following disadvantages:

1. The gravitational field must be computed explicitly.

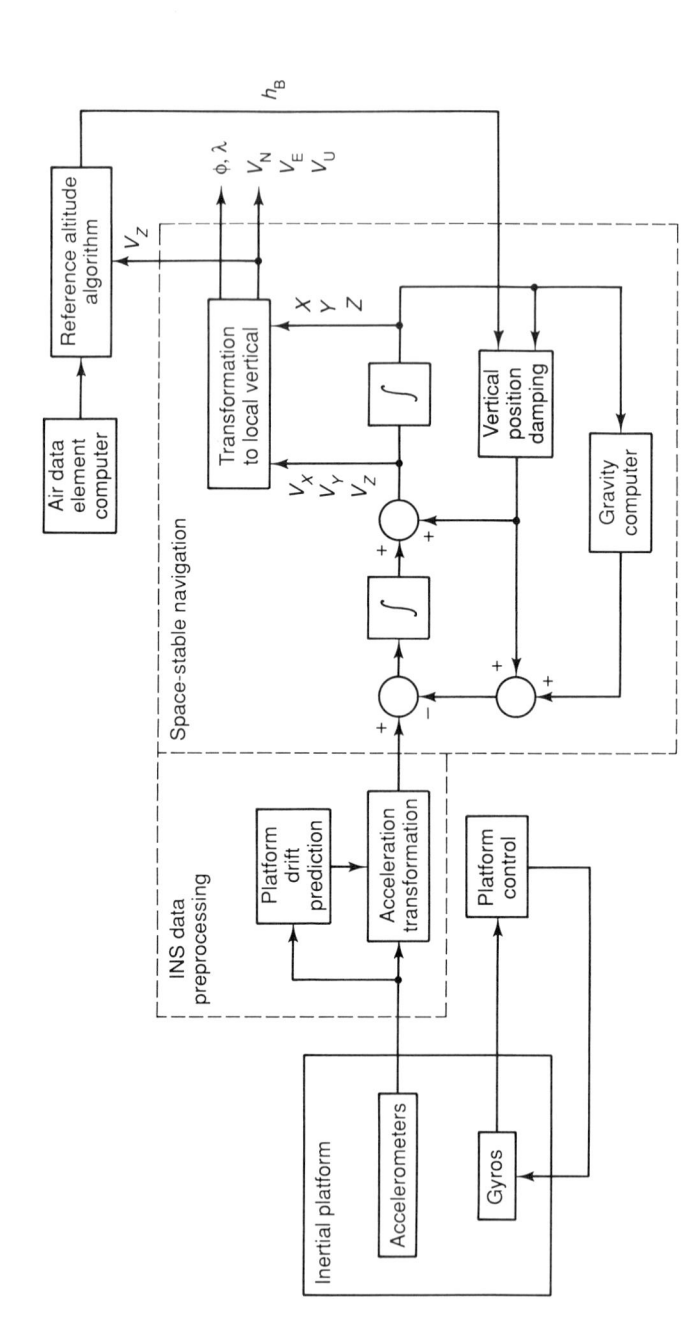

FIGURE 4-10
Generic space-stable navigation block diagram.

2. The system does not give earth-referenced velocity (ground speed) as an output.
3. The system requires considerably more computation than local-vertical systems.
4. Position, velocity, and heading errors become unbounded.
5. The system requires vertical position damping, since vertical velocity and position integrators may not be actually present.

Local-level systems have a bounded latitude error, but the space-stable system develops a latitude error whose amplitude is characteristic of the 24-h mode increasing linearly with time (i.e., the product of angular velocity and time). The discussion of this subsection can be summarized by noting that for a pure or free (one that is not aided by external navaids) inertial navigation system, the following hold:

1. Navigation calculations are a basic space-stable mechanization.
2. The platform does drift; however, its rate is predictable.
3. An alignment matrix is needed that indicates the position of the platform coordinate frame relative to a space-stable coordinate system.
4. The accelerometer data is processed through a transformation matrix which transforms it to the computational inertial frame.
5. A barometric altitude is needed to damp the Schuler oscillations.
6. The space-stable navigation system error dynamics are identical to those of the local-level navigation system. However, it should be noted that the dynamics of the inertial sensor errors which drive both systems are different, that is, resulting in different system error propagation characteristics.

4.4.3.1. Error Equations for the Space-Stable System

We have seen in Section 4.3 [Eq. (4.12b)] that the specific force measured by the accelerometers is

$$\mathbf{A} = [d^2\mathbf{R}/dt^2]_I - \mathbf{g}_m(\mathbf{R})$$

From this relationship, one obtains the perturbation or error equation

$$\Delta \mathbf{A} = \Delta \ddot{\mathbf{R}} - \Delta \mathbf{g} \tag{4.83}$$

where $\Delta \mathbf{R}$ and $\Delta \mathbf{g}$ are the errors in computed position and gravitational acceleration, respectively. Now assume an inertial coordinate system (x, y, z) with origin at the earth's center. Furthermore, it will be assumed that the z axis is the earth's spin axis and the x, y-axes define the equatorial plane. The gravitational acceleration, assuming a spherical earth model, is given by

$$\mathbf{g} = -g_0 \frac{R_0^2}{|\mathbf{R}|^3} \mathbf{R} = -g_0 \left(\frac{R_0^2}{R^3}\right) \begin{bmatrix} x \\ y \\ z \end{bmatrix} \tag{4.84}$$

where $R = R_0 + h$ (R_0 is the radius of the earth and h is the altitude) and g_0 is the gravitational acceleration at the surface of the earth. Since R has components (x, y, z), then

$$R^2 = \mathbf{R}^T\mathbf{R} = [x \quad y \quad z]\begin{bmatrix} x \\ y \\ z \end{bmatrix} = x^2 + y^2 + z^2$$

The error in the computed gravitational acceleration due to an error in computed position $\Delta\mathbf{R}$ is

$$\Delta\mathbf{g} = -g_0 \frac{R_0^2}{R^3} \Delta\mathbf{R} + 3g_0 \frac{R_0^2}{R^5}(\mathbf{R} \cdot \Delta\mathbf{R})\mathbf{R} \qquad (4.85)$$

where $\Delta\mathbf{R} = [\Delta x \quad \Delta y \quad \Delta z]^T$. Similarly, the error in the measured acceleration $\Delta\mathbf{A}$ is $\Delta\mathbf{A} = [\Delta A_x \quad \Delta A_y \quad \Delta A_z]^T$. Substituting Eq. (4.85) into Eq. (4.83) and rearranging, we obtain [8]

$$\Delta\ddot{\mathbf{R}} + g_0 \frac{R_0^2}{R^3} \Delta\mathbf{R} - 3g_0 \frac{R_0^2}{R^5}(\mathbf{R} \cdot \Delta\mathbf{R})\mathbf{R} = \Delta\mathbf{A} \qquad (4.86)$$

In component form, Eq. (4.86) can be written as [9]

$$\Delta\ddot{x} + g_0\left(\frac{R_0^2}{R^3}\right)\Delta x - 3g_0\left(\frac{R_0^2}{R^5}\right)x(x\Delta x + y\Delta y + z\Delta z) = \Delta A_x \qquad (4.87a)$$

$$\Delta\ddot{y} + g_0\left(\frac{R_0^2}{R^3}\right)\Delta y - 3g_0\left(\frac{R_0^2}{R^5}\right)y(x\Delta x + y\Delta y + z\Delta z) = \Delta A_y \qquad (4.87b)$$

$$\Delta\ddot{z} + g_0\left(\frac{R_0^2}{R^3}\right)\Delta z - 3g_0\left(\frac{R_0^2}{R^5}\right)z(x\Delta x + y\Delta y + z\Delta z) = \Delta A_z \qquad (4.87c)$$

Equations (4.87a)–(4.87c) represent a set of three simultaneous linear second-order differential equations which may be solved to yield the relationship between velocity errors $[\Delta\dot{x} \quad \Delta\dot{y} \quad \Delta\dot{z}]$ and position errors $[\Delta x \quad \Delta y \quad \Delta z]$ and the errors in sensed acceleration. The accelerometer errors ΔA_x, ΔA_y, ΔA_z represent forcing functions in the system equations of motion. Therefore, the complete set of Eqs. (4.87a)–(4.87c) will contain forcing functions due to gyroscopic drift, accelerometer errors, and platform misalignments.

Finally, it can be shown that for white noise (i.e., flat power spectral density) the horizontal channel errors will grow with $\sqrt{t}$. This is less severe than the linear error growth associated with a constant gyroscope drift rate bias.

4.5. THE VERTICAL CHANNEL

In pure inertial navigation systems using three accelerometers, vertical velocity and position are commonly derived by integrating the vertical channel accelerometer output. It is well known that this output is inherently divergent (unstable). This characteristic is commonly corrected by using an externally measured altitude reference (e.g., barometric or radar altimeter) to stabilize the INS vertical channel. Specifically, the vertical velocity V_z cannot be obtained directly from Eq. (4.25c) since this equation is basically unstable, necessitating a baroinertial loop to damp this equation. Thus, because of the unbounded errors in latitude and longitude possible in such systems, recourse is frequently made to the use of external inputs to update the position information. A series of external fixes may also be used to provide damping of the error-induced oscillations characteristic in all inertial navigation systems.

The vertical channel is damped with measurements of elevation above mean sea level. The measured elevation is obtained from the Central Air Data Computer (CADC) outputs. Subsequently, the prime CADC free air temperature (T_{FAT}) and static pressure (P_S) are used with the data from the prime system in order to compute a reference altitude. This algorithm is then provided to the INS. The system altitude can be initialized either manually or automatically with altitude derived in the CADC from a standard atmosphere model. After initialization, CADC static pressure (P_S) and free-air temperature (T_{FAT}) are used to adjust the altitude output for non-standard-atmosphere conditions. This is implemented with an algorithm that computes a corrected altitude by numerically integrating the physical relation for a column of air as a function of pressure, gravity, and absolute temperature. Barometric and inertial altitude systems are complementary to each other. The former provides good altitude rate information in nearly level flight [subject only to the slope of the isobars in the vertical plane, which is typically 0.3048 m per kilometer (or 1 ft per mile) standard deviation] but is poor in climbs or dives or in the presence of significant vertical acceleration. On the other hand, the INS needs to be bounded by an external reference for sustained periods, but provides direct information about vertical acceleration and a good short-term reference for use during climbing or diving in high-performance (fighter) aircraft. In a prolonged ascent and/or descent, significant error develops in the barometric data with the INS closely following it. Specifically, climbing or diving through a non-standard atmosphere will result in height errors of up to 8%. Also, the time constant associated with a barometric altimeter is very large, which means that through prolonged ascent or turns, the vertical channel inherits an error that can persist for as long as 2 min. This error can degrade weapon delivery, especially on reattack.

Consequently, the vertical error is critical during weapon delivery because this error greatly affects the miss distance of the weapon on a target. In cases where the external velocity is of good quality, the inertial navigation system's velocities may be slaved to the external data. In order to avoid the vertical channel instability problem, the INS vertical channel is frequently not implemented, such as in the Aeronautical Radio Incorporated 561 (or ARINC 561) commercial aircraft systems. Therefore, as stated above, in systems for high-performance aircraft, the vertical channel is usually slaved to barometric height data. Typically, in conventional local-level mechanizations, the difference between the inertial and altimeter indications of altitude is, as we shall presently see, fed back to the input of the vertical velocity and position integrators. Therefore, by combining independent altimeter data with inertial altitude data, vertical channel errors can be damped and bounded.

Consideration of the implementation of the vertical INS channel is important because of the impact on the propagation of other sources of errors. In fact, altitude is a significant factor in the earth radii used in computing latitude and longitude rates. Also, altitude rates can contribute significant Coriolis acceleration terms in combination with northerly torque rates, particulary where the aircraft has significant easterly velocity. In cases where the vertical inertial channel is not mechanized, barometric altitude is commonly used. Some systems assume a fixed cruising altitude, or an altitude determined as a function of speed.

As we shall discuss in Chapter 6, in certain mechanizations the atmosphere must be modeled stochastically. In arriving at a stochastic mathematical model of the atmosphere, the following error sources must be considered: (1) random instrument noise, (2) instrument lags (e.g., in the CADC), and (3) effects of winds. The most important source of error in the vertical channel is the fluctuation in the altimeter bias (due to scale factor error and nonzero vertical velocity). The second most significant source of error is the short correlation time acceleration error (e.g., due to specific force measurement error during a maneuver). Theoretically, the vertical velocity error can be reduced by optimizing the loop gains. More specifically, the baroaltimeter exhibits a scale factor error due to atmosphere not having a standard-day temperature–altitude profile. In climbs and dives the scale factor induces significant vertical velocity error. Since the beginning of flight, one of the most important parameters to the pilot has been the measure of the aircraft's height above the ground. Therefore, at this point, certain definitions relating to the altitude are in order [14]:

Absolute altitude (H_{abs}) The height above the surface of the earth at any given surface location. The vertical clearance between an airplane and a mountain top is an example of absolute altitude (see Fig. 4-11).

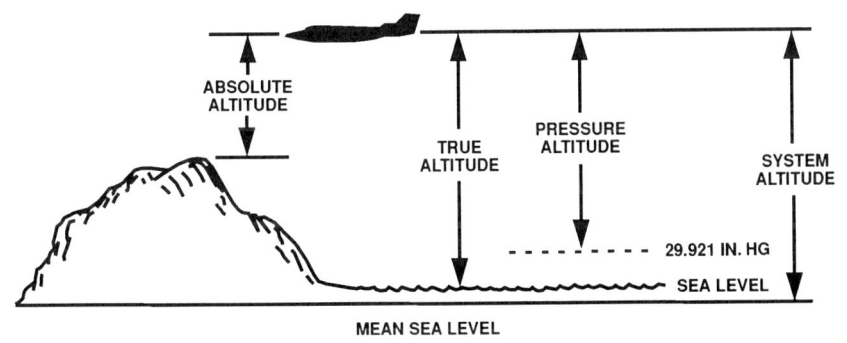

4.5. The Vertical Channel

FIGURE 4-11
Types of altitude measurement.

True altitude (H_{true}) The actual height above the standard sea level. True altitude is the sum of absolute altitude and the elevation above sea level of the ground below the aircraft.

Pressure altitude (H_p) The height in a model atmosphere above the standard pressure datum plane (sea level) of 29.921 in. (760 mm) of mercury. Pressure altitude differs from the true altitude in that pressure altitude does not consider the high and low pressure or temperature variations which occur constantly on the surface of the earth due to changing weather phenomena. When flying in a high-pressure area, the indicated pressure altitude will be less than the true altitude.

System altitude (H_s) System altitude above mean sea level (MSL) is obtained by correcting CADC pressure altitude for non-standard-day variations of temperature with altitude. This refined estimate of system altitude is computed using CADC and INS data combined through a weighted mixing function. The system altitude solution then achieves the long-term stability of the air data system with the short-term accuracy of the INS. The weighting function is adjusted in order to minimize the air data computer sensor lag and angle of attack effects during maneuvers. The air data computation uses an open-loop integration on pressure, while the INS computations involve straightforward integration of inertially sensed vertical velocity.

These concepts are illustrated in Fig. 4-11. Before we proceed with the discussion of the vertical channel, certain preliminaries are necessary.

4.5.1. Physics of the Atmosphere

Because the barometric altimeter is influenced to a large degree by the atmospheric physics, this section introduces and reviews briefly the basic laws of the atmosphere. The atmosphere is commonly described by tabular data representing the altitude, density, and temperature. The equations solve for static equilibrium of the column of air, given the input temperature profile, while the density profile is included as input data only for convenience. Historically, a standard (or model) atmosphere (SA) which was an attempt to model a typical "column of atmosphere," was first defined in the 1920s in both the United States and Europe to satisfy a need for aircraft instrumentation standardization and weather forecasting. These two independently developed models differed slightly. These differences were resolved on November 7, 1952, when the International Civil Aviation Organization (ICAO) adopted a new standard atmosphere. The standard is still in use today, although data above 15,140 m (50,000 ft) has undergone several revisions. The latest revision is ICAO Standard Atmosphere 1976. Figure 4-12 shows the characteristics of the ICAO 1976 Standard Atmosphere (SA). It should be pointed out that the Standard Atmosphere depicted in Fig. 4-12 is the result of extensive worldwide measurements of temperature versus altitude and represents the "average" condition. The standard sea-level conditions are for 45° north latitude. In fact, the actual temperature rarely matches the standard condition, so that the measured altitude H_p is not the actual height above sea level. However, the vertical separation between aircraft in a local vicinity is maintained because all nations have adopted the ICAO atmosphere model. For general purposes, the SA model is adequate and a valuable tool. However, as anyone who appreciates the constantly changing weather, the atmosphere is never truly in a "standard" state. This can cause problems in the use of modern navigation and fire-control systems that use SA-based algorithms.

The earth's atmosphere is not perfectly homogeneous. In fact, it is made up of several layers. A computer program was written to model the atmosphere to an altitude of 90 km. This altitude range contains four atmospheric layers. From the surface up, they are (1) the troposphere, (2) the stratosphere, (3) the mesosphere, and (4) the thermosphere. The various layers are differentiated by their temperature characteristics, or lapse rate. (The lapse rate is the change in air temperature with altitude.) These atmospheric layers will now be described in some more detail:

Troposphere The troposphere is the lowest layer of the atmosphere. It varies in altitude, from 8 km at the poles to 18 km at the equator, expanding in the summer. The troposphere contains most of the atmosphere's moisture, air mass, and therefore most of the earth's weather phenomena. The surface of the earth acts as a collector of a solar radiation, thereby turning this radiation into heat, and warms the air immediately above it. The air temperature decreases with altitude. This negative lapse rate continues to the upper boundary of the stratosphere, the tropopause.

4.5. The Vertical Channel

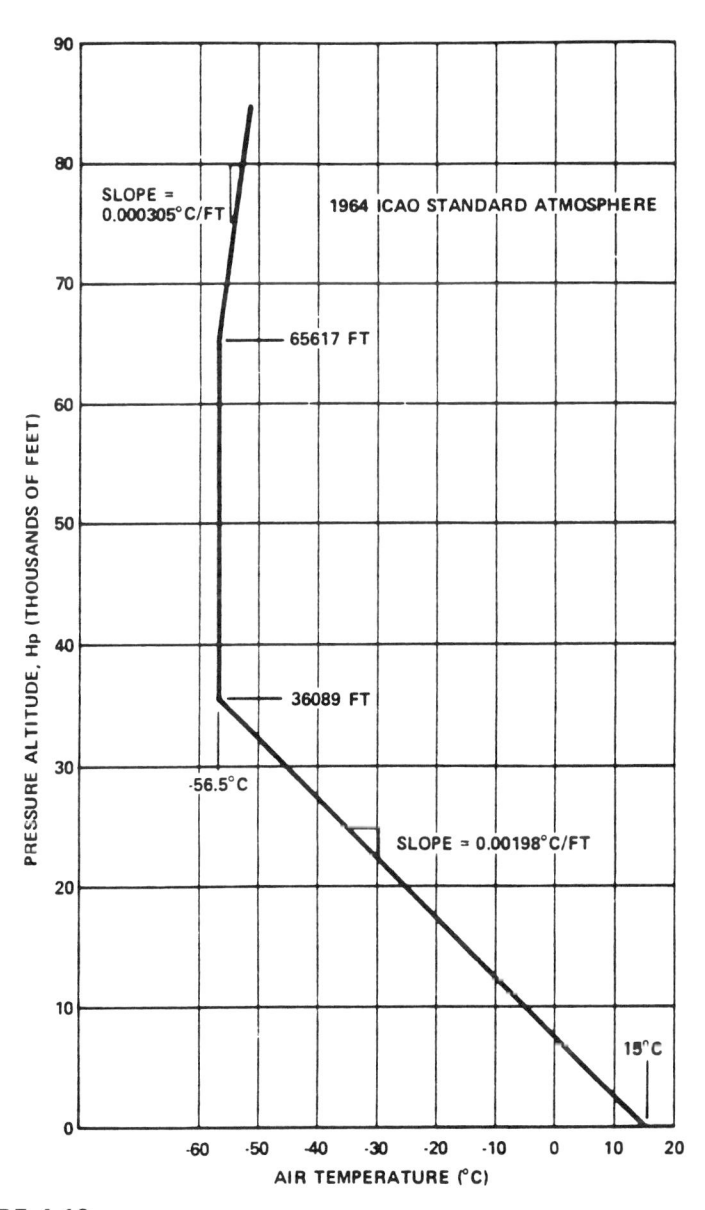

FIGURE 4-12
The 1976 ICAO standard atmosphere.

Stratosphere The lapse rate changes its direction at the tropopause. This is where the stratosphere begins, and the temperature increases with altitude. The stratosphere is also characterized by a marked decrease in water vapor. With the exception of occasional thunderstorm tops and jet-stream effects, weather phenomena are not observed.

Mesosphere At the top of the stratosphere, the lapse rate reverses itself and temperature decreases with increasing altitude. In this atmospheric layer, there is no water vapor and no weather effects are observed.

Thermosphere Above 80 km, the air temperature increases with altitude. The gas molecules at this altitude are separated by relatively large distances. Collisions among them are less frequent and they maintain high velocity, and thus higher temperatures. Ionic atmospheric phenomena such as Aurora Borealis occur in the thermosphere.

The incremental change in pressure dP for an incremental change in height dZ is governed by the hydrostatic equation [14]

$$dP = -\rho g \, dZ \qquad (4.88)$$

where ρ is the atmospheric density, g is the acceleration due to gravity, and Z is geometric altitude. Making use of the inverse-square law of gravitation, an expression for g as a function of altitude, with sufficient accuracy for most model atmosphere computations, is as follows:

$$g = \frac{g_0 R_e^2}{(R_e + Z)^2}$$

where g_0 is the acceleration due to gravity at sea level (9.78032677 m s^{-2} at the equator) and R_e is an effective earth's radius at a specific latitude.

The geopotential altitude H_g can be calculated from the expression

$$H_g = \frac{H R_e}{R_e + H}$$

where H is the altitude above the reference ellipsoid. In the range of altitudes of our interest, g may be assumed constant at standard sea-level value (g_0) with little error. From the Ideal Gas Law, the density is expressed as

$$\rho = \frac{MP}{R^* T} \qquad (4.89)$$

where M is the mean molecular weight of air (assumed constant), R^* is the universal gas constant, and T is the absolute temperature. Equations (4.88)

4.5. The Vertical Channel

and (4.89) form the basis for modeling the atmosphere. Combining Eqs. (4.88) and (4.89) results in

$$d(\ln P) = \left(-\frac{gM}{R^*T}\right) dZ \qquad (4.90)$$

If one neglects the variation of gravity and temperature with altitude, this results in the so-called exponential atmospheric pressure. Thus, integrating Eq. (6.90) yields

$$P = P_0 \exp\left(-\frac{gM}{R^*T} Z\right) \qquad (4.91)$$

where P_0 is the pressure at sea level (e.g., at the bottom of an isothermal layer). Similarly, the density for an isothermal layer is given by

$$\rho = \rho_0 \exp\left(-\frac{gM}{R^*T} Z\right) \qquad (4.92)$$

If, in a given layer, the temperature varies linearly with Z, then for a temperature T_0 at the bottom of the layer and a lapse rate of Γ,

$$T = T_0 - \Gamma Z$$

$$P = P_0 \left(\frac{T}{T_0}\right)^{(gM/R^*T)-1}$$

$$\rho = \rho_0 \left(\frac{T}{T_0}\right)^{(gM/R^*T)-1} \qquad (4.93)$$

$$\Gamma = \frac{\partial T}{\partial Z}$$

(*Note*: Γ is defined to be positive when the temperature decreases with height.) Also, the solutions for Eqs. (4.88) and (4.89) for T equal to $\Gamma(H_p - H_0) + T_0$ are given by [1]

$$H_p = H_0 + \frac{T_0\left[\left(\frac{P}{P_0}\right)^{-\frac{R^*\Gamma}{g_0 M}} - 1\right]}{\Gamma}, \qquad \Gamma \neq 0 \qquad (4.94a)$$

$$H_p = H_0 - \frac{R^*T_0}{g_0 M} \ln\left(\frac{P}{P_0}\right), \qquad \Gamma = 0 \qquad (4.94b)$$

Substituting the appropriate constants for Γ, T_0, H_0, and P_0 from the U.S. Standard Atmosphere model, Eqs. (4.94a) and (4.94b) represent the algorithm used to obtain pressure altitude. Variations in the troposphere due to season and latitude have been known for years. Pressure altitude and true altitude can differ significantly. For example, in Ref. 14 it is shown that the 75°N-latitude January model of the atmosphere differs from the SA model (i.e., pressure altitude) in a linear fashion in the 0–7315.2-m (or 24,000-ft) range by an amount equal to approximately 10% of the true altitude. This difference continues to increase above 7,315.2 m, but at a lower rate. The 15°N-altitude annual model differs from pressure altitude by about 5% for the same range. Therefore, differences between true altitude and pressure altitude of between 5 and 10% can be considered typical. Some other useful relationships which are compatible with air data computation techniques are

Temperature ratio

$$\frac{T}{T_0} = 1.0 - 6.87535 \times 10^{-6} Z$$

Pressure altitude H_p

$$H_p = 145{,}542 \left[1 - \left(\frac{P_S}{29.9213} \right)^{0.19026} \right]$$

for $H_p < 36{,}089$ ft (10,999.93 m)

$$H_p = 36.089 - 20.806 \ln \left(\frac{P_S}{6.68324} \right)$$

for $36{,}089 \leq H_p \leq 65{,}617$ ft (20,000.06 m)

Static pressure P_S

$$P_S = 29.9213 \left[1 - \frac{H_p}{145{,}542} \right]^{5.2561}$$

(*Note*: P_S is given in inches of mercury.)

True freestream air temperature T_{FAT}
For standard day

$T_{FAT} = 288.15 - 0.0003048 H_p$ (K)   for $H_p < 36{,}089$ ft

$\phantom{T_{FAT}} = 216.120$ (K)   for $36{,}089 \leq H_p \leq 65{,}617$ ft

Density ratio σ

$$\sigma = 17.3344979 \left(\frac{P_S}{T_{FAT}} \right)$$

[*Note*: P_S is given in inches mercury and T_{FAT} in °R (degrees Rankine).]

4.5. The Vertical Channel

TABLE 4-1

Relationship between Reference Altitude, Pressure Ratio, and Temperature

Reference altitude Z (ft)	Pressure ratio	Temperature (°R)
5,000	0.8320	500.86
10,000	0.6877	483.03
15,000	0.5643	465.20
20,000	0.4595	447.37
25,000	0.3711	429.53
30,000	0.2970	411.70
35,000	0.2353	393.87
40,000	0.1851	389.99

True pressure ratio δ

$$\delta = \frac{P_S}{29.9212598}$$

The ICAO atmosphere values at sea level are as follows:

$T_0 = 518.67°R$
$P_0 = 2116.8 \text{ lb/ft}^2$
$\rho_0 = 2.3769 \times 10^{-3} \text{ slugs/ft}^3$
1 standard atmosphere = 14.7 lb/in.^{2} = 1013.25 mb (millibars)
1 mb = 1.4504×10^{-2} lb/in.^{2}

Table 4-1 gives the pressure ratio using Eq. (4.91) and temperature as a function of reference altitude. The temperature remains constant starting at approximately 37,000 ft (11,277.6 m).

4.5.2. The Central Air Data Computer (CADC)

The CADC continuously measures and computes altitude and airspeed information. The primary pneumatic and thermal inputs to the CADC consist of static air pressure (P_S) and total pressure (P_t) from the pitot–static system and free-air temperature (T_{FAT}) from the temperature probe, converting this information into electrical signals representing barometric altitude, indicated airspeed (IAS), true airspeed (TAS), and Mach number. More specifically, these input functions may be defined as follows:

Static pressure (P_S) This is the ambient atmospheric pressure and is the force unit area exerted by the atmosphere on the surface of a body at rest relative to the air mass.

Total pressure (P_t)   This is the sum of the static pressure and the impact pressure, and is the total force per unit area exerted by the air mass on the surface of a body; P_t is sometimes called *pitot pressure*.

Total temperature (T_t)   This is the temperature of a sample of the compressed air mass that is brought up to the speed of the moving body. The total temperature is greater than the ambient temperature by the amount of heat associated with the adiabatic compression of the air sample.

The CADC makes direct measurement of the static and total pressures via the aircraft's pitot–static system. In block diagram form, the CADC can be represented as illustrated in Fig. 4-13.

The CADC interfaces with several instruments. Some of these CADC signals provide information directly to cockpit indicators for pilot readout, while others provide information to instruments computing navigation and weapon delivery solutions. We will discuss four of these signals: (1) When the automatic flight control system is in the "altitude hold" mode, the CADC signals of barometric altitude are used to maintain the aircraft at the designated altitude; (2) the TAS indicator, which continuously displays computed true airspeed, is the only instrument dependent entirely on the CADC for operation; (3) signals of barometric altitude are sent by the CADC to the "forward-looking radar" for sweep delay corrections, to convert slant range information into true ground range, in the ground mapping mode; (4) in military applications, the barometric altitude, Mach number, and TAS signals sent by the CADC to the tactical computer control panel, "head-up-display" (HUD), "projected map display set," "horizontal situation indicator" (HSI), and "attitude direction indicator" (ADI).

In the study of air data computers, one encounters several different airspeed terms which often result in confusion unless the exact type of

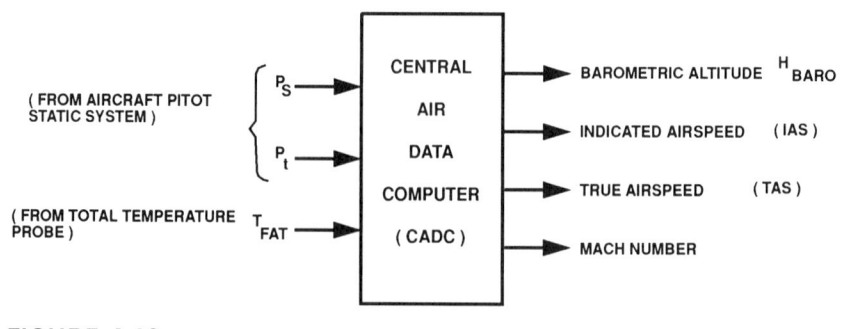

FIGURE 4-13
Block diagram of the CADC.

4.5. The Vertical Channel

airspeed is specified and understood. For convenience, the following definitions of airspeed and/or aerodynamic parameters most commonly used in avionics are given:

Speed of sound (a) This is the propagation of sound waves in ambient air. The speed of sound is proportional to the square root of the static temperature T_s in the atmosphere through which an aircraft is flying; $a^* = K\sqrt{T_s}$, where K is a constant. The speed of sound is also given by

$$a = \sqrt{\gamma\left(\frac{P_S}{\rho}\right)}$$

where γ is the specific heat ratio for air (constant), ρ is the gas density (air), and P_S is the static atmospheric pressure.

Mach number (M) Is the ratio of the TAS to the speed of sound at that flight condition. The Mach number is the chief criterion of airflow pattern, and is usually represented by the freestream, steady-state value:

$$M = \frac{V_t}{a}$$

Dynamic pressure (q) This is the force per unit area required to bring an ideal (incompressible) fluid to rest; $q = \frac{1}{2}\rho V_t^2$ where V_t is TAS (see definition below) and ρ is the mass density of the fluid.

Impact pressure (Q_c) This is the force per unit area required to bring moving air to rest. Specifically, it is the pressure exerted at the stagnation point on the surface of a body in motion relative to the air. Mathematically, it is equal to the total pressure minus the static pressure ($P_t - P_S$). At speeds under $M = 0.3$ the impact pressure and dynamic pressure are nearly identical. For higher speeds, the two differ because of the compressibility of the air.

Indicated airspeed The indicated airspeed (IAS) is the speed indicated by a differential-pressure airspeed indicator calibrated to the standard formulas relating airspeed to pitot–static pressures. The instrument's displayed airspeed is computed from the pitot and static system pressures and is not corrected for installation and position errors.

True airspeed V_t The true airspeed (TAS) is the speed of the airplane's center of mass with respect to the ambient air through which it is passing. That is, TAS is defined as the speed of the aircraft with respect

---

*The speed of sound at sea level is 661.478599 knots or 1,116.576 ft/s. (*Note:* The conversion of knots to m/s is knots × 0.514 = m/s.)

to the air mass that it is passing:

$$V_t = 29.0449233 M \sqrt{T_{FAT}} \quad \text{(knots)}$$

where T_{FAT} is in units of Rankine temperature (°R).

Computed airspeed V_c The computed airspeed (CAS) is an air data function related to the impact pressure (Q_c). Moreover, computed airspeed is the airspeed related to the differential pressure by the formulas for Q_c given below. At standard sea-level conditions, the calibrated airspeed and TAS are equal. The computed airspeed may be thought of as the indicated airspeed, corrected for instrument installation errors. For this reason it is sometimes called *true indicated airspeed*. For subsonic speeds, V_c is derived from Bernoulli's standard adiabatic formula for compressible flow:

$$P_t = P_s \left(1 + \frac{\gamma - 1}{2\gamma} \frac{\rho}{P_s} V_t^2 \right)^{\gamma/(\gamma - 1)} \quad V_t \leq a$$

where γ is the specific heat ratio for air (constant). Since $Q_c = P_t - P_s$, and the fact that V_c is based on standard sea-level conditions (where $V_c = V_t$), the above equation can be rewritten as

$$Q_C = P_0 \left[\left(1 + \frac{\gamma - 1}{2} \left(\frac{V_c}{a_0} \right)^2 \right)^{\gamma/(\gamma - 1)} - 1 \right] \quad V_c \leq a$$

where the subscript 0 denotes the standard sea-level values defined by the 1964 ICAO Standard Atmosphere.

For the supersonic range ($M > 1$) where a normal shock wave occurs ahead of the pitot tube, the relation for the total pressure behind the normal shock wave is

$$P_t = \frac{1+\gamma}{2\gamma} \rho V_t^2 \left[\frac{\frac{(\gamma+1)^2}{\gamma} \cdot \frac{\rho}{P_s} \cdot V_t^2}{\frac{4\rho}{P_s} V_t^2 - 2(\gamma - 1)} \right]^{1/(\gamma - 1)} \quad V_t \geq a$$

Using the relationships $Q_c = P_t - P_s$, $a = (\gamma P_s/\rho)^{1/2}$, and assuming standard sea-level conditions so that $V_c = V_t$, we can rewrite the above equation as

$$Q_c = \frac{1+\gamma}{2} \left(\frac{V_c}{a_0} \right)^2 P_0 \left[\frac{(\gamma+1)^2}{4\gamma - 2(\gamma-1)(a_0/V_c)^2} \right]^{(1/(\gamma-1))} - P_0 \quad V_c \geq a_0$$

4.5. The Vertical Channel

Since the Mach number is also a function of P_t and P_S, we have

$$\frac{P_t}{P_S} = \begin{cases} (1+0.2M^2)^{3.5} & \text{for } M \leq 1.0 \\ \dfrac{166.92158M^7}{(7M^2-1)^{2.5}} & \text{for } M > 1.0 \end{cases}$$

where P_t and P_S are expressed in inches of mercury. In summary, we note the following points:

1. Central air data computer—primary data source
   - Inputs: freestream air temperature (T_{FAT}); static pressure (P_S)
   - Provides accurate altitude during periods of low vertical velocity
2. Inertial navigation system—secondary data source
   - Inputs vertical velocity
   - Provides accurate altitude during periods of high vertical velocity

4.5.3. Vertical Channel Damping

The goal of vertical channel mechanization is to minimize altitude and vertical velocity output errors, and to achieve a satisfactory system response time to aircraft disturbances. The inertial navigation system mechanization accepts input accelerations from the inertial measurement unit, via a processing module, and generates velocity and position outputs by two vector integrations. As stated earlier, the vertical component of this mechanization is inherently unstable because of the need for gravity subtraction. The gravity correction is the cause of instability. An air data altitude is used to stabilize the vertical position component. Altitude data from the INS is used to transform the three-axis acceleration data into the local horizontally referenced coordinate frame in order to obtain measurements of the aircraft's vertical and horizontal acceleration components. The difference between the indicated system altitude derived from baroinertial mixing and the measured system altitude can be used to correct the baroinertial altitude. This type of altitude mechanization is used for radar bombing. Furthermore, errors in the local geoid height will be common between altitude and target deviation; thus, the need for an accurate gravity model will be required.

Historically, first- and/or second-order vertical channel mechanizations have been used because of computer limitations. Today, however, most systems use third-order mechanizations for two reasons: (1) they reduce errors originating from accelerometer (or other sources) biases, and (2) the third-order vertical channel mechanizations achieve better response characteristics. We will begin this section by analyzing a simple first-order vertical channel. Consequently, in order to gain some insight into the vertical channel error behavior, consider the simplistic error model shown in Fig. 4-14. From

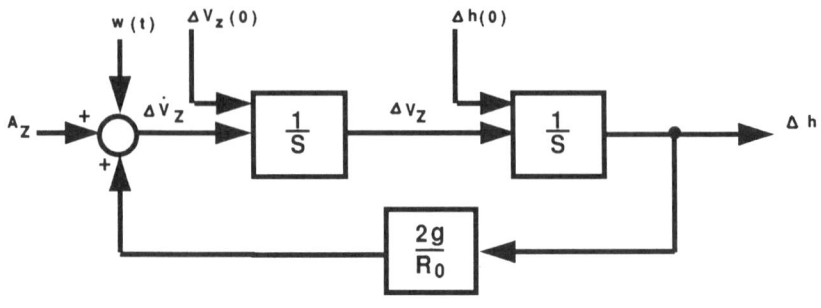

FIGURE 4-14
Simplified diagram of the vertical channel error model.

Fig. 4-14, the system error model equations can be written as

$$\Delta \dot{V}_z = \left(\frac{2g}{R_0}\right)\Delta h + A_z + w(t)$$

$$\Delta \dot{h} = \Delta V_z$$

where $w(t)$ represents random noise (or disturbances). The positive feedback with gain $(2g/R_0)$ is the destabilizing effect on the loop. Note that $2g/R_0 = 2\omega_s^2$, where ω_s is the Schuler frequency. Neglecting the noise term, and using state–space notation, the closed-loop solution for this system is

$$\Delta h = \frac{1}{\omega_s}(\sinh \omega_s t)\Delta V_z(0) + (\cosh \omega_s t)\Delta h(0)$$

$$+ \frac{A_z}{\omega_s^2}(\cosh \omega_s t - 1) \tag{4.95a}$$

$$\Delta V_z = (\cosh \omega_s t)\Delta V_z(0) + (\omega_s \sinh \omega_s t)\Delta h(0)$$

$$+ \frac{A_z}{\omega_s}(\sinh \omega_s t) \tag{4.95b}$$

where $\Delta V_z(0)$ and $\Delta h(0)$ are the initial velocity and altitude errors, respectively. A program listing in BASIC for computing Δh and ΔV_z for $A_z = 0.0001$ m/s^2 and $t = 0$–120 s is given below. For g and R_0, use the following values from WGS-84:

Equatorial radius: $R_0 = 6378,137$ m
Equatorial gravity: $g = 9.7803267714$ m/s^2

```
10    INPUT 'INITIAL VELOCITY ERROR (m/s) ="; IVELERR
20    LPRINT "INITIAL VELOCITY ERROR (m/s) ="; IVELERR
30    INPUT "INITIAL ALTITUDE ERROR (m)  ="; IALTERR
40    LPRINT "INITIAL ALTITUDE ERROR (m)  ="; IALTERR
50    LPRINT
```

4.5. The Vertical Channel

```
 60  INPUT 'START TIME (sec) ="; STIME
 70  INPUT 'FINISH TIME (sec) = "; FTIME
 80  INPUT "TIME STEP (sec) ="; STEPT
 90  EQRAD = 6378137#
100  EQGRAV = 9.7803268#
110  AZ = .0001
120  OMEGA = SQRT(2*EQGRAV/EQRAD)
130  LPRINT "TIME", "ALT ERR (m)", "VEL ERROR (m/s)"
140  PRINT "TIME", "ALT ERR (m)", "VEL ERROR (m/s)"
150  FOR ITIME = (STIME) TO (FTIME) STEP (STEPT)
160  OMEGAT = OMEGA*ITIME
170  HSIN = .5*(EXP(OMEGAT) - EXP(-(OMEGAT)))
180  HCOS = .5*(EXP(OMEGAT) + EXP(-(OMEGAT)))
190  DELH = ((1/OMEGA)*HSIN)*IVELERR + HCOS*IALTERR +
          + (AZ/(OMEGA*OMEGA))*(HCOS - 1)
200  DELV = (HCOS*IVELERR) + (OMEGA*HSIN*IALTERR) +
          + (AZ/OMEGA)*HSIN
210  LPRINT ITIME , DELH, DELV
220  PRINT ITIME, DELH, DELV
230  IALTERR = DELH
240  IVELERR = DELV
250 NEXT ITIME
```

INITIAL VELOCITY ERROR (m/s) = .5
INITIAL ALTITUDE ERROR (m) = 30.0

TIME [s]	ALT ERROR [m]	VEL ERROR [m/s]
0	30.00000	.5
5	32.50243	.5009793
10	37.52245	.5030530
15	45.09330	.5064531
20	45.09330	.5115306
25	68.14868	.5187607
30	83.85775	.5287511
35	102.59450	.5422536
40	124.63420	.5601838
45	150.35710	.5836471
50	180.27860	.6139777
55	215.08780	.6527920
60	255.69610	.7020611
65	303.29850	.7642101
70	359.45440	.8422517
75	426.19160	.9399653
80	506.14500	1.0621380
85	602.74040	1.2148890
90	720.44140	1.4061020
95	865.08350	1.6460190
100	1044.32800	1.9480390
105	1268.28500	2.3298150
110	1550.36400	2.8147330
115	1908.46400	3.4341000
120	2366.61500	4.2298300

The results are plotted in Fig. 4-15.

These figures illustrate the divergent nature of the undamped vertical channel. The derivation for altitude divergence will be given in Section 4.5.4. Next, we will consider the second-order vertical channel damping mechanization. Vertical channel damping of an inertial navigation system may be implemented by using a barometric altimeter to generate a vertical position error signal, which is fed back to the input of the vertical velocity and position integrators. Stated in other words, the inertially derived altitude minus a generated barometric altitude constitutes the feedback signal. However, the baroaltimeter output is not used in the gravity calculation. Viewing the vertical channel as uncoupled, then a typical second-order vertical channel damping-loop mechanization is represented in Fig. 4-16.

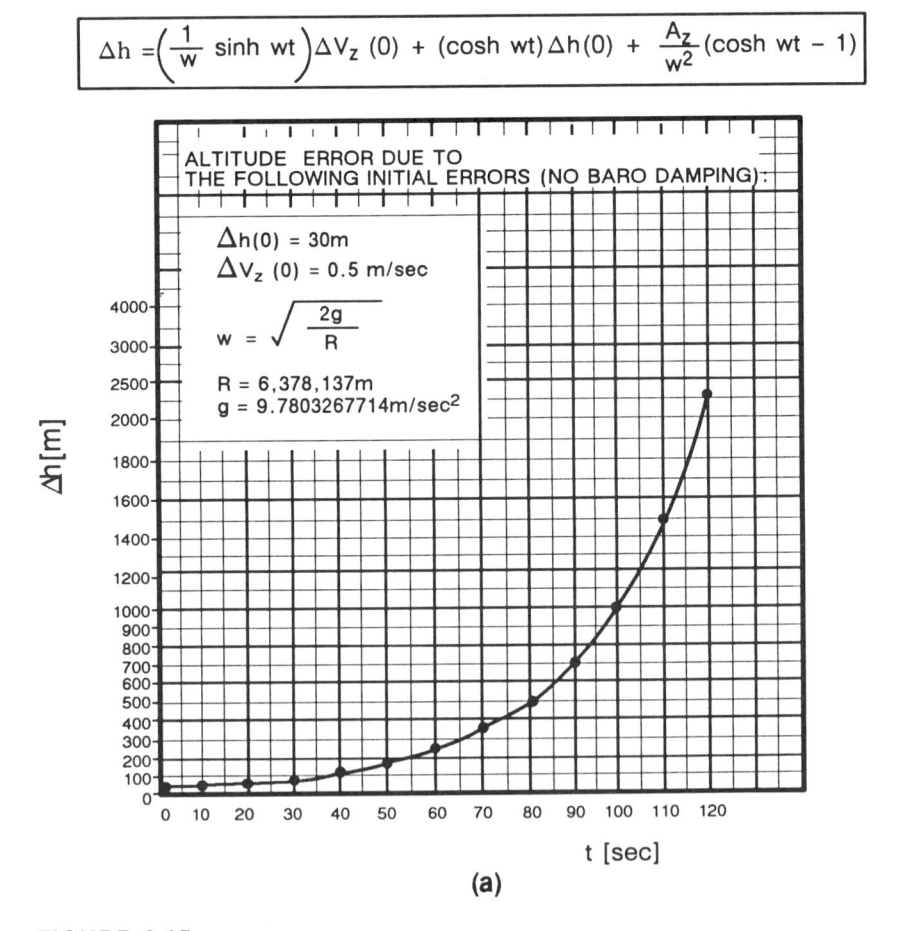

$$\Delta h = \left(\frac{1}{w} \sinh wt\right)\Delta V_z(0) + (\cosh wt)\Delta h(0) + \frac{A_z}{w^2}(\cosh wt - 1)$$

ALTITUDE ERROR DUE TO THE FOLLOWING INITIAL ERRORS (NO BARO DAMPING):

$\Delta h(0) = 30$ m
$\Delta V_z(0) = 0.5$ m/sec

$$w = \sqrt{\frac{2g}{R}}$$

$R = 6,378,137$ m
$g = 9.7803267714$ m/sec^2

(a)

FIGURE 4-15

First-order undamped vertical channel: (a) altitude error; (b) velocity error.

4.5. The Vertical Channel

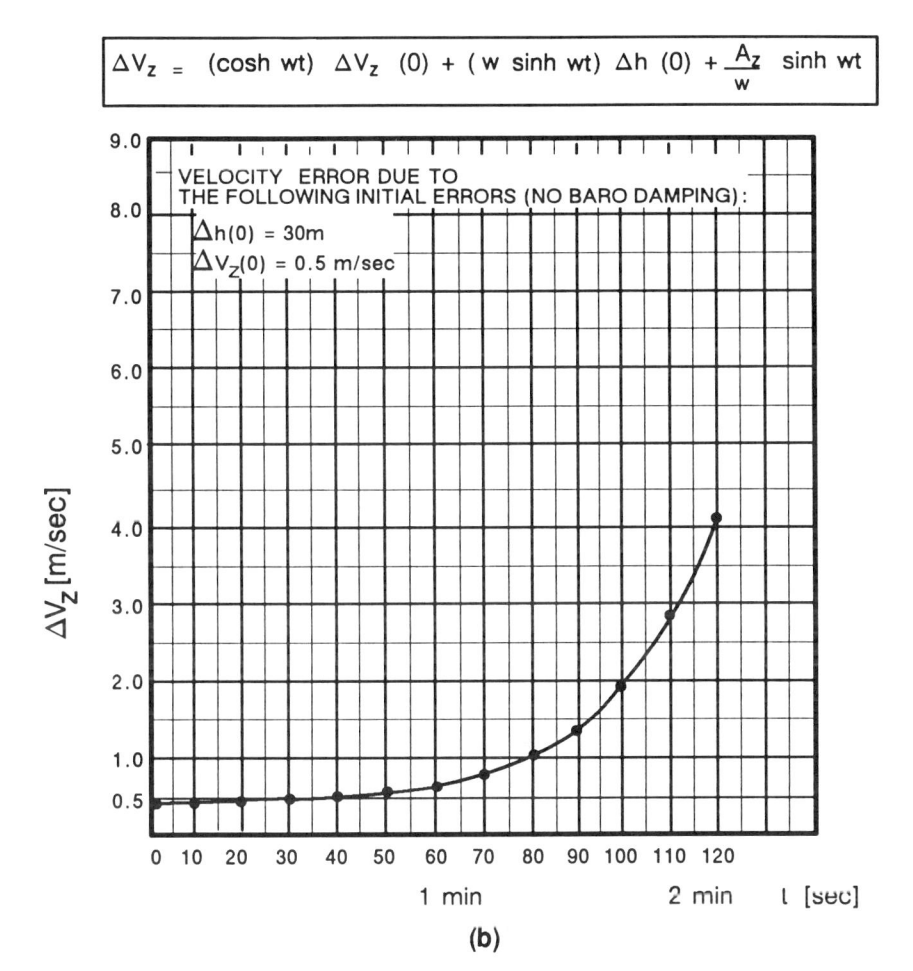

$$\Delta V_z = (\cosh wt) \Delta V_z(0) + (w \sinh wt) \Delta h(0) + \frac{A_z}{w} \sinh wt$$

VELOCITY ERROR DUE TO THE FOLLOWING INITIAL ERRORS (NO BARO DAMPING):
$\Delta h(0) = 30m$
$\Delta V_z(0) = 0.5$ m/sec

(b)

FIGURE 4.15
(*continued*)

From Fig. 4-16, we note that the barometric position measurements h_B are compared with the inertial navigation system's computed position X_z, and the difference is fed back through the loops C_1 and C_2 in order to adjust the computed (or indicated) vertical velocity V_z, and the measured specific force A_z. It should be pointed out, however, that this model is based on the assumption that the inertial navigation system processes the altitude measurements and feeds back the altitude deviations for every computation cycle of the inertial navigation system's computer.

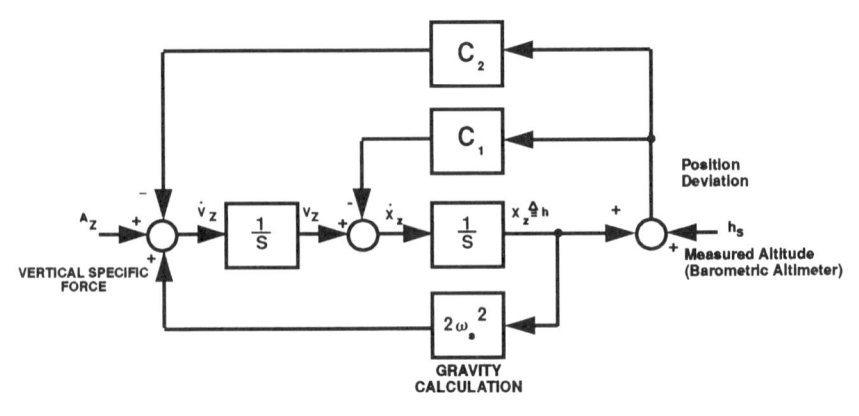

FIGURE 4-16
Second-order vertical channel damping-loop mechanization.

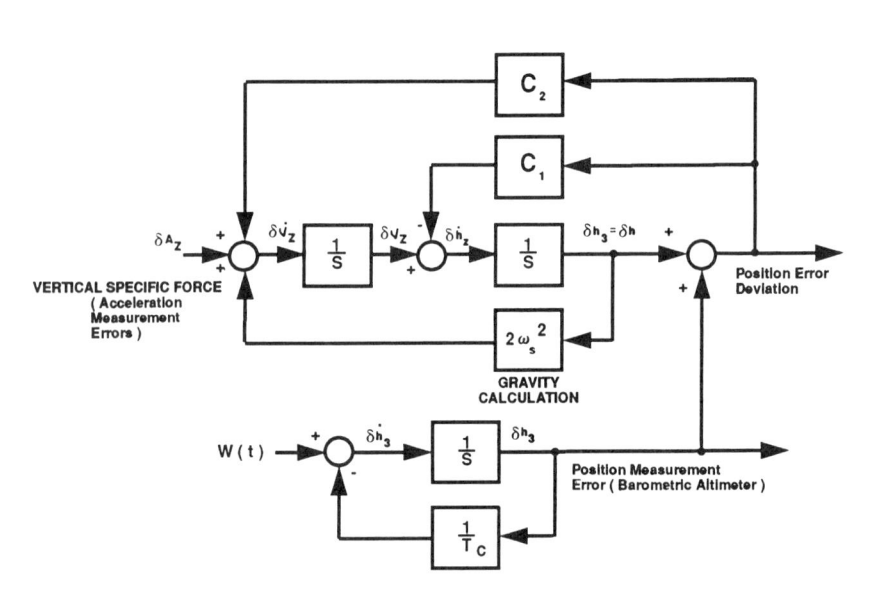

FIGURE 4-17
Error model for the vertical channel damping loop.

4.5. The Vertical Channel

An error model derived from the above mechanization is shown in Fig. 4-17. This model assumes that the barometric altimeter error is modeled as a first-order Markov process driven by an uncorrelated (white) noise process. The vertical channel error dynamics can be written directly from Fig. 4-17 as follows:

$$\delta \dot{X}_z = \delta V_z - C_1[\delta X_z + \delta h_B] \tag{4.96a}$$

$$\delta \dot{V}_z = \delta A_z + 2\omega_s^2 \delta X_z - C_2[\delta X_z + \delta h_B] \tag{4.96b}$$

$$\delta \dot{h}_B = -\frac{1}{\tau_c} \delta h_B + w(t)$$

where $w(t)$ = white-noise driving process
 τ_c = correlation time

Moreover, it can be shown that the characteristic equation of this loop is

$$s^2 + C_1 s + \left(C_2 - \frac{2g}{R_0}\right) = 0$$

The loop will be stable provided C_1 is positive and C_2 is greater than $2g/R_0$. Further, if one wishes the system to have two real poles at

$$s = -\frac{1}{\tau}\left[\left(s^2 + \frac{1}{\tau}\right)^2 = 0\right]$$

one chooses the gains to be [17]

$$C_1 = \frac{2}{\tau}$$

$$C_2 = 2\omega_s^2 + \frac{1}{\tau^2}$$

These constants are determined from specifications and/or requirements and time constant of the system τ. For many applications, the system time constant is $\tau = 100$ s. A significant performance problem of the second-order loop is that a steady vertical acceleration error causes a steady-state vertical velocity error. In order to illustrate this point, consider a second-order system with a 100-s loop time constant. Then, a 100 µg accelerometer bias produces a 0.2-m/s (0.656-ft/s) bias in velocity. Mathematically, the velocity bias is approximated by

$$\delta V_b = 2\tau \, \delta A_b = 2 \times 100 \text{ s} \times 100 \times 10^{-6} \times 9.8 \text{ m/s}^2 \approx 0.2 \text{ m/s}$$

This is a significant velocity error for weapon delivery applications and must be reduced. Vertical velocity is critical during weapon delivery because this error greatly affects the miss distance of the weapon on the target. More specifically, in delivery of ordnance, the two largest contributions to miss distance would be due to (1) altitude navigation error and (2) vertical velocity navigation error; both these errors are computed at the instant of weapon release. The altimeter output and vertical position are defined with their positive directions being opposite; that is, the altimeter output is positive up, while vertical position is positive down, assuming a NED frame, which accounts for the sum $(\delta X_z + \delta h_B)$ being fed back in the error equations. Also, note that since the inertial navigation system obtains a measurement of specific force in the platform "p" axes, these axes in general do not coincide with the geographic or navigation "n" axes. A coordinate transformation is therefore required. Thus, $\mathbf{A}^n = C_p^n \mathbf{A}^p$. Equations (4.96a) and (4.96b) are the equations that model the vertical channel in an error analysis computer program. A note of caution is in order at this point. Since the reference altitude h_B comes from a slowly reacting barometric instrument, the feedback through C_1 in Fig. 4-16 will include a large error in the event of a rapid change in altitude, as in high-performance military aircraft. For this reason, V_z is commonly taken after the first integrator rather than after the summation point $(\delta \dot{X}_z)$. The first integrator acts as a filter during a rapid change in altitude.

In order to obtain adequate and/or improved performance, a third-order loop is commonly used in many applications. Figure 4-18 illustrates a standard third-order vertical channel damping technique for mixing baro-altimeter and inertially derived data [1, 17]. An analysis of this system follows. First, from Eq. (4.24) we note that the inertial navigation system's output is A_z, which is the accelerometer-measured specific force, plus $-(\omega_y + \Omega_y)V_x + (\omega_x + \Omega_x)V_y$, the Coriolis compensation, plus g_z, the acceleration due to gravity. Next, the air-data computer's input to Fig. 4-18 is reference altitude h_B. The outputs are vertical velocity V_z and system altitude h. As shown in Fig. 4-18, changes in barometric derived altitude and inertially derived altitude are combined in a baroinertial mixer that emphasizes barometric information at low vertical velocity rates and inertial information at high vertical velocity rates. Thus, the result of mixing the barometric and inertial changes in altitude provides the total change in system altitude. These changes are continuously summed to provide an updated system altitude at all times. Moreover, system altitude computed by the baroinertial mixer represents the system's best estimate of true altitude above mean sea level (MSL). The positive feedback path through $2\omega_s^2$ is required to compensate g_z for changes in altitude, but as we have seen earlier, it causes the instability in the unaided vertical channel. The feedback paths through C_1, C_2, and C_3 damp and control this instability using the error signal $(h - h_B) - C_4 V_z$. The

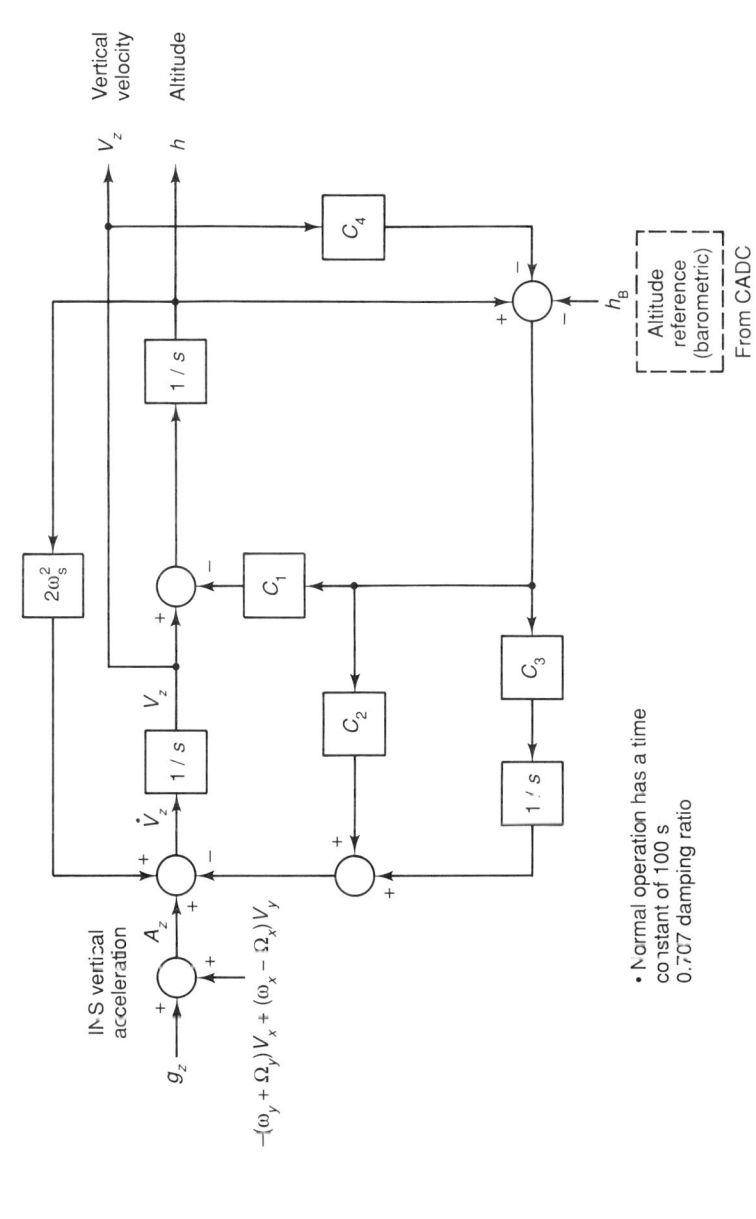

FIGURE 4-18
Third-order vertical channel damping-loop mechanization.

feedforward path through C_4 compensates for altimeter lag. The main source of altimeter error due to dynamic lag is the pressure transducer. Typical values for C_4 range between 0.5 and 0.8 s [1]. If the integral feedback were omitted, an INS vertical acceleration error would cause a steady-state difference between the system and reference altitudes ($h - h_B$).

For the purposes of the present analysis, the values of the three gains (C_1, C_2, C_3) have been chosen so that the system characteristic equation yields three poles at the complex frequency $s = -(1/\tau)$ as follows:

$$\left(s + \frac{1}{\tau}\right)^3 = 0$$

where τ is the vertical channel time constant. Therefore, the values of these gains that achieve these system poles are as follows:

$$C_1 = \frac{2\omega_s^2 C_4 \tau^3 + 3\tau^2 + 3C_4\tau + C_4^2}{(1 - 2\omega_s^2 C_4^2)\tau^3} \tag{4.97a}$$

$$C_2 = \frac{2\omega_s^2(\tau^3 + 3C_4\tau^2) + 3\tau + C_4}{(1 - 2\omega_s^2 C_4^2)\tau^3} \tag{4.97b}$$

$$C_3 = \frac{1}{\tau^3} \tag{4.97c}$$

Typically, depending on the system design, τ is chosen to be 100–200 s. Once the appropriate values for τ and C_4 are chosen, these relations yield a time invariant linear system equation. If $C_4 = 0$, the gains given by Eqs. (4.97a)–(4.97c) reduce to

$$C_1 = \frac{3}{\tau}$$

$$C_2 = 2\omega_s^2 + \frac{3}{\tau^2}$$

$$C_3 = \frac{1}{\tau^3}$$

The third-order vertical channel mechanization improves significantly the system's performance. If the system gains are assumed constant, the baroinertial equation to be mechanized is Eq. (4.76). Thus, the vertical velocity V_z and h can be approximated by solving Eq. (4.76). An error model for the vertical channel can be designed similar to the one shown in Fig. 4-18, with the various parameters replaced by error states (δh, δV_z, etc.). The error state δA_z can be taken to be a random walk modeling any bias or slowly varying error in the vertical acceleration due to accelerometer bias, gravity anomaly, or error in the Coriolis terms.

4.5. The Vertical Channel

The value of h_B provided by the air data computer can be calculated in the standard manner, or by means of the technique developed by Blanchard [1, 2]. Blanchard develops an improved algorithm for computing h_B in a "nonstandard" atmosphere. Rather than calculating pressure altitude based on a standard atmosphere (SA) model (and possibly compensating later for nonstandard conditions), this method computes a dynamically corrected altitude by numerically integrating the physical relation for a column of air as a function of pressure gradients resulting from weather changes during flight, gravity, and absolute temperature. This algorithm has been flight tested and has demonstrated good performance. Figure 4-19 illustrates the error between standard and non-standard-day conditions. The difference between baroaltitude and true altitude is computed by the navigation (or fire-control) computer.

Using the trapezoidal integration method, we can write this algorithm as follows:

$$Z_n = Z_{n-1} - \frac{C}{g_z}\left[\frac{T_n}{P_n} + \frac{T_{n-1}}{P_{n-1}}\right](P_n - P_{n-1}) \tag{4.98a}$$

$$C = \frac{R^*}{2M} = 858.277 \text{ ft}^2/(°\text{R-s}^2)$$

$$g_z = 32.087437359 + 0.169944938 \sin^2\phi - 3.0877321 \times 10^{-6} Z^*_{n-1}$$
$$+ 4.39404 \times 10^{-9} \sin^2\phi Z^*_{n-1} \quad \text{ft/s}^2$$

$$\Delta Z_n = \Delta Z_{n-1} + \frac{\Omega \sin\phi}{g_z}(V_{gn} V_{wcn} + V_{gn-1} V_{wcn-1})\Delta t \tag{4.98b}$$

$$P_{n-1} = P_n; \quad T_{n-1} = T_n; \quad Z_{n-1} = Z_n; \quad \Delta Z_{n-1} = \Delta Z_n$$

$$V_{wc} = w_N \sin\psi_v - w_E \cos\psi_v$$

$$V_{gn-1} = V_{gn}; \quad V_{wcn-1} = V_{wcn}$$

$$Z^*_n = Z_n + \Delta Z_n$$

$$Z^*_{n-1} = Z^*_n$$

$$H_D = Z^*_n + C_4 V_Z$$

The algorithm is initialized by setting

$$Z^*_{n-1} = h_0; \quad Z_{n-1} = h_0; \quad \Delta Z_{n-1} = 0$$

$$P_{n-1} = P_s; \quad T_{n-1} = T_{FAT}$$

$$V_{gn-1} = \text{initial ground-track velocity}$$

$$V_{wcn-1} = \text{initial cross-track wind velocity}$$

$$\Delta t = 0.0625 \text{ s}; \quad C_4 = 0.075 \text{ s}$$

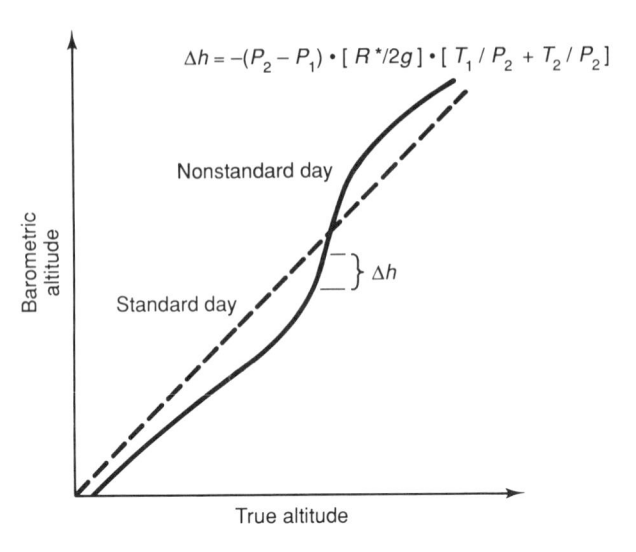

FIGURE 4-19
Altitude error due to nonstandard day atmospheric conditions.

$$\Delta h = -(P_2 - P_1) \cdot [R^*/2g] \cdot [T_1/P_2 + T_2/P_2]$$

where Δt = computer iteration time = 0.0625 (s)
Z_n = altitude reference without gradient correction (non-standard atmosphere altitude reference) (ft)
Z_n^* = altitude reference with gradient correction (ft)
ΔZ_n = pressure gradient correction term (ft)
C = precomputed constant = 858.277 [ft^2/(°R-s^2)]
C_4 = altimeter lag correction constant (0.075 s)
V_Z = INS vertical velocity (ft/s)
H_D = reference altitude for damping the vertical channel (ft)
ϕ = geodetic latitude (rad)
Ω = earth rate of rotation (7.292115 × 10^{-5} rad/s)
g_Z = radial (or normal) component of local gravity (ft/s^2)
M = atmosphere molecular weight (dimensionless)
R^* = universal gas constant (°R)
T_n = current freestream air temperature T_{FAT} (°R)
P_n = current static pressure P_S (inches Hg)
V_{gn} = current ground-track velocity (ft/s)
V_{wcn} = current value of wind component perpendicular to ground track (ft/s)
w_E, w_N = east and north wind components, respectively (ft/s)
$T_{n-1} = T_n$ at previous iteration cycle step
$P_{n-1} = P_n$ at previous iteration cycle step

4.5. The Vertical Channel

$V_{gn-1} = V_{gn}$ at previous iteration cycle step
$V_{wcn-1} = V_{wcn}$ at previous iteration cycle step
ψ_v = velocity heading (rad)

The initial altitude h_0 is obtained from the navigator's entry of runway altitude, when $V_g = 0$. After algorithm initialization, these equations are iterated in the order in which they appear. The reference altitude H_D is made available to the INS in order to damp its divergent velocity and altitude. In most applications, the algorithm would be iterated at a fairly high rate, such as 16 Hz, in order to keep the altitude information current for all onboard systems. However, this iteration rate must be chosen to be compatible with the software in the navigation computer. Figure 4-20 shows the functional relationship between the CADC-derived data and inertial system outputs.

The discussion of the third-order vertical channel mechanization will now be summarized. In order to implement the third-order vertical channel mechanization with minimum altitude and vertical velocity output errors, while at the same time achieving a satisfactory response time to aircraft disturbances, the following features must be considered in the design:

1. An optimal set of gains must be found by minimizing the system's disturbance with appropriate response time.
2. Air data reference (true) altitude must include
   - Non-standard-atmosphere model
   - Horizontal pressure gradient (wind) effect correction
   - Lag time of temperature and pressure sensors as functions of altitude

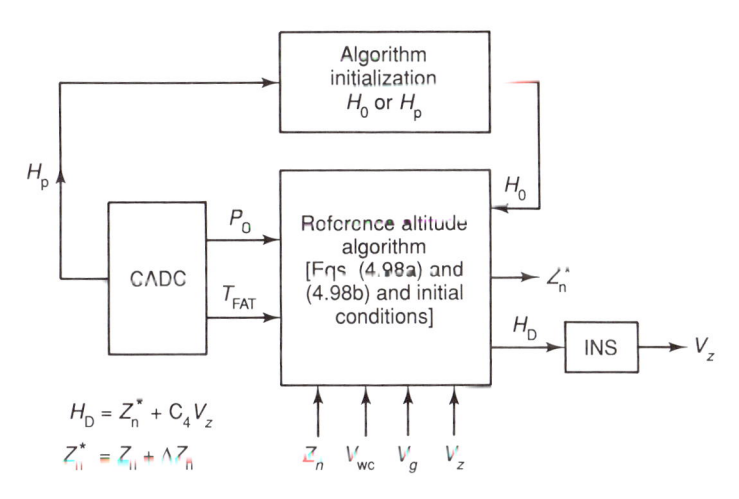

FIGURE 4-20
Block diagram for altitude determination.

3. Altitude calibrations must be implemented through a feedforward technique. This minimizes the output disturbances caused by calibration.
4. Third-order mechanization inherently can achieve better accuracy and better response characteristics than second-order mechanization.
5. In order to get the best performance and accuracy in vertical velocity and altitude outputs, the use of a highly stable and accurate inertial navigation system will be required.

The third-order vertical channel mechanization, when properly designed, will reduce errors originating from the accelerometer (or other sources) biases, and achieve better time response characteristics. Therefore, a third-order mechanization is preferred for the reasons of optimum balance between performance and mathematical tractability. In some applications, the difference between indicated system altitude derived from baroinertial mixing and the measured system altitude is used to correct the baroinertial altitude. This type of altitude mechanization is ideal for military aircraft applications such as radar bombing systems, since an altitude calibration performed in the vicinity of the target removes mapping elevation errors from the height above target computation.

For the reader's convenience, Table 4-2 gives selected numerical values adopted as exact for the various computations. These values are also published in the ICAO Standard Atmosphere. In addition, presented below are certain conversion factors that should be of help in converting from one system of units into another.

TABLE 4-2
Adopted Primary Constants[a]

Symbol	Name	Metric system (mks)	English units (ft-lb-s)
P_0	Sea-level pressure	1.01325×10^5 N m^{-2}	2116.8 lb$_f$ ft^{-2}
ρ_0	Sea-level air density	1.2250 kg m^{-3}	0.07647 lb ft^{-3}
T_0	Sea-level temperature	15°C	59.0°F
T_i	Temperature of the ice point	273.15 K	491.67°R
R^*	Universal gas constant	8.31432 J(K)$^{-1}$ mol^{-1}	1545.31 ft-lb(lb-mol)$^{-1}$(°R^{-1})
g_0	Sea-level acceleration due to gravity	9.80665 m s^{-2}	32.1741 ft s^{-2}
M	Molecular weight of air (constant up to 90 km)	28.96(−)	28.96(−)

[a] Key: N = Newtons, J = Joules, lb$_f$ = pounds force, (−) = dimensionless.

4.5. The Vertical Channel

Conversion Factors

1 ft = 0.3048 m
1 m = 3.2808399 ft
1 nm = 6,076.1155 ft
1 nm = 1,852.0 m
1 lb = 0.45359237 kg
1 kg = 2.2046226 lb
1 slug = 32.174049 lb = 14.594 kg

The absolute temperature (in kelvins) of the melting point of ice under a pressure of 101,325.0 (newtons)/m^2 is

$$T_i = 273.15 \text{ K} = 491.67°\text{R}$$

Temperatures in the metric absolute scale (K) are given by

$$T(\text{K}) = T_i + t$$

where t is the temperature on the Celsius scale (°C); in the English system of units, this is

$$T(°\text{R}) = 1.8[t(°\text{C}) + 273.15(°\text{C})]$$

Therefore, for $t = 15°\text{C} = 59.0°\text{F}$, we have

$$T(\text{K}) = 273.15 + 15.0 = 288.15 \text{ K}$$

$$T(°\text{R}) = 1.8[15 + 273.15] = 518.67°\text{R}$$

For the melting point of ice stated above, T_i in the English system of units is obtained from the relation

$$T_i(°\text{R}) = 1.8 T_i(\text{K})$$

4.5.4. Altitude Divergence

In Section 4.5.3 we noted that the vertical channel of an inertial navigation system is unstable. The vertical velocity and position errors diverge, that is to say, these errors increase exponentially with the passage of time. Therefore, in this section we will discuss the altitude divergence problem, since it affects the mechanization of all inertial navigation systems. Now consider a simple vertical accelerometer with its input axis along the Z axis. The measured specific force A is given by

$$A = \ddot{Z} + g \tag{4.99a}$$

where $\ddot{Z}$ is the altitude above the earth and g is the acceleration due to gravity. From Newton's Law of Gravitation

$$g = g_0 \left(\frac{Z_0}{Z}\right)^2$$

where Z_0 is an arbitrary initial point and g_0 is the gravity at Z_0. Now let $Z = Z_0 + h$. Hence

$$g = g_0 \left(1 + \frac{h}{Z_0}\right)^{-2} \tag{4.99b}$$

Expansion into a Taylor series yields

$$g = g_0 \left[1 - \frac{2h}{Z_0} + \cdots \right] \tag{4.99c}$$

Retaining only the first two terms of Eq. (4.99c), we have

$$g \approx g_0 - \left(\frac{2g_0}{Z_0}\right) h \tag{4.99d}$$

Substituting g into the specific force equation [Eq. (4.99a)], one obtains

$$A = \ddot{Z} + g = \ddot{Z} + g_0 - \left(\frac{2g_0}{Z_0}\right) h$$

or

$$\ddot{h} - \left(\frac{2g_0}{Z_0}\right) h = A - g_0 \tag{4.99e}$$

since

$$Z = \frac{d^2}{dt^2}(Z_0 + h) = \ddot{h}$$

Equation (4.99e) is the equation that must be mechanized in order to compute h. Because of the negative sign on the left-hand side of Eq. (4.99e), it will be recognized that this equation diverges, that is, one term in the solution is a positive exponential function of time. Let us now consider what happens if there is an error ΔA in A. Since Eq. (4.99e) is linear, we can write

$$\Delta \ddot{h} - \left(\frac{2g_0}{Z_0}\right) \Delta h = \Delta A \tag{4.99f}*$$

---

*Note that Eq. (4.99f) has a pole in the right-hand plane giving rise to the instability discussed in this section.

4.5. The Vertical Channel

where Δh is the corresponding error in h. The general solution of Eq. (4.99f) is

$$\Delta h = A \cosh \sqrt{\frac{2g_0}{Z_0}} \cdot t + B \sinh \sqrt{\frac{2g_0}{Z_0}} \cdot t - \frac{\Delta A}{(2g_0/Z_0)} \quad (4.99\text{g})$$

Assuming the initial conditions $\Delta h(0) = \Delta \dot{h}(0)$, we have the final result

$$\Delta h = \frac{Z_0}{2}\left(\frac{\Delta A}{g_0}\right)\left[\cosh \sqrt{\frac{2g_0}{Z_0}} \cdot t - 1\right] \quad (4.99\text{h})$$

where Z_0 corresponds to the radius of the earth, R_0.

The FORTRAN program listing shown below, computes Eq. (4.99h) assuming the following values:

$$Z_0 = R_0 = 6{,}378{,}137 \text{ m}; \qquad g_0 = 9.7803267714 \text{ m/s}^2; \qquad \frac{\Delta A}{g_0} = 10^{-5}$$

```
C   CALCULATE ALTITUDE DIVERGENCE AS A FUNCTION OF TIME

C   CONSTANTS

      R0     = 6378137.0              ! meters
      G0     = 9.7803267714           ! meters/s**2
      CONST1 = 1.0E-5
      CONST2 = SQRT(2.0*G0/R0)

C   WRITE HEADER FOR OUTPUT

            WRITE(*,1)
1           FORMAT(5X,'  T[sec]      Dh[m]',/,
     *             5X,'                    ')

C   CALCULATE DeltaH(t)

            DO 10 T = 0.0, 5000.0, 100.0
            DELTAH = 0.5*R0*CONST1*(COSH(CONST2*T) - 1.0)
            WRITE(*,2) T,DELTAH
2           FORMAT(2,(5X,F10.2))
10          CONTINUE

            END
```

T[s]	Dh[m]	T[s]	D[m]
0.00	0.00	2600.00	1482.08
100.00	0.49	2700.00	1771.79
200.00	1.98	2800.00	2116.95
300.00	4.50	2900.00	2528.18

400.00	8.15	3000.00	3018.13
500.00	13.03	3100.00	3601.85
600.00	19.29	3200.00	4297.30
700.00	27.12	3300.00	5125.86
800.00	36.76	3400.00	6113.00
900.00	48.52	3500.00	7289.08
1000.00	62.75	3600.00	8690.25
1100.00	79.89	3700.00	10359.61
1200.00	100.46	3800.00	12348.47
1300.00	125.11	3900.00	14717.98
1400.00	154.58	4000.00	17541.00
1500.00	189.79	4100.00	20904.33
1600.00	231.81	4200.00	24911.38
1700.00	281.94	4300.00	29685.36
1800.00	341.72	4400.00	35373.04
1900.00	412.99	4500.00	42149.32
2000.00	497.93	4600.00	50222.53
2100.00	599.17	4700.00	59840.87
2200.00	719.81	4800.00	71300.16
2300.00	863.56	4900.00	84952.63
2400.00	1034.84	5000.00	101218.16
2500.00	1238.93		

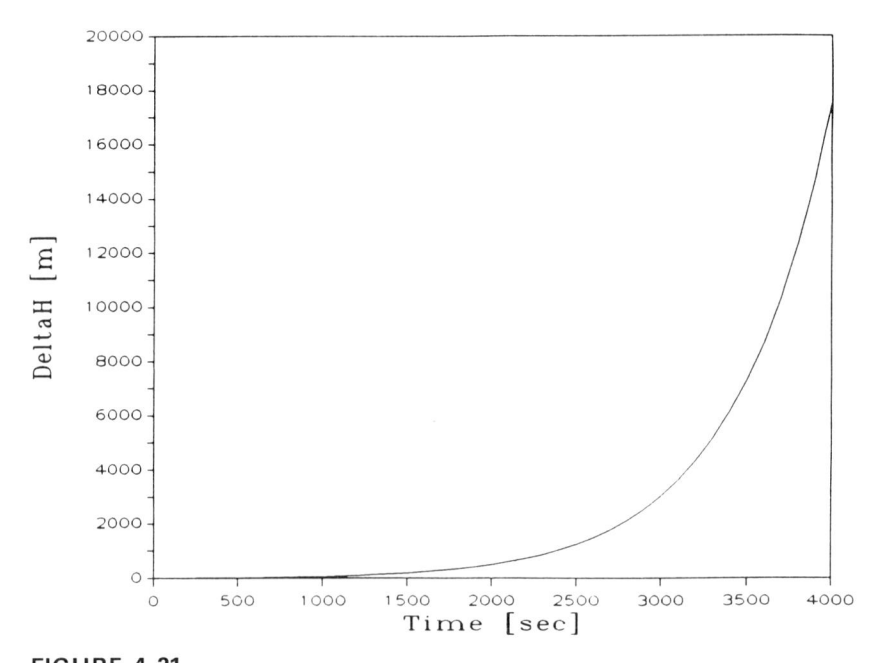

FIGURE 4-21
Altitude divergence.

The preceding table and plot (Fig. 4-21) indicate that an inertial navigation system can be used for measurement of vertical distance only for short periods. If long periods are desired, the pure inertial system cannot be used to provide altitude information. This is because inertial navigation requires prior knowledge of gravity. The proof of divergence given here is general, since there is no restriction on the choice of Z_0 or the location of the Z axis on the earth. Consequently, because of the divergence problem, all inertial navigation systems must be aided with external altitude information.

4.5.5. Altitude Calibration

Altitude calibration may be performed periodically to correct for errors in system altitude caused by sensor errors or unpredictable changes in atmospheric conditions. Altitude calibrations are performed over terrain of known elevation and may be accomplished either on the ground or in flight. The calibration mechanization uses radar ranging to determine relative height above terrain of known elevation to compute true altitude above mean sea level. If a radar altimeter system (RAS) is used, a low-altitude calibration is accomplished. A radar altimeter is used to measure the height above ground level, and is limited to a maximum of 5000 ft. The largest error in using a radar altimeter as an altitude reference is the unmodeled variation in the local terrain height above sea level. In addition, radar altimeters exhibit a scale factor error, which can be as much as 2–3% of height above the terrain.

REFERENCES

1. Blanchard, R. L.: "A New Algorithm for Computing Inertial Altitude and Vertical Velocity," *IEEE Transactions on Aerospace and Electronic Systems*, **AES-7**(6), 1143–1146, November 1971.
2. Blanchard, R. L.: "An improvement to an Algorithm for Computing Aircraft Reference Altitude," *IEEE Transactions on Aerospace and Electronic Systems*, **AES-8**(5), 685–686 (Correspondence), September 1972.
3. Britting, K. R.: *Inertial Navigation Systems Analysis*, Wiley-Interscience, New York, 1971.
4. Brockstein, A. J., and Kouba, J. T.: "Derivation of Free-Inertial, General Wander Azimuth Mechanization Equations," Litton Systems, Inc., Publication No. 15960, rev. ed., June 1981.
5. Carlson, N. A.: "Fast Geodetic Coordinate Transformations," AIAA Guidance and Control Conference, August 11–13, 1980, Paper No. AIAA-80-1771-C8.
6. Goldstein, H.: *Classical Mechanics*, Addison-Wesley, Reading, Mass., 1965.
7. Huddle, J. R.: "Inertial Navigation System Error-Model Considerations in Kalman Filtering Applications," in *Control and Dynamic Systems*, C. T. Leondes (ed.), Academic Press, New York, 1983, pp. 293–339.
8. Kayton, M., and Fried, W. R. (eds.): *Avionics Navigation Systems*, Wiley, New York, 1969.
9. Macomber, G. R., and Fernandez, M.: *Inertial Guidance Engineering*, Prentice-Hall, Englewood Cliffs, N.J., 1962.

10. Nash, R. A., Jr., Levine, S. A., and Roy, K. J.: "Error Analysis of Space-Stable Inertial Navigation Systems," *IEEE Transactions on Aerospace and Electronic Systems*, **AES-7**(4), 617–629, July 1971.
11. O'Donnell, C. F.: *Inertial Navigation Analysis and Design*, McGraw-Hill, New York, 1964.
12. Pinson, J. C.: "Inertial Guidance for Cruise Vehicles," in *Guidance and Control of Aerospace Vehicles*, C. T. Leondes (ed.), McGraw-Hill, New York, 1963, pp. 113–187.
13. Pitman, G. R., Jr. (ed.): *Inertial Guidance*, Wiley, New York, 1962.
14. *U.S. Standard Atmosphere Supplements, 1966*, U.S. Government Printing Office, Washington, D.C. 20402.
15. Webster, A. G.: *Dynamics of Particles and Rigid, Elastic, and Fluid Bodies*, 2nd ed., Dover, New York, 1959.
16. Widnall, W. S., and Grundy, P. A.: "Inertial Navigation System Error Models," TR-03-73, Intermetrics, Inc., May 11, 1973.
17. Widnall, W. S, and Sinha, P. K.: "Optimizing the Gains of the Baro-Inertial Vertical Channel," *AIAA Journal of Guidance and Control*, **3**(2), 172–178, March–April 1980.
18. Wrigley, W., Hollister, W. M., and Denhard, W. G.: *Gyroscopic Theory, Design, and Instrumentation*, MIT Press, Cambridge, Mass., 1969.

5 Error Analysis

5.1. INTRODUCTION

Error analysis is an integral part in the design and development of inertial navigation systems. Generally, it is through error analysis that requirements (or "error budgets") on instrument accuracies are specified and system performance evaluated. The errors in an inertial navigation system are due to several sources. Some of these errors are accelerometer and gyroscope inaccuracies, initial errors in platform alignment, misalignment of the inertial sensors on the platform, computational errors, and approximations in the mechanization of the system equations. Specifically, the objective of error analysis is twofold: (1) to show how to obtain appropriate equations for programming the onboard navigation computer and to depict the error behavior of specific system mechanizations and (2) to introduce enough additional assumptions so that the behavior of the system may be analytically portrayed. For vehicles utilizing inertial navigation systems, it is desirable that the system computes position and velocity with respect to the earth with high accuracy. However, as will be discussed shortly, several types of errors occur in inertial navigation systems. For this reason, analysis and propagation of errors is one of the major analytical activities supporting the design, development, and fielding of inertial navigation systems. Here, the underlying assumption is that the error dynamics are described by linear differential equations driven by white noise. Moreover, the prime concern is to determine variances of the resulting errors at the output. If the system dynamics are nonlinear, then Monte Carlo simulation techniques are most likely to be used. In some cases, a simplified system model will be sufficient to analyze the problem. Insight into the error propagation can be gained if the dominant characteristics and parameters of the system are identified, and approximate answers can be provided to check the problem against a more sophisticated computer solution. As discussed in the literature [2, 7], the Kalman filter development involves covariance propagation as an error analysis tool. Errors normally fall into two categories: (1) deterministic errors and (2) stochastic (or random) errors. Deterministic errors are usually simple in form and are easy to describe mathematically. The characteristics of such errors are that they are constant, have constant coefficients, or may be sinusoidal in nature. Furthermore, these errors are easily compensated for, that is, they can be effectively subtracted out of the system. For instance, the zero calibration of an instrument is such an error. Regression analysis may be used in such error sources. Stochastic or random errors, on the other hand, are treated statistically, based on a mathematical specification. Therefore, in stochastic analysis, the error sources can be time-varying. Parameter or state estmation is used in this case. However, it should be pointed out that both of these categories require very good error models. In general, some of these errors do not consist of pure white noise but are correlated in time. This problem can be solved by adding states to the state vector, and simulating the errors as being the outputs of stochastic processes with white-noise inputs. The main sources of errors in an inertial navigation system are due to the following facts:

1. The time rates of change of the velocity errors are driven chiefly by accelerometer errors and gravity anomalies.
2. The attitude error rates are driven chiefly by gyroscope errors.

When dealing with errors in error analysis, it is necessary to describe these errors mathematically, in order to study their propagation. Inertial navigation and guidance system performance requirements are usually specified in terms of such statistical quantities as the rms position or velocity error, or the time "rate of growth" of rms position errors. The covariance matrix $P(t)$ is a measure of filter performance, since its diagonal elements are the filter's estimate of the mean-square errors in the state-vector estimate. The covariance matrix gets smaller as more measurements are made. Furthermore, error covariance analysis can be used as a tool for determining the design requirements of inertial sensors. It is a way of estimating overall system performance as a function of the performance of subsystems and components that contribute to system errors. Error budgeting requires solving the inverse problem, which is to find an allocation of allowable errors (i.e., error budget) among the subsystems such that the system meets a specified set of performance criteria. Finally, in addition to the two main sources of error in an inertial navigation system, three other basic classes of errors must be considered:

1. *Physical component errors*—errors that are usually deviations of the inertial sensors from their design behavior. The main errors are gyroscope drifts, accelerometer bias, and scale factor errors of the gyroscope torquers and accelerometers.
2. *Construction errors*—errors in the overall system construction such as mechanical alignment errors of the inertial sensors mounted on the platform.
3. *Initial conditions*—errors that arise from imperfect determination of the initial position vector, initial velocity vector, and initial platform misalignment.

5.2. DEVELOPMENT OF THE BASIC INS SYSTEM ERROR EQUATIONS

Generally, in inertial navigation systems, "error" is defined as the difference between the value of the variable as it physically exists in the system computer, and the value of the variable as interpreted in the reference local-level* navigation coordinate system when determined without error. Specifically, the numerical value of a measured quantity as registered by any instrument is not necessarily correct and must be considered to be approximate. The degree of approximation between the numerical value and the real value depends on measurement errors. Expanding on the classes of errors discussed in Section 5.1, we can define as systematic errors those errors that during repeated measurement of the same quantity either remain constant or change according to a defined law. For instance, the zero calibration of an instrument is a systematic error. It is possible to eliminate systematic errors if a good mathematical model of the phenomenon is available. The random errors, on the other hand, are those of undetermined value and source and result from both intervals of measuring instruments and nonnominal environmental conditions. These type of errors cannot, by definition, be reduced by experimental means, but can be reduced by multiple measurements.

The development of error models for inertial navigation systems is of the utmost importance in the analysis and evaluation of systems using the INS. The reason for this is that, unlike position fixing systems, INSs are essentially dead reckoning systems similar to compass/airspeed navigation. Note that the "model" is the mathematical description of the physical system which is used in the Kalman filter design. Commonly, a Kalman filter is implemented within an INS which accepts the bounding source (e.g., a

---

*A locally level (or "local vertical," as it is sometimes called) coordinate frame is one in which two axes of the frame are always normal to $g(\mathbf{R})$.

secondary sensor such as Doppler, GPS, Omega, TACAN, etc.) inputs and provides a best estimate of the vehicle velocity, position, attitudes, and heading, as well as estimates of the modeled error sources and the inertial system errors.

The INS must be initialized with position and velocity information, so that in the free inertial mode the INS must determine subsequent vehicle accelerations relative to a reference frame and integrate these to maintain current velocity and position. It must be noted here that anomalies in the gravitational field cause errors in computing acceleration which must be included in the system model and corrected. Because of the difficulty in controlling the initialization and subsequent errors, aided navigation systems using external sources of position and velocity are routinely used. Errors induced into the INS propagate in a characteristic manner (i.e., Schuler oscillation, etc.) due to operation in the gravitational field of the earth. The propagation equations are dependent on the system position and velocity. However, since error propagation cannot be generally described by closed form solutions, simulation techniques must be used.

In order to determine how the computed position and velocity deviate from the true (or ideal), position and velocity require an error model encompassing the navigation parameters, inertial instruments, and noninertial external measurement sensors. The principal navigation error parameters are vehicle position, velocity, and heading. A number of good textbooks discuss error analysis equations and error propagation for an INS [1, 5, 8]. Three sets of coordinate axes are necessary in defining and deriving the system error equations:

True set (X_T, Y_T, Z_T) Specifies the ideal (or error-free) platform orientation relative to the earth.
Platform set (X_p, Y_p, Z_p) Specifies the actual platform alignment (IMU "tilt" error).
Computer set (X_c, Y_c, Z_c) Specifies the computer's perception of platform alignment (i.e., angular position error).

The most adaptable convention, in terms of both conventional error analysis and state variable description of system error behavior, employs the angle error vectors $\delta\theta$, ψ, and ϕ. In an error-free system, the three sets of coordinate axes are coincident. In an actual system, which is subject to navigation sensor and initial condition errors, the three sets are angularly divergent. For small-divergence angles, we define the three vectors as follows:

$$\phi = \begin{bmatrix} \phi_x \\ \phi_y \\ \phi_z \end{bmatrix} \quad \text{Angle from ideal (or true) to platform axes (i.e., platform misalignment or attitude error)}$$

5.2. Basic INS System Error Equations

$$\delta\theta = \begin{bmatrix} \delta\theta_x \\ \delta\theta_y \\ \delta\theta_z \end{bmatrix}$$
Angle from ideal (or true) to computer axes (i.e., error in computed position)

$$\psi = \begin{bmatrix} \psi_x \\ \psi_y \\ \psi_z \end{bmatrix}$$
Angle from computer to platform axes (this corresponds to platform drift)

The relationship between these three sets of error angles is

$$\phi = \psi + \delta\theta \qquad (5.1)$$

Error analysis for inertial navigation systems can be performed by using the nine-state "Pinson" error model [5]. This involves three-position errors, three-velocity errors, and three-misalignment or attitude errors. The nine-state Pinson error model can readily be derived by application of the "psi" (ψ) equations to the tilt angle errors to obtain the position and tilt error state equations. Therefore, we will use the Pinson model that is parameterized with the error angles ψ. Figure 5-1 is a two-dimensional illustration of the relationships between the error Euler angles defined above.

The Euler angles between the three coordinate systems are very small, and thus vector addition is valid since the theory of infinitesimal rotations applies. Position errors can be defined by the three misalignment angles $\delta\theta_x$, $\delta\theta_y$, and $\delta\theta_z$, which represent the angular displacement or misorientation in position and heading between the airborne computer's "conception" of where the navigation axes are with respect to the earth and where an errorless (or ideal) set of navigation axes lie. If a wander–azimuth angle is mechanized,

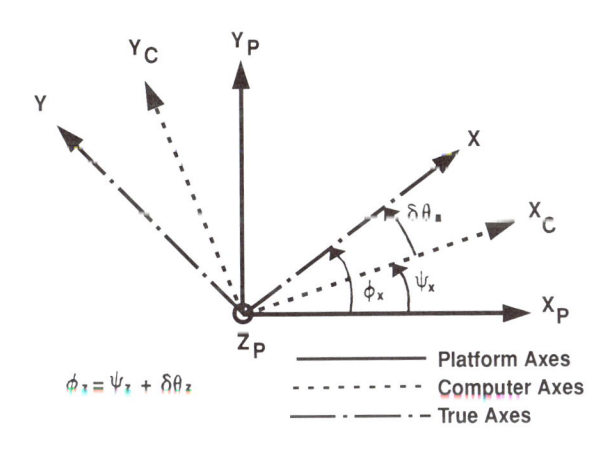

FIGURE 5-1

Two-dimensional illustration of INS error Euler angles.

it can be shown (Euler transformation) that $\delta\theta_x$, $\delta\theta_y$, and $\delta\theta_z$ are related to the latitude ($\delta\phi$), longitude ($\delta\lambda$), and wander angle ($\delta\alpha$) errors by the following expressions:

$$\delta\phi = \delta\theta_x \sin\alpha + \delta\theta_y \cos\alpha$$

$$\delta\lambda = \delta\theta_x \cos\alpha - \delta\theta_y \left(\frac{\sin\alpha}{\cos\phi}\right)$$

$$\delta\alpha = \delta\theta_z - \delta\lambda \sin\phi$$

In addition, the error in computed position (defined as computed position minus ideal position) is defined by the relation

$$\delta \mathbf{R} = \delta\boldsymbol{\theta} \times \mathbf{R} \tag{5.2}$$

where $\mathbf{R}$ is the position vector defining the location of the vehicle with respect to the center of the earth. The error equations will be developed in terms of $\delta\boldsymbol{\theta}$, $\boldsymbol{\psi}$, and $\delta\mathbf{V}$ (velocity errors), with occasional reference to Eqs. (5.1) and (5.2) to assist in their reduction to a simplified form. Pitman [8] deduces that the behavior of $\boldsymbol{\psi}$ is governed by the equation

$$\dot{\boldsymbol{\psi}} = \boldsymbol{\varepsilon} \tag{5.3}$$

where the differentiation refers to an inertially fixed reference frame and $\boldsymbol{\varepsilon}$ is the vectorial platform drift due to gyroscope drift and torquer errors. Equation (5.3) may be written in any coordinate system. By application of the Coriolis equation, we obtain

$$\dot{\boldsymbol{\psi}} = \boldsymbol{\varepsilon} + \boldsymbol{\psi} \times \boldsymbol{\omega} \tag{5.4}$$

or we can write it as in Eq. (4.25)

$$\dot{\boldsymbol{\psi}} + \boldsymbol{\omega} \times \boldsymbol{\psi} = \boldsymbol{\varepsilon}$$

Equation (5.4) is usually referred to as the "psi" equation and is of significant value since, with its use, the error equations become basically uncoupled. The vector error equations are linear equations, but in general they have time-varying coefficients. The psi equations are driven by the gyroscope drift rate vector $\boldsymbol{\varepsilon}$ and are solved first. Its solution then becomes a driving function for the position error equation. In particular, the misalignment $\boldsymbol{\psi}$ is caused by four principal factors: (1) initial misalignment ($\boldsymbol{\psi}_0$), (2) gyroscope drift, (3) torquing errors, and (4) computational errors. (Note that computational errors are commonly small in gimbaled systems; however, they become pronounced in strapdown configurations.)

Equation (5.4) can also be written in component form along the computed navigation axes X, Y, Z (here we assume that the X axis points east, the Y axis north, and the Z axis upward and nominally along the normal to the reference ellipsoid). Therefore, the psi equation can be expanded into

5.2. Basic INS System Error Equations

three first-order differential equations:

$$\dot{\psi}_X = \varepsilon_X + \omega_Z \psi_Y - \omega_Y \psi_Z \tag{5.5a}$$

$$\dot{\psi}_Y = \varepsilon_Y + \omega_X \psi_Z - \omega_Z \psi_X \tag{5.5b}$$

$$\dot{\psi}_Z = \varepsilon_Z + \omega_Y \psi_X - \omega_X \psi_Y \tag{5.5c}$$

The state variable description of system error propagation may be computed by reference to the basic mechanization equations of an incrtial navigator discussed in Chapter 4. The basic INS mechanization is [6]

$$\left[\frac{d\mathbf{V}}{dt}\right]_p = \mathbf{A} - (\mathbf{\Omega} + \boldsymbol{\omega}_{\mathrm{IC}}) \times \mathbf{V} - \mathbf{g}(\mathbf{R}) \tag{4.22}$$

or in terms of error quantities

$$\frac{d\,\delta\mathbf{V}}{dt} = \delta\mathbf{A} - (\mathbf{\Omega} + \boldsymbol{\omega}_{\mathrm{IC}}) \times \delta\mathbf{V} - \delta\mathbf{g}(\mathbf{R})$$

where $\boldsymbol{\omega}_{\mathrm{IC}}$ is the angular rate of the computer frame with respect to the inertial space. The term $\delta\mathbf{A}^*$ measured in the computer frame consists of errors due to the acceleration measuring instruments, $\delta\mathbf{A}_p$ (in the platform frame), plus errors due to transformation from the platform frame to the computer frame through the angle, $-\boldsymbol{\psi}$. Thus

$$\delta\mathbf{A} = \delta\mathbf{A}_p - \boldsymbol{\psi} \times \mathbf{A} \tag{5.6}$$

where products of errors have been neglected. Moreover, since $\mathbf{g}$ is accurately defined in ideal coordinates,

$$\delta\mathbf{g} = \delta\boldsymbol{\theta} \times \mathbf{g} \tag{5.7}$$

in the computer frame. Substituting Eqs. (5.6) and (5.7) into the error equation [Eq. (4.22)] yields

$$\delta\dot{\mathbf{V}} = \delta\mathbf{A}_p - \boldsymbol{\psi} \times \mathbf{A} - (\mathbf{\Omega} + \boldsymbol{\omega}) \times \delta\mathbf{V} + \mathbf{g} \times \delta\boldsymbol{\theta} \tag{5.8}$$

From the Coriolis equation [Eq. (6.15)] we can write

$$\mathbf{V} = \left[\frac{d\mathbf{R}}{dt}\right]_I = \left[\frac{d\mathbf{R}}{dt}\right]_E + \boldsymbol{\omega}_{\mathrm{EC}} \times \mathbf{R} = \mathbf{v} + \mathbf{\Omega} \times \mathbf{R}$$

where $\boldsymbol{\omega}_{\mathrm{EC}}$ is the angular rate of the computer frame with respect to the earth ($\boldsymbol{\omega}_{\mathrm{EC}} = \boldsymbol{\rho} = \boldsymbol{\omega}_{\mathrm{IC}} - \mathbf{\Omega}$). We can write the Coriolis equation in component

---

*δ represents a perturbation.

form in computed navigation axes (X, Y, Z) as follows:

$$\delta \dot{R}_X = \delta V_X - \rho_Y \delta R_Z + \rho_Z \delta R_Y$$
$$\delta \dot{R}_Y = \delta V_Y - \rho_Z \delta R_X + \rho_X \delta R_Z \qquad (5.9)$$
$$\delta \dot{R}_Z = \delta V_Z - \rho_X \delta R_Y + \rho_Y \delta R_X$$

Assuming that either an altitude constraint is placed on the vehicle (surface applications) or assistance from an external altitude reference (aircraft applications) maintains δR_Z and $\delta \dot{R}_Z$ close to zero. Then

$$\delta \dot{R}_X = \delta V_X + \rho_Z \delta R_Y$$
$$\delta \dot{R}_Y = \delta V_Y - \rho_Z \delta R_X \qquad (5.10)$$

Utilizing Eq. (5.2) in its component form

$$\delta R_X = \delta \theta_Y R_Z - \delta \theta_Z R_Y = \delta \theta_Y R_Z$$
$$\delta R_Y = \delta \theta_Z R_X - \delta \theta_X R_Z = -\delta \theta_X R_Z \qquad (5.11)$$

Since $R_X \approx R_Y \approx 0$, and differentiating, yields

$$\delta \dot{R}_X = R_Z \delta \dot{\theta}_Y$$
$$\delta \dot{R}_Y = -R_Z \delta \dot{\theta}_X \qquad (5.12)$$

for constant-altitude flight. Substitution of Eqs. (5.11) and (5.12) into (5.10) finally yields

$$\delta \dot{\theta}_X = -\frac{\delta V_Y}{R_Z} + \rho_Z \delta \theta_Y = -\frac{\delta V_Y}{R_Z} + \rho_Z \frac{\delta R_X}{R_Z}$$
$$\delta \dot{\theta}_Y = \frac{\delta V_X}{R_Z} - \rho_Z \delta \theta_X = \frac{\delta V_X}{R_Z} - \rho_Z \left(-\frac{\delta R_Y}{R_Z}\right) \qquad (5.13)$$

For clarity, Eqs. (5.8) and (5.13) will be written in component form in the computer navigation axes (X, Y, Z) under the assumed conditions

$$\delta \dot{\theta}_X = -\frac{\delta V_Y}{R_Z} + \rho_Z \delta \theta_Y$$

$$\delta \dot{\theta}_Y = \frac{\delta V_X}{R_Z} - \rho_Z \delta \theta_X \qquad (5.14)$$

$$\delta \dot{V}_X = \delta A_X + \psi_Z A_Y - \psi_Y A_Z + (\Omega_Z + \omega_Z) \delta V_Y - g_Z \delta \theta_Y$$
$$\delta \dot{V}_Y = \delta A_Y - \psi_Z A_X + \psi_X A_Z - (\Omega_Z + \omega_Z) \delta V_X + g_Z \delta \theta_X$$

Considerable simplification is possible for cruise conditions when $A_Z = g_Z = \mathbf{g}$. Then, Eqs. (5.5) and (5.14) under these conditions yield the final

5.2. Basic INS System Error Equations

state variable equations describing the error model of the subject inertial system:

$$\left.\begin{aligned}
\delta\dot{V}_X &= \delta A_X + \psi_Z A_Y - g\psi_Y + (\Omega_Z + \omega_Z)\delta V_Y - g\,\delta\theta_Y \\
\delta\dot{V}_Y &= \delta A_Y - \psi_Z A_X + g\psi_X + (\Omega_Z + \omega_Z)\delta V_X + g\,\delta\theta_X \\
\delta\dot{\theta}_X &= \frac{\delta V_Y}{R} + (\omega_Z - \Omega_Z)\delta\theta_Y \\
\delta\dot{\theta}_Y &= \frac{\delta V_X}{R} - (\omega_Z - \Omega_Z)\delta\theta_X \\
\dot{\psi}_X &= \omega_Z \psi_Y - \omega_Y \psi_Z + \varepsilon_X \\
\dot{\psi}_Y &= \omega_X \psi_Z - \omega_Z \psi_X + \varepsilon_Y \\
\dot{\psi}_Z &= \omega_Y \psi_Y - \omega_X \psi_Y + \varepsilon_Z
\end{aligned}\right\} \quad (5.15)$$

Equation (5.15) represents the basic inertial system error model used as a basis for the generation of the system dynamics **F**-matrix, and state-transition matrix.

Using vector notation, the general nine-state error model can be summarized for a local-level (NEU) coordinate system as follows:

$$\delta\dot{\mathbf{R}} = -\boldsymbol{\rho} \times \delta\mathbf{R} + \delta\mathbf{V} \qquad (5.16)$$

$$\delta\dot{\mathbf{V}} = \begin{bmatrix} -\omega_s^2 & 0 & 0 \\ 0 & -\omega_s^2 & 0 \\ 0 & 0 & 2\omega_s^2 \end{bmatrix} \delta\mathbf{R} - (\boldsymbol{\rho} + 2\boldsymbol{\Omega}) \times \delta\mathbf{V} - \boldsymbol{\psi} \times \mathbf{A} + C\boldsymbol{\varepsilon}_a \quad (5.17)$$

$$\boldsymbol{\psi} = -(\boldsymbol{\rho} + \boldsymbol{\Omega}) \times \boldsymbol{\psi} + C\boldsymbol{\varepsilon}_g \qquad (5.18)$$

where
  $\mathbf{R}$ = position vector of vehicle $\simeq [0, 0, r_e + h]^T$
  $\boldsymbol{\rho}$ = angular velocity of the local-level frame relative to the earth $\simeq (\mathbf{v} \times \mathbf{R})/R^2$
  $\boldsymbol{\Omega}$ = earth's angular velocity vector relative to an inertial frame
  $\omega_s^2 = g/R \simeq \mu/R^3 \simeq$ Schuler rate
  $\boldsymbol{\psi}$ = small-angle error in the body–earth transformation
  C = transformation matrix from the body frame to the local-level frame = C_b^L
  $\mathbf{v}$ = vehicle velocity relative to the earth; error $\simeq \delta\mathbf{v} - \delta\mathbf{V} - \boldsymbol{\varepsilon} \times \mathbf{v}$
  $\mathbf{g}$ = plumb-bob gravity vector = $[0\ \ 0\ \ -g]^T$
    (g is assumed to lie entirely along the vertical direction)
  $\boldsymbol{\varepsilon}_a, \boldsymbol{\varepsilon}_g$ = accelerometer and gyro error terms excited by sensed acceleration (A_s): these error terms would be zero if the instruments were error-free

The relationship to the Pinson error model is indicated below.

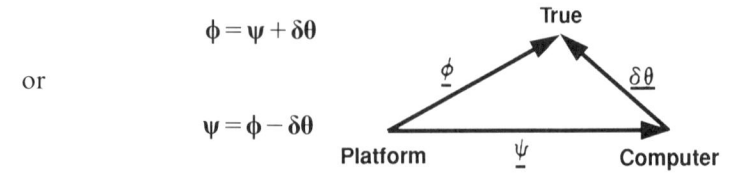

$$\phi = \psi + \delta\theta$$

or

$$\psi = \phi - \delta\theta$$

In scalar form, the above error model can be derived as follows. First, the matrix ρ is a skew-symmetric realization of the vector ρ, defined by its components as

$$\rho = \begin{bmatrix} 0 & -\rho_z & \rho_y \\ \rho_z & 0 & -\rho_x \\ -\rho_y & \rho_x & 0 \end{bmatrix} \quad \text{with} \quad \rho \simeq \mathbf{v} \times \mathbf{R}/R^2 = \begin{bmatrix} v_y/R \\ -v_x/R \\ 0 \end{bmatrix}$$

The level components of ρ, as defined above, maintain the level axes perpendicular to the gravity vector (i.e., the radius vector $\mathbf{R}$). In gimbaled, platform systems, it is common to have the vertical component of ρ equal and opposite to the vertical component of Ω. This is referred to as the "free azimuth" case (see Chapter 2), and has a definite benefit in that the vertical axis of the platform need not be physically precessed, and, consequently, less hardware is required to mechanize the system. Here we will assume that the position errors are small, so that $\delta \mathbf{v} = \delta \mathbf{V}$. Next, we note that

$$\rho \times \delta \mathbf{R} = \begin{vmatrix} \mathbf{i} & \mathbf{j} & \mathbf{k} \\ \dfrac{V_y}{R} & -\dfrac{V_x}{R} & 0 \\ \delta R_x & \delta R_y & \delta R_z \end{vmatrix} = \left[-\dfrac{V_x}{R} \delta R_z \right] \mathbf{i} - \left[\dfrac{V_y}{R} \delta R_z \right] \mathbf{j} + \left[\dfrac{V_y}{R} \delta R_y + \dfrac{V_x}{R} \delta R_x \right] \mathbf{k}$$

Therefore

$$-\rho \times \delta \mathbf{R} + \delta \mathbf{V} = \begin{bmatrix} \dfrac{V_x}{R} \delta R_z \\ \dfrac{V_y}{R} \delta R_z \\ -\dfrac{V_y}{R} \delta R_y - \dfrac{V_x}{R} \delta R_x \end{bmatrix} + \begin{bmatrix} \delta V_x \\ \delta V_y \\ \delta V_z \end{bmatrix}$$

so that

$$\delta \dot{R}_x = \dfrac{V_x}{R} \delta R_z + \delta V_x \tag{5.19a}$$

$$\delta \dot{R}_y = \dfrac{V_y}{R} \delta R_z + \delta V_y \tag{5.19b}$$

5.2. Basic INS System Error Equations

$$\delta \dot{R}_z = -\frac{V_y}{R}\delta R_y - \frac{V_x}{R}\delta R_x + \delta V_z \tag{5.19c}$$

Assuming now that the earth rate vector $\Omega = [\Omega \cos \phi \quad 0 \quad -\Omega \sin \phi]^T$, the $(\rho + 2\Omega) \times \delta V$ term becomes

$$(\rho + 2\Omega) \times \delta V = \begin{vmatrix} \mathbf{i} & \mathbf{j} & \mathbf{k} \\ \frac{V_y}{R} + 2\Omega \cos \phi & -\frac{V_x}{R} & -2\Omega \sin \phi \\ \delta V_x & \delta V_y & \delta V_z \end{vmatrix}$$

$$= \left[-\frac{V_x}{R}\delta V_z + 2\Omega \sin \phi \, \delta V_y\right]\mathbf{i}$$

$$- \left[\left(\frac{V_y}{R} + 2\Omega \cos \phi\right)\delta V_z + 2\Omega \sin \phi \, \delta V_x\right]\mathbf{j}$$

$$+ \left[\left(\frac{V_y}{R} + 2\Omega \cos \phi\right)\delta V_y + \frac{V_x}{R}\delta V_x\right]\mathbf{k}$$

Also

$$\psi \times \mathbf{A} = \begin{vmatrix} \mathbf{i} & \mathbf{j} & \mathbf{k} \\ \psi_x & \psi_y & \psi_z \\ A_x & A_y & A_z \end{vmatrix}$$

$$= (\psi_y A_z - \psi_z A_y)\mathbf{i} - (\psi_x A_z - \psi_z A_x)\mathbf{j} + (\psi_x A_y - \psi_y A_x)\mathbf{k}$$

The last term, $C\varepsilon_a$ becomes

$$C\varepsilon_a = \begin{bmatrix} C_{11} & C_{12} & C_{13} \\ C_{21} & C_{22} & C_{23} \\ C_{31} & C_{32} & C_{33} \end{bmatrix}\begin{bmatrix} \varepsilon_{a_x} \\ \varepsilon_{a_y} \\ \varepsilon_{a_z} \end{bmatrix} = \begin{bmatrix} C_{11}\varepsilon_{a_x} + C_{12}\varepsilon_{a_y} + C_{13}\varepsilon_{a_z} \\ C_{21}\varepsilon_{a_x} + C_{22}\varepsilon_{a_y} + C_{23}\varepsilon_{a_z} \\ C_{31}\varepsilon_{a_x} + C_{32}\varepsilon_{a_y} + C_{33}\varepsilon_{a_z} \end{bmatrix}$$

Combining all the above terms, we have the following velocity error equations:

$$\delta \dot{V}_x = -\omega_s^2 \delta R_x + \left(-\frac{V_x}{R}\right)\delta V_z + (2\Omega \sin \phi)\delta V_y - \psi_y A_z + \psi_z A_y$$

$$+ C_{11}\varepsilon_{a_x} + C_{12}\varepsilon_{a_y} + C_{13}\varepsilon_{a_z} \tag{5.20a}$$

$$\delta \dot{V}_y = -\omega_s^2 \delta R_y + \left(-\frac{V_y}{R} - 2\Omega \cos \phi\right)\delta V_z - (2\Omega \sin \phi)\delta V_x + \psi_x A_z$$

$$- \psi_z A_x + C_{21}\varepsilon_{a_x} + C_{22}\varepsilon_{a_y} + C_{23}\varepsilon_{a_z} \tag{5.20b}$$

$$\delta \dot{V}_z = 2\omega_s^2 \, \delta R_z + \left(\frac{V_y}{R} + 2\Omega \cos \phi\right) \delta V_y + \left(\frac{V_x}{R}\right) \delta V_x - \psi_x A_y + \psi_y A_x$$
$$+ C_{31}\varepsilon_{a_x} + C_{32}\varepsilon_{a_y} + C_{33}\varepsilon_{a_z} \qquad (5.20c)$$

The equations for $\delta \dot{V}_x$ and $\delta \dot{V}_y$ are commonly called the *level velocity error equations*. The vertical velocity error equation, δV_z, cannot be obtained as indicated, since it is basically unstable. However, it can be damped by a baroinertial loop to affect a reasonable solution for δV_z. Following the above procedure, it is easy to derive the components of the ψ equation. These components are as follows [solution of $\psi(t)$ for the platform drift]:

$$\dot{\psi}_x = \left(-\frac{V_x}{R}\right)\psi_z + (\Omega \sin \phi)\psi_y + C_{11}\varepsilon_{a_x} + C_{12}\varepsilon_{a_y} + C_{13}\varepsilon_{a_z} \qquad (5.21a)$$

$$\dot{\psi}_y = -\left(\frac{V_y}{R} + \Omega \cos \phi\right)\psi_z + (\Omega \sin \phi)\psi_x + C_{21}\varepsilon_{a_x}$$
$$+ C_{22}\varepsilon_{a_y} + C_{23}\varepsilon_{a_z} \qquad (5.21b)$$

$$\dot{\psi}_z = \left(\frac{V_y}{R} + \Omega \cos \phi\right)\psi_y + \left(\frac{V_x}{R}\right)\psi_x + C_{31}\varepsilon_{a_x} + C_{32}\varepsilon_{a_y} + C_{33}\varepsilon_{a_z} \qquad (5.21c)$$

In some applications, where only horizontal navigation is desired, the above equations can be simplified. Thus, for level navigation, $\delta V_z = 0$, and if we assume error free instruments, the equations become

$$\begin{aligned}
\delta \dot{R}_x &= \delta V_x \\
\delta \dot{R}_y &= \delta V_y \\
\delta \dot{V}_x &= -\omega_s^2 \, \delta R_x + \left(-\frac{V_x}{R}\right)\delta V_z + (2\Omega \sin \phi) \, \delta V_y - \psi_y A_z + \psi_z A_y \\
\delta \dot{V}_y &= -\omega_s^2 \, \delta R_y + \left(-\frac{V_y}{R} - 2\Omega \cos \phi\right)\delta V_z - (2\Omega \sin \phi) \, \delta V_x \\
&\quad + \psi_x A_z - \psi_z A_x \\
\dot{\psi}_x &= \left(-\frac{V_x}{R}\right)\psi_z + (\Omega \sin \phi)\psi_y \\
\dot{\psi}_y &= -\left(\frac{V_y}{R} + \Omega \cos \phi\right)\psi_z + (\Omega \sin \phi)\psi_x \\
\dot{\psi}_z &= \left(\frac{V_y}{R} + \Omega \cos \phi\right)\psi_y + \left(\frac{V_x}{R}\right)\psi_x
\end{aligned} \qquad (5.22)$$

Finally, for time periods short compared to the earth rate, $1/\Omega$ (about 3.9 h), and neglecting the Coriolis term, Eqs. (4.87a)–(4.87c) can be simplified to three uncoupled equations as follows [making use of Eq. (5.8)]:

$$\delta \ddot{R}_x + \frac{g_0}{R_0} \delta R_x = -\psi_y A_z + \psi_z A_y + \nabla_x \qquad (5.23a)$$

$$\delta \ddot{R}_y + \frac{g_0}{R_0} \delta R_y = -\psi_z A_x + \psi_x A_z + \nabla_y \qquad (5.23b)$$

$$\delta \ddot{R}_z - \frac{2g_0}{R_0} \delta R_z = -\psi_x A_y + \psi_y A_x + \nabla_z \qquad (5.23c)$$

where $\nabla_{(x,y,z)}$ is the accelerometer bias error. The last equation has a pole in the right-hand plane, giving rise to the well-known cosh instability in the vertical channel, while the first two equations have two poles on the imaginary axis and therefore act as undamped oscillators, with resonant frequency $\sqrt{g_0/R_0}$, the familiar Schuler frequency.

5.3. GENERAL INS ERROR EQUATIONS

In order to determine how the computed position and velocity deviate from the true (or ideal) position and velocity, an error model is required, encompassing the navigation parameters, inertial instruments, and noninertial external measurement sensors. The principal navigation error parameters are vehicle position, velocity, and heading. Inertial platform (off-level) tilt is also of interest in beam pointing or tracking applications. A commonly used error model for the navigation parameters associated with inertial systems is the Pinson model, discussed in the previous section. The Pinson model is based on velocity and craft rate vector equations which may be evaluated in any coordinate system. Expressed in true coordinates, these equations are as follows:

$$\delta \mathbf{V}^T = \mathbf{V}_c^T - \mathbf{V}_T^T$$

$$\delta \mathbf{\rho}^T = \mathbf{\rho}_c^T - \mathbf{\rho}_T^T$$

where the subscript c denotes computed values and T denotes the true or error free values. That is, the navigation system errors are defined as the difference between computed values and true values. Similarly, the navigation direction cosine errors are defined as

$$\delta C = \ddot{C}_c - \ddot{C}_T$$

where the C_c are computed direction cosine elements and the C_T are ideal direction cosines. Because of its simplicity, the Pinson error model is probably best suited to system mechanizations in which the angle measurements are made with a star tracker. There are three error angle sets that must be defined in order to study error propagation in inertial systems. Section 5.2 defines these error angles. Consequently, the manner in which navigation system errors are defined is basic to an error dynamics formulation. In the discussion that follows, the symbol $\delta(\,\cdot\,)$ will denote sets of numbers obtained by differencing the computed values in computer coordinates with the true values (in ideal coordinates). In Section 4.3 we noted that inertial navigation system mechanizations generallly fall into categories of local-level, space-stable, wander–azimuth, and strapdown systems. Moreover, since generally one is interested in earth-referenced quantities, the accelerometer outputs are usually either transformed to the local-level frame, or the accelerometer outputs are integrated in the inertial frame, and then transformed in open-loop configuration to yield latitude, longitude, velocity, and so on. Britting [1] has shown that the basic error differential equations for any INS may be written in standard coordinates, regardless of the physical mechanization or internal navigation variables. Furthermore, under certain very broad assumptions, the homogeneous (unforced) portion of these differential equations is identical for any arbitrary configured terrestrial INS. Consequently, INS error propagation is, to a large extent, completely independent of system mechanization. The error propagation equations for an inertial navigation system can be derived in the form of a set of first-order linear differential equations. Using state–space notation, the standard form of these equations is [7]

$$\dot{\mathbf{x}}(t) = \mathbf{F}(t)\mathbf{x}(t) + \mathbf{G}(t)\mathbf{u}(t) + \mathbf{w}(t), \quad \mathbf{x}(0) \text{ given} \quad (5.24)$$

where $\mathbf{x}$ is the state vector; $\mathbf{w}$ is a white, Gaussian forcing function compensating for approximations in $\mathbf{F}(t)$; and $\mathbf{u}$ is the rate-of-correction vector due to noninertial aiding (when available). The **F**-matrix defines the dynamical coupling among the position, velocity, and attitude errors themselves, while the **G**-matrix, if used, defines the coupling between the navigation parameter errors and the inertial sensors. The state vector $\mathbf{x}$ consists of position and velocity errors coordinatized in the inertial frame, platform tilt, states in the altitude and damping filters, and states that represent the various error sources driving the system such as gyroscope drift rate, gravity anomalies, and reference velocity error. In general, it is assumed that the error sources are random in nature.

5.3.1. Position Error Equations

Accurate navigation systems must compensate for the changes in the radius of curvature of the earth, whereby the earth's shape is approximated as an oblate spheroid (i.e., reference ellipsoid). As a result, a small change in

5.3. General INS Error Equations

latitude is defined as the radius of curvature r of the path

$$r = \left|\frac{ds}{d\phi}\right|$$

From Section 4.3, the radius of curvature in the north–south direction is

$$R_\phi = \frac{R_0(1-\varepsilon^2)}{(1-\varepsilon^2 \sin^2 \phi)^{3/2}}$$

while in the east–west direction, the radius of curvature is

$$R_\lambda = \frac{R_0}{(1-\varepsilon^2 \sin^2 \phi)^{1/2}}$$

In the present discussion, we will consider a north–east–up (NEU) coordinate system. The equations governing geographic latitude* ϕ, longitude λ, and altitude h, above the reference ellipsoid are as follows:

Latitude

$$\phi = \phi_0 + \int_0^t \dot\phi \, dt$$

$$\dot\phi = \frac{V_N}{R_\phi + h}$$

(5.25)

Longitude

$$\lambda = \lambda_0 + \int_0^t \dot\lambda \, dt$$

$$\dot\lambda = \frac{V_E}{(R_\lambda + h)\cos\phi} = \frac{V_E}{R_0 \cos\phi}\left[1 - f\sin^2\phi - \frac{h}{R_0}\right]$$

(5.26)

Altitude

$$\dot h = V_z$$

(5.27)

Equations (5.25)–(5.27) can be used by the inertial navigator to compute latitude, longitude, and altitude; however, the computed latitude, longitude, and altitude are estimates (because of approximations made in these equations). The estimate of a state variable is commonly defined as

$$\hat x = x + \delta x$$

(5.28)

where $\hat x$ is the output (estimation) from the INS, x is the true value, and δx is the error in the estimation, defined to be positive if the estimate is larger

---

*In many navigation systems, latitude and longitude are the desired outputs; therefore, the system should be mechanized so as to yield those outputs directly.

than the true state. Similarly, the time derivative of the error in estimation is

$$\delta\dot{x} = \dot{\hat{x}} - \dot{x} \tag{5.29}$$

Assume that the inertial navigator computes its estimate $\hat{\lambda}$ of longitude by integration of the same equation using estimated variables as required. Then

$$\dot{\hat{\lambda}} = \frac{V_E}{R_0 \cos \phi} \left[1 - f \sin^2 \phi - \frac{h}{R_0} \right] + u_\lambda \tag{5.30}$$

The variable u_λ is used in aided navigation systems and is the rate of change of the indicated longitude due to the adding equations. (Note that in pure inertial navigation systems this variable is zero.) The estimate of longitude and the time derivative of longitude error are as follows:

$$\begin{aligned} \hat{\lambda} &= \lambda + \delta\lambda \\ \delta\dot{\lambda} &= \dot{\hat{\lambda}} - \dot{\lambda} \end{aligned} \tag{5.31}$$

From Eq. (5.31) the longitude error rate is

$$\delta\dot{\lambda} = \frac{\hat{V}_E [1 - f \sin^2 \hat{\phi} - \hat{h}/R_0]}{R_0 \cos \hat{\phi}} + u_\lambda - \dot{\lambda} \tag{5.32}$$

Using Eq. (5.28) for $\hat{V}_E$, $\hat{\phi}$, $\hat{h}$ and expanding the trigonometric functions to first-order in $\delta\phi$, we obtain

$$\delta\dot{\lambda} = \frac{(V_E + \delta V_E)[1 - f \sin^2 \phi - 2f \sin \phi \cos \phi \, \delta\phi - (h + \delta h)/R_0]}{(R_0 \cos \phi - R_0 \sin \phi \, \delta \phi)} + u_\lambda - \dot{\lambda} \tag{5.33}$$

In general, the squares and/or cross-products of $\delta\phi$, $\delta\lambda$, δh, δV_E, δV_N, δV_U, f, and h/R_0 can be neglected. Therefore, the term including $f \, \delta\phi$ in Eq. (5.33) is neglected. Equation (5.33) can therefore be written as

$$\delta\dot{\lambda} = \frac{V_E}{R_0 \cos \phi} \left[1 - f \sin^2 \phi - \frac{h}{R_0} + \tan \phi \, \delta\phi - \frac{\delta h}{R_0} + \frac{\delta V_E}{V_E} \right] + u_\lambda - \dot{\lambda} \tag{5.34}$$

where again second-order terms have been neglected. The first three terms are recognized as being the true longitude rate $\dot{\lambda}$ from Eq. (5.26). Hence, the linearized error equation is

$$\delta\dot{\lambda} = \delta\phi \left(\frac{V_E}{R_0} \right) \frac{\tan \phi}{\cos \phi} - \frac{\delta h V_E}{R_0^2 \cos \phi} + \left(\frac{\delta V_E}{R_0} \right) \frac{1}{\cos \phi} + u_\lambda \tag{5.35}$$

5.3. General INS Error Equations

Making use of the equations for ρ_E, ρ_N, ρ_U, R_ϕ, and R_λ from Section 4.3, Eq. (5.35) can be written in the form

$$\delta\dot{\lambda} = \left(\frac{\rho_U}{\cos\phi}\right)\delta\phi - \left(\frac{\rho_N}{R_0\cos\phi}\right)\delta h + \left(\frac{1}{R_0\cos\phi}\right)\delta V_E + u_\lambda \quad (5.36)$$

From Eq. (5.25) the latitude of the aircraft can be expressed approximately as

$$\dot{\phi} = \frac{V_N}{R_0}\left[1 + 2f - 3f\sin^2\phi - \frac{h}{R_0}\right] \quad (5.37)$$

Proceeding as in the case of the derivation for the longitude error, the differential equation for the latitude error assumes the form

$$\delta\dot{\phi} = \delta h\left(\frac{\rho_E}{R_0}\right) + \left(\frac{\delta V_N}{R_0}\right) + u_\phi \quad (5.38)$$

where u_ϕ is used in the case of aided inertial navigation systems. Finally, from Eq. (5.27), the error in the indicated altitude is

$$\delta\dot{h} = \delta V_U + u_h \quad (5.39)$$

where u_h is the rate of correction of altitude due to the barometric altimeter or other altitude aiding source.

5.3.2. Velocity Error Equations

In order to derive the inertial navigation system velocity error equations, the reference frames must first be defined. Let (x, y, z) be the vehicle's local-level geographic, "g," coordinate system, which is located at or near the surface of the earth. As in the previous section, this is a north–east–up (NEU) triad, where the x axis is identified with the east direction, the y axis with the north direction, and the z axis up (or vertical). Furthermore, let the (X, Y, Z) triad be inertially fixed with the Z axis in the polar direction and the (X, Y) axes define the equatorial plane; as we noted in Chapter 2, this coordinate system is known as *earth-centered inertial* (ECI) (having an origin that is at the center of the earth and that is motionless with respect to the fixed stars). The equations of motion simplify for this ECI coordinate frame in the sense that no additional Coriolis cross-product terms normally associated with a moving coordinate frame need be considered. These coordinate systems are illustrated in Fig. 5-2.

FIGURE 5-2

Coordinate systems for position and velocity error equations.

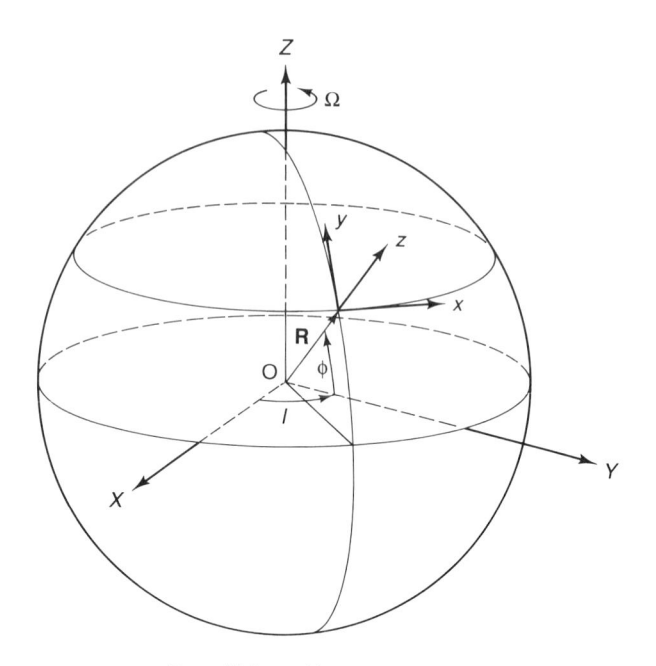

R = vehicle position vector
ϕ = geographic latitude
l = celestial longitude
  = $\lambda + \Omega t$ (when $t = 0$, $l = \lambda$)

The generalized navigation equation expressed in component form along the platform (x, y, z) axes can be written as follows (see Section 4.3):

$$\dot{V}_x = A_x - (\rho_y + 2\Omega_y)V_z + (\rho_z + 2\Omega_z)V_y + g_x \quad (4.25a)$$
$$\dot{V}_y = A_y - (\rho_z + 2\Omega_z)V_x + (\rho_x + 2\Omega_x)V_z + g_y \quad (4.25b)$$
$$\dot{V}_z = A_z - (\rho_x + 2\Omega_x)V_y + (\rho_y + 2\Omega_y)V_x + g_z \quad (4.25c)$$

where the x axis corresponds to east (E), the y axis north (N), and the z axis up (U); A_x, A_y, A_z are the components of the specific force. Since the inertial angular rate of the geographic frame (or platform) can be written as $\omega = \rho + \Omega$, then the vehicle velocity with respect to the rotating earth, expressed in the local geographic coordinates, can be written from Eqs. (4.25a)–(4.25c) in the form

$$\dot{V}_E = A_E - (\omega_N + \Omega_N)V_z + (\omega_z + \Omega_z)V_N + g_E \quad (5.40a)$$
$$\dot{V}_N = A_N - (\omega_z + \Omega_z)V_E + \omega_E V_z + g_N \quad (5.40b)$$
$$\dot{V}_z = A_z + (\omega_N + \Omega_N)V_E - \omega_E V_N + g_z \quad (5.40c)$$

since $\Omega_E = \rho_E$ and $\Omega_x = \Omega_E = 0$ (*Note*: $\Omega_N = \Omega_y = \Omega \cos \phi$ and $\Omega_z = \Omega \sin \phi$.)

5.3. General INS Error Equations

From the above discussion, we can now develop the velocity error equations. The error in the east velocity can be obtained from Eq. (5.40a). The time derivative of the east velocity error is

$$\delta \dot{V}_E = \hat{\dot{V}}_E - \dot{V}_E \tag{5.41}$$

To a first-order approximation, the velocity error can be written as

$$\delta \dot{V}_E = \delta A_E + \delta g_E - (\delta \omega_N + \delta \Omega_N) V_z - (\omega_N + \Omega_N) \delta V_z$$
$$+ (\delta \omega_z + \delta \Omega_z) V_N + (\omega_z + \Omega_z) \delta V_N + u_{V_E} \tag{5.42}$$

where the error in the specific force is the driving term and is given by $\delta A_E = \hat{A}_E - A_E$ while

$$\delta \omega_N + \delta \Omega_N = -\delta \phi (2\Omega \sin \phi) - \delta h \left(\frac{V_E}{R_0^2}\right) + \frac{\delta V_E}{R_0} \tag{5.43}$$

and

$$\delta \omega_z + \delta \Omega_z = \delta \phi \left[2\Omega \cos \phi + \left(\frac{V_E}{R_0}\right)\frac{1}{\cos^2 \phi}\right] - \delta h \left(\frac{V_E \tan \phi}{R_0^2}\right) + \delta V_E \left(\frac{\tan \phi}{R_0}\right) \tag{5.44}$$

Combining these results, and after some simplification, the resulting velocity error differential equation is [3]

$$\delta \dot{V}_E = \left[2(\Omega_N V_N + \Omega_z V_z) + \frac{\rho_N V_N}{\cos^2 \phi}\right]\delta \phi + \left(\rho_z \rho_E + \rho_N \frac{V_z}{R_0}\right)\delta h$$
$$+ \left(-\rho_E \tan \phi - \frac{V_z}{R_0}\right)\delta V_E + (\omega_z + \Omega_z) \delta V_N$$
$$+ (-\omega_N - \Omega_N) \delta V_z + \delta A_E + \delta g_E + u_{V_E} \tag{5.45}$$

Following the procedure outlined above for the derivative of $\delta \dot{V}_E$, and making use of the expressions

$$\delta \hat{A}_N = \hat{A}_N - A_N$$

$$\delta \omega_N - \left(\frac{V_N}{R_0^2}\right)\delta h - \frac{\delta V_N}{R_0},$$

the differential equation for the computed north velocity error is obtained from Eq. (5.40b) as

$$\delta \dot{V}_N = \left[-2\Omega_N V_E - \frac{\rho_N V_E}{\cos^2 \phi}\right]\delta \phi + \left(\rho_N \rho_z - \rho_E \frac{V_z}{R_0}\right)\delta h + (-2\omega_z) \delta V_E$$
$$+ \left(-\frac{V_z}{R_0}\right)\delta V_N + \delta V_z \rho_E + \delta A_N + \delta g_N + u_{V_N} \tag{5.46}$$

Finally, from Eq. (5.40c), the linearized differential equation for the error in the vertical velocity channel is

$$\delta \dot{V}_z = (-2\Omega_z V_E) \delta\phi - (\rho_N^2 + \rho_E^2) \delta h + (2\omega_N) \delta V_E$$
$$+ (-2\rho_E) \delta V_N + \delta A_z + \delta g_z + u_{V_z} \quad (5.47)$$

where again the specific force error driving function is $\delta A_z = \hat{A}_z - A_z$. Next, the error terms for the specific force and gravity computation are needed. From the above discussion, we note that since the specific force is being measured by the accelerometers in the platform frame, "p," a transformation matrix C_p^g from the platform to the geographic frame g is required. However, since the specific force is measured by a nonorthogonal set of instruments, the transformation matrix is an estimate that contributes to the specific force error. The transformation of the specific force from the platform to the geographic frame is

$$\hat{A}^g = \hat{C}_p^g \hat{A}^p \quad (5.48)$$

The error in transforming from the accelerometer output to the geographic frame is termed the "attitude error" or inertial system misalignment; since the angles involved are small, the attitude error may be treated as a vector Ψ. The skew-symmetric matrix form of the attitude error is

$$\Psi^g = \begin{bmatrix} 0 & -\psi_z & \psi_N \\ \psi_z & 0 & -\psi_E \\ -\psi_N & \psi_E & 0 \end{bmatrix} \quad (5.49)$$

The relationship between the computed transformation and the true transformation is [1]

$$\hat{C}_p^g = (I - \Psi^g) C_p^g \quad (5.50)$$

Recognizing that the specific force error in platform coordinates is defined by

$$\hat{A}^p = A^p + \delta A^p \quad (5.51)$$

substituting Eqs. (5.50) and (5.51) into Eq. (5.48) results in

$$\hat{A}^g = (I - \Psi^g) A^g + C_p^g \delta A^p \quad (5.52)$$

where second-order effects have been neglected. Consequently, the specific force error is

$$A^g = -\Psi^g A^g + C_p^g \delta A^p \quad (5.33)$$

Equation (5.53) can be written in component form as follows:

$$\delta A_E = -\psi_N A_z + \psi_z A_N + w_{A_E} \quad (5.54a)$$

$$\delta A_N = \psi_E A_z - \psi_z A_E + w_{A_N} \quad (5.54b)$$

$$\delta A_z = -\psi_E A_N + \psi_N A_E + w_{A_z} \quad (5.54c)$$

5.3. General INS Error Equations

where the terms $w_{A_E}, w_{A_N}, w_{A_z}$ represent the error due to truncation in Eq. (5.53); that is, $w^g = C_p^g \delta A^p$.

The errors in computed gravity δg_E, δg_N, and δg_z are by definition [see Eq. (5.28)]

$$\delta g_E = \hat{g}_E - g_E \qquad (5.55a)$$

$$\delta g_N = \hat{g}_N - g_N \qquad (5.55b)$$

$$\delta g_z = \hat{g}_z - g_z \qquad (5.55c)$$

Specifically, and as discussed in Chapter 2, the errors due to computing gravity arise mainly from two sources: (1) an imperfect model of the earth's gravitational field and (2) evaluating the gravity model at the estimated position rather than the true position. For error modeling purposes, the gravity vector is assumed to act entirely in the radial direction. Therefore, the horizontal components δg_E and δg_N are approximately zero everywhere. The vertical component of gravity, on the other hand, varies with changes in altitude, consequently an error in altitude will cause an error δg_z. From Section 4.4.2, the error δg_z is approximately $(2g/R_0)\,\delta h$. Thus, the components of the gravity error can be written in the form [4]

$$\delta g_E = w_{g_E} \qquad (5.56a)$$

$$\delta g_N = w_{g_N} \qquad (5.56b)$$

$$\delta g_z = (2g/R_0)\,\delta h + w_{g_z} \qquad (5.56c)$$

5.3.3. Attitude Error Equations

The final step in the development of the navigation error equations is the attitude error, Ψ^g. The differential equations governing the INS platform attitude errors can be developed by the use of transformation matrices between the inertial "i," geographic, and platform reference frames. The transformation matrix from the platform to the geographic frame is given by Eq. (5.48). This transformation can be divided into the transformation from the platform to the inertial frame and the transformation from the inertial to the geographic frame as follows:

$$\hat{C}_p^g = \hat{C}_i^g \hat{C}_p^i \qquad (5.57)$$

Next, the errors in the transformation from the platform to the inertial frame are given by [10]

$$\hat{C}_p^i = (\mathbf{I} - \mathbf{B}^i) C_p^i \qquad (5.58)$$

where

$$\mathbf{B}^i = \begin{bmatrix} 0 & -\beta_z & \beta_y \\ \beta_z & 0 & -\beta_x \\ -\beta_y & \beta_x & 0 \end{bmatrix} \qquad (5.59)$$

and in the inertial to geographic frame by

$$\hat{C}_i^g = (I - N^g) C_i^g \tag{5.60}$$

where

$$\mathbf{N}^g = \begin{bmatrix} 0 & -v_z & v_N \\ v_z & 0 & -v_E \\ -v_N & v_E & 0 \end{bmatrix} \tag{5.61}$$

Substituting Eqs. (5.50), (5.58), and (5.60) into Eq. (5.57), neglecting second-order terms and simplifying, we obtain [10]

$$\mathbf{\Psi}^g = \mathbf{N}^g + C_i^g \mathbf{B}^i C_g^i \tag{5.62}$$

or in vector form

$$\boldsymbol{\psi}^g = \mathbf{v}^g + \boldsymbol{\beta}^g \tag{5.63}$$

The time derivative of Eq. (5.63) is

$$\dot{\boldsymbol{\psi}}^g = \dot{\mathbf{v}}^g + \dot{\boldsymbol{\beta}}^g \tag{5.64}$$

In order to solve for $\dot{\mathbf{v}}^g$ and $\dot{\boldsymbol{\beta}}^g$, we take the time derivatives of Eqs. (5.58) and (5.60)

$$\dot{\hat{C}}_p^i = (\mathbf{I} - \mathbf{B}^i) \dot{C}_p^i - \dot{\mathbf{B}}^i C_p^i \tag{5.65}$$

$$\dot{\hat{C}}_i^g = (\mathbf{I} - \mathbf{N}^g) \dot{C}_i^g - \dot{\mathbf{N}}^g C_i^g \tag{5.66}$$

The time rate of change of a direction cosine matrix is related to the angular velocity matrix by [see Chapter 4, Eq. (4.50)]

$$\dot{C}_p^i = C_p^i \Omega_{ip} \tag{5.67}$$

$$\mathbf{\Omega}_{ip} = \begin{bmatrix} 0 & -\omega_z & \omega_y \\ \omega_z & 0 & -\omega_x \\ -\omega_y & \omega_x & 0 \end{bmatrix} \tag{5.68}$$

In order to solve for $\dot{\boldsymbol{\beta}}$, the time derivative of the platform to inertial angular error, apply Eq. (5.67) to Eq. (5.65):

$$\hat{C}_p^i \Omega_{ip} = (\mathbf{I} - \mathbf{B}^i) C_p^i \Omega_{ip}^p - \dot{\mathbf{B}}^i C_p^i \tag{5.69}$$

The indicated platform angular velocity $\hat{\Omega}_{ip}^p$ is defined as

$$\hat{\Omega}_{ip}^p = \Omega_{ip}^p - \delta\Omega_{ip}^p - \mathbf{u}_\psi^p \tag{5.70}$$

where $\delta\Omega_{ip}$ is the platform angular velocity error and $\mathbf{u}_\psi^p$ is the alignment correction rate. Substituting Eqs. (5.58) and (5.70) into Eq. (5.69) and solving for $\dot{\mathbf{B}}^i$ yields

$$\dot{\mathbf{B}}^i = C_p^i (\delta\Omega_{ip}^p - \mathbf{u}_\psi^p) C_i^p \tag{5.71}$$

5.3. General INS Error Equations

Writing Eq. (5.71) in column vector form, we obtain

$$\dot{\boldsymbol{\beta}}^i = C_p^i(\boldsymbol{\omega}_{ip}^p - \mathbf{u}_\psi^p) \tag{5.72}$$

In order to write the derivative of $\dot{\boldsymbol{\beta}}^i$ in the geographic frame, we apply the theorem of Coriolis in matrix form to $\dot{\boldsymbol{\beta}}^i$

$$\dot{\boldsymbol{\beta}}^i = C_g^i(\dot{\boldsymbol{\beta}}^g + \boldsymbol{\Omega}_{ig}^g \times \boldsymbol{\beta}^g) \tag{5.73}$$

Substituting Eq. (5.73) into Eq. (5.72) and solving for $\dot{\boldsymbol{\beta}}^g$ results in

$$\dot{\boldsymbol{\beta}}^g = \boldsymbol{\Omega}_{ig}^g \boldsymbol{\beta}^g + C_p^g(\delta\boldsymbol{\omega}_{ip}^p + \mathbf{u}_\psi^p) \tag{5.74}$$

Next, a solution to $\dot{\mathbf{v}}^g$, the time derivative of the inertial to geographic angular error, needs to be derived. First, an alternative form of Eq. (5.67) can be written

$$\dot{C}_i^g = \boldsymbol{\Omega}_{ig}^g C_i^g \tag{5.75}$$

Applying Eq. (5.75) to Eq. (5.66) results in [10]

$$\hat{\boldsymbol{\Omega}}_{ig}^g \hat{C}_i^g = (\mathbf{I} - \mathbf{N}^g)(\boldsymbol{\Omega}_{ig}^g C_i^g) - \dot{\mathbf{N}}^g C_i^g \tag{5.76}$$

Now, define the error in the computed geographic angular velocity as

$$\hat{\boldsymbol{\Omega}}_{ig}^g = \boldsymbol{\Omega}_{ig}^g + \delta\boldsymbol{\Omega}_{ig}^g \tag{5.77}$$

Substituting Eqs. (5.60) and (5.77) into Eq. (5.76) and solving for $\dot{\mathbf{N}}^g$, neglecting second-order terms, gives

$$\dot{\mathbf{N}}^g = \mathbf{N}^g \boldsymbol{\Omega}_{ig}^g - \boldsymbol{\Omega}_{ig}^g \mathbf{N}^g + \delta\boldsymbol{\Omega}_{ig}^g \tag{5.78}$$

Again, writing Eq. (5.78) in column vector form,

$$\dot{\mathbf{v}}^g = \boldsymbol{\Omega}_{ig}^g \mathbf{v}^g + \delta\boldsymbol{\omega}_{ig}^g \tag{5.79}$$

Substituting Eqs. (5.74) and (5.79) into Eq. (5.64), we have

$$\dot{\boldsymbol{\psi}}^g = \boldsymbol{\Omega}_{ig}^g \mathbf{v}^g + \delta\boldsymbol{\omega}_{ig}^g - \boldsymbol{\Omega}_{ig}^g \boldsymbol{\beta}^g + C_p^g(\delta\boldsymbol{\omega}_{ip}^p + \mathbf{u}_\psi^p) \tag{5.80}$$

Using Eq. (5.63), this can be written as

$$\dot{\boldsymbol{\psi}}^g = \boldsymbol{\Omega}_{ig}^g \boldsymbol{\psi}^g + \delta\boldsymbol{\omega}_{ig}^g + \mathbf{w}_\psi^g + \mathbf{u}_\psi^g \tag{5.81}$$

where $\mathbf{w}_\psi^g$ is the vector that accounts for gyroscope errors and is given by

$$\mathbf{w}_\psi^g = C_p^g \delta\boldsymbol{\omega}_{ip}^p \tag{5.82}$$

Expanding Eq. (5.81) and writing the component forms results in

$$\dot{\psi}_E = -\omega_z \psi_N - \omega_N \psi_z + \delta\omega_E + w_{\psi_E} + u_{\psi_E} \tag{5.83a}$$

$$\dot{\psi}_N = \omega_z \psi_E + \omega_E \psi_z + \delta\omega_N + w_{\psi_N} + u_{\psi_N} \tag{5.83b}$$

$$\dot{\psi}_z = \omega_N \psi_E - \omega_E \psi_N + \delta\omega_z + w_{\psi_z} + u_{\psi_z} \tag{5.83c}$$

In order to resolve $\delta\omega_{ig}$, we first write the components of the angular velocity in the geographic frame ω with respect to inertial space [see Eqs. (5.84a)–(5.84c)] [4]:

$$\omega_E = -\dot{\phi} \tag{5.84a}$$

$$\omega_N = (\Omega + \dot{\lambda}) \cos \phi \tag{5.84b}$$

$$\omega_z = (\Omega + \dot{\lambda}) \sin \phi \tag{5.84c}$$

Using Eq. (5.28), simplifying, and neglecting higher-order terms, we can write the error equations for Eqs. (5.84a)–(5.84c) as follows:

$$\delta\omega_E = -\delta\dot{\phi} \tag{5.85a}$$

$$\delta\omega_N = \delta\dot{\lambda}(\cos \phi) - \delta\phi(\Omega + \dot{\lambda}) \sin \phi \tag{5.85b}$$

$$\delta\omega_z = \delta\dot{\lambda}(\sin \phi) + \delta\phi(\Omega + \dot{\lambda}) \cos \phi \tag{5.85c}$$

Finally, using Eqs. (5.26) and (5.39), and the equations for ρ_E, ρ_N, ρ_z from Section 4.3, we can write the linearized platform error Eqs. (5.83a)–(5.83c) in the form

$$\dot{\psi}_E = -\left(\frac{\rho_E}{R_0}\right)\delta h - \left(\frac{1}{R_0}\right)\delta V_N + \omega_z\psi_N - \omega_N\psi_z + w_{\psi_E} + u_{\psi_E} \tag{5.86a}$$

$$\dot{\psi}_N = -\Omega_z\,\delta\phi - \left(\frac{\rho_N}{R_0}\right)\delta h + \left(\frac{1}{R_0}\right)\delta V_E - \omega_z\psi_E + \omega_E\psi_z + w_{\psi_N} + u_{\psi_N} \tag{5.86b}$$

$$\dot{\psi}_z = (\omega_N + \rho_z \tan \phi)\,\delta\phi - \left(\frac{\rho_z}{R_0}\right)\delta h + \frac{\delta V_E}{R_0}\tan \phi$$
$$+ \omega_N\psi_E - \omega_E\psi_N + w_{\psi_z} + u_{\psi_z} \tag{5.86c}$$

Equations (5.36), (5.38), (5.39), (5.45), (5.46), (5.47), and (5.86a)–(5.86c) represent the linearized INS error equations. These equations can be placed in the standard form

$$\dot{\mathbf{x}}(t) = \mathbf{F}(t)\mathbf{x}(t) + \mathbf{G}(t)\mathbf{u}(t) + \mathbf{w}(t) \tag{5.24}$$

where the state vector $\mathbf{x}$ consists of the various errors such that

$$\mathbf{x} = [\delta\phi \quad \delta\lambda \quad \delta h \quad \delta V_E \quad \delta V_N \quad \delta V_z \quad \psi_E \quad \psi_N \quad \psi_z]^T$$

The matrix $\mathbf{F}(t)$ is the system dynamics matrix; $\mathbf{w}$ is a white, Gaussian forcing function compensating for approximations made in deriving the error equations; and $\mathbf{u}$ is the rate-of-correction vector of the INS errors from any noninertial aiding equations, when the system is implemented in the aided configuration. Consider now a simple horizontal navigation example [as in Eq. (5.22)] operating in the pure (unaided) inertial mode. Further-

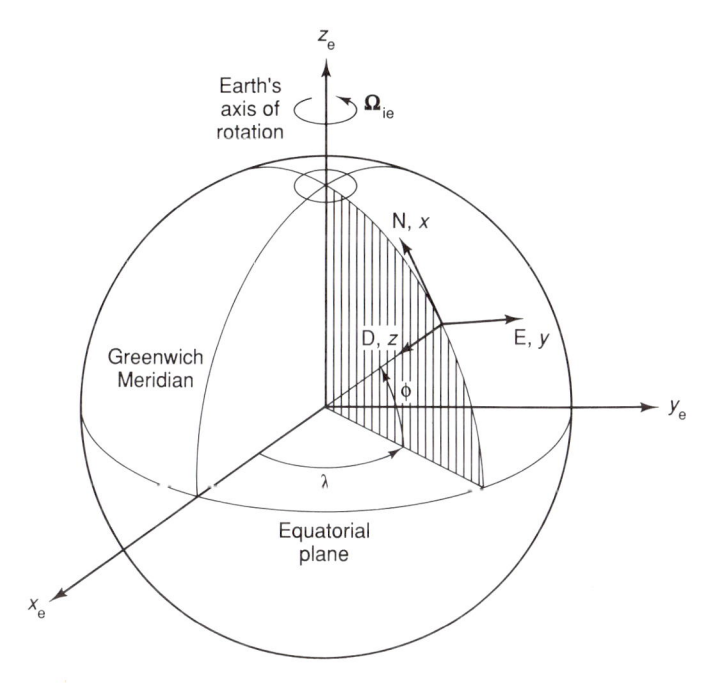

5.3. General INS Error Equations

FIGURE 5-3
Locally level coordinate frame.

more, a locally level north–east–down (NED) coordinate frame as illustrated in Fig. 5-3 is assumed.

The aircraft inertial navigation system error equations are as follows:

$$\delta\dot{x} = \delta V_x$$
$$\delta\dot{y} = \delta V_y + \delta x(\dot{\lambda}\sin\phi) - \delta\lambda(\dot{\phi}\tan\phi)$$
$$\delta\dot{V}_x = \delta A_x - 2\delta V_y(\Omega + \dot{\lambda})\sin\phi - \delta x\dot{\lambda}(2\Omega\cos^2\phi + \dot{\lambda}) + g\psi_y + g\xi$$
$$\delta\dot{V}_y = \delta A_y + \delta V_x(2\Omega + \dot{\lambda})\sin\phi + \delta x\dot{\phi}(2\Omega\cos\phi + \dot{\lambda}\sec\phi)$$
$$\quad + \delta V_y\dot{\phi}\tan\phi - g\psi_x + g\eta$$
$$\dot{\psi}_x = \frac{\delta V_y}{R} - \frac{\delta x}{R}\Omega\sin\phi - \psi_y(\Omega + \dot{\lambda})\sin\phi + \dot{\phi}\psi_z + \varepsilon_x$$
$$\dot{\psi}_y = -\frac{\delta V_x}{R} + \psi_x(\Omega + \dot{\lambda})\sin\phi + \psi_z(\Omega + \dot{\lambda})\cos\phi + \varepsilon_y$$
$$\dot{\psi}_z = -\frac{\delta V_y}{R}\tan\phi - \frac{\delta x}{R}(\dot{\lambda}\sec\phi + \Omega\cos\phi) - \psi_y(\Omega + \dot{\lambda})\cos\phi - \psi_x\dot{\phi} + \varepsilon_z$$

where λ = vehicle longitude
ϕ = operating latitude (depends on flight path)
$\dot\phi$ = vehicle latitude rate (positive in northward direction)
$\dot\lambda$ = vehicle longitude rate (positive in eastward direction)
Ω = earth's rate of rotation
g = acceleration due to gravity
R = radial distance from the earth's center to the aircraft.

Inputs   [input forcing functions (INS error sources)]:

$\delta A_x = x$—accelerometer error
$\delta A_y = y$—accelerometer error
$\varepsilon_x = x$—gyro drift rate
$\varepsilon_y = y$—gyro drift rate
$\varepsilon_z = z$—gyro drift rate
$\xi = x$—channel vertical deflection
$\eta = y$—channel vertical deflection

Outputs

$\delta x = x$—channel position (latitude) error (also denoted $\delta\phi$)
$\delta y = Y$—channel position (longitude) error (also denoted $\delta\lambda$)
$\delta V_x = x$—(north) velocity error
$\delta V_y = y$—(east) velocity error
ψ_x = platform tilt about the x (north) axis
ψ_y = platform tilt about the y (east) axis
ψ_z = azimuth error (measured clockwise from north)

As another example, let us consider an aircraft flying along a great circle. For flights along a great circle, the heading changes throughout the flight (except in north and south flights). The coordinate frame chosen for this example is a rotating, wander–azimuth, local-level frame, with the x axis pointing north and the y axis west at $t = 0$. Thus, $V_y = -V$ for a 90° heading, $V_y = V$ for 270° heading, and $V_x = 0$ at $t = 0$. Again, as in the previous example, we will consider only horizontal navigation. The platform misalignment term ψ_z indicates how the difference in heading (east vs. west) causes a difference in navigation performance. The angular velocity of the geographic frame ω in inertial space is given in component form as follows:

$$\omega_x = \Omega \cos\phi - \frac{V_y}{R}; \quad \omega_y = \frac{V_x}{R}; \quad \omega_z = \Omega \sin\phi - \frac{V_y}{R}\tan\phi$$

where Ω is the angular velocity of the earth, ϕ is the geographic latitude, and R is the distance from the center of the earth to the vehicle. (For an altitude h above the reference ellipsoid, the curvature of the earth must be accounted for, so that $R = R_\phi + h$ for east–west and $R = R_\lambda + h$ for north–south directions, respectively.) From Eq. (5.24), we can write the system

5.3. General INS Error Equations

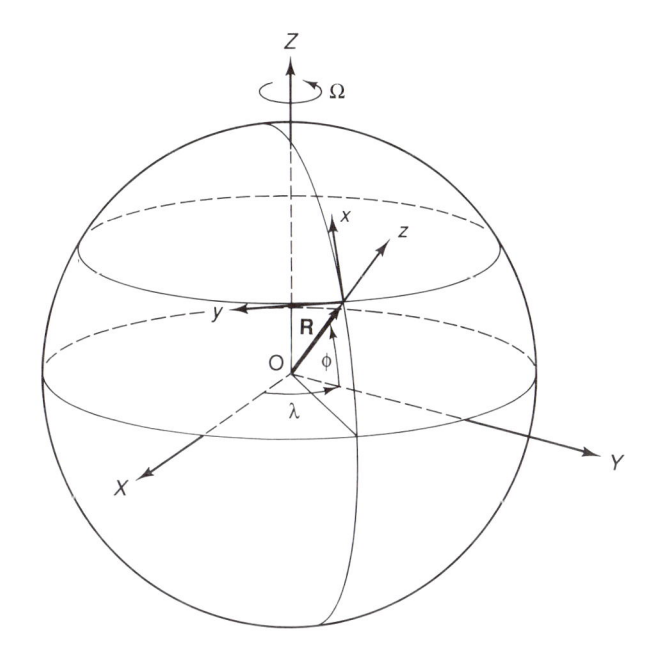

dynamics matrix **F** as follows:

$$F = \left[\begin{array}{ccc|cc|ccc} 0 & 0 & 1 & 0 & 0 & 0 & 0 \\ 0 & 0 & 0 & 1 & 0 & 0 & 0 \\ \hline -g/R & 0 & 0 & 2\Omega_z & 0 & -A_z & A_y \\ 0 & -g/R & -2\Omega_z & 0 & A_z & 0 & -A_x \\ \hline & & & & 0 & \omega_z & -\omega_y \\ & 0_{3\times 2} & & 0_{3\times 2} & -\omega_z & 0 & \omega_x \\ & & & & \omega_y & \omega_x & 0 \end{array} \right] \begin{array}{l} \delta x \\ \delta y \\ \delta V_x \\ \delta V_y \\ \psi_x \\ \psi_y \\ \psi_z \end{array} \begin{array}{l} \text{Position} \\ \text{Position} \\ \text{Velocity} \\ \text{Velocity} \\ \\ \text{Platform misalignment} \\ \\ \end{array}$$

From the **F**-matrix, we note that

$$\dot{\psi}_z = \omega_y \psi_x - \omega_x \psi_y$$

To this expression, the azimuth gyroscope drift ε_z must be added:

$$\dot{\psi}_z = \omega_y \psi_x - \omega_x \psi_y + \varepsilon_z$$

where $\omega_x = -(1/R)V_y + \Omega_x = \rho_x + \Omega_x$
$\omega_y = (1/R)V_x + \Omega_y = \rho_y + \Omega_y$
$\Omega_x = \Omega \cos \phi \cos \alpha$
$\Omega_y = -\Omega \cos \phi \sin \alpha$

In these equations, the ρ values are the craft rates (V/R) and α is the wander-azimuth angle measured positive clockwise from north. The complete equation for $\dot{\psi}_z$ is therefore

$$\dot{\psi}_z = \{(1/R)V_x + \Omega_y\}\psi_x - \{-(1/R)V_y + \Omega_x\}\psi_y + \varepsilon_z$$

Since $V_x \approx 0$ and $\Omega_y \approx 0$, this equation reduces to

$$\dot{\psi}_z \approx -\{-(1/R)V_y + \Omega_y\}\psi_y + \varepsilon_z$$

Now, in order to compare for the east–west azimuth rate, we substitute the respective velocities as follows:

$$\dot{\psi}_z(\text{west}) \approx -\{-(1/R)V + \Omega_x\}\psi_y + \varepsilon_z$$
$$\dot{\psi}_z(\text{east}) \approx -\{-(1/R)V + \Omega_x\}\psi_y + \varepsilon_z$$

Therefore, taking the difference

$$\Delta\dot{\psi}_z = \dot{\psi}_z(\text{west}) - \dot{\psi}_z(\text{east}) = 2(V/R)\psi_y$$

which indicates that the rates of change of the azimuth misalignment differ by the factor $2(V/R)\psi_y$. Furthermore, this result indicates that the platform azimuth error grows faster when flying west. To get a feeling for the error magnitude, consider an aircraft flying at 440 knots. The value of the y-platform tilt ψ_y after 6 min of Doppler leveling is approximately 0.018° (or 1 arcmin). Therefore, the difference in platform drifts is approximately 0.0046 deg/h.

Some inertial system designers use a different convention to designate the earth axes. Specifically, the X axis is taken to be directed along the earth's polar axis, and the (Y, Z) axes in the equatorial plane, with the Z axis in the Greenwich Meridian plane. The direction cosines between the earth X

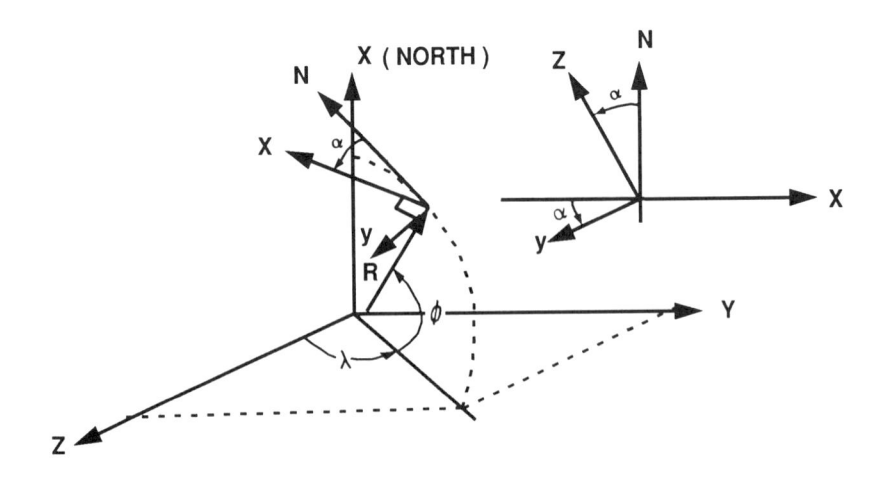

5.3. General INS Error Equations

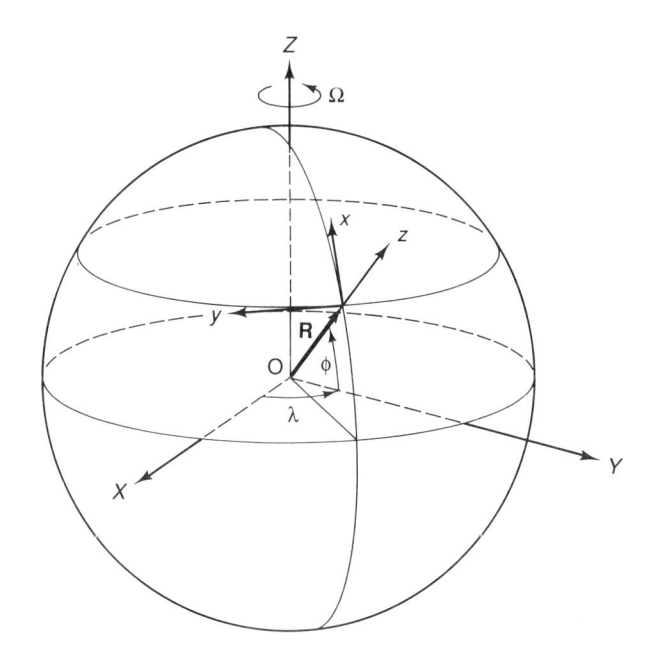

dynamics matrix **F** as follows:

$$\mathbf{F} = \left[\begin{array}{ccc|cc|ccc} 0 & 0 & 1 & 0 & 0 & 0 & 0 \\ 0 & 0 & 0 & 1 & 0 & 0 & 0 \\ \hline -g/R & 0 & 0 & 2\Omega_z & 0 & -A_z & A_y \\ 0 & -g/R & -2\Omega_z & 0 & A_z & 0 & -A_x \\ \hline & & & & 0 & \omega_z & -\omega_y \\ & 0_{3\times 2} & & 0_{3\times 2} & -\omega_z & 0 & \omega_x \\ & & & & \omega_y & -\omega_x & 0 \end{array}\right] \begin{array}{l} \delta x \\ \delta y \\ \delta V_x \\ \delta V_y \\ \psi_x \\ \psi_y \\ \psi_z \end{array} \begin{array}{l} \text{Position} \\ \text{Position} \\ \text{Velocity} \\ \text{Velocity} \\ \\ \text{Platform misalignment} \\ \\ \end{array}$$

From the **F**-matrix, we note that

$$\dot{\psi}_z = \omega_y \psi_x - \omega_x \psi_y$$

To this expression, the azimuth gyroscope drift ε_z must be added:

$$\dot{\psi}_z = \omega_y \psi_x - \omega_x \psi_y + \varepsilon_z$$

where $\omega_x = -(1/R)V_y + \Omega_x = \rho_x + \Omega_x$
$\omega_y = (1/R)V_x + \Omega_y = \rho_y + \Omega_y$
$\Omega_x = \Omega \cos \phi \cos \alpha$
$\Omega_y = -\Omega \cos \phi \sin \alpha$

In these equations, the ρ values are the craft rates (V/R) and α is the wander-azimuth angle measured positive clockwise from north. The complete equation for $\dot{\psi}_z$ is therefore

$$\dot{\psi}_z = \{(1/R)V_x + \Omega_y\}\psi_x - \{-(1/R)V_y + \Omega_x\}\psi_y + \varepsilon_z$$

Since $V_x \approx 0$ and $\Omega_y \approx 0$, this equation reduces to

$$\dot{\psi}_z \approx -\{-(1/R)V_y + \Omega_y\}\psi_y + \varepsilon_z$$

Now, in order to compare for the east–west azimuth rate, we substitute the respective velocities as follows:

$$\dot{\psi}_z(\text{west}) \approx -\{-(1/R)V + \Omega_x\}\psi_y + \varepsilon_z$$
$$\dot{\psi}_z(\text{east}) \approx -\{-(1/R)V + \Omega_x\}\psi_y + \varepsilon_z$$

Therefore, taking the difference

$$\Delta\dot{\psi}_z = \dot{\psi}_z(\text{west}) - \dot{\psi}_z(\text{east}) = 2(V/R)\psi_y$$

which indicates that the rates of change of the azimuth misalignment differ by the factor $2(V/R)\psi_y$. Furthermore, this result indicates that the platform azimuth error grows faster when flying west. To get a feeling for the error magnitude, consider an aircraft flying at 440 knots. The value of the y-platform tilt ψ_y after 6 min of Doppler leveling is approximately 0.018° (or 1 arcmin). Therefore, the difference in platform drifts is approximately 0.0046 deg/h.

Some inertial system designers use a different convention to designate the earth axes. Specifically, the X axis is taken to be directed along the earth's polar axis, and the (Y, Z) axes in the equatorial plane, with the Z axis in the Greenwich Meridian plane. The direction cosines between the earth X

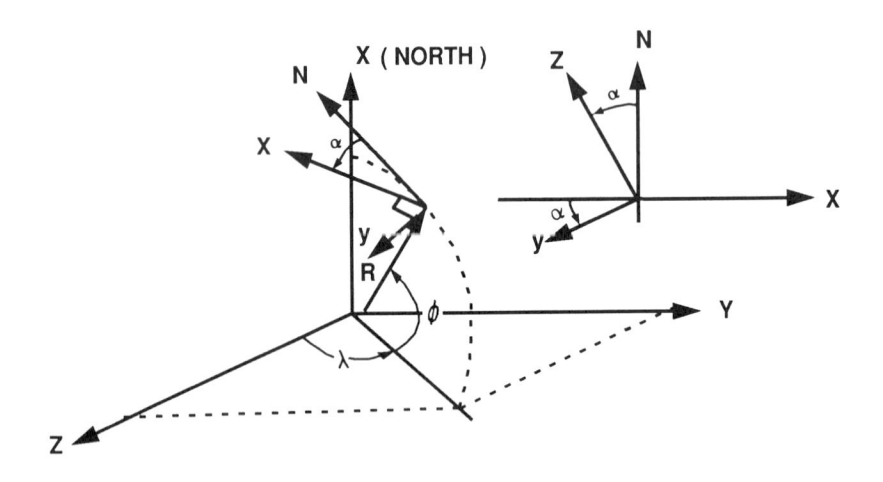

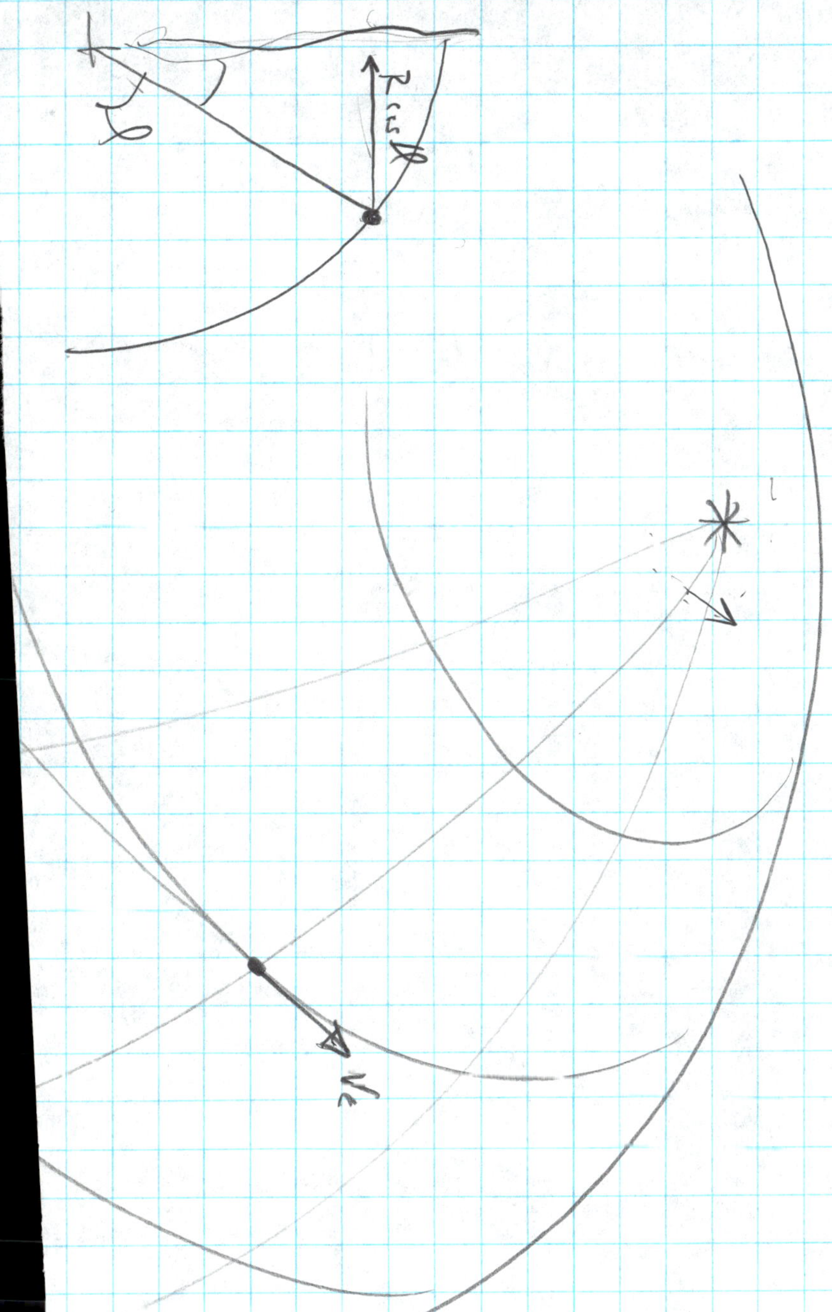

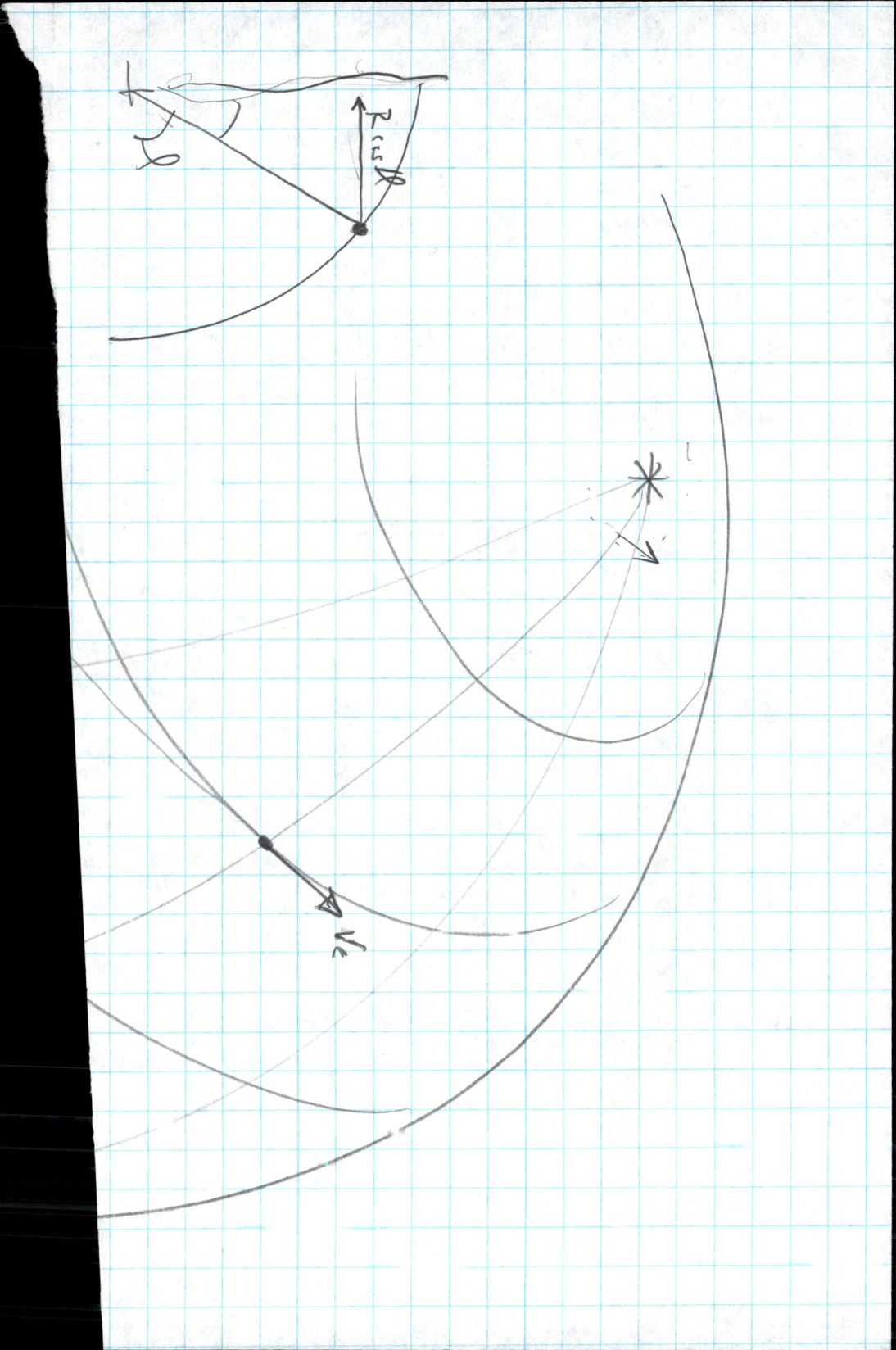

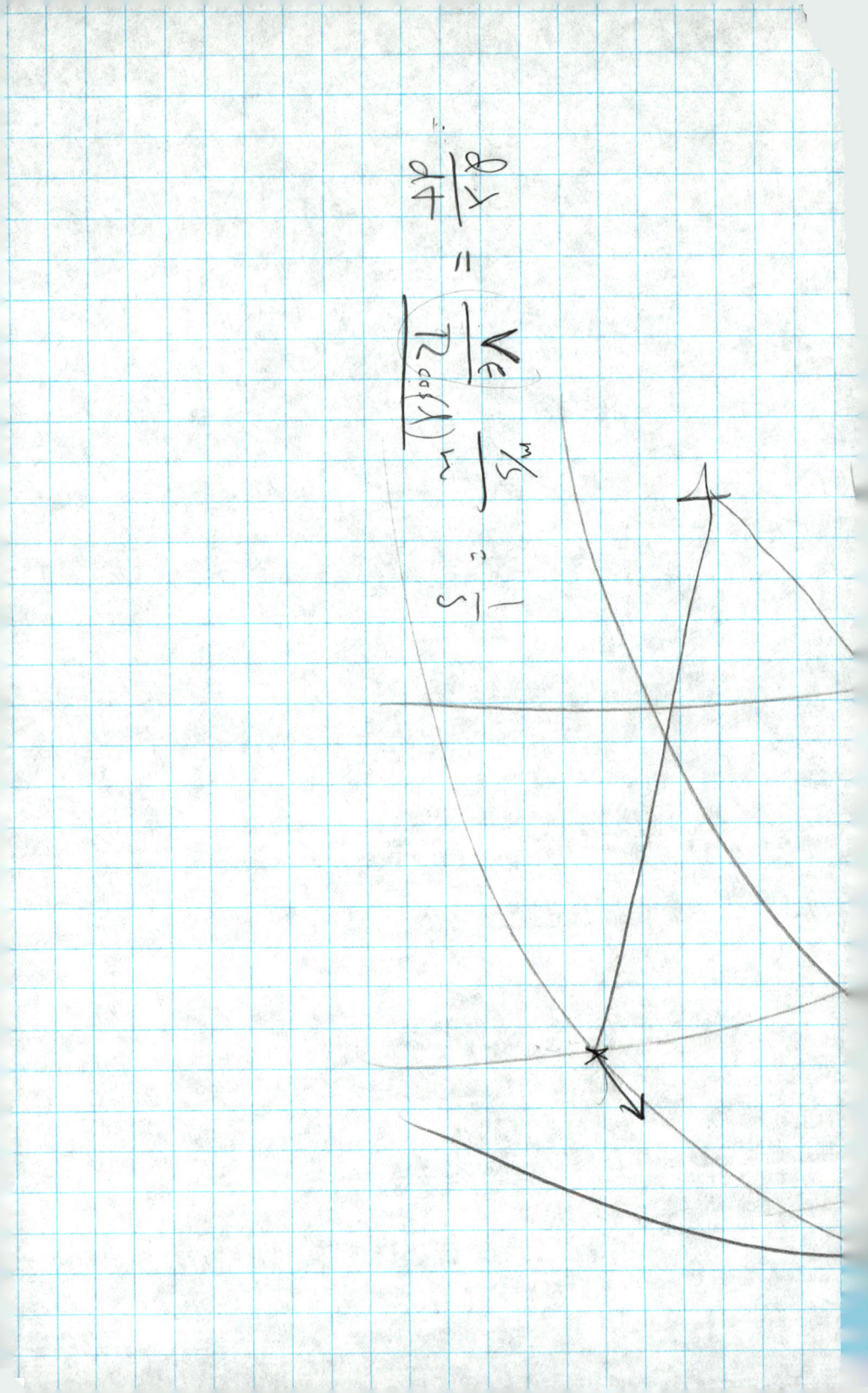

$$\frac{1}{S} = \frac{1}{\sum_{m} R_{eff}(\lambda)} = \frac{V_e}{V_T} = \frac{d\theta}{dt}$$

5.3. General INS Error Equations

axis and the platform (x, y, z) axes are given by

$$C_{Xx} = \cos \phi \cos \alpha$$
$$C_{Xy} = -\cos \phi \sin \alpha$$
$$C_{Xz} = \sin \phi$$

Therefore, the north and east components of the position error can be expressed in terms of the platform x and y components of the error as follows:

$$\begin{bmatrix} \Delta N \\ \Delta E \end{bmatrix} = \begin{bmatrix} \cos \alpha & -\sin \alpha \\ -\sin \alpha & -\cos \alpha \end{bmatrix} \begin{bmatrix} \Delta x \\ \Delta y \end{bmatrix}$$

The wander azimuth angle error, in terms of the error in position, is given by

$$\Delta \alpha = -\Delta \lambda \sin \phi$$

Also

$$\Delta E = R \cos \phi \cdot \Delta \lambda$$

or

$$\Delta \lambda = \frac{\Delta E}{R \cos \phi}$$

Making use of these relations, and substituting for

$$\Delta E = -\Delta x \sin \alpha - \Delta y \cos \alpha$$

we obtain

$$\Delta \alpha = \frac{\tan \phi}{R} (\Delta x \sin \alpha + \Delta y \cos \alpha)$$

A more comprehensive example will now be discussed. Consider a global positioning system (GPS)-aided inertial navigation system (INS). (The INS/GPS integration and/or aiding is discussed in detail in Section 6.2.5.6). As in the previous development of the error equations, the present error analysis will be based on a set of linearized INS and GPS error equations. No specific INS accuracy and/or performance is proposed for this general example. The accuracy and error budget allocation will depend, to a large extent, on the intended use and mission requirements. Furthermore, the INS modeled in this example is a local-level, NEU, wander–azimuth system, as illustrated in Fig. 5-4.

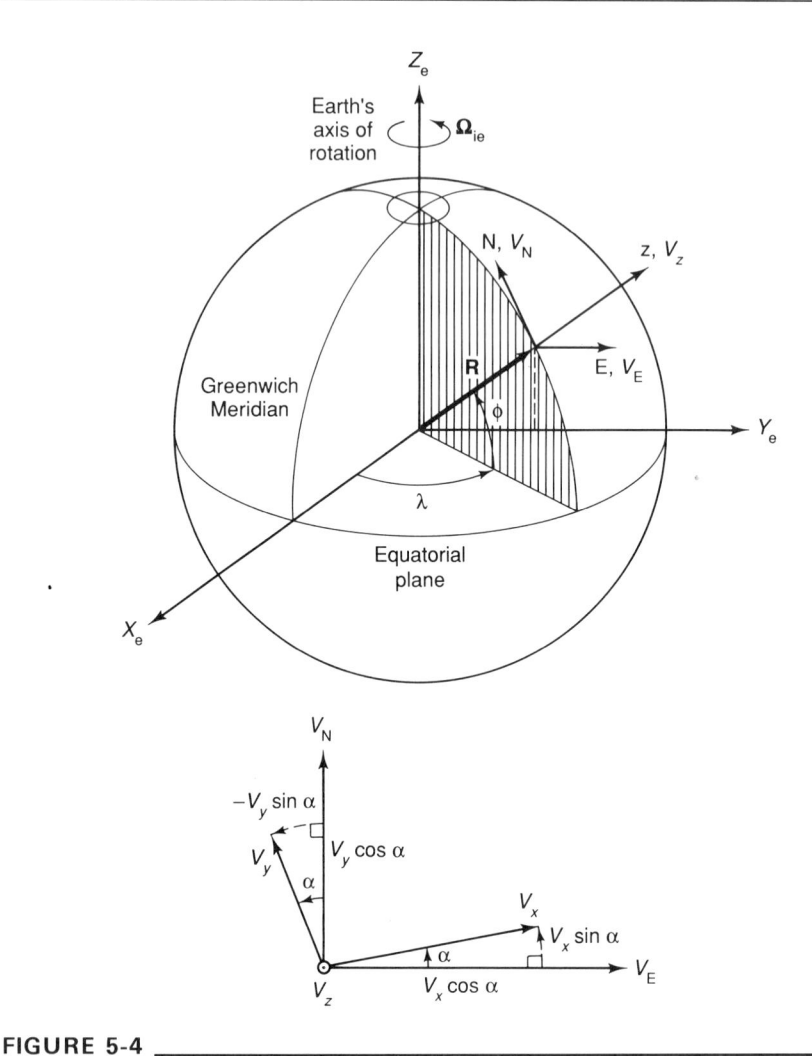

FIGURE 5-4
Coordinate system for the example.

Gyroscopes and accelerometers are mounted on a gimbaled, stabilized platform. The platform is kept aligned to the wander–azimuth frame, defined as a local-level frame with the x and y axes displaced from the east and north axes by the wander–azimuth angle α by a pair of 2-degree-of-freedom gyroscopes. Three accelerometers measure the specific force in the x, y, and z directions (components of the wander–azimuth frame). A baroinertial altimeter is also used to aid the INS by bounding the inherently unstable INS errors in the vertical channel. Moreover, the INS modeled has a position

5.3. General INS Error Equations

error growth of approximately 1.852 km/h due to Schuler and 24-h mode effects, instrument errors (i.e., instrument biases, drifts, misalignments, and scale factor errors), and environmental errors such as gravity uncertainty and variations in pressure. Consequently, a 54-state system error model (or "truth model") will be developed in the form of a stochastic, linear, vector differential equation as given by

$$\dot{\mathbf{x}}(t) = \mathbf{F}(t)\mathbf{x}(t) + \mathbf{G}(t)\mathbf{w}(t) \tag{5.87}$$

where $\mathbf{x}(t)$ = the 54-state vector (54×1)
$\mathbf{F}(t)$ = the system dynamics matrix (54×54)
$\mathbf{G}(t)$ = input matrix (54×12)
$\mathbf{w}(t)$ = vector of white noise forcing functions (12×1)

In Eq. (5.87), $\dot{\mathbf{x}}(t)$ represents a set of 54 first-order linear differential equations, which model the errors and error sources in the GPS-aided INS. The error equations, Eqs. (5.36), (5.38), (5.39), (5.45)–(5.47), and (5.86a)–(5.86c), will be used in a slightly modified form to account for the baroinertial altimeter loop, gyroscope, and accelerometer errors. For convenience, these error equations are summarized below. As before, these equations are expressed relative to the north, east, and up frames [9]:

$$\delta\dot{\lambda} = (\rho_z/\cos\phi)\,\delta\phi - (\rho_N/R\cos\phi)\,\delta h + \delta V_E/R\cos\phi \tag{5.88}$$

$$\delta\dot{\phi} = (\rho_E/R)\,\delta h + \delta V_N/R \tag{5.89}$$

$$\delta\dot{h} = \delta V_z - K_1\,\delta h = \delta V_z - K_1(h - h_{\text{ref}}) \tag{5.90}$$

$$\delta\dot{V}_E = (2(\Omega_N V_N + \Omega_z V_z) + \rho_N V_N/\cos^2\phi)\,\delta\phi$$
$$+ (\rho_N\rho_E + \rho_N K_z)\,\delta h$$
$$- (\rho_E \tan\phi + K_z)\,\delta V_E + (\omega_z + \Omega_z)\,\delta V_N$$
$$- (\omega_N + \Omega_N)\,\delta V_z + \delta f_E + \delta g_E \tag{5.91}$$

$$\delta\dot{V}_N = -(2\Omega_N V_E + \rho_N V_E \cos^2\phi)\,\delta\phi + (\rho_N\rho_z - \rho_E K_z)\,\delta h$$
$$- 2\omega_z\,\delta V_E - K_z\,\delta V_N + \rho_E\,\delta V_z + \delta f_N + \delta g_N \tag{5.92}$$

$$\delta\dot{V}_z = -2\Omega_z V_E\,\delta\phi - (\rho_N^2 + \rho_E^2)\,\delta h$$
$$+ 2\omega_N\,\delta V_E - 2\rho_E\,\delta V_N + \delta f_z + \delta g_z - K_2\,\delta h \tag{5.93}$$

$$\dot{\psi}_E = (-\rho_E/R)\,\delta h - \delta V_N/R + \omega_z\psi_N - \omega_N\psi_z + \delta\omega_E \tag{5.94}$$

$$\dot{\psi}_N = -\Omega_z\,\delta\phi - (\rho_N/R)\,\delta h + \delta V_E/R - \omega_z\psi_E + \omega_E\psi_z + \delta\omega_N \tag{5.95}$$

$$\dot{\psi}_z = (\omega_N + \rho_N \tan\phi) - (\rho_z/R)\,\delta h + (\tan\phi/R)\,\delta V_E + \omega_N\psi_E$$
$$- \omega_E\psi_N + \delta\omega_z \tag{5.96}$$

$$\delta\dot{a} = K_3\,\delta h \tag{5.97}$$

where
$\psi_{E,N,z}$ = platform misalignment angles
$\delta f_{E,N,z}$ = accelerometer errors
$\delta\omega_{E,N,z}$ = gyroscope errors
$\delta\hat{a}$ = estimated vertical acceleration error in altitude channel
λ = longitude
ϕ = geographic latitude
α = wander–azimuth angle
$f_{x,y,z}$ = measured specific forces in wander–azimuth coordinates
$g_{x,y,z}$ = components of gravity in wander–azimuth coordinates
h, h_{ref} = INS and baroinertial altimeter indicated altitudes
K_1, K_2, K_3 = gains in the damping loop on the INS vertical channel
$\Omega_{y,z}$ = components of earth angular velocity in wander–azimuth coordinates
R_λ = radius of curvature of the earth reference ellipsoid in east–west direction
R_ϕ = radius of curvature of the earth reference ellipsoid in north–south direction
V_E, V_N = east and north velocities
V_x, V_y, V_z = velocity in wander–azimuth coordinates
$\omega_x, \omega_y, \omega_z$ = angular velocity of wander–azimuth frame with respect to inertial space
$V_E = V_x \cos\alpha - V_y \sin\alpha$
$V_N = V_x \sin\alpha + V_y \cos\alpha$
$\dot{\alpha} = -V_E \tan\phi / R$
$\dot{\lambda} = V_E / (R_\lambda + h) \cos\phi$
$\rho_E = -V_N / (R_\phi + h)$
$\rho_N = V_E / (R_\lambda + h)$
$\rho_z = \rho_N \tan\phi$
$K_z = V_z / R$

The 54-state variables are summarized in Table 5-1. The system dynamics matrix $\mathbf{F}(t)$ is represented by Figs. 5-5–5-11. Table 5-2 gives the elements appearing in Fig. 5-6. Before an attempt at system simulation is made, the inertial system analyst or designer must specify the initial conditions and statistics for the state variables. In addition to the basic error equations, error equations emanating from accelerometer and gyroscope errors, gravity uncertainties, GPS receiver errors, and baroinertial altimeter errors are derived. These errors are modeled as random constants, random walks, and first-order Markov processes. A random constant is modeled as the output of an integrator with zero input and an initial condition that has a zero-mean and variance P_0. The model is suitable for an instrument bias that

TABLE 5-1
Error Model State Variables

State variable	State variable
Basic inertial navigation errors 1. $\delta\lambda$ Error in east longitude 2. $\delta\phi$ Error in north latitude 3. δh Error in altitude 4. δV_E Error in east velocity 5. δV_N Error in north velocity 6. δV_z Error in vertical velocity 7. ψ_E East attitude error 8. ψ_N North attitude error 9. ψ_z Vertical attitude error	32. YG_z Y-gyro input axis misalignment about Z 33. ZG_x Z-gyro input axis misalignment about X 34. ZG_y Z-gyro input axis misalignment about Y
Barointertial altimeter error 10. e_{SF} Altimeter scale factor error	Accelerometer biases 35. AB_x X-accelerometer bias 36. AB_y Y-accelerometer bias 37. AB_z Z-accelerometer bias
Vertical channel error variable 11. δa Vertical acceleration error variable in altitude channel	Accelerometer scale factor errors 38. ASF_x X-accelerometer scale factor error 39. ASF_y Y-accelerometer scale factor error 40. ASF_z Z-accelerometer scale factor error
Clock errors 12. δt_u User clock phase error 13. δt_{bu} User clock frequency error	
g-Insensitive gyro drifts 14. DX_f X-gyro drift rate 15. DY_f Y-gyro drift rate 16. DZ_f Z-gyro drift rate	Accelerometer input axis misalignments 41. θ_{xY} X-accelerometer input axis misalignment about Y 42. θ_{xZ} X-accelerometer input axis misalignment about Z 43. θ_{yX} Y-accelerometer input axis misalignment about X 44. θ_{yZ} Y-accelerometer input axis misalignment about Z 45. θ_{zX} Z-accelerometer input axis misalignment about X 46. θ_{zY} Z-accelerometer input axis misalignment about Y
g-Sensitive gyro drift 17. DX_x X-gyro input axis g-sensitivity 18. DX_Y X-gyro spin axis g-sensitivity 19. DY_x Y-gyro input axis g-sensitivity 20. DY_y Y-gyro spin axis g-sensitivity 21. DZ_y Z-gyro input axis g-sensitivity 22. DZ_z Z-gyro spin axis g-sensitivity	
g^2-Sensitive gyro drift coefficients 23. DX_{xy} X-gyro spin-input g^2-sensitivity 24. DY_{xy} Y-gyro spin-input g^2-sensitivity 25. DZ_{yz} Z-gyro spin-input g^2-sensitivity	Barointertial altimeter errors 47. e_{p0} Error due to variation in altitude of a constant-pressure surface
Gyro scale factor errors 26. GSF_x X-gyro scale factor error 27. GSF_y Y-gyro scale factor error 28. GSF_z Z-gyro scale factor error	Gravity uncertainties 48. δg_E East deflection of gravity 49. δg_N North deflection of gravity 50. δg_z gravity anomaly
	Clock errors 51. δt_{bu} Clock aging bias 52. δt_{ru} Clock r and m frequency bias (τ_c is the correlation time)
Gyro input axis misalignments 29. XG_y X-gyro input axis misalignment about Y 30. XG_z X-gyro input axis misalignment about Z 31. YG_x Y-gyro input axis misalignment about X	Additional barointertial altimeter errors 53. C_{sp} Coefficient of static pressure measurements 54. τ_b Altimeter lag

	1–11	12–13	14–22	23–34	35–46	47–50	51–52	53–54
1–11	F_{11}	0	F_{13}	F_{14}	F_{15}	F_{16}	0	F_{18}
12–13	0	$\begin{matrix}0 & 1\\ 0 & 0\end{matrix}$	0	0	0	0	$\begin{matrix}0 & 1\\ 0 & 0\end{matrix}$	0
14–22	0	0	0	0	0	0	0	0
23–34	0	0	0	0	0	0	0	0
35–46	0	0	0	0	F_{55}	0	0	0
47–50	0	0	0	0	0	0	0	0
51–52	0	0	0	0	0	$\begin{matrix}0 & 1\\ 0 & -1/\tau_c\end{matrix}$	0	0
53–54	0	0	0	0	0	0	0	0

FIGURE 5-5
System dynamics matrix, $\mathbf{F}(t)$.

normally changes each time the instrument is turned on, but remains constant while the instrument is turned on. The random-walk model is the output of an integrator driven by zero-mean, white, Gaussian noise. Random-walk models are useful for errors that grow without bound or change slowly with time. The first-order Markov model is a first-order lag driven by a zero-mean white Gaussian noise. The first-order Markov model is used to represent exponentially time correlated noises (i.e., band limited

	$\delta\lambda$ 1	$\delta\phi$ 2	δh 3	δV_E 4	δV_N 5	δV_z 6	ψ_E 7	ψ_N 8	ψ_z 9	e_{SF} 10	δa 11
$\delta\dot\lambda$ 1	0	$\rho_z/\cos\phi$	$-\rho_N/R\cos\phi$	$1/R\cos\phi$	0	0				0	0
$\delta\dot\phi$ 2	0	0	ρ_E/R	0	$1/R$	0	\multicolumn{3}{c}{$0_{3\times 3}$}	0	0
$\delta\dot h$ 3	0	0	$-K_1$	0	0	1				$K_1 h$	0
$\delta\dot V_E$ 4	0	A_{42}	A_{43}	A_{44}	$(\omega_z+\Omega_z)$	$-(\omega_N+\Omega_N)$	0	$-f_z$	f_N	0	0
$\delta\dot V_N$ 5	0	A_{52}	A_{53}	$-2\omega_z$	$-K_z$	ρ_E	f_z	0	$-f_E$	0	0
$\delta\dot V_z$ 6	0	$2\Omega_z V_E$	A_{63}	$2\omega_n$	$-2\rho_E$	0	$-f_N$	f_E	0	$K_2 h$	-1
ψ_E 7	0	0	$-\rho_E/R$	0	$-1/R$	0	0	ω_z	$-\omega_N$	0	0
ψ_N 8	0	$-\Omega_z$	$-\rho_N/R$	$1/R$	0	0	$-\omega_z$	0	ω_E	0	0
ψ_z 9	0	A_{92}	$-\rho_N/R$	$\tan\phi/R$	0	0	ω_N	$-\omega_E$	0	0	0
e_{SF} 10	0	0	0	\multicolumn{3}{c}{$0_{2\times 3}$}	\multicolumn{3}{c}{$0_{2\times 3}$}	0	0
δa 11	0	0	K_3						$-K_3 h$	0	0

FIGURE 5-6
F_{11}, upper 9×9 partition of $\mathbf{F}(t)$.

5.3. General INS Error Equations

TABLE 5-2
Notation Used in Fig. 5-6

$\Omega_N = \Omega \cos \phi$
$\Omega_z = \Omega \sin \phi$ } Components of earth rate

$\rho_E = -V_N/R$
$\rho_N = V_E/R$ } Components of angular velocity of $E-N-z$ frame with respect to earth
$\rho_z = V_E \tan \phi/R$

$\omega_E = \rho_E$
$\omega_N = \rho_N + \Omega_N$ } Components of angular velocity of $E-N-z$ frame with respect to inertial space
$\omega_z = \rho_z + \Omega_z$

$K_z = V_z/R$
$A_{42} = 2(\Omega_N V_N - \Omega_z V_z) + \rho_N V_N/\cos^2 \phi$
$A_{43} = \rho_z \rho_E - \rho_N K_z$
$A_{44} = -\rho_E \tan \phi - K_z$
$A_{52} = -2\Omega_N V_E - \rho_N V_E/\cos^2 \phi$
$A_{53} = \rho_N \rho_z - \rho_E K_z$
$A_{63} = 2g/R - (\rho_N^2 + \rho_E^2)$
$A_{92} = \omega_N \quad \rho_z \tan \phi$

	g-Insensitive gyro drifts					g-Sensitive gyro drifts			
	14	15	16	17	18	19	20	21	22
1–6	$\mathbf{0}_{6\times 3}$					$\mathbf{0}_{6\times 6}$			
7	C	$-S$	0	$f_x C$	$f_y C$	$-f_x C$	$-f_y S$	0	0
8	S	C	0	$f_x S$	$f_y S$	$f_x C$	$f_y C$	0	0
9	0	0	1	0	0	0	0	f_y	f_z
10–11	$\mathbf{0}_{2\times 3}$					$\mathbf{0}_{2\times 6}$			

Note: $C = \cos \alpha$, $S = \sin \alpha$.

FIGURE 5-7
F_{13}, partitions of g-insensitive and g-sensitive gyro error state variables.

	g^2-Sensitive gyro drifts			Gyro scale factor errors			Gyro input axis misalignments					
	23	24	25	26	27	28	29	30	31	32	33	34
1–6	$\mathbf{0}_{6\times 3}$			$\mathbf{0}_{6\times 3}$			$\mathbf{0}_{6\times 6}$					
7	$f_x f_y C$	$-f_x f_y S$	0	$\omega_x C$	$-\omega_y S$	0	$\Omega_z C$	$-\omega_y S$	$\Omega_z S$	$-\omega_x S$	0	0
8	$f_x f_y S$	$f_x f_y C$	0	$\omega_x S$	$\omega_y C$	0	$\Omega_z S$	$-\omega_y S$	$-\Omega_z C$	$\omega_x C$	0	0
9	0	0	$f_y f_z$	0	0	Ω_z	0	0	0	0	ω_y	$-\omega_x$
10–11	$\mathbf{0}_{2\times 3}$			$\mathbf{0}_{2\times 3}$			$\mathbf{0}_{2\times 6}$					

Note: $C = \cos \alpha$, $S = \sin \alpha$.

FIGURE 5-8
F_{14}, partition of g^2-sensitive, gyro scale factor, and gyro input axis misalignment state variables.

	Accelerometer biases			Accelerometer scale factor errors			Accelerometer input axis misalignments					
	35	36	37	38	39	40	41	42	43	44	45	46
1–3	$0_{3 \times 3}$			$0_{3 \times 3}$			$0_{3 \times 6}$					
4	C	$-S$	0	$f_x C$	$-f_y S$	0	$-f_z C$	$f_y C$	$-f_z S$	$f_x S$	0	0
5	S	C	0	$f_x S$	$f_y C$	0	$-f_x S$	$f_y C$	$f_z C$	$-f_x C$	0	0
6	0	0	1	0	0	f_z	0	0	0	0	$-f_y$	f_x
7–11	$0_{5 \times 3}$			$0_{5 \times 3}$			$0_{5 \times 6}$					

Note: $C = \cos \alpha$, $S = \sin \alpha$.

FIGURE 5-9

F_{15}, partition of accelerometer error state variables.

	e_{p_0}	Gravity uncertainties			C_{sp}	b
	47	48	49	50	53	54
1–2	$0_{2 \times 1}$	$0_{2 \times 3}$			$0_{2 \times 2}$	
3	K_1	0	0	0	$K_1 V^2$	$-K_1 V_z$
4	0	1	0	0	0	0
5	0	0	1	0	0	0
6	K_2	0	0	1	$K_2 V^2$	$-K_2 V_z$
7–10	$0_{4 \times 1}$	$0_{4 \times 3}$			$0_{4 \times 2}$	
11	$-K_3$	$0_{1 \times 3}$			$-K_3 V^2$	$K_3 V_z$

Note: V = vehicle speed; V_z = vertical velocity.

FIGURE 5-10

F_{16} and F_{18}, partition of gravity uncertainty and additional baroinertial altimeter error state variables.

	35–46	e_{p_0}	Gravity uncertainty		
		47	48	49	50
47		$-V/d_{\text{alt}}$	0	0	0
48	**0**	0	$-V/d_{g_E}$	0	0
49		0	0	$-V/d_{g_N}$	0
50		0	0	0	$-V/d_{g_z}$

Note: V = vehicle speed; V_z = vertical velocity.

FIGURE 5-11

F_{55}, partition of pressure variation and gravity uncertainty error state variables.

5.3. General INS Error Equations

errors). All modeled errors are assumed to be independent with initial covariance given by

$$P_{ij}(0) = 0, \quad i \neq j \quad (5.98a)$$

$$P_{ii}(0) = X_i^2 \quad (5.98b)$$

where X_i is the initial condition of the ith truth model state at t_0. The accelerometer errors modeled in the truth model are accelerometer input axis misalignment, biases, and scale factor errors. Accelerometer misalignment can be obtained by considering the following diagram. The angle θ_{xZ}

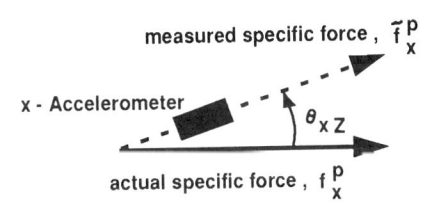

Accelerometer Misalignment

(which is a rotation of the x accelerometer about the z axis) is a small-angle misalignment induced when the accelerometer is mounted on the platform. In essence, there are six misalignment angles that cause the accelerometer measurement error f^P as shown by the matrix

$$\delta \mathbf{f}^P = \begin{bmatrix} 0 & \theta_{xZ} & -\theta_{xY} \\ -\theta_{yZ} & 0 & \theta_{yX} \\ \theta_{zY} & -\theta_{zX} & 0 \end{bmatrix} \mathbf{f}^P \quad (5.99a)$$

where

$$\mathbf{f}^P = [f_E \quad f_N \quad f_z]^T \quad (5.99b)$$

Each misalignment angle is modeled as a random constant. Next, accelerometer bias induces a constant specific force measurement error, which can be adequately modeled by a random walk. Scale factor errors induce a measurement error when a measured voltage is translated into a specific force reading. The scale factor is assumed to be linear over the range of the accelerometer and is modeled as a random constant.

Gyroscope errors modeled include gyroscope drift (g-insensitive, g-sensitive, and g^2-sensitive), scale factor errors, and input axis misalignments. Gyroscope drift consists of a g insensitive random component that exhibits growth in time. It is modeled by a random walk and is one of the most significant error sources. The g-sensitive gyroscope drifts, on the other hand, induce a constant drift proportional to the specific force in each gyroscope.

These drifts are modeled as random constants. The g^2-sensitive gyroscope drifts are induced by anisoelastic torques and are also modeled as random constants. Scale factor errors and input axis misalignments are similar to those defined for the accelerometers and are modeled as random constants.

The gravity uncertainties modeled are gravity deflections in the east and north directions and gravity anomaly; they are modeled as first-order Markov processes. The baroinertial altimeter errors modeled in the truth model are scale factor errors modeled as first-order Markov processes, scale factor error modeled as a random constant, static pressure measurement error modeled as a random constant, and time lag modeled also as a random constant. Finally, since the assumption was made that the satellites' clocks are synchronized and operate with negligible error, and that the broadcast ephemerides are accurate, then the GPS measurement errors can be attributed to the user receiver clock. The clock error is assumed to be the sum of three processes: (1) white noise, (2) a slowly shifting bias, and (3) a first-order Markov process. Figure 5-12 shows a block diagram of the simulation processes.

In Section 5.1 we noted that inertial navigation and guidance system performance requirements are usually specified in terms of statistical quantities, such as rms position or velocity errors. A conventional approach for determining such statistics for a particular error budget is to integrate the matrix differential equations defining error covariance dynamics for the modeled system and mission, using an error budget to define the initial covariances and covariance dynamics. In order to define an error budget that meets the specified requirements, this process is repeated, using engineering

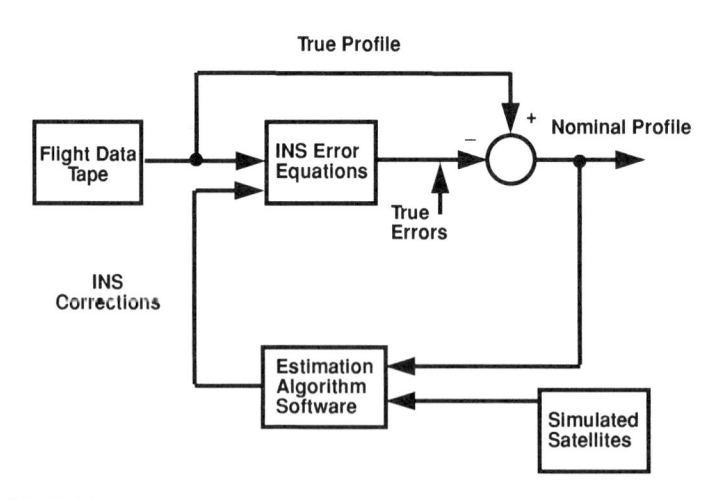

FIGURE 5-12
Feedback simulation for the example.

5.3. General INS Error Equations

judgment to adjust individual budget allocations until a satisfactory error budget is obtained. It should be noted here that an error budget is essentially a special case of sensitivity analysis whereby the inertial system designer determines from the outset the individual effects of a group of error sources. Furthermore, if the error sources are uncorrelated, the total system errors can be found by taking the root sum square (RSS) of all contributions to the system's performance.

The navigation accuracy performance can be represented by the following performance parameters:

1. Position CEP (circular error probable)
2. Position error (north, east, and radial components)
3. Velocity error (north, east, and radial components)
4. Ground track angle error
5. True heading angle error
6. Drift angle error
7. Azimuth error
8. Magnetic variation error

In addition, the error sources inherent to an inertial navigation system can be thought to consist of the following errors:

1. Initial condition errors
2. Gyroscope drift rates
3. Accelerometer errors
4. Platform misalignments
5. Position and velocity errors
6. Altitude reference (or indicated) errors
7. Velocity reference errors
8. Vertical deflections
9. Gravity anomalies

Figure 5-13 depicts an approximate breakdown of the error contribution to INS performance due to the inertial sensors. These errors must be identified by the system designer before a sensitivity error covariance analysis is carried out.

Error covariance analysis is a commonly used tool for determining the design requirements of inertial sensor systems. More specifically, it is a method of estimating the overall system performance as a function of the performance of subsystems and components that contribute to system errors. Error budgeting, on the other hand, requires solving the inverse problem; that is, finding an allocation of allowable errors (error budget) among the subsystems, such that the system meets a set of performance criteria specified by the user (e.g., the military services or the airlines). In inertial navigation and guidance systems, these performance requirements are usually specified

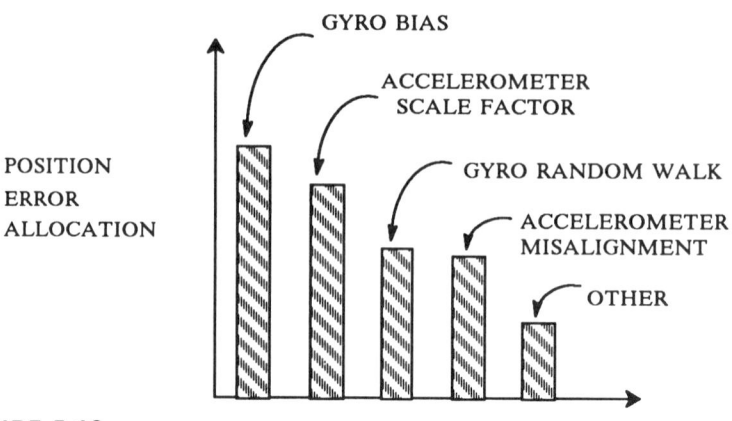

FIGURE 5-13
Error budgets (sensitivities).

in terms of statistical quantities such as the root-mean-squared (RMS) position or velocity errors.

The gyroscope and accelerometer errors consist of the following errors:

Gyroscope errors: Bias, scale factor, misalignment, g-dependent errors, and rotation induced errors.

Accelerometer errors: Bias, scale factor, axis misalignment, nonlinearities, cross-coupling, and rotation induced errors.

REFERENCES

1. Britting, K. R.: *Inertial Navigation Systems Analysis*, Wiley-Interscience, New York, 1971.
2. Chen, G., and Chui, C. K.: *Kalman Filtering with Real-Time Applications*," Springer Series in Information Sciences, Vol. 17, 2nd ed., Springer-Verlag, Berlin–Heidelberg–New York, 1991.
3. Huddle, J. R.: "Inertial Navigation System Error-Model Considerations in Kalman Filtering Applications," in *Control and Dynamic Systems*, C. T. Leondes (ed.), Academic Press, New York, 1983, pp. 293–339.
4. Kayton, M., and Fried, W. (eds.): *Avionics Navigation Systems*, Wiley, New York, 1969.
5. Leondes, C. T. (ed.): *Guidance and Control of Aerospace Vehicles*, McGraw-Hill, New York, 1963.
6. Macomber, G. R., and Fernandez, M.: *Inertial Guidance Engineering*, Prentice-Hall, Englewood Cliffs, N.J., 1962.
7. Maybeck. P. S.: *Stochastic Models, Estimation, and Control*, Vol. I, Academic Press, New York, 1979.
8. Pitman, G. R., Jr. (ed.): *Inertial Guidance*, Wiley, New York, 1962.
9. Siouris, G. M.: "Navigation, Inertial," *Encyclopedia of Physical Science and Technology*, Vol. 8, Academic Press, San Diego, 1987, pp. 668–717.
10. Widnall, W. S., and Grundy, P. A.: "Inertial Navigation System Error Models," Intermetrics, Inc., TR-03-73, May 11, 1973.

6 Externally Aided Inertial Navigation Systems

6.1. INTRODUCTION

Modern commercial and military aircraft use inertial navigation systems principally for one of two reasons: (1) the INS's redundancy offers an economically viable alternative to the cost of human navigator crews on long-range transports, such as in the Boeing 747 and C-5 (or C-141); and (2) the INS's unique characteristics of self-contained, nonjammable position, velocity, and acceleration insensitive attitude data. Furthermore, the INS drift angle and cross-track velocity have proved useful in autonomous landing for reducing lateral touchdown dispersion caused by cross-wind gusts and ILS (instrument landing system) beam degradation. INS data have been used successfully to provide velocity trend vectors and along-track acceleration to pilot's displays. Accuracy of terminal area navigation and guidance using radio data, such as DME (distance measurement equipment) or TACAN (tactical air navigation), and CARP (computed air release point) is improved using INS data. More importantly, the combination of INS velocity and CAD (central air data) airspeed can be used to warn the pilot of dangerous vertical wind-shear conditions during approach.

Inertial navigation systems exhibit position errors that tend to increase with time in an unbounded manner. This degradation is due, in part, to errors in the initialization of the inertial measurement unit and inertial sensor imperfections such as accelerometer bias and gyroscope drift. One way of mitigating this growth and bounding the errors is to update the inertial system periodically with external position fixes. Specifically, the effect of position fixes is to reset or correct the position error in the inertial system to the same level of accuracy inherent in the position fixing technique. In between position fixes, the inertial system error grows at a rate equal to the velocity error in the system. For this reason, external sources of data are routinely used in air and marine navigation for inertial navigation systems operating for extended periods of time (several hours). Therefore, when noninertial sensors are available on board the navigating vehicle, the accuracy of the navigation process can be greatly improved through external position and velocity measurements. Kalman filter integration of the inertial and noninertial navigation sensor information provides a method of utilizing these updates in an optimal manner. For instance, in overwater operation the normal buildup of position errors due to Schuler oscillations can be minimized by making measurements of range-to-sea surface with, say, a sophisticated radar system at several beam positions, establishing aircraft attitude with respect to the local vertical, thereby resulting in an order of magnitude improvement in performance over free (unaided) inertial navigation. Navigation-aided (Navaid) systems, also known as "hybrid" systems, use one or more different sources of navigation information integrated into a single system. An integrated system configuration eliminates redundancy in display hardware by utilizing a programmable multisensor display processor to integrate data from the navigation, radar, and other avionics systems for display on a common display unit. Furthermore, the motivation for the integrated system approach is to take advantage of size, weight, and power savings. Therefore, aided navigation systems involve an inertial navigation system plus one or more additional navaid sensor subsystems such as GPS, TACAN, Loran, Omega, VOR/DME, and ILS to bound or damp the navigation errors. The various radio navigation aids will be discussed in the next section. This synergistic effect, as it will be discussed later, is obtained through external measurements updating the inertial navigation system, commonly using a Kalman filter software package residing in the navigation computer, thereby optimizing system performance. It is therefore natural to demand and expect that the INS data and that of the external aids be combined in an optimal and efficient manner. As mentioned above, position errors grow with time in the free inertial mode. In order to get a feeling of the free inertial mode behavior, consider the navigation errors along the north, east, and down (or up) axes. An examination of the state equations [Eqs. (5.16)–(5.18)] that describe the behavior of the system state in the absence of updates reveals that there should be the following behavior:

1. Diverging azimuth errors (in the north and east coordinates) with an 84.4-min period (Schuler oscillation)
2. An exponentially increasing vertical channel error with a 10-min time constant.

6.1. Introduction

Consequently, the north and east coordinates are expected to grow linearly in time with sinusoidal motion around that linear growth. Figure 6-1 depicts these errors in the free inertial mode.

As discussed in Chapter 5, as attitude errors accumulate in the system, there is cross-coupling of the azimuth with the vertical channel and vice versa. As a result, there will be some spillover of the azimuth errors into the vertical channel and vice versa. The Schuler oscillations and resulting growth in azimuth and vertical channel arises from the double integration of accelerations sensed along wrong axes due to attitude errors. In the free inertial mode, no measurement of attitude is available, so that the errors continue to grow, while position errors grow according to the behavior discussed above. Using the barometric measurement (e.g., in 10-s intervals), the behavior of the vertical channel should increase linearly with time. The azimuth channel, on the other hand, is not directly affected by the barometric measurement. However, with the accumulation of attitude errors, the cross-coupling of the azimuth with the vertical channel means that in the barometrically updated free inertial mode there will be a linear growth of errors with sinusoidal motion superimposed on all channels. Note that the growth

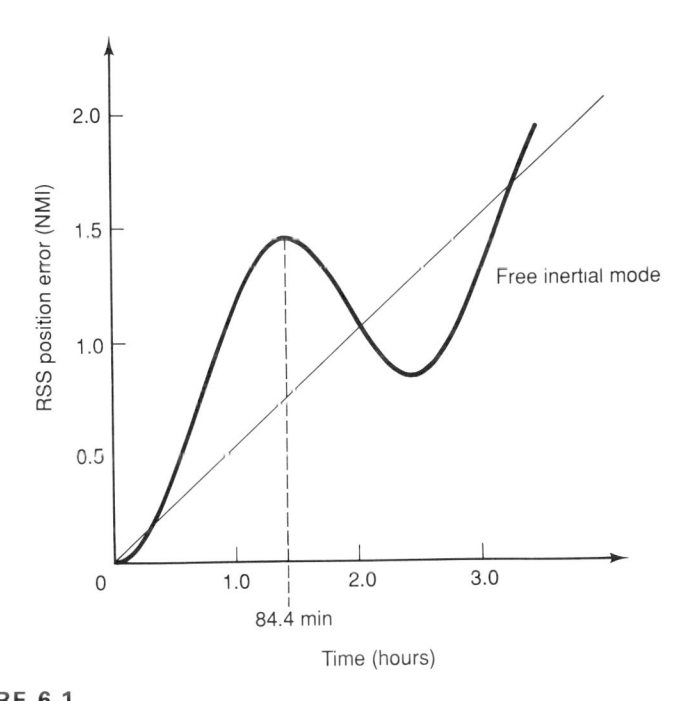

FIGURE 6-1

Root sum square of north and east errors in the free inertial mode.

will decrease in the down (up) component because of the measurements in that channel. Oscillations can be noted at 84.4 min, the Schuler frequency (see Fig. 6-1).

6.2. NAVAID SENSOR SUBSYSTEMS

For several years now, inertial navigation system designers have successfully integrated and/or automated online handling of avionics navigation system information involving several sensor subsystems. As discussed in the previous section, such navigation configurations usually include an INS, in conjunction with the use of simultaneous alternative navaid sensor subsystems such as TACAN, Loran, GPS, and JTIDS RelNav. In such a multisensor navigation system, the sensors are integrated via a mission computer. In military applications, navigation systems must be capable of operating over hostile terrain, and in a jamming environment, after some or all ground or satellite navigation systems have been disabled. Under these conditions, aircraft operation, sensor management, and information handling will increase the operator tasking, so that a fully automatic navigation function will become necessary. Therefore, in order to achieve mission accuracy requirements, future inertial navigation systems will most likely include more navigation sensors and subsystem modes. The problem associated with studying navigation errors associated with externally aided inertial navigation systems involves the following:

1. Modeling external measurement errors as well as inertial sensor errors
2. Error propagation between updates
3. Updating the inertial system when external measurements become available
4. The generation of a nominal mission (or trajectory) profile

Consequently, since inertial navigation errors seriously degrade over long periods of time, but are relatively small for short periods of time, external sensor aiding is important for long-range missions.

Figure 6-2 illustrates a typical strapdown multisensor navigation system. Multisensor navaids are classified in accordance with the INS updating function they perform. Specifically, external information sources that will aid an INS are position and velocity data. Listed below are the navaid sensors that provide position and velocity information for updating the INS.

1. Position data
    - Radar [including multimode radar (MMR)]
    - Radio navigation aids: TACAN, Loran, Omega, GPS, VOR, DME, VOR/TACAN, and JTIDS RelNav
    - Position fixes (e.g., Flyover, FLIR, TERCOM, and star sightings)

6.2. Navaid Sensor Subsystems

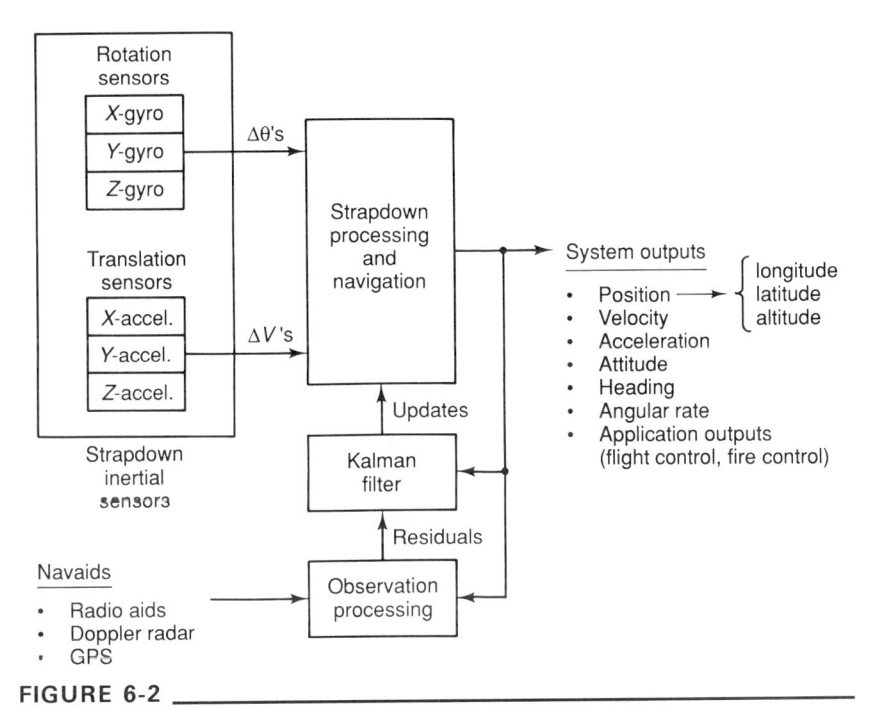

FIGURE 6-2
Typical aided strapdown system configuration.

2. Velocity data
    - Doppler radar
    - IAS
    - Global Positioning System (GPS)
    - EM-Log or Speedlog (used in marine applications)
3. Attitude/heading
    - Star tracker
    - Multi-antenna GPS

Since, as we have noted earlier in this book, the vertical channel of an INS is unstable, external information data must be used to damp the altitude channel. Strictly speaking, the vertical (altitude) channel must be stabilized regardless of whether the system is operating in the free or aided navigation mode. Conventional altitude data can therefore be summarized here as consisting of (1) barometric altimeter, (2) radar altimeter, and (3) laser altimeter. It should be pointed out here that, as opposed to an inertial navigation system, the navaids provide data which is good at low frequencies; however, they are subject to high-frequency noise, atmospheric effects, and ground effects [12]. Figure 6-3 shows a partial list of the navigation functions.

In Chapter 5 it was discussed that the error mechanism of an aircraft (or surface vehicle) may be described by a set of nine linear first-order differential equations. Two general types of error sources are identifiable: (1) those associated with the inertial sensors (gyroscopes and accelerometers) and (2) those associated with the external measurement processes (e.g., Doppler radar, air data–dead reckoning, position fix equipment). These errors will, in general, exhibit complex statistical characteristics, such as exponentially correlated or "colored" noise, and bias effects [10]. The avionics system provides flexibility and can be partitioned to take advantage of autonomous subsystem capabilities and fault detection. In the subsections that follow, the various navaids that make up the avionics system will be discussed in some detail.

6.2.1. Radar

Basically, conventional airborne radar for navigation provides range and bearing to a previously known point. Airborne radar, which is a self-contained air-based navigation device, has unlimited coverage over land areas of the globe. Its useful position fixing range to a checkpoint is limited to about 92.6 km (50 NM); the primary accuracy limitation for navigation is the azimuth error [9]. However, the azimuth error has been overcome through the use of synthetic aperture radar (SAR). The synthetic aperture radar provides a high-resolution image of a ground patch. In the high-resolution ground mapping (GM) mode, an artificial aperture is created that overcomes the physical limits of antenna size. Many of the modern radars, especially those for military applications, are coherent, multimode, digital, two-channel, monopulse sensors that employ phased array technology, mode interleaving, and programmable processing. Four basic capabilities of this radar type are (1) generating data for navigation, (2) penetration, (3) weapon delivery, and (4) ancillary functions such as aerial refueling.

The navigation function requires that the radar obtain an accurate present position update, which is then fed to the INS. The navigation function consists of the following modes: high-resolution mapping, real-beam mapping, weather detection, velocity update, and altitude update. Penetration requires automatic terrain following and manual terrain avoidance. Weapon delivery demands not only the ability to locate and identify an obscure target area but also extremely accurate velocity and altitude information. The weapon delivery modes are precision position fix, ground moving track, and ground beacon track. The ancillary functions consist of aerial refueling and rendezvous with airborne beacon-equipped aircraft. The multimode radar (MMR) described above, used in military applications,

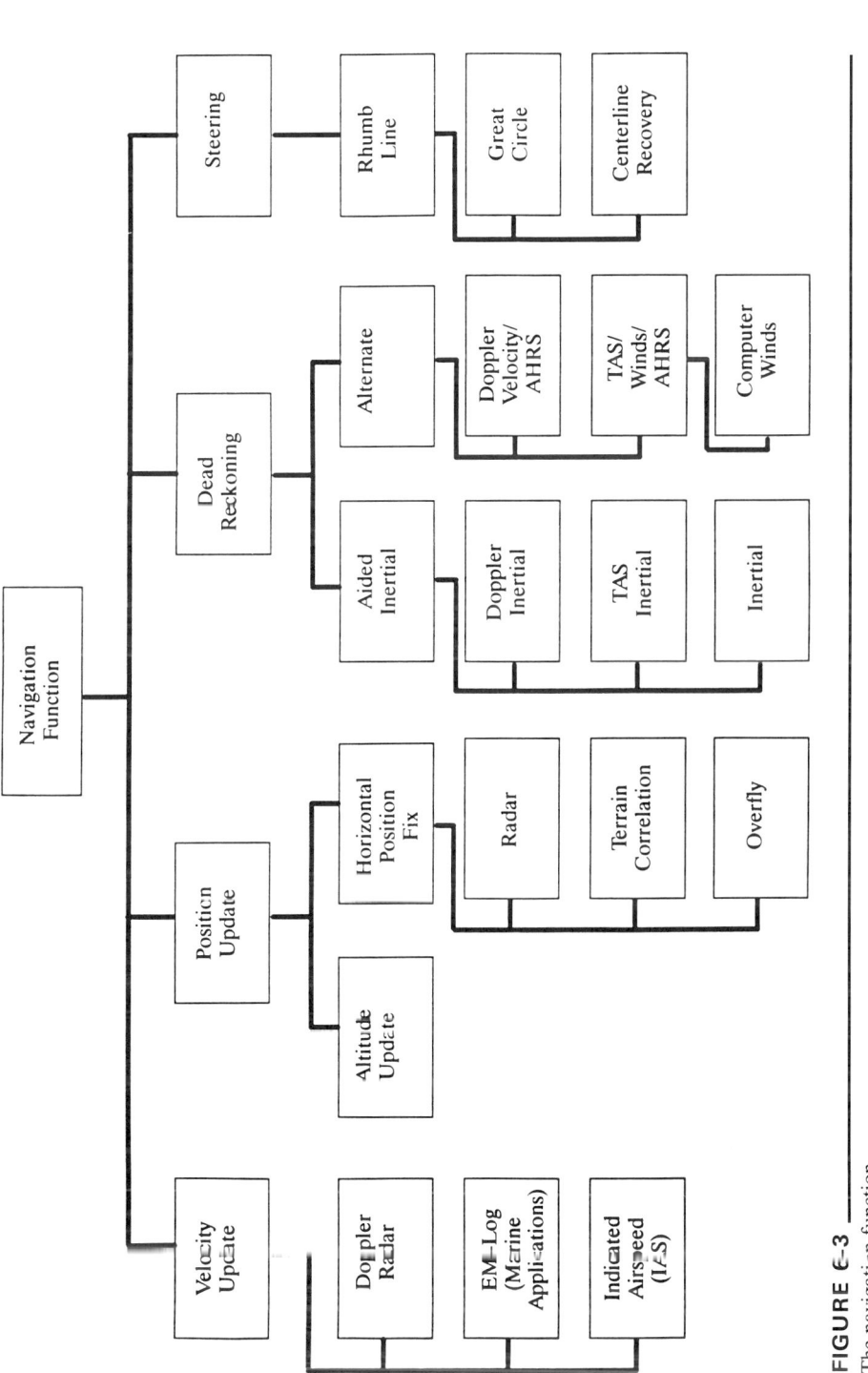

FIGURE 6-3
The navigation function.

provides many modes of operation. For example, two modes are used to obtain navigation information. In the terminal phase of flight, the navigator can, through a display, sight on landmark patterns whose position relative to the desired supply and/or personnel drop location is known. Slant range and bearing data is provided by the MMR during the terminal phase. The navigator can also use the MMR for sighting on landmarks whose absolute position (i.e., latitude and longitude) is known. In this mode, measurements of range, bearing, and landmark location go directly to the Kalman filter. The basic GM radar equations used in the measurement of the error equations in the z (up), north, and east directions are given below.

$$\boldsymbol{\rho}_s = -H \cdot \mathbf{1}_U + \sqrt{R_s^2 - H^2} \cdot [\sin \alpha \cdot \mathbf{1}_E + \cos \alpha \cdot \mathbf{1}_N]$$

with components

$$\rho_U = -H$$
$$\rho_E = \sqrt{R_s^2 - H^2} \sin \alpha$$
$$\rho_N = \sqrt{R_s^2 - H^2} \cos \alpha$$

where $\boldsymbol{\rho}_s$ = slant range from the aircraft to the ground checkpoint
R_s = slant range distance = $|\boldsymbol{\rho}_s|$
H = measured baroinertial system altitude
α = measured radar azimuth angle

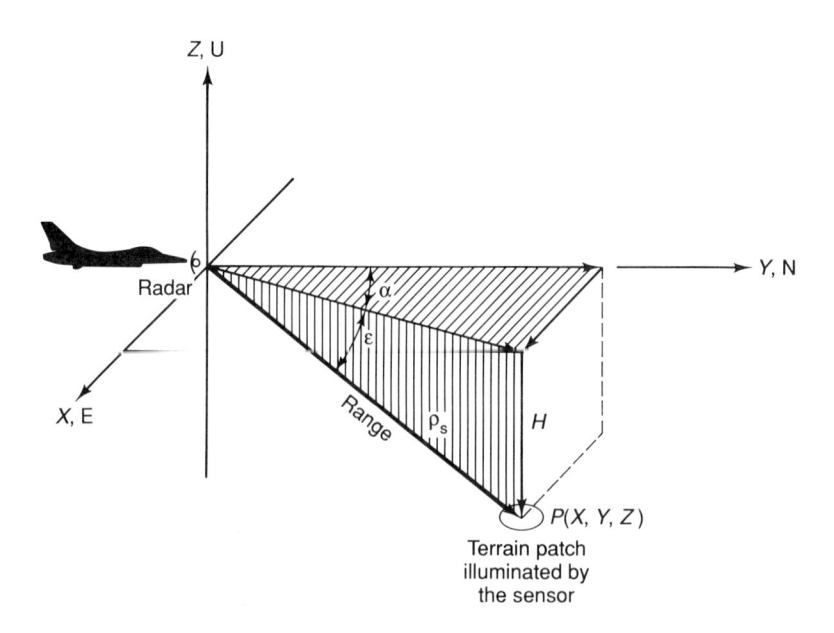

Terrain patch illuminated by the sensor

Position update accuracy is dependent on distance from the checkpoint, so that the CEP (circular error probable) position error increases with range from the checkpoint or position fix. Typical position update sighting accuracies are

Range = 100 m or 0.5% of range (whichever is greater)
Azimuth = 0.5° (rms)

Note that accuracy degrades with aircraft bank angle. For example, the average update radial error while the aircraft is in a, say, 30° bank angle may be 350 m, while it is 100 m at level flight. Azimuth angle error varies inversely with slant range.

6.2.2. Tactical Air Navigation (TACAN)

Simple operation, small-size, modest power requirements, and low cost make TACAN the primary method of aircraft navigation in the United States. Airborne TACAN provides the pilot and flight control systems slant-range and relative bearing information about a ground station. It also provides an audio identification of the ground station. By transmitting pulses to a ground station and timing the delay before receiving response pulses, the system measures the distance between aircraft and beacon. The system also displays bearing from the beacon to the aircraft based on variations in the beacon's rotating antenna pattern. As a function of bearing, and in conjunction with the horizontal situation indicator (HSI), TACAN provides TO/FROM and course deviation information relative to aircraft heading. TACAN has a range of 722.3 km [390 nautical miles (NM)] from a ground station and 370.4 km (200 NM) from another aircraft. TACAN coverage is limited to LOS. Moreover, TACAN navigation involves polar geometry. As mentioned above, the airborne equipment displays the aircraft slant-range and bearing from a ground beacon.

Beacon locations are displayed on navigation charts, thereby making it simple to determine the aircraft's position with a chart, a pair of dividers, and a protractor [9]. Onboard navigation systems typically access commands entered by the pilot and use this information when estimating aircraft trajectory. In order to simplify en route navigation using TACAN, the FAA (Federal Aviation Administration) has established airways and jet routes between TACAN beacons. Aircraft fly along these routes from one TACAN to another using the range measurement to indicate their position in the route and to provide an estimate of how long until the next course change.

A TACAN system model can be developed consisting of nine states: three position, three velocity, and three acceleration states. For a scalar, one-dimensional case, a simple model can be written as

$$\dot{P}(t) = V(t) \tag{6.1}$$

$$\dot{V}(t) = A(t) \tag{6.2}$$

$$\dot{A}(t) = -\frac{1}{T} A(t) + w(t) \tag{6.3}$$

where $P(t)$ is the position, $V(t)$ is the velocity, $A(t)$ is the acceleration, T is the time constant correlating input accelerations ($T = 1/\tau$; τ is correlation time), and $w(t)$ is a white Gaussian noise. Next, consider an earth-centered earth-fixed (ECEF) orthogonal reference frame (x, y, z), with the x axis pointing out along the equator at the prime meridian, the z axis pointing through the North Pole, and the y axis pointing out along the equator at 90° east longitude to complete a right-handed (x, y, z) coordinate system. For navigation purposes, the earth is normally modeled as an ellipsoid with semimajor axis a and eccentricity ε. On the basis of this reference ellipsoid, transforming geodetic coordinates expressed in latitude, longitude, and altitude (ϕ, λ, h) to ECEF coordinates used by the Kalman filter (P_x, P_y, P_z) results in

$$P_x = \left[\frac{a}{\sqrt{1 - \varepsilon^2 \sin^2 \phi}} + h\right] \cos \phi \cos \lambda \tag{6.4}$$

$$P_y = \left[\frac{a}{\sqrt{1 - \varepsilon^2 \sin^2 \phi}} + h\right] \cos \phi \sin \lambda \tag{6.5}$$

$$P_z = \left[\frac{a(1 - \varepsilon^2)}{(1 - \varepsilon^2 \sin^2 \phi)^{3/2}} + h\right] \sin \phi \tag{6.6}$$

where the altitude is given in this frame as

$$h = \frac{P_z}{\sin \phi} - \frac{a(1 - \varepsilon^2)}{(1 - \varepsilon^2 \sin^2 \phi)^{3/2}} \tag{6.7}$$

which is a nonlinear function of position in this ECEF coordinate system. The Kalman filter knows a priori the ECEF coordinates for the TACAN beacons designated by (X_B, Y_B, Z_B). Using TACAN position and aircraft coordinates, the following equation gives slant range from the beacon to the aircraft:

$$R = \sqrt{(P_x - X_B)^2 + (P_y - Y_B)^2 + (P_z - Z_B)^2} \tag{6.8}$$

The filter also uses bearing measurements from the TACAN to the aircraft. These measurements are magnetic bearings based on the local tangent plane defined at the beacon. The range vector between the beacon and the aircraft,

6.2. Navaid Sensor Subsystems

$[(P_x - X_B)(P_y - Y_B)(P_z - Z_B)]^T$, is projected into a north, east, and down coordinate system based on the beacon's geodetic latitude (ϕ_B) and longitude (λ_B). The north, east, and down (N, E, D) components of the range vector are given by the standard coordinate transformation as follows:

$$\begin{bmatrix} N \\ E \\ D \end{bmatrix} = \begin{bmatrix} -\sin \phi_B \cos \lambda_B & -\sin \phi_B \sin \lambda_B & \cos \phi_B \\ -\sin \lambda_B & \cos \lambda_B & 0 \\ -\cos \phi_B \cos \lambda_B & -\cos \phi_B \sin \lambda_B & -\sin \phi_B \end{bmatrix} \begin{bmatrix} P_x - X_B \\ P_y - Y_B \\ P_z - Z_B \end{bmatrix} \quad (6.9)$$

After calculating the north and east range vector components, using the inverse tangent gives the true heading from the beacon as shown in Fig. 6-4. The magnetic heading reported by the TACAN receiver will also have the local magnetic variation ξ at the beacon added to the true heading. It should be noted that east magnetic variation is added, while west magnetic variation is subtracted from the true heading β to give the magnetic heading γ.

The state vector of a representative Kalman filter TACAN model would consist of nine error states: seven navigation states consisting of two position error states, two velocity error states, and three misalignment error states, while the two TACAN error states would consist of a TACAN receiver bearing bias error and one receiver range bias error. Figure 6-5 is a composite figure, depicting the position fix for a single TACAN position fix, and the true position used in determining the positioning error.

Polar geometry causes the TACAN accuracy to decrease as distance to the station increases. Inaccurate bearing information, measurement noises, and biases constitute the major error components of TACAN signals. Combining several more equipment components displays the aircraft slant range and bearing from a ground beacon. Accurate range measurements can determine an aircraft's position without using bearing data. Figure 6-6 illustrates the position fix when more than one TACAN station is available.

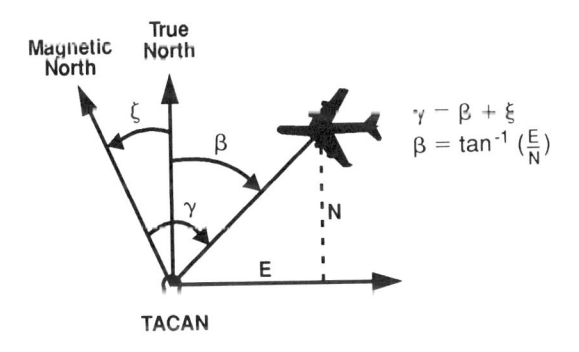

FIGURE 6-4
True bearing TACAN to aircraft

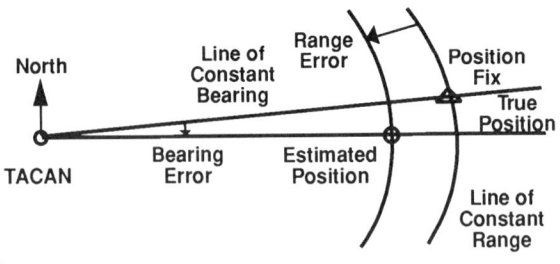

FIGURE 6-5
Single TACAN position fix and positioning error.

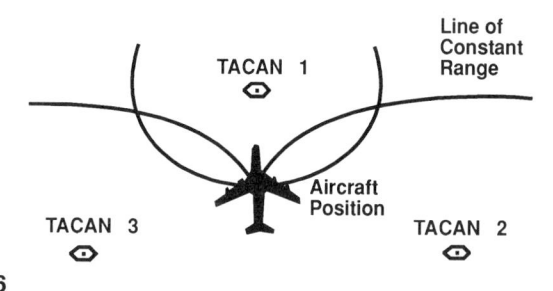

FIGURE 6-6
Ideal triple range fix.

Following is a list of state-of-the-art digital TACAN system characteristics:

Range (or readout capability): 722.28 km (390 NM)
Absolute range accuracy: ±0.1852 km (0.1 NM)
Bearing accuracy: ±1.0°
Frequency range: 962–1213 MHz
Number of channels available: 252 (126 X-channels, 126 Y-channels)
Power requirements: 115 V ac, 26 V ac, 28 V dc

The TACAN determines whether a signal is invalid and informs the pilot by either garbling the station identifier or showing warning flags. Moreover, velocity memory keeps the bearing and range indications tracking for the pilot during temporary signal loss, while the TACAN reacquires and locks on to the ground station when the signal returns. Distance and bearing can be acquired in less than 1 and 3 s, respectively. Finally, the unit's output is compatible with INS, R-NAV (area navigation), and ANS (automatic navigation system), providing full Rho–Theta real-time update information to these systems. Range rate information is also available for R-NAV.

6.2.3. Long Range Navigation (Loran)

The word "Loran" is an acronym for Long range navigation. Loran is a low-frequency, pulsed, hyperbolic radio-navigation position-fixing system. Historically, the system was developed in the early 1940s by the Radiation Laboratory of the Massachusetts Institute of Technology, in response to a requirement by the U.S. Army Signal Corps for "precision navigational equipment for guiding airplanes," as an aid to navigation. More specifically, the requirement was for a system that had the following characteristics: (1) the capability of guiding aircraft in all weather, (2) receivable at a range of 926 km (500 NM) from the transmitter, and (3) receivable at an altitude of 10,668 m (35,000 ft). The first system, consisting of two stations operated by the U.S. Coast Guard, became fully operational in January 1943. By mid-1943, stations in Labrador, Greenland, and Newfoundland also became operational. In addition, Loran stations were also built and operated in the Pacific area; in particular, stations were built in the Aleutians, the Central Pacific area, and in the China–Burma–India theater. Known then as "Loran-A," it consisted of two stations transmitting at the same pulse rate, from which a receiver could determine its location. Loran-A had 83 stations located 463–1111.2 km (250–600 NM) apart, operating at frequencies of 1850, 1900, and 1950 kHz. The coverage areas were the northern Atlantic, northern Pacific, North Sea, and the Gulf of Mexico. Predictable accuracies were 2.8 km at 926 km and 12.9–13.7 km at 2222.4 km. Following the war, its use was extended by the U.S. Coast Guard to aid marine navigation. Loran-A was phased out in 1980 because of station obsolescence and increasing operating costs. "Loran-B" was developed in an attempt to improve the accuracy of Loran-A. Loran-B used three stations, instead of two, transmitting at the same pulse rate. The intended purpose of Loran-B was to provide precision navigation in harbors and bays; however, technical problems prevented the deployment of an operational Loran-B system. In 1958, Loran-C became operational and was used commercially for water navigation. However, the development of modern "Loran-C" for commercial use dates to the early 1970s, from a commitment of the U.S. government to provide accurate navigation for the navigable waters of the United States to such users as commercial shippers and fishermen, and recreational boaters. More importantly, the area under concern was the Coastal Confluence Zone (CCZ), which was defined as extending seaward from a harbor entrance to 92.6 km (50 NM), or the edge of the continental shelf. Consequently, in May 1974, the Secretary of Transportation established Loran-C as the official radio navigation system for the coastal waters of the United States. Loran-C is a long-range (>1850 km), low-frequency (100-kHz) hyperbolic radio-navigation system which uses the difference in time of arrival of signals from two synchronized transmitters to establish position. The lower frequency allows the ground wave to follow the earth's curvature for longer distances, and the multiple pulses allow the receiver to distinguish the sky wave that bounces off the ionosphere from the ground wave. Depending on geometry, receiver time measurement accuracy, and propagation conditions, Loran-C provides predictable accuracies of between 100 and 200 m. "Loran-D," developed by the U.S. Air Force in the mid-1960s, is a short-range tactical version of the Loran-C system. It is a low-power transportable system, whose pulse sequencing consists of 16 phase-coded pulses 500 s apart in each group. Loran-D's range is limited to 1100 km. From the preceding discussion, it is noted that Loran has changed four times since its inception. When completely phased out, Loran-D will be replaced by "Loran-C/D." Loran-C/D is an advanced hyperbolic radio-navigation system for tactical deployment, operating within the low-frequency band of 90–110 kHz, and with ground-wave coverage of about 1100 km. Under optimum conditions, Loran-C/D can provide accuracies of better than 100 m, depending on position in the grid. For more details on Loran, Refs. 22, 27, 28, and 29 are recommended.

6.2.3.1. Hyperbolic Navigation

The most common form of Loran used today is the hyperbolic mode, as opposed to the circular mode. The reason for this is that the time difference in time of arrival of signals from the various transmitters can be measured precisely using a simple clock. In the circular mode (i.e., direct ranging), a line of position (LOP) represents a constant range from a transmitter. Radar, TACAN, and DME (distance measurement equipment) use range (circular) coordinates. Therefore, the word "hyperbolic" originates from the fact that, as signals are received from two different sources, position is established between the sources along a line that is in the form of a hyperbola. Seen from a geometric point of view, in a hyperbolic navigation system an LOP represents a constant range from two transmitters. For instance, if a flat earth model is assumed, the transmitter locations form the loci for a family of hyperbolic LOPs. For this reason, the lines of constant time difference marked on Loran charts are hyperbolas. As already mentioned, Loran is based on the concept that the amount of time it takes for a radio pulse to travel a certain distance is a measurement of that distance. Theoretically, if the radio waves were traveling in free space, then the time difference between reception at the vehicle of signals transmitted from two separate stations would be proportional to the difference the distance traveled by the two waves. Mathematically, the model for the (observed) time difference can be expressed as

$$\Delta T = \frac{\rho_S - \rho_M}{c} + \varepsilon \qquad (6.10)$$

6.2. Navaid Sensor Subsystems

where c is the speed of light in vacuum, ε is the error in the emission delay between master and slave stations, and ρ_M and ρ_S are the geodesic distances traveled by the waves transmitted from the master and slave stations, respectively. In reality, however, the signals do not travel in free space. Instead, they are trapped in the space between the earth and the ionosphere. Consequently, the velocity of propagation is not c, but a function of the nature of the waveguide. Loran can be used as a stand alone system or as an integrated part of the navigation system. Figure 6-7 illustrates how Loran is used for position determination using the hyperbolic navigation principle.

Consider Fig. 6-7, where h_1, h_2, h_3, h_5, h_6, and h_7 are hyperbolas of constant time difference (TD). Two transmitters at points M and S transmit synchronized signals. The "master" station (M) transmits a pulse at time $t = 0$, and a "slave" or secondary station (S) transmits a similar pulse signal, δ microseconds (µs) after receiving the master pulse signal. The slave station receives the master signal at time $t = \beta$ microseconds, and transmits its pulse at $t = \beta + \delta$ microseconds. In hyperbolic navigation, as we saw earlier, an LOP represents a constant range difference from two transmitters. As before, if we assume a flat earth model, then a receiver (e.g., an aircraft or other vehicle) R will measure the difference in time of arrival of the signal from the slave and master stations. In equation form, this time difference

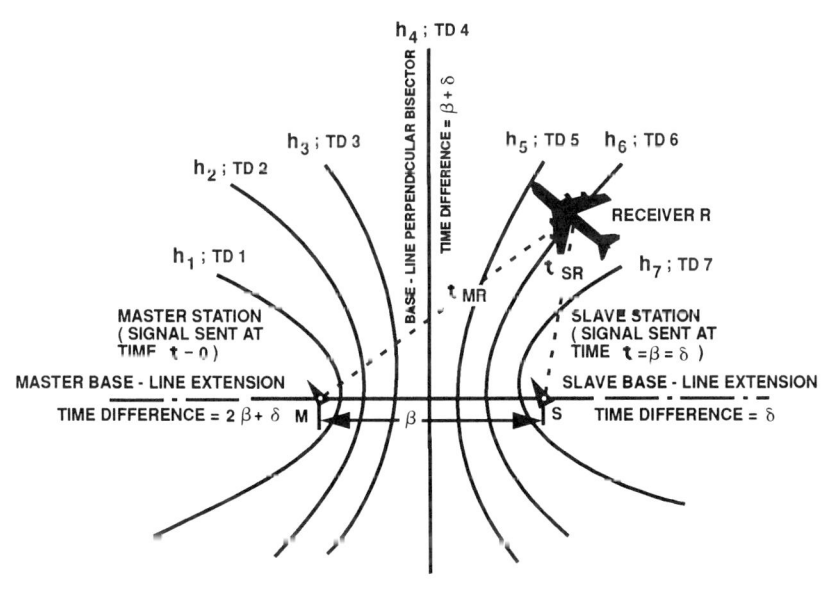

FIGURE 6-7
Loran time-difference grid.

(TD) is given by

$$TD = (\beta + \delta) + t_{SR} - t_{MR} \qquad (6.11)$$

where t_{SR} = time for a radio wave to travel between S and R
t_{MR} = time for a radio wave to travel between M and R
β = time for the radio wave to travel from M to S
δ = time after receiving the master pulse signal, before the slave station transmits its own pulse signal

Note that if the receiver were on the slave baseline extension, the hyperbolas would become great circles on the earth and represent the locus of all points with the following time differences: slave baseline extension $TD = \delta$; perpendicular bisector of baseline $TD = \beta + \delta$; master baseline extension $TD = 2\beta + \delta$. Thus, each hyperbola is uniquely defined by a time difference, such that $2\beta + \delta \geq TD \geq \delta$.

As can be seen from Fig. 6-7, two transmitting stations are not enough to obtain a position fix. A position fix requires another hyperbolic grid. Hence, a third transmitting station must be coupled with the master transmitter to generate the second set of LOPs. With both of these grids generated, navigators are able to determine their exact location based on the intersection of the two different hyperbolic LOPs. The coverage area of a Loran chain depends on the geometry of the chain. A Loran chain consists of three or more transmitter stations, in which one station is the master and the remaining stations are secondaries. Thus, the addition of a secondary station would increase the size of the coverage. There are three general types of chain-station configurations used in the Loran-C system: (1) triad, (2) wye, and (3) star [27]. The triad and star configurations are the most common transmitter layouts. A master and two slave stations or secondaries is called a *triad*; a master and three slave stations is referred to as the *wye* configuration; the *star* configuration consists of four slave stations and one master station, making a total of five stations that provide optimal coverage. The star configuration can be considered as four triads. The navigator uses the master and the two slave stations that provide the best coverage, as illustrated in Fig. 6-8. If two time-difference measurements TD1 and TD2 are made using either a triad or star transmitting station configuration, then the intersection of the two hyperbolas so defined provides a position fix for the receiver R.

In Fig. 6-8 the solid hyperbolas correspond to the triad configuration formed by the master station and the slave stations 1 and 2. By adding the slave station 3, the complete array of hyperbolas corresponds to the wye configuration. Included in the Loran chain is a monitor station that monitors the synchronization of the chain. Therefore, by monitoring the chain, one can maintain the accuracy and stability of the hyperbolic grid. Each chain is designed to give coverage to a particular area. Other configurations may

6.2. Navaid Sensor Subsystems

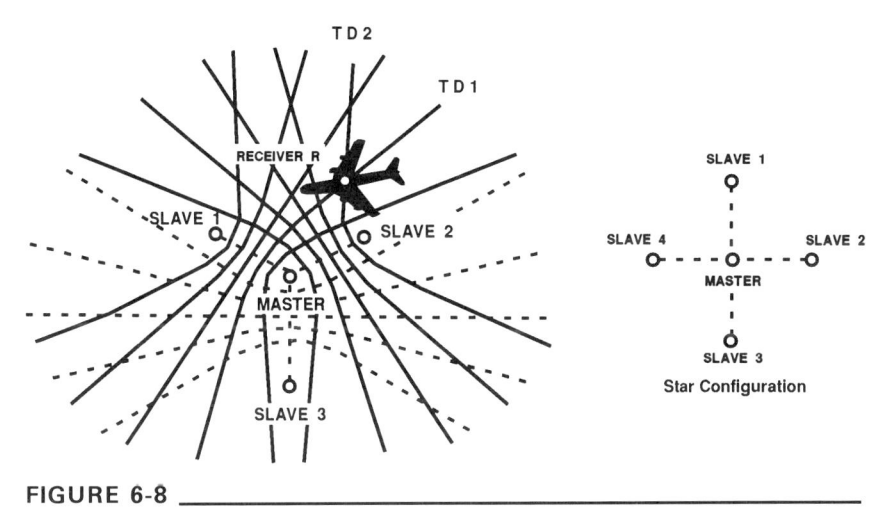

FIGURE 6-8
Triad, wye/star Loran chain configuration.

be dictated by special operational requirements or by practical considerations, such as the availability of a suitable transmitter site. Loran-C currently has 45 stations in 12 chains. Many Loran receivers today can receive two chains at once, and give the two time differences at once. Some also convert the time differences to latitude and longitude automatically, and with the addition of a plotter will actually plot the course of the vehicle. In addition to the chain's geometry, the coverage area is affected by the following factors: (1) ground conductivity, (2) signal strength, (3) atmospheric noise, and (4) dielectric constants. Even with all these factors perturbing the hyperbolic grid, a Loran-C chain has demonstrated predictable accuracies of 0.1852–0.5556 km (0.1–0.3 NM) at 1852-km (1000-NM) range and 1.852–9.26 km (1–5 NM) at 4630-km (2500-NM) range and a repeatable accuracy of 91.44 m (300 ft).

One final comment with regard to system geometry is in order. When the hyperbolic Loran navigation technique is used, the accuracy of the position solution is affected by the vehicle's (receiver) position relative to the transmitters. This geometric effect consists of two separate effects: (1) acuteness of the crossing angle of the hyperbolic LOPs and (2) proximity to the baseline extension. In regions sufficiently far from baseline extensions, the rms position error is approximately proportional to the rms time-difference measurement error. When the time measurements are converted to distance by multiplying them by the ground-wave velocity, a dimensionless proportionality constant is obtained, called the *geometric dilution of precision* (GDOP). When the hyperbolas intersect in a small acute angle, the GDOP becomes very large, approaching infinity as the crossing is along the bisector of the acute crossing angle (see Fig. 6-9). In regions near a baseline extension, the position error does not vary linearly with the measurement errors. This is because the gradient of the time difference associated with the baseline in question is zero on the baseline extension and very small near the baseline extension. Furthermore, the useful coverage area of a chain is inherently limited by the GDOP.

From Fig. 6-9 it is noted that a Loran-C position fix occurs at the intersection of two LOPs, each resulting from a time difference (TD) measurement. However, various factors, such as transmitter/timer instability and propagation effects, would cause random errors in the LOPs. Assuming that the distribution of errors is normal, and that the errors associated with

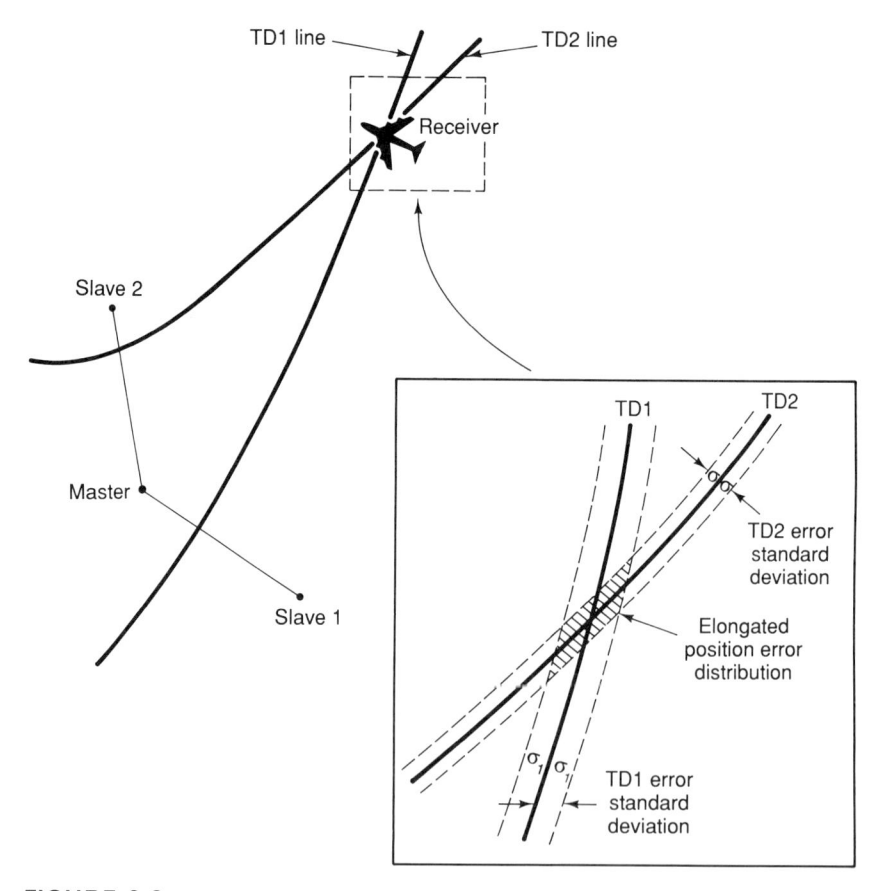

FIGURE 6-9
Errors due to hyperbolic crossing angles.

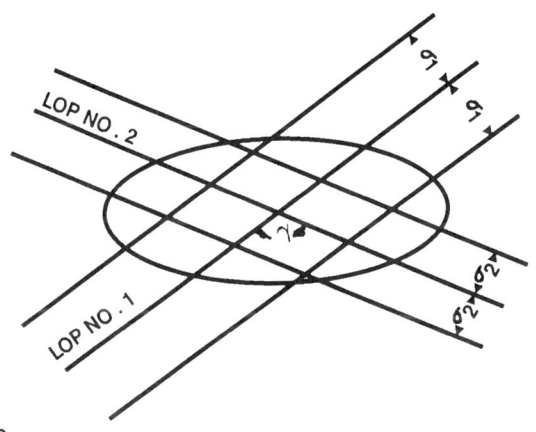

FIGURE 6-10
LOP error ellipse.

the two LOPs are independent, then a region of equal probability of errors about the fix would be an ellipse. Figure 6-10 illustrates the intersection of two lines of position, the standard deviation of each LOP, and the ellipse defining the region of equal probability [22]. The angle of crossing of the two lines of position is designated by the letter γ.

A standard measure of error for an LOP is its standard deviation (rms error). Numerically, the standard deviation σ corresponds to 68% of the distribution. That is, if a large number of measurements is made of a given quantity, 68% will be within the range $\pm 1\sigma$ of the mean value. Similarly, 95% will be within the range $\pm 2\sigma$, and 99.6% within $\pm 3\sigma$ of the mean value. At this point, it is noted that three types of accuracies are commonly used in connection with Loran-C [28]:

Predictable accuracy Accuracy of position with respect to the geographic or geodetic coordinates of the earth (i.e., latitude and longitude).

Repeatable accuracy Accuracy with which a user can return, again and again, to a position whose coordinates have been measured at a previous time and with the same navigation system.

Relative accuracy Accuracy with which a user can measure position relative to another user or to some reference point such as a beacon or a buoy. It may also be expressed as a function of the distance between two users.

6.2.3.2 The Loran-C Pulse System

As mentioned in Section 6.2.3, the carrier frequency for the Loran-C is 100 kHz. This frequency was selected because of its ground-wave characteristics and since it was the only one available where the Loran pulse would not interfere with existing broadcasting stations. The frequency bandwidth is 90–110 kHz; 99% of the energy radiated must be within this region. Since the system uses ground-wave transmissions, sky-wave contamination is avoided by using pulse techniques. Moreover, since in a Loran-C chain the pulse is the same for both the master and slave stations, a technique of pulse sequencing is used to distinguish the master stations from the slave stations. A slave station's pulse sequence consists of eight pulses that are 1 µs apart. Each master station transmits its pulses in groups of nine at a repetition rate of 10–25 groups per second. In normal operation, the pulses of each group are transmitted for a designated period of time, 200–300 µs. Pulse sequencing increases the average power of the transmitter. With eight pulses transmitted, phase coding is possible. Pulse coding provides a means to reject synchronous jamming signals. Spacing between the pulses in a group is 1000 µs. The selection of 100 µs between pulses was arbitrarily chosen to allow enough time for the passing of the sky wave from the previous pulse before transmitting the next pulse. Figure 6-11 shows a typical Loran-C pulse.

In order to meet these specifications, the pulse must have a short rise time and a slow decay time. A short rise time has the advantage of minimizing sky-wave interference. Most sky-wave interference begins at approximately 30 µs after the start of the pulse. Finally, at the 30-µs zero crossing of the Loran-C pulse, 50% of the pulse amplitude is available for receiver detection and tracking. It should be noted that the hyperbolic radio navigation system is equally valid for a continuous-wave system. However, in this case, a given phase difference does not represent a given LOP unambiguously, but represents an LOP in each of many lanes, with the width of a lane, and hence the number of lanes between a master and a slave station, depending on the

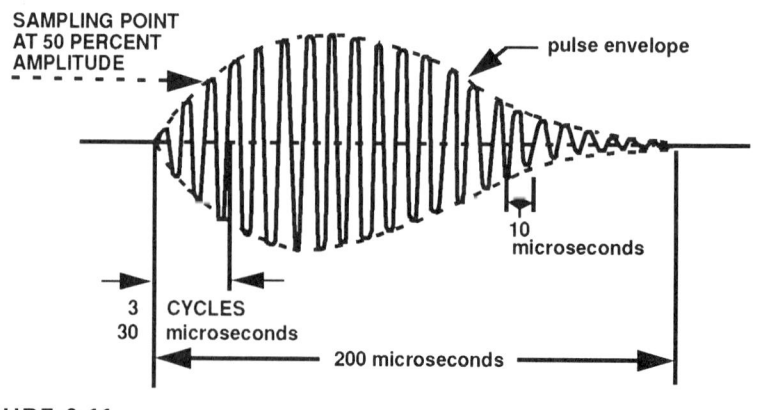

FIGURE 6-11
The Loran-C pulse.

carrier frequencies. Since a line of position is defined as "a locus of points which, when connected, form a line," an LOP will take one or more of the following forms:

Constant-difference LOP [Rho (ρ)]   A circle with the transmitting station antenna at the center.

Constant azimuth LOP [Theta (θ)]   A straight line whose angle is measured from a true north reference.

Constant distance/time-difference LOP   A hyperbolic line along the constant distance differential between two transmitting stations. This form of LOP is used in Loran-C.

The advantages of Loran are that it provides the best accuracy for a given area of coverage, there is no ambiguity on a position fix, and the receivers are affordable. Its disadvantages are as follows: (1) it does not cover a large area, (2) it does not provide altitude information, (3) its transmitters cannot be placed for best coverage, and (4) its signal can be jammed.

6.2.3.3. Loran–Inertial System Mechanization

In a Loran/INS system mechanization, the Kalman filter processes the Loran receiver time difference data, in order to estimate the errors in the INS solutions. Figure 6-12 [12] illustrates a typical integrated Loran/INS. The function of the Loran receiver system on board the aircraft (or ship) is to receive the pulsed RF (radio-frequency) Loran signals transmitted by a triad (one master and two slaves) of Loran stations and measure precisely the differences in the times of arrival between the master and each slave signal. As stated earlier, the Loran receiver time difference measurements (TD1, TD2) establish a position measurement at the intersection of the two hyperbolic LOPs defined by the time differences. Furthermore, the Loran time-difference measurements are used in the Loran/INS Kalman filter innovation or residual ($\{\mathbf{H}[\mathbf{\Phi}(\mathbf{X}_{k-1/k-1})] - \mathbf{Z}_k\}$) vector. The Loran time differences are processed at, say, the time instants T_j, and between successive

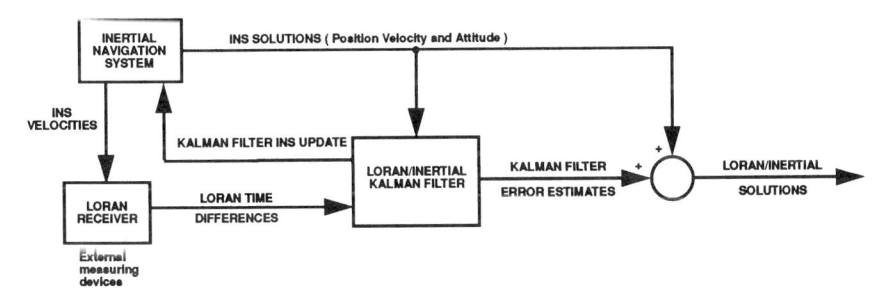

FIGURE 6-12
Integrated Loran/INS.

observation times, the state vector estimate is time updated at the times τ_i. The innovation components, which are observations of the INS position error, are the differences between the Loran time differences (TD1, TD2) and the Kalman filter's estimates of the time difference $(\hat{TD}1, \hat{TD}2)$, obtained from the Kalman filter estimates of latitude $(\hat{\phi})$ and longitude $(\hat{\lambda})$.

As seen in Fig. 6-12, the Loran/INS Kalman filter differences the INS position solution and the Loran position information (i.e., the Loran time differences) and estimates errors in the INS. Consequently, the INS error estimates are added to the INS navigation solutions to obtain the Loran/INS Kalman filter solutions for position, velocity, and heading. Therefore, if the Loran data being processed by the Kalman filter are valid, the INS is periodically updated by the Kalman filter to correct the INS navigation solutions. Furthermore, Fig. 6-12 also shows the Loran receiver using the INS velocity solutions to obtain better dynamic performance in its time difference measurements [1].

When Loran-C is used in an integrated mode, the master–slave time difference errors can be modeled as the sum of two independent random processes. For instance, one may model the master station's time difference errors by a stochastic differential equation driven by a Wiener process.

6.2.4. Omega

Omega is a long-range, worldwide, all-weather, day-and-night radio-navigation system. The Omega concept was pioneered by Professor J. A. Pierce in the late 1940s. Subsequently, the system underwent several stages of development, and in 1965 the U.S. Navy established the Omega project office, developing the system to the point where today the system provides worldwide coverage from a network of eight transmitter stations strategically located around the world. Two of these stations are controlled by the U.S. government, while the remaining six stations are under the control of foreign governments. Therefore, the Omega system includes the vehicle receiver and eight radio transmitters operating in the very-low-frequency (VLF) range. Omega receivers employ automatic station selection algorithms to select the three or four Omega stations from which the signals will be processed for use in the navigation solution. At least three of the Omega transmitter signals can be received any place on earth. The eight transmitter stations transmit at precise times and for exact intervals. Therefore, cesium clocks are used for synchronizing station transmissions. A signal from any given transmitter can be received up to 14,816 km (8000 NM). Users of Omega are the commercial airlines, ships, and land vehicles. Omega, like Loran, is a hyperbolic, very-low-frequency, navigation system using phase difference of continuous-wave (cw) radio signals, operating in the electromagnetic spectrum between

6.2. Navaid Sensor Subsystems 291

10 and 14 kHz. Since Omega operates at these very low frequencies (10–14 kHz), propagation can be considered as taking place within a spherical waveguide formed by the earth and the ionosphere. Therefore, since signals are propagated within the waveguide formed by the earth and ionosphere, changes in propagation velocity of the signal occur as a result of changes in the ionosphere or differences on the earth's surface such as mountains, oceans, ice, and plains. Also, navigationally, the most obvious undesirable effect is the daily phase change from day to night. Solar conditions also alter the propagation velocity. Each station broadcasts four omnidirectional, time-multiplexed 10-kW signals on 10.2, 11.05, 11.33, and 13.6 kHz during each 10-second transmission period. That is, the time-multiplexing permits one station to be on at a time on one frequency. The period is divided into eight transmission slots, an equal time of 0.2 s separating the slots. As a result, this transmission format allows easy identification of both station and phase. If a measurement with a phase difference of 10.2 kHz is transmitted from two stations, the wavelength is nearly 25.74 km (16 miles), repeating every 12.87 km (8 miles). Each of these intervals are called "lanes." Figure 6-13 illustrates the signal transmission format [23].

Navigation information is obtained in the form of pseudorange (phase) measurements from, as mentioned above, a minimum of three stations. Specifically, the receiver synchronizes to the transmitter frequency and measures the phase relationship of the receiver's location. The phase-difference relationship between two received signals defines an LOP. Two or more LOPs define the receiver location. Therefore, phase measurements from three or more stations provide position fixing (latitude and longitude) information to the user with an accuracy of 1.85–3.70 km during daytime operation, degrading to 3.70–7.40 km during night operation due to diurnal effects. It should be pointed out, however, that the accuracy depends on geographic location of the vehicle, stations used, accuracy of propagation corrections, day or night, and receiver characteristics. Moreover, signal propagation effects can degrade the system accuracy, and the signals are not denied to potential hostile forces and/or governments. Furthermore, the network is vulnerable to jamming, attack, and natural disturbances.

Selection of the above frequencies prevents the "lane ambiguity," yielding at 133-km (72-NM)-wide unambiguous lane. Lane ambiguity exists because a series of concentric circles emanates from the transmitting source, where each circle is a distance of a wavelength from an adjacent circle. Consequently, during transition from daylight to darkness or vice versa, an Omega receiver can experience "lane slip" of about 18.5 km (10 NM). Consider now two transmitters located on the baseline as shown in Fig. 6-7, where we will designate the master (M) station to be transmitter 1 and the slave (S) to be transmitter 2. The location of a receiver in the Omega

292 CHAPTER 6 • *Externally Aided Inertial Navigation Systems*

TIME SEGMENT	T 1	T 2	T 3	T 4	T 5	T 6	T 7	T 8
TRANSM INTERVAL	0.9	1.0	1.1	1.2	1.1	0.9	1.2	1.0
STATIONS								
NORWAY	10.2	13.6	11.33	12.1	12.1	11.05	12.1	12.1
TRINIDAD/LIBERIA	12.0	10.2	13.6	11.33	12.0	12.0	11.05	12.0
HAWAII	11.8	11.8	10.2	13.6	11.33	12.0	12.0	11.05
NORTH DAKOTA	11.05	12.1	13.1	10.2	13.6	11.33	12.1	13.1
LA REUNION ISLAND	12.3	11.05	12.3	12.3	10.2	13.6	11.33	12.3
ARGENTINA	12.9	12.9	11.05	12.9	12.9	10.2	13.6	11.33
AUSTRALIA	11.33	13.0	13.0	11.05	13.0	13.0	10.2	13.6
JAPAN	13.6	11.33	12.8	12.8	11.05	12.8	12.8	10.2

FIGURE 6-13
The Omega signal transmission format.

environment can be accomplished by the phase measurement technique. Define the phase difference for a pair of transmitting stations as follows:

$$\phi_1 = \phi_{10} + 2\pi f \left(\frac{\rho_1}{c}\right) \tag{6.12a}$$

$$\phi_2 = \phi_{20} + 2\pi f \left(\frac{\rho_2}{c}\right) \tag{6.12b}$$

where $\Delta\phi$ = change in phase = $\phi_1 - \phi_2$
  c = speed of light
  f = frequency of transmitted signal
  ϕ_{10} = initial phase at transmitter 1
  ϕ_{20} = initial phase at transmitter 2
  ρ_1 = distance from transmitter 1 to receiver
  ρ_2 = distance from transmitter 2 to receiver

Therefore

$$\Delta\phi = \phi_1 - \phi_2 = (\phi_{10} - \phi_{20}) + 2\pi f \frac{1}{c}(\rho_1 - \rho_2) \tag{6.13}$$

6.2. Navaid Sensor Subsystems

Because of the precise synchronization of the eight Omega ground stations resulting from the use of cesium clocks, $\phi_{10} = \phi_{20}$; therefore, Eqs. (6.12a) and (6.12b) simplify to

$$\Delta \phi = 2\pi f \frac{1}{c} (\rho_1 - \rho_2) \qquad (6.14)$$

This is the expression that defines the hyperbolic LOP distance difference from two fixed points. Now assume that the receiver is located on the baseline between the two transmitters. The phase-difference reading repeats on the baseline at integer multiples of half a wavelength. If σ_b is the phase repetition interval on the baseline, then $\sigma_b = \lambda/2$, where the wavelength is given by $\lambda = c/f$. Furthermore, if σ_c is the phase repetition interval anywhere in the coverage area, then

$$\sigma_c = \tfrac{1}{4}\lambda \csc \gamma \qquad (6.15)$$

where γ is the crossing angle of the LOPs. In order to reduce the ambiguity of the LOPs, several techniques such as the utilization of an external reference (i.e., satellite) or counting of lanes from the initial point via an automatic lane counter are used to resolve exactly in which lane the receiver is located. The Omega system was originally designed to be a hyperbolic system. However, many of the Omega receiver systems today are operating using the "ranging mode." One of the main reasons for this change is that the ranging mode allows greater flexibility to the user. In the hyperbolic system, at least three ground stations are required to obtain a position fix, while in the ranging mode only two ground stations are required to obtain a position fix. The basic operation of the ranging mode, often called "Rho–Rho" navigation, is the measurement of the total time it takes for each of the two signals to travel from the individual ground station transmitters to the receiver. These time measurements for each of the transmitters define a circular LOP. The intersection of two circular LOPs defines a position fix. The ambiguity of this position fix can be resolved by knowledge of approximate location. The main disadvantage of the Rho–Rho navigation technique is the requirement for a very accurate and stable clock or oscillator, against which the time measurements are made. However, with the advancement of technology, very accurate clocks are becoming more available and economical, thus decreasing the overall cost to the user. Any clock drift will directly affect the range measurement due to the induced clock error. In the hyperbolic system, the clock bias is not a factor, since time differences between signals are not measured. If a third station is introduced to complement the two range measurements, three circular LOPs are generated. This is known as "Rho–Rho–Rho" navigation. These three LOPs will intersect at some common point, only if there are no clock errors. However, if the range measurements are adjusted until a common intersection point is achieved, the clock drift error can be estimated and applied to subsequent range measurements. As a result, the use of this method allows for the use of a less accurate and expensive clock, but requires the use of a third Omega transmitter. For the hyperbolic system, clock initialization errors have very little effect on position accuracy due to the fact that only the phase difference between the incoming signals is measured, resulting in the cancellation of initialization errors. Initialization errors in the ranging mode are more significant because they add a constant error to the position fix.

The Omega error does not accumulate with elapsed time, as is the case with inertial navigation systems. For this reason, some airlines use two INSs and one Omega receiver. Omega requires heading and velocity information, which is available from the INS. Prior to entering the Omega navigation mode, the system must be initialized. A hybrid Omega–inertial system uses the Omega to damp (or limit) the error accumulation on the inertial system, while using the INS stability to correct for lane slip errors. Thus, in a hybrid Omega/INS-aided navigation system, the mixing of Omega and inertial data takes advantage of the long-term accuracy of Omega position measurements and the short-term accuracy of the inertial system. The inputs of Omega and INS are combined in a Kalman filter to yield the best estimate of present position. Moreover, the long-term velocity and position drifts in the inertial data must be corrected by continuously comparing the inertial position to the Omega position. As is the case with all radio navaids, the largest source of error in position determination with Omega is due to signal propagation effects such as (1) diurnal effects, (2) seasonal and sunspot activity (11-year cycle) variations of the ionosphere, (3) earth geometry, (4) earth's magnetic field, (5) ground conductivity, and (6) latitude effects. Propagation uncertainties are reduced with a predicted propagation correction computed using simplified models of the ionosphere. The nominal effective ionosphere height is approximately 70 km (43 miles) during daytime and 90 km (56 miles) at night. Also, synchronization errors are reduced either by using hyperbolic navigation or by estimation techniques involving modeling of the receiver's internal time source (e.g., clock or precision oscillator).

6.2.4.1. Differential Omega

An improvement over the conventional Omega is the "Differential Omega." The differential Omega system concept was developed to enhance the accuracy of the original Omega system. The differential Omega system consists of a ground unit with a monitor (or reference) receiver at a fixed, precisely known location with an uplink transmitter. Figure 6-14 depicts in block diagram form the differential Omega concept. The ground reference receiver, whose location is precisely known, measures the actual Omega signal phases and compares them with the nominal phase characteristics of the known receiver location. The differences between the nominal and actual

6.2. Navaid Sensor Subsystems

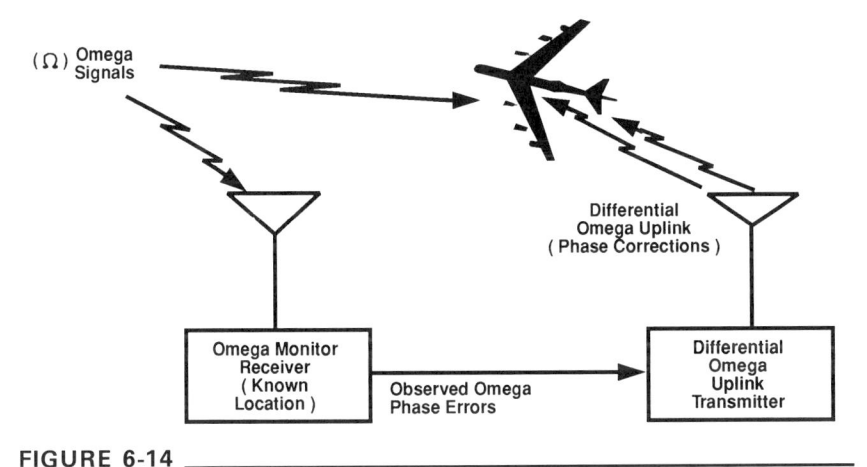

FIGURE 6-14
Differential Omega concept.

phase measurements are used to generate signal correction data, which are uplinked (or broadcast) to the airborne differential receiver within the area of coverage. Consequently, the airborne receiver decodes the correction data from the uplink transmitter and uses these to correct the Omega signals measured by the Omega receiver. For ranges of <370.4 km (<200 NM) there exists a good correlation between the Omega signal errors measured by the monitor station and by the user equipment. Flight tests have shown that the differential Omega system can provide a substantial accuracy improvement for the Omega system. This improvement is based on having reasonably good Omega coverage over the area of interest. It should be noted that the signal reliability and data content of the differential Omega are essentially the same as that of the conventional Omega. With good Omega coverage, the differential Omega can reduce the errors resulting from propagation phase prediction errors but cannot correct for poor phase measurements due to poor signal/noise (SNR) ratios or poor Omega station receiver geometry [7].

The Omega system is a relatively accurate long-range navigation system. The system can be used as a stand-alone system for this type of navigation or may be used as a backup for other navaids. Omega is being used by both military and commercial aviation in a large number of aircraft as one of the navaids. Because of its higher accuracy, the differential Omega finds extensive use in such applications as terminal or approach navigation, coastal and harbor navigation for ships, and military applications (e.g., inflight refueling, and precision air-drop activities). With improved electronics and more efficient algorithms, it is conceivable that the Omega system will provide day and night, all-weather, worldwide navigation with position errors of better than 1.852 km (1 NM). The disadvantages of the differential Omega system is that it requires the installation and maintenance of a large number of surveyed reference stations, and the requirement for the establishment of a communication system for transmission of the compensation signals from the reference stations.

On completion of the global positioning system (GPS), to be discussed next, Omega will slowly be phased out with an expected date of termination to be approximately the year 2000.

6.2.5. The Navstar–Global Positioning System (GPS)

The Navstar*–GPS is a space-based, pseudoranging navigation satellite system that will provide worldwide, nearly continuous, three-dimensional position, velocity, and coordinated universal time to the suitably equipped user, and is designed primarily for global navigation of a terrestrial or near-earth user. In particular, the system broadcasts continuously the information needed for a GPS receiver to compute its own position and velocity. For missions where an autonomous navigation is needed, a hybrid of the GPS and an INS can be used, which has several advantages. The GPS can update, align, and calibrate the inertial system continuously until the GPS signals are lost. At that time, the inertial system would have a much more recent update and would provide precise positioning through its short-term stability. This combination will reduce position accuracy degradation during high-energy–high-dynamic maneuvers, and provide navigation redundancy should one system fail. The GPS is being developed as a DoD joint service program, for such military applications as guidance, rendezvous, reconnaissance, targeting operations, command and control, weapon delivery, anti-submarine warfare, and inertial guidance systems. Furthermore, civil applications as well can benefit from the total worldwide coverage, all-weather operations, and the unlimited number of passive users that the system can support. Consequently, potential commercial uses for aircraft, ships, and land vehicles are practically unlimited. Precision airline or general aviation navigation anywhere in the world will be possible when the system is fully operational. In this regard, the GPS would offer far-reaching benefits for both the military and civilian community. For instance, GPS accuracies would provide full worldwide coverage of all airports in a common grid, thereby allowing for closer air traffic densities, reduced holding times, and reduced airport congestion, resulting in substantial savings to the airlines in terms of fuel and other resources. Other civilian uses of the GPS will be in the areas of search-and-rescue, mineral exploration, geodesy, geology, mapping, and surveying.

---

*NAVigation Signal Time and Range.

The idea that navigation and positioning could be accomplished using radio signals transmitted from satellites has been actively pursued by the U.S. Navy since the 1950s, with the TRANSIT program, and the U.S. Air Force since the early 1960s. At that time, each service separately established its own concept of such a system through an extensive program of studies and tests designed to demonstrate the feasibility of a space-based positioning and navigation system. The success of the Navy Navigation Satellite System (NNSS), better known as TRANSIT, stimulated both the Navy and the Air Force to develop a more advanced system that would provide enhanced capabilities and global coverage. The Navy's concept, TIMATION (TIMe and navigATION), was essentially a two-dimensional system that could not provide position updates in a high-dynamic aircraft environment. The TRANSIT system was made available to nonmilitary users in July 1967, and the first commercial TRANSIT receivers became available in 1968. The Air Force concept, known as "Program 621B," could provide the high dynamic capability, but had its own shortcomings as well, particularly from a survivability standpoint. On April 17, 1973, the Deputy Secretary of Defense issued a memorandum entitled "Defense Navigation Satellite Development Program," designating the Air Force as the executive service to coalesce the Pos/Nav satellite concepts that were being developed by the Navy and the Air Force into a Defense Navigation Satellite System (DNSS). On December 22, 1973, the Secretary of Defense approved the DNSS proposed by the Air Force in Paper No. 133, and renamed the program as the Navstar/Global Positioning System [4].

When first conceived, the design for the fully operational system consisted of a constellation of 24 satellites, deployed with eight satellites uniformly distributed in each of three orbital planes, providing continuous three-dimensional global coverage with predicted accuracies in the 10-m range. During 1978 and 1979 Defense budgetary constraints forced a reduction in funding for the GPS program. As a result, in 1980, the Navstar/ GPS program was restructured and the number of satellites for the fully operational system was reduced from the 24 originally planned to 18. Extensive studies have been made since that time to establish the optimum orbital configuration for these remaining 18 satellites. System accuracy, survivability, satellite visibility, ease of buildup, location and duration of outages, ease of sparing and replacement, and growth potential of the constellation to 24 satellites were considered in these studies. After the mid-1980s, the baseline constellation was reconfigured, presently consisting of 21 satellites and 3 active spares, again forming a constellation of 24 satellites. Basically, as it will be discussed in the next subsection, the GPS consists of a space segment, a control segment, and a user segment. The space segment will be a 24-satellite constellation (21 active + 3 spares) continuously broadcasting signals containing time and satellite positioning information to an unlimited number of passive users. The control segment consists of monitor stations located on U.S. territory and a master control station located in the CONUS (CONtinental U.S.). Satellite positioning and time information will be updated through the control segment. The user segment consists of appropriate GPS user equipment (receivers) sets mounted in aircraft or other vehicles. Each GPS user set receives information broadcast by the GPS satellites and derives the set's position and velocity; each set also receives and decodes the time signals broadcast by each GPS satellite.

6.2.5.1. GPS System Overview

The Navstar/GPS, hereinafter called simply GPS, is divided into three major segments: (1) the space vehicle (SV) segment, which is the constellation of earth-orbiting satellites; (2) the ground-control segment, which monitors the orbits of all satellites and provides them with updated information several times each day; and (3) the user segment, which is all the air-, land-, sea-, and space-based users equipped with GPS receivers. Figure 6-15 illustrates these three segments. These three segments will now be discussed in some more detail.

Space Segment The space vehicle segment consists of all the GPS satellites. The orbital constellation consisting of 21 satellites plus 3 active spares will be deployed in subsynchronous circular orbits, at an altitude of approximately 20,183–20,187 km (10,898–10,900 NM) with an orbital period of 12 h, in six orbital planes, each orbital plane inclined 55° to the equator with three or four operational satellites in each plane. Moreover, the orbital planes will be separated from each other in longitude by 60°, with a nonuniform phasing* of the satellites within the planes. These orbits are phased so that any user on earth can acquire, at any time, at least four satellites. Each satellite is equipped with a highly accurate atomic cesium clock with a known or predictable offset from GPS time. Figure 6-16 shows how an optimized 21 satellite constellation may be arranged [4].

Each satellite will transmit its information on two L-band frequencies designated L_1 and L_2: L_1 is 1575.42 MHz and L_2 is 1227.60 MHz. Two frequencies are required to correct for ionospheric delay uncertainties in the transmission. The L_1 frequency will be modulated in quadrature by two pseudorandom codes, a coarse/acquisition (C/A) code and a precision (P) code. The C/A code has a frequency of 1.023 MHz and repeats itself every millisecond. The P code is a classified code sequence, which is created from a product of two pseudorandom codes and modulated at a frequency of 10.23 MHz, making it difficult to acquire, and is one week long. The C/A

---

*For uniform spacing of the satellites within a plane, assuming a constellation of 24 satellites, it will be designated as 24/6/1 (24 sats/6 planes/relative phasing $360° \times 1/24 = 15°$).

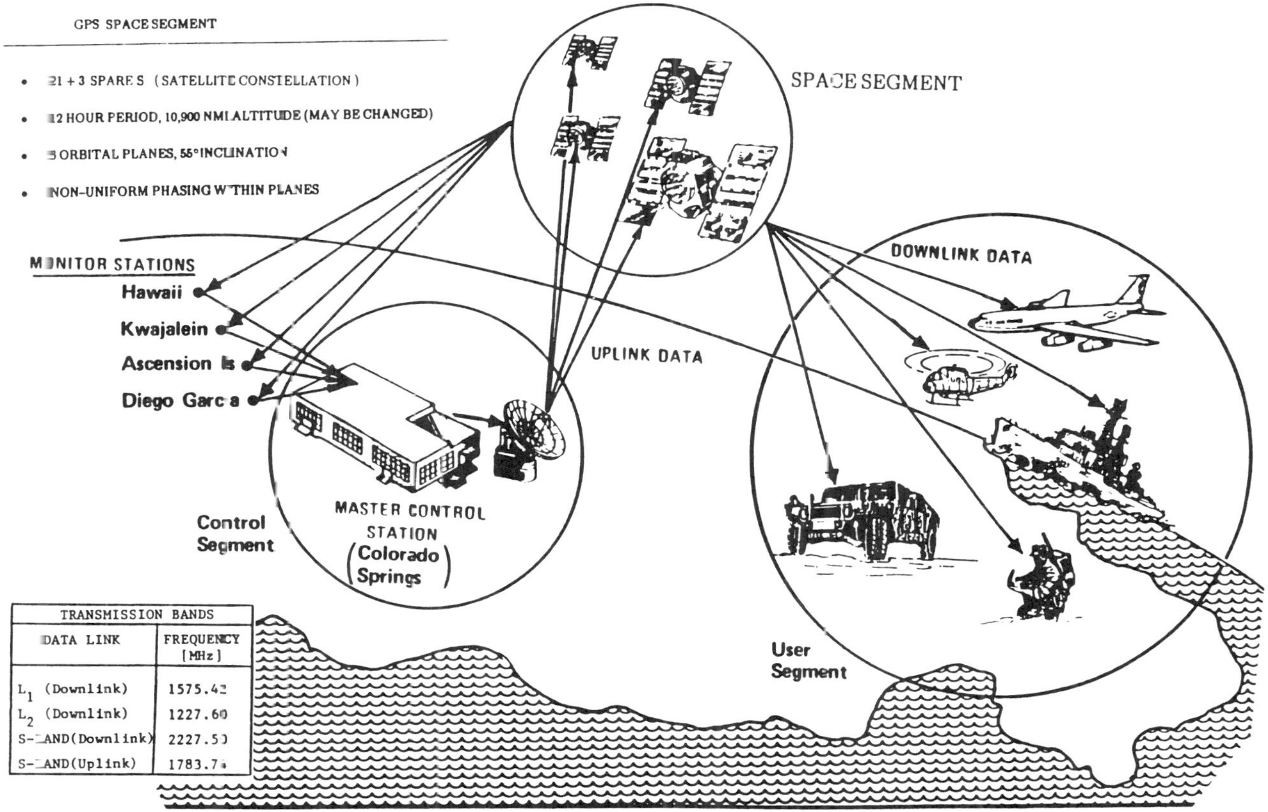

TRANSMISSION BANDS	
DATA LINK	FREQUENCY [MHz]
L_1 (Downlink)	1575.42
L_2 (Downlink)	1227.60
S-BAND (Downlink)	2227.50
S-BAND (Uplink)	1783.74

FIGURE 6-15

Navstar/GPS major segments.

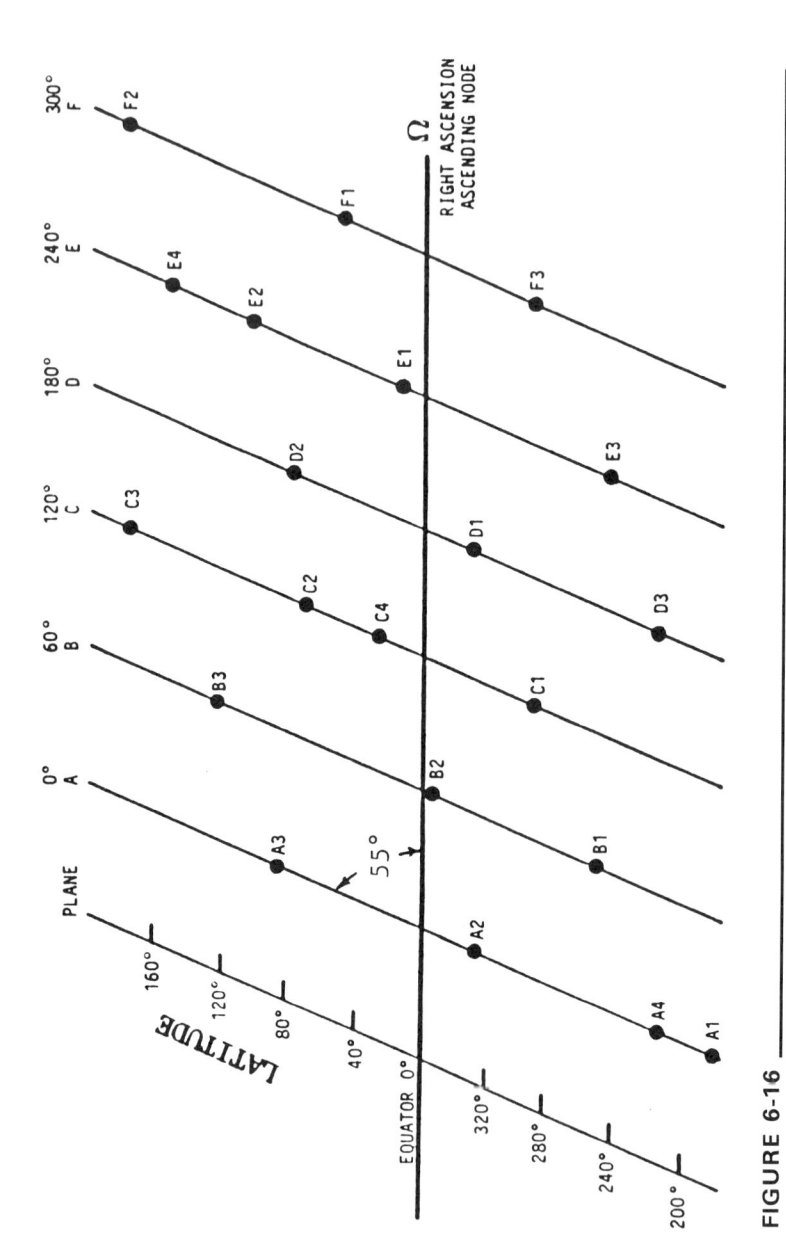

FIGURE 6-16
Optimized 21-satellite constellation phasing.

code is acquired first, and then the transfer is made to the precision code by using a handover word contained in the C/A code. The L_2 frequency is modulated by the P code, but not by the C/A code. Moreover, the specific code sequences broadcast by each satellite are different and are used by the GPS receiver in order to distinguish the satellites from each other. Each frequency (L_1 and L_2) is further modulated by the navigation message. The transmissions are sent in five 1500-bit data blocks in a bit stream of 50-bit/s rate.

When fully operational, two classes of the GPS service will be available: (1) the Standard Positioning Service (SPS), and (2) the Precise Positioning Service (PPS). The SPS, utilizing the C/A-code signal, will be available to the general public. It will provide a horizontal position accuracy of 40 m CEP (100 m 2-D RMS). The PPS, utilizing the P-code signal, is a highly accurate positioning, velocity, and timing service, which will only be available to authorized users. The PPS will provide a 3-D position RMS accuracy of 10–16 m SEP, 0.1 m/s RMS velocity accuracy, and 100 ns (1-sigma) Universal Coordinated Time (UCT) time transfer to authorized users. In a "stand-alone" GPS installation, the GPS position and velocity outputs are calculated by using passive trilateration. Specifically, the GPS user equipment measures the pseudorange and delta pseudorange to four satellites, computes the position of the four satellites using the received ephemeris data, and estimates the user's position, velocity, and time.

The navigation message contains GPS time, satellite ephemeris data, atmospheric propagation correction data, system almanac data, and any other information needed by the GPS receivers. As mentioned above, the use of two signals permits the user's equipment to compensate for the ionospheric group delay or electromagnetic disturbances in the atmosphere that may alter the affected signals. Each satellite has a mean mission duration of 6.2 years and a design life of 7.5 years.

Control Segment The ground-control segment monitors the satellite broadcast signals and uplinks corrections to ensure predefined accuracies. This segment is also responsible for monitoring and controlling the orbits of the satellites, for maintaining the GPS system time, and for uploading necessary information to the satellites three times each day. The operational control segment will consist of five monitor stations, a master control station, and three uplink antennas. The master control station is located near Colorado Springs, Colorado. The widely separated monitor stations, positioned worldwide, will allow simultaneous tracking of the full satellite constellation and will relay orbital and clock information to the master control station. The ranging data accumulated by the monitor stations will be processed by the Navstar Operations Center (master control station) for use in satellite determination and systematic error elimination. The master control station then forms corrections that are uploaded to the satellites by the uplink antennas.

User Segment The GPS user segment is intended for both military and civilian users of the system. Using GPS requires a GPS receiver, which consists of three major functional divisions: (1) an antenna to capture the GPS signals, (2) an amplifier to increase the power level of the received signal, and (3) a digital computer to process the information contained in the signal. The primary computer output is the position and velocity of the GPS receiver. The user segment selects the four optimally positioned satellites from those visible and, using the navigation signals passively received from each of these satellites, the user's receiver measures four independent pseudoranges and pseudorange rates to the satellites. The receiver-processor then converts these signals to three-dimensional position, velocity, and system time. However, in situations where one or more of the satellites are temporarily obscured from the receiver antenna's view, the receiver will have to acquire additional satellite signals in order to generate a continuous position-velocity-time solution. The aforementioned solution degrades until the new satellites are acquired. Because the GPS information is available in a common reference absolute grid, the WGS-84 coordinate system, or any desired mapping or coordinate system, civil and military position data can be standardized on a worldwide basis. The user equipment can transform navigation information into other commonly used datums as well. That is, a coordinate system can be converted to any other coordinate system the user desires. Frequently, in many applications the position solution is in the WGS-84, earth-centered earth-fixed (ECEF) coordinate system. In essence, the GPS navigation system will perform three major tasks:

1. Acquire and measure the signal of four geometrically optimum satellites (if less than four satellites are in sight, then the aircraft altitude, heading, and airspeed are used with the satellite data, to calculate the vehicle fix; also, currently if less than four satellites are available, the GPS receiver provides no output on the MIL-STD-1553 databus).
2. Process the satellite data, determine the position of the receiver, and transform that information into a coordinate system that is familiar to the operator (i.e., latitude, longitude, and altitude).
3. Create an interface between the user and the vehicle by providing a means to receive signals from other vehicle systems in both digital and analog form, a command output to the user's vehicle for such functions as steering signals, and an interaction with the operator through a control-and-display unit (CDU).

Presently, four main receiver models are produced: (1) a five-channel aircraft model, (2) a five-channel marine model, (3) a two-channel model, and (4) a one-channel model. It should be pointed out, however, that GPS receivers with up to 12 channels are available from various manufacturers using the SPS or PPS service. An example of a one-channel receiver is the portable battery powered "manpack." Each channel can acquire and track only one satellite at a time. Thus, the five-channel receiver parallel processes the range information from four satellites. The five-channel air model, which is referred to as the "Standard GPS IIIA receiver," operates in three different modes, depending on the type of aiding information available. The first mode, the Doppler radar system (DRS) mode, operates when Doppler radar aiding is available. The INS mode is employed when INS aiding is present and provides a generic INS error model to emulate the actual errors that the INS experiences. The last mode, the position velocity acceleration (PVA) mode, occurs when no form of aiding is available. Since solving the navigation equations requires data from four different satellites, receivers with fewer than four channels must listen to the different satellites sequentially. In particular, there are three basic types of GPS receiver architectures used to perform satellite tracking. These are as follows:

Sequential tracking receivers A sequential receiver tracks the necessary satellites by using one or two channels. In this case, the receiver will track one satellite at a time, time tag the measurements, and combine them when all four satellite pseudoranges have been measured. Moreover, the four pseudorange measurements are made on both L_1 and L_2 frequencies in order to determine position and compensate for ionospheric delay.

Continuous receivers A continuous tracking receiver will normally have four or more channels to track four satellites simultaneously. If a five-channel receiver is used, the receiver uses the fifth channel to read the navigation message of the next satellite to be used when the receiver changes the satellite selections. The continuous receiver is ideal for high dynamic environment vehicles, such as fighter aircraft.

Multiplex receivers The multiplex receiver switches at a fast rate (e.g., 50 Hz), between the satellites being tracked, while continuously collecting sampled data to maintain two to eight signal processing algorithms in the software. The navigation message is read continuously from all the satellites.

Finally, a GPS receiver implements a phase-lock loop and delay-lock loop to acquire the satellite transmissions and extract the pseudorange and delta-range to each satellite. The phase-locked loop locks on the phase of the carrier frequency and is therefore able to detect any phase change of the carrier. The delay-lock loop locks on to the time sequence of a signal and is therefore able to detect any delay in signal acquisition.

Typical GPS receiver inputs and outputs are presented below. Note that receiver input and output data may be managed through the GPS dedicated control and display unit (CDU), or as part of an integrated system function.

Inputs

- Receiver aiding signals
- Initialization inputs (e.g., position, time)
- Waypoint navigation data (if the system is used as a navigator)
- Self-test command

Outputs

- Position, velocity, and time
- 3-D area navigation and steering data
- Receiver status [e.g., figure of merit, jamming detection, automatic performance monitoring, and built-in-test (BIT)]

6.2.5.2. The Navigation Solution

Navigation—whether it is air, land, or sea—using the GPS is accomplished by means of passive trilateration. Users determine their position by measuring the range between their antenna and four satellites. As mentioned earlier, each satellite contains its own highly accurate atomic cesium clock. These clocks are kept synchronized to GPS time by the master control station. The signal transmitted by each satellite contains the time of the start of the transmission. Since the transmission propagates from the satellite to the GPS receiver at the speed of light, which is constant, the distance traveled by the signal can be determined by noting the time the signal was received, subtracting the time it was transmitted, and multiplying this difference by the speed of light.* The measurement of the signal's time of arrival will therefore be in error by an amount equal to the difference between the atomic time standard maintained by the satellite and the time maintained by the GPS receiver. Because of factors such as cost and portability, the GPS receiver employs a less accurate clock (i.e., crystal), which causes a significant time difference between the satellite and the receiver clock, thereby resulting in an inaccurate range measurement. As a result of the user clock error, this range measurement is referred to as "pseudorange." Therefore, pseudorange is the term used to describe the measured distance between the GPS receiver and a particular satellite. Mathematically, the pseudorange may be expressed

---

*Since the velocity of light is 3×10^8 m/s, 1 ns of transmission time error corresponds to 0.3048 m of range error.

6.2. Navaid Sensor Subsystems

by the equation

$$R_p = R_a + c\Delta T_b = R_a + c(T_r - T_t) \tag{6.16}$$

where R_p = pseudorange
  R_a = actual or true range
  c = speed of light
  T_r = time of signal reception
  T_t = time of signal transmission
  ΔT_b = receiver's clock bias or clock offset

At this point, it is appropriate to recapitulate the functions of the user equipment stated in the previous section. The GPS user equipment performs the following functions: (1) measures the pseudorange to four satellites in view, (2) computes the position of these four satellites using the received ephemeris data, and (3) processes the pseudorange measurements and satellite positions to estimate the three-dimensional user position and time.

Now consider Fig. 6-17. At time zero, the X_E axis passes through the North Pole, and the Y_E axis completes the right-handed orthogonal system. Because of the earth's rotation, the user x and y axes are constantly changing in longitude with time.

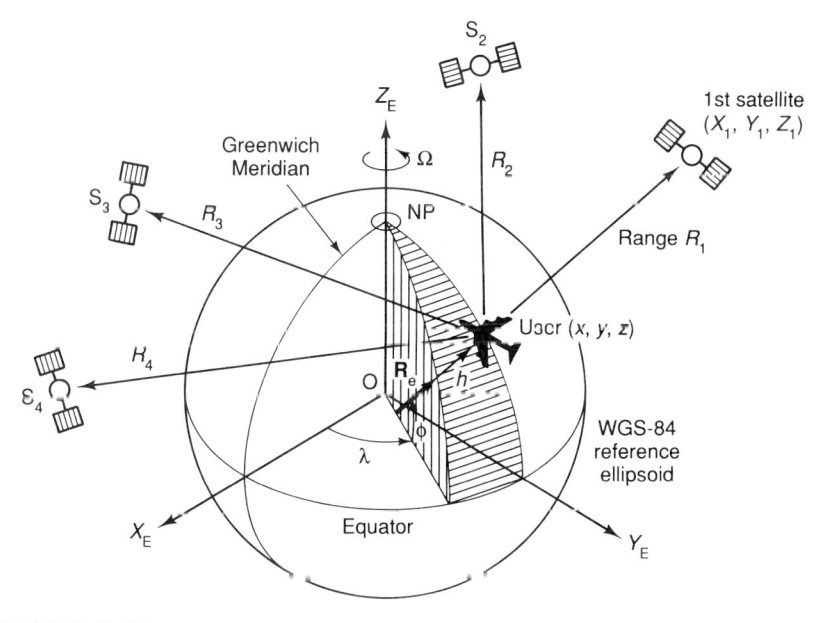

FIGURE 6-17
Earth-centered, earth-fixed (ECEF) coordinate system.

The GPS receiver calculates its position in an ECEF Cartesian coordinate system. These coordinates may be converted to some other system such as latitude, longitude, and altitude if desired. Specifically, the GPS receiver calculates precisely the position of each satellite using the information transmitted by the satellite. As stated earlier, the GPS receiver uses one-way transmissions to measure pseudorange and delta pseudorange (or range rate) between the user and the selected satellite. Note that pseudorange is the measured signal delay expressed in units of distance. Normally, the navigation data frame is 6 s long; however, most systems provide a code epoch pulse every 1.5 s. Now let (x_i, y_i, z_i) be the position of the ith satellite being tracked. Therefore, if the ith satellite is being tracked, a pseudorange estimate R_i can be obtained every 1.5 s. The four unknown quantities that must be solved for are the three coordinates of the GPS receiver location (x, y, z) and the GPS receiver clock bias ΔT_b. Using four satellites, these four quantities may be found from the simultaneous solution of four coupled, nonlinear equations of the form

$$R_i = \sqrt{(x-x_i)^2 + (y-y_i)^2 + (z-z_i)^2} \\ + I_i(f) + c\Delta T_i + c\Delta T_b + c\delta_i + \gamma_i + \varepsilon_i \quad (6.17)$$

where x_i, y_i, z_i = position of the ith satelite (known); $i = 1, 2, 3, 4$
x, y, z = user (or receiver) position (unknown)
ΔT_b = receiver clock bias (unknown)
ΔT_i = ith-satellite clock offset from GPS time
R_i = pseudorange measurement to the ith satellite
$I_i(f)$ = ionospheric delay
δ_i = receiver clock drift
γ_i = term which accounts for any other biases in the system (e.g., antennas, cables)
ε_i = statistical error in the measurement

The largest error in the measurement is by far the ionospheric dispersive delay. In order to remove this error, the GPS satellites transmit identical signals at the two frequencies, L_1 and L_2. However, since the ionospheric delay error varies inversely with the frequency squared, this term can be eliminated. The error ΔT_i can also be eliminated since it can be precisely measured by the monitor stations. By using simultaneous multiple satellite observations, the δ_i, γ_i, and ε_i terms can be eliminated. This, however, requires a multiple-channel receiver designed specifically for navigation applications. Consequently, Eq. (6.17) assumes the simpler form

$$R_i = \sqrt{(x-x_i)^2 + (y-y_i)^2 + (z-z_i)^2} + \Delta T_b \quad (i=1, 2, 3, 4) \quad (6.18)$$

Note that the units in this equation have been so chosen so that the speed of light is unity. These four equations are called the *navigation equations*. Their solution requires the measurement of the pseudoranges to four different satellites. The GPS receiver's computer may be programmed to solve directly the navigation equations in the form given above. However, the computation time required to solve them, even if only a few seconds, may be too long for many applications. As an alternate way, these equations may be approximated by a set of four linear equations that the GPS receiver can solve using a much faster and simpler algorithm. Linearization of Eq. (6.18) can proceed as follows [8,15,18]. Let

x_n, y_n, z_n, T_n = nominal (a priori best-estimate) values of $x, y, z,$ and T
$\Delta x, \Delta y, \Delta z, \Delta T$ = corrections to the nominal values
R_{ni} = nominal pseudorange measurement to the ith satellite
ΔR_i = residual (difference) between actual and nominal range measurements

Thus, the following incremental relationships are obtained:

$x = x_n + \Delta x$
$y = y_n + \Delta y$
$z = z_n + \Delta z$
$T = T_n + \Delta T$
$R_i = R_{ni} + \Delta R_i$

and

$$R_{ni} = \sqrt{(x_n - x_i)^2 + (y_n - y_i)^2 + (z_n - z_i)^2} + T_n$$

Substituting these incremental relationships into Eq. (6.18) yields

$$[(x_n + \Delta x - x_i)^2 + (y_n + \Delta y - y_i)^2 + (z_n + \Delta z - z_i)^2]^{1/2}$$
$$= R_{ni} + \Delta R_i - T_n - \Delta T_i \qquad (i = 1, 2, 3, 4) \qquad (6.19)$$

Working with the left-hand side of the equation and expanding terms results in

$$[(x_n + \Delta x - x_i)^2 + (y_n + \Delta y - y_i)^2 + (z_n + \Delta z - z_i)^2]^{1/2}$$
$$= [(x_n - x_i)^2 + 2\Delta x(x_n - x_i)$$
$$+ (\Delta x)^2 + (y_n - y_i)^2$$
$$+ 2\Delta y(y_n - y_i) + (\Delta y)^2$$
$$+ (z_n - z_i)^2 + 2\Delta z(z_n - z_i)^2$$
$$+ (\Delta z)^2]^{1/2} \qquad (6.20)$$

Rearranging terms and eliminating second-order terms, we can write Eq. (6.20) in the form

$$\sqrt{a+2b} \qquad (6.21)$$

where

$$a = [(x_n - x_i)^2 + (y_n - y_i)^2 + (z_n - z_i)^2]$$
$$b = [\Delta x(x_n - x_i) + \Delta y(y_n - y_i) + \Delta z(z_n - z_i)]$$

Equation (6.21) can be expanded using the binomial series expansion obtaining

$$(a+2b)^{1/2} = (a)^{1/2}[1 + 2b/a]^{1/2}$$
$$= (a)^{1/2}[1 + b/a + \text{higher-order terms}]$$

By noting that all higher-order terms containing second-order and higher terms of Δx, Δy, or Δz can be ignored, this equation reduces to

$$(a)^{1/2}[1 + b/a] = (a)^{1/2} + (a)^{1/2}[b/a]$$
$$= (a)^{1/2} + b/(a)^{1/2}$$
$$= R_{ni} + \Delta R_i - T_n - \Delta T \qquad (6.22)$$

Substituting the incremental relationship for R_{ni} into this equation and simplifying, we have

$$(a)^{1/2} + b/(a)^{1/2} = [(a)^{1/2} + T_n] - T_n - \Delta T + \Delta R_i \qquad (6.23)$$
$$b/(a)^{1/2} = \Delta R_i - \Delta T \qquad (6.24)$$

Furthermore, we note that

$$R_{ni} = (a)^{1/2} + T_n \rightarrow (a)^{1/2} = R_{ni} - T_n$$

Substituting this expression and the expression for b into Eq. (6.24), we obtain a set of linearized equations ($i = 1, 2, 3, 4$) that relate the pseudorange measurements to the desired user navigation information as well as the user's clock bias:

$$\left[\frac{x_u - x_i}{R_{ni} - T_n}\right]\Delta x + \left[\frac{y_u - y_i}{R_{ni} - T_n}\right]\Delta y + \left[\frac{z_u - z_i}{R_{ni} - T_n}\right]\Delta z + \Delta T = \Delta R_i \qquad (6.25)$$

The quantities on the right-hand side are known; they are simply the differences between the actual measured pseudoranges and the predicted measurements, which are supplied by the user's computer, based on knowledge of the satellite position and current estimate of the user's position and clock bias. Therefore, the quantities to be computed (Δx, Δy, Δz and ΔT) are the corrections that the user will make to the current estimate of position and

6.2. Navaid Sensor Subsystems

clock time bias. The coefficients of these quantities on the left-hand side represent the direction cosines of the line-of-sight (LOS) vector from the user to the satellite as projected along the (x, y, z) coordinate system.

The system of equations, Eq. (6.18), can be solved by using a standard least-squares procedure to form a new estimate and the entire procedure repeated. Section 6.2.5.4 discusses the method of least squares in more detail. For test data with noise but no biases, the procedure works very well since the initial estimate is not critical, and three to five iterations will reproduce the true coordinates within a few meters. The most difficult problem is controlling the bias owing to the clock drift.

Returning our attention now to the four linearized equations represented by Eq. (6.25), we note that these equations can be expressed in matrix notation as

$$\begin{bmatrix} \beta_{11} & \beta_{12} & \beta_{13} & 1 \\ \beta_{21} & \beta_{22} & \beta_{23} & 1 \\ \beta_{31} & \beta_{32} & \beta_{33} & 1 \\ \beta_{41} & \beta_{42} & \beta_{43} & 1 \end{bmatrix} \times \begin{bmatrix} \Delta x \\ \Delta y \\ \Delta z \\ \Delta T \end{bmatrix} = \begin{bmatrix} \Delta R_1 \\ \Delta R_2 \\ \Delta R_3 \\ \Delta R_4 \end{bmatrix} \quad (6.26)$$

where β_{ij} is the direction cosine of the angle between the LOS to the ith satellite and the jth coordinate. This equation can be written more compactly by letting

$B =$ the 4×4 solution matrix (i.e., a matrix of coefficients of the linear equations)

$\mathbf{x} =$ user position and time correction vector

$\mathbf{r} =$ the four element pseudorange measurement difference vector (i.e., the difference between the expected and measured pseudoranges)

$$B = \begin{bmatrix} \beta_{11} & \beta_{12} & \beta_{13} & 1 \\ \beta_{21} & \beta_{22} & \beta_{23} & 1 \\ \beta_{31} & \beta_{32} & \beta_{33} & 1 \\ \beta_{41} & \beta_{42} & \beta_{43} & 1 \end{bmatrix}$$

$$\mathbf{x} \equiv [\Delta x \quad \Delta y \quad \Delta z \quad \Delta T]^T \quad (6.27)$$

$$\mathbf{r} \equiv [\Delta R_1 \quad \Delta R_2 \quad \Delta R_3 \quad \Delta R_4]^T$$

Thus, this equation becomes simply [8]

$$B\mathbf{x} = \mathbf{r} \quad (6.28)$$

The position error and clock bias can be solved by taking the inverse of the matrix B. That is, premultiplying both sides of this linear matrix equation by the inverse of B, yields

$$\mathbf{x} = B^{-1}\mathbf{r}$$

This equation expresses the relationship between pseudorange measurements and user position and clock bias in a compact form. In order to understand how the geometry of the satellites at a point in time can result in a system outage, we need only to examine the solution matrix, B. If the ends of the unit vectors from the user to the four satellites selected are in a common plane, the direction cosine of the four unit vectors along a direction perpendicular to this plane are equal. When this occurs, the determinant of the 4×4 solution matrix becomes zero (i.e., singular) and no solution is possible from the four equations. Consequently, the navigation equations "blow up," and what is known as a "system outage" occurs as a result of poor geometry. Figure 6-18 illustrates the concept of pseudorange and its associated errors.

The receiver (or user equipment) errors are given in Table 6-1. The user range error is defined at the phase center of the satellite antenna. It should be noted that the values given for ionospheric delay compensation error are

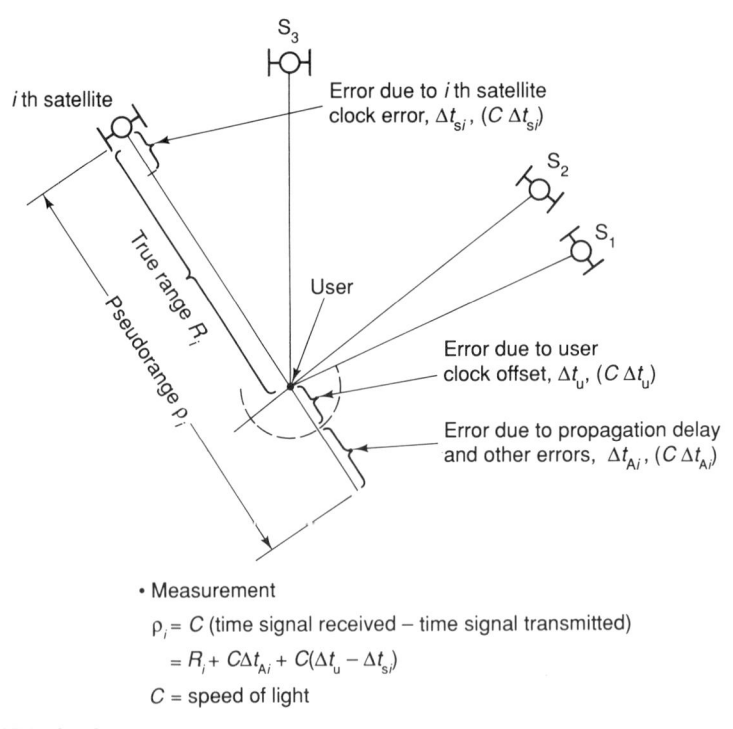

- Measurement

$\rho_i = C$ (time signal received − time signal transmitted)

$\quad = R_i + C\Delta t_{Ai} + C(\Delta t_u - \Delta t_{si})$

C = speed of light

FIGURE 6-18
GPS pseudorange concept and error components.

TABLE 6-1
GPS User Equipment Range-Error Budget

Error source	P-Code pseudorange error 1σ (m)	C/A-Code pseudorange error 1σ (m)
Multipath	1.24	1.24
Receiver noise and resolution	1.48	7.10
Tropospheric delay compensation	1.92	1.92
Ionospheric delay	2.33	5.00–10.21

based on dual-frequency delay measurements for the P-code and the single-frequency ionospheric delay model for the C/A-code. Furthermore, the budgeted value for the C/A-code receiver noise and resolution can be improved by digital phase locking techniques.

6.2.5.3. Geometric Effects

In the previous section the pseudorange equation to the various satellites was developed. However, ranging error alone does not determine position fix error. The relative geometry of the four satellites and the user also affects fix accuracy. Therefore, in order to determine the accuracy available from the four satellites selected as a function of their geometry, we must calculate the dilution of precision (DOP) values available from the four satellites selected. Since the overall position accuracy is a product of this value and other system errors, small DOP values are highly desirable. Moreover, since the navigation Eq. (6.28) is a linear relationship, it can also be used to express the relationship between errors in pseudorange measurement and the errors in user position and clock bias. Mathematically, this relationship can be expressed as

$$\varepsilon_x = B^{-1}\varepsilon_r$$

where ε_x = error in user position and clock bias
ε_r = pseudorange measurement errors

From Refs. 4 and 12, we note that

$$C_x = \mathcal{E}\{\varepsilon_x \ \varepsilon_x^T\}$$

$$C_r = \mathcal{E}\{\varepsilon_r \ \varepsilon_r^T\}$$

where C_r = the covariance matrix of errors in pseudorange measurements
C_x = the covariance matrix of the resulting errors in the three components of user position and clock bias

The relationship between the two covariance matrices can be written as

$$C_x = B^{-1} C_r B^{-T} \tag{6.29}$$

where B is the same 4×4 matrix of coefficients of the unknowns derived previously and is a function only of the direction cosines of the LOS unit vectors from the user to the four satellites and the user's clock bias. Thus, the error relationships are functions only of the satellite geometry, which leads to the concept of geometric dilution of precision (GDOP), a measure of how satellite geometry degrades navigation accuracy. To a good approximation, Eq. (6.29) can be written as

$$C_x = (B^T B)^{-1} \tag{6.30}$$

Assuming sufficient signal strength, this covariance matrix depends only on the direction and is in no way dependent on the distances between the user and each satellite. Furthermore, since the diagonal elements of this covariance matrix are actually the variances of user position and time, various DOP values can be obtained from looking at appropriate elements of this matrix. Consequently, these values can be used to compare position accuracies (from a geometric standpoint) of different orbital configurations as well as to measure the overall effect of geometry to errors in the user's position and time bias. Expressing the diagonal elements of the covariance matrix of errors in user position and clock bias as V_x, V_y, V_z, and V_t, then the four-dimensional GDOP is obtained by taking the square root of the trace (tr) of the matrix:

$$\begin{aligned} \text{GDOP} &= \sqrt{\text{tr}[(B^T B)^{-1}]} \\ &= [V_x + V_y + V_z + V_t]^{1/2} \end{aligned} \tag{6.31}$$

where V_x = variance of x position (or velocity) estimate
V_y = variance of y position (or velocity) estimate
V_z = variance of z position (or velocity) estimate
V_t = variance of the clock bias (or clock drift rate) estimate

From the above discussion, geometric dilution of precision (GDOP) is a measure of the error contributed by the geometric relationships of the GPS as seen by the receiver. In modeling the GPS receiver, the Kalman filter will contain the covariance matrix of the estimates of the pseudorange measurement errors. As stated above, the diagonal of the covariance matrix contains the variances of the position errors and the receiver clock bias error. Mathematically, the GDOP is given by the following expression:

$$\text{GDOP} = \sqrt{\sigma_{xx}^2 + \sigma_{yy}^2 + \sigma_{zz}^2 + \sigma_{tt}^2}$$

where $\sigma_{xx}, \sigma_{yy}, \sigma_{zz} = 1 - \sigma$ errors in position
$\sigma_{tt} = 1 - \sigma$ error in time

6.2. Navaid Sensor Subsystems

Equation (6.31) includes all four unknowns (three dimensions of position and time) and is the conventional measure of overall geometric performance. A more frequently used measure of geometric performance is the three-dimensional position dilution of precision (PDOP), which relates only to the three components of position error. PDOP is an error amplification factor. The product of PDOP and ranging error determines fix error. The numerical value of PDOP ranges from 1.5 to infinity. Thus, it is a strong factor of position error. The receiver normally attempts to track the set of four among n visible satellites that has the smallest PDOP. Computation of the PDOP involves taking the inverse of the square of a 4×4 matrix, which is not a trivial task. The PDOP is also invariant with the coordinate system and is used because the most important consideration in any navigation system is position accuracy; knowing time is a secondary by-product. Mathematically, the PDOP can be expressed as

$$PDOP = [V_x + V_y + V_z]^{1/2}$$

or (6.32)

$$PDOP = \sqrt{\sigma_{xx}^2 + \sigma_{yy}^2 + \sigma_{zz}^2}$$

PDOP is useful in aircraft weapon delivery applications. The position accuracy available from the GPS can be divided into two multiplicative factors: PDOP and other "system" errors. The "system" errors depend on the accuracy of the ephemeris data and transmitted time from the satellites, ionospheric and atmospheric effects, and various other error sources. Since the PDOP factors depend predominantly on the user/navigation satellite geometries, they can be analyzed independently of system errors, which depend, as we have seen earlier, on a multitude of factors. This characteristic allows separate analyses of alternative orbital configurations, user motion, and satellite losses sustained for the purposes of comparison and choosing the optimal constellation. Several other alternative DOP values are occasionally used in evaluating satellite constellations and relate only some of the variances of user position and time: (1) horizontal dilution of precision (HDOP), (2) altitude or vertical dilution of precision (VDOP), (3) time dilution of precision (TDOP), and (4) the larger component of horizontal position error, the maximum dilution of precision (MDOP). Mathematically, these various DOPs are defined as follows [21]:

$$HDOP = [V_x + V_y]^{1/2} \tag{6.33}$$
$$= \sqrt{\sigma_{xx}^2 + \sigma_{yy}^2}$$
$$VDOP = [V_z]^{1/2} \tag{6.34}$$
$$= \sqrt{\sigma_{zz}^2}$$

$$\text{TDOP} = [V_t]^{1/2} = \sqrt{\sigma_{tt}^2} \qquad (6.35)$$

$$\text{MDOP} = \max[(V_x)^{1/2}, (V_y)^{1/2}] \qquad (6.36)$$

HDOP is mostly used in surface and aircraft navigation, and TDOP is used in time transfer applications. Geometric dilution of precision can also be defined as the ratio of $1-\sigma$ measurement error normal to the line-of-position (LOP) to the $1-\sigma$ error in position determination. Therefore,

$$\text{GDOP} = \frac{\sqrt{\sigma_{xx}^2 + \sigma_{yy}^2 + \sigma_{zz}^2 + \sigma_{tt}^2}}{\sigma_{\text{TOA}}}$$

where $\sigma_{xx}, \sigma_{yy}, \sigma_{zz} = 1 - \sigma$ errors in position
  $\sigma_{tt} = 1 - \sigma$ error in time
  $\sigma_{\text{TOA}} = 1 - \sigma$ error in time-of-arrival (TOA) measurement
    (note that σ_{TOA} can be considered as a scaling factor)

Similarly, in this case the equations for HDOP, VDOP, and TDOP must be divided by σ_{TOA}.

In order to obtain the most accurate user position, it would be highly desirable to utilize those four satellites with the most favorable geometry (lowest DOP values) with respect to the user at any instant of time. This presents no problem, should there be only four visible satellites to choose from, as all four must be used to determine the user's three-dimensional position. The majority of the time, however, there will be six or more satellites in view by an earth based user and even more by a low altitude satellite user, and the computational time required to compute PDOP values for all the possible combinations of satellites is excessive. Figure 6-19 depicts what constitutes good or bad satellite geometry [13].

The results of many computer runs and analytical studies have demonstrated an almost total correlation between PDOP and the volume of a tetrahedron formed by lines connecting the tips of the four unit vectors from the user toward the four satellites. Usually, the larger the volume of this tetrahedron, the smaller the corresponding PDOP* value will be for this same set of satellites. Therefore, the system designer must identify the "best four" satellites, which yield the largest tetrahedron volumes. The problem of satellite selection is not only a problem to those conducting global analyses of orbital constellations for evaluation but is just as significant a problem to the designers of user equipment, as the equipment must be designed to operate quickly in a dynamic environment, and large computer resources are not available.

The results of this section can be summarized by noting that the GDOP is a figure of merit for selecting the best four satellite set with the best

---

*PDOP = 6 is the threshold. Therefore, the larger the PDOP, the less the accuracy.

6.2. *Navaid Sensor Subsystems* **315**

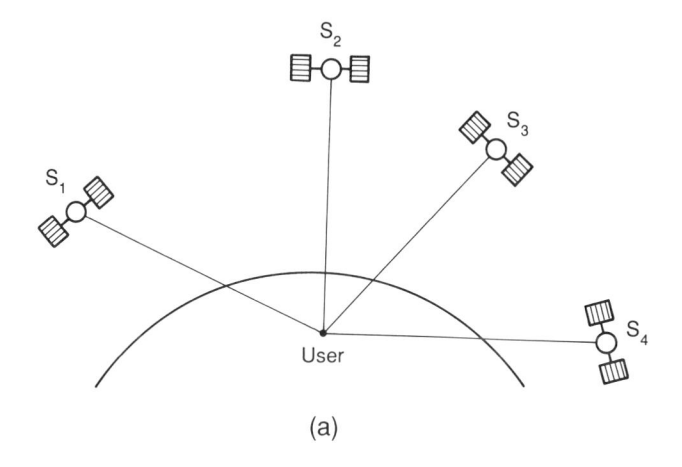

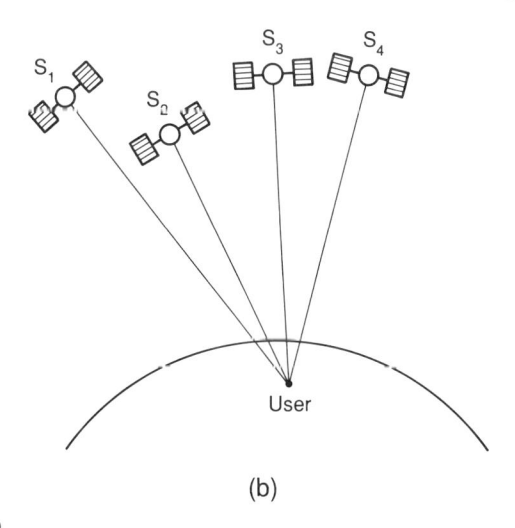

FIGURE 6-19

Navstar geometry for the determination of PDOP: (a) good (low) PDOP; (b) poor (high) PDOP.

geometry for tracking. The PDOP is used extensively in satellite constellation design and analysis, without having to specify the user's position and trajectory. Horizontal dilution of precision (HDOP) is a horizontal plane accuracy descriptor, similar to the CEP. VDOP is valuable in TF/TA (terrain following/terrain avoidance). Finally, new signal processing techniques and currently developing technology have the potential for significantly improving

GPS receivers. These improvements are primarily lower production cost, smaller size, and higher jamming resistance in military applications. The GPS performance characteristics are summarized below:

Navigation signals	
Precise positioning service (PPS)	$L_1 = 1575.42$ MHz
P-code (military) ranging signal	10.23 mbps[a]
Standard positioning service (SPS)	$L_2 = 1227.60$ MHz
C/A-code (civil) ranging signal	1.023 mbps
Earth model	WGS-84
Accuracy	
Position	40-m CEP (or 100-m 2-D rms)
Velocity	0.1 m/s (1σ)
Time	100 ns (1σ)
P-code position	10–15-m 3-D SEP
Dynamic capability	
Velocity	600 m/s
Acceleration	40 m/s^2 ($\approx 4\,g$)
Allowable initialization uncertainty	
Position	100 km
Velocity	300 m/s
Time to first fix	
Stationary	150 s
Dynamic	470 s at 300 m/s
Power	<15 W

[a]Megabits per second.

6.2.5.4. GPS UE Satellite Selection and Least-Squares Navigation Solution

The ultimate objective of any satellite selection scheme is to select from the available satellites the set which minimizes the user's navigation errors. In a normal environment (no jamming), the optimum set of four satellites to track is determined by the geometry between the user and the satellites in view. The standard procedure for user equipment (UE) satellite selection is based solely on the geometrical dilution of precision (GDOP). Navigation measurements are dependent on three factors (external to the UE signal processor): (1) geometry, (2) incoming signal, and (3) noise. The UE set uses a figure of merit known as the GDOP to select from the available satellites the four satellite set with the best geometry for tracking. The GDOP, however, does not address the incoming signal and noise. An optimum satellite selection process must use every available piece of information available to determine the satellite set which will provide the smallest navigation errors over a particular period of time. From Eq. (6.31), the GDOP is given by

$$\text{GDOP} = [\text{tr}(H^\mathrm{T} H)^{-1}]^{1/2} \qquad (6.37)$$

6.2. Navaid Sensor Subsystems

The user–satellite geometry matrix H, also called the *visibility matrix*, is formed by calculating the direction cosines of the LOS unit vector from the user to each satellite in view (expressed in user local-level coordinates). Each row of the H-matrix is composed of the 3-direction cosines of the LOS unit vectors and the fourth element for all satellites is a one, to account for the equal uncertainty in range due to satellite-user clock time bias for each satellite. Thus

$$H = \begin{bmatrix} h_E(1) & h_N(1) & h_u(1) & 1 \\ h_E(2) & h_N(2) & h_u(2) & 1 \\ \vdots & \vdots & \vdots & \vdots \\ h_E(j) & h_N(j) & h_u(j) & 1 \end{bmatrix} \tag{6.38}$$

where $h_E(\cdot), h_N(\cdot), h_u(\cdot)$ = the respective direction cosines of the LOS unit vectors to each satellite expressed in user local-level (east, north, up) coordinates
j = number of available satellites in view

Because of variations in geographic terrain, all satellites below 5° elevation from the horizon are deleted from possible satellite selection. The 5° angle is known as the "masking angle." The GDOP figure of merit [Eq. (6.37)] calculates a scalar value for each possible satellite set, and, as discussed earlier, the satellite set with the lowest GDOP is the best user–satellite geometry to provide the smallest navigation errors.

In this section, we will discuss the navigation solution treated in Section 6.2.5.2, from the point of view of the least-squares estimation. This can be considered as an extension of the aforementioned section. Now consider a vector of measurement components, $\mathbf{z}$, corrupted by additive noise $\mathbf{v}$ and linearly related to the state $\mathbf{x}$. This measurement vector, $\mathbf{z}$, is given by

$$\mathbf{z} = H\mathbf{x} + \mathbf{v} \tag{6.39}$$

where H is the matrix that relates the state to the measurements. An estimate of $\mathbf{x}$, called $\hat{\mathbf{x}}$, is sought from the given $\mathbf{z}$ vector. One approach to this problem is the least-squares estimator, $\hat{\mathbf{x}}$. That is, select $\hat{\mathbf{x}}$ so as to minimize the "sum squared" of the components of the difference $(\mathbf{z} - H\hat{\mathbf{x}})$. This means that $\hat{\mathbf{x}}$ should minimize

$$J_1 = (\mathbf{z} - H\hat{\mathbf{x}})^T (\mathbf{z} - H\hat{\mathbf{x}}) \tag{6.40}$$

Differentiating J_1 with respect to $\hat{\mathbf{x}}$ and setting the result to zero gives the well-known least-squares estimate as [16]

$$\hat{\mathbf{x}} = (H^T H)^{-1} H^T \mathbf{z} \tag{6.41}$$

It should be pointed out that, in the deterministic case, that is, when the additive noise $\mathbf{v} = 0$, the equation $\mathbf{z} = H\mathbf{x}$ corresponds to Eq. (6.28), where B

corresponds to the measurement (or observation) matrix H, $\mathbf{x}$ represents the state vector, and $\mathbf{r}$ corresponds to the measurement vector $\mathbf{z}$. The quality of this estimate can be assessed by forming the error

$$\tilde{\mathbf{x}} = \hat{\mathbf{x}} - \mathbf{x} \qquad (6.42)$$

Making use of Eqs. (6.39) and (6.40) results in

$$\tilde{\mathbf{x}} = (H^T H)^{-1} H^T H \mathbf{x} + (H^T H)^{-1} H^T \mathbf{v} - \mathbf{x}$$
$$= (H^T H)^{-1} H^T \mathbf{v} \qquad (6.43)$$

If the noise $\mathbf{v}$ has zero mean, taking the expected value $\mathcal{E}\{\ \}$ of Eq. (6.43) shows that the estimator error $\tilde{\mathbf{x}}$ also has a zero mean. The covariance of the estimation error is related to the covariance of the noise $\mathbf{v}$ as follows:

$$\mathcal{E}\{\tilde{\mathbf{x}}\tilde{\mathbf{x}}^T\} = (H^T H)^{-1} H^T \mathcal{E}\{\mathbf{v}\mathbf{v}^T\} H (H^T H)^{-1} \qquad (6.44)$$

Furthermore, if all components of $\mathbf{v}$ are pairwise uncorrelated and have unit variance, then

$$\mathcal{E}\{v_i v_j\} = \delta_{ij} = \begin{cases} 1 & \text{if } i = j \\ 0 & \text{if } i \neq j \end{cases}$$

where

$$\mathcal{E}\{\mathbf{v}\mathbf{v}^T\} = I$$

so that

$$\mathcal{E}\{\tilde{\mathbf{x}}\tilde{\mathbf{x}}^T\} = (H^T H)^{-1} H^T H (H^T H)^{-1}$$
$$= (H^T H)^{-1} \qquad (6.45)$$

The square root of $[\text{tr}(H^T H)^{-1}]^{1/2}$ will be recognized as the GDOP [Eq. (6.31)]. From the preceding discussion, it is clear that all GDOP-related performance measures indicate the error is an estimated navigation quantity "per unit of measurement noise" covariance. Moreover, all the above GDOP-related measures depend solely on the geometry matrix H. Smaller GDOP values indicate stronger or more robust geometric solutions to the estimation problem. For these reasons, when some freedom exists in the choice of the measurements, good (i.e., small) GDOP is often used as the selection criterion. It should be noted that the "per unit noise" concept mentioned above, inherent in the GDOP, is not so useful when certain measurement components are noisier than others. If a choice exists between two possible sets of measurements, the set with the poorer GDOP may be preferable, if they are of sufficiently higher accuracy.

If the measurement noise covariance is not exactly the unit matrix, but rather $\mathcal{E}\{\mathbf{v}\mathbf{v}^T\} = I\sigma^2$, then the GDOP matrix increases by the scalar σ^2.

6.2. Navaid Sensor Subsystems

Moreover, if all potential measurements have this same variance σ^2, then the choice of the best measurement set will still be the best GDOP set. The multiplication scalar σ^2 will not affect the relative rankings. More specifically, the measurement noise is called nonuniform when different measurements have different noise levels, as indicated by their variances. In this case, a weighted least-squares approach to estimation is normally used. The quadratic form of Eq. (6.40) is modified by inserting a weighting matrix W. Thus

$$J_2 = (\mathbf{z} - H\hat{\mathbf{x}})^T W (\mathbf{z} - H\hat{\mathbf{x}}) \tag{6.46}$$

Commonly, the weighting matrix is selected as the inverse of the noise covariance matrix. That is, $W = R^{-1}$ where $R = \mathcal{E}\{\mathbf{v}\mathbf{v}^T\}$. This choice weights the accurate measurements more and the noisy ones less. Minimization of Eq. (6.46) results in

$$\hat{\mathbf{x}} = (H^T R^{-1} H)^{-1} H^T R^{-1} \mathbf{z} \tag{6.47}$$

which is a modification of Eq. (6.41). Similarly, Eq. (6.45) becomes

$$P = \mathcal{E}\{(\hat{\mathbf{x}} - \mathbf{x})(\hat{\mathbf{x}} - \mathbf{x})^T\} = \mathcal{E}\{\tilde{\mathbf{x}}\tilde{\mathbf{x}}^T\} = (H^T R^{-1} H)^{-1} \tag{6.48}$$

From the above discussion and the discussion of Section 6.2.5.2, the user determines its position by measuring the range between its antenna and four satellites. The range is determined by measuring the time the signal needs to propagate from the satellite to the user position scaled by the speed of light. Assuming for simplicity an ECEF coordinate system, and referring to Fig. 6-20, the user position vector $\mathbf{R}_u$ is determined by measuring the

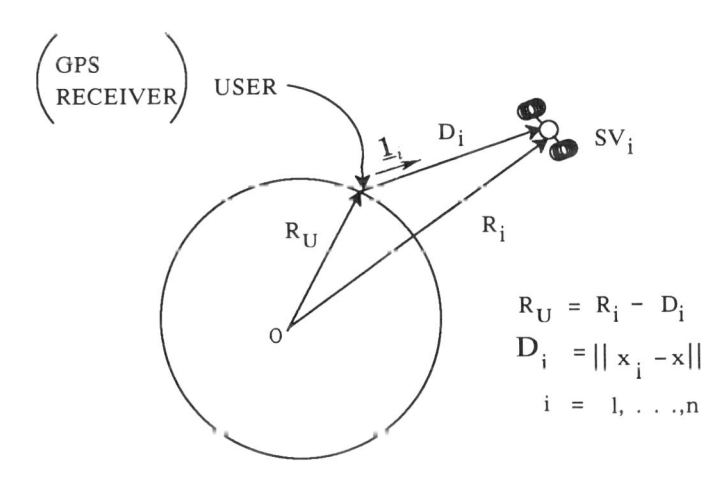

FIGURE 6-20
GPS–user position solution vectors.

magnitude of the vector $\mathbf{D}_i$, which is the difference between the user position vector and the ith satellite position vector $\mathbf{R}_i$. More specifically, both the user's position $\mathbf{R}_u$ and the satellite position $\mathbf{R}_i$ will be calculated in the WGS-84 ECEF reference frame.

The ith satellite transmits $\mathbf{R}_i$ (known), with elements (x_i, y_i, z_i) and the time of that position. The propagation time scaled by the speed of light yields the measured pseudorange ρ_i to the ith satellite, which consists of the magnitude D_i and the user clock offset error ΔT_u. The user components of $\mathbf{R}_u$ (x, y, z), which are unknown, are determined by solving Eq. (6.16) for each satellite. Therefore, based on Eq. (6.16), the measured pseudorange ρ_i to the ith satellite can be written in a more complete form as (see also Fig. 6-18)

$$\rho_i = R_i + c\Delta t_{Ai} + c(\Delta t_u - \Delta t_{Si}) \qquad (i \geq 4) \qquad (6.16)$$

where
ρ_i = measured pseudorange to the ith satellite
R_i = the true range = $\sqrt{(x-x_i)^2 + (y-y_i)^2 + (z-z_i)^2}$
c = the speed of light
Δt_{Si} = satellite i clock offset from the GPS system time
Δt_u = the user clock offset bias from the GPS system time
Δt_{Ai} = propagation delays and other errors

In order to obtain the navigation solution, certain preliminaries need to be addressed. Referring to Fig. 6-20, the user vector is expressed as [4,13]

$$\mathbf{R}_u = \mathbf{R}_i - \mathbf{D}_i, \qquad \text{where } D_i = \|\mathbf{x}_i - \mathbf{x}\| \qquad (6.49)$$

Now we define a coordinate system with respect to the user. Defining $\mathbf{1}_i$ as the LOS unit vectors from the user to the ith satellite, and noting that $\mathbf{1}_i \cdot \mathbf{D}_i = |\mathbf{D}_i| = D_i$, Eq. (6.49) becomes

$$\mathbf{1}_i \cdot \mathbf{R}_u = \mathbf{1}_i \cdot \mathbf{R}_i - D_i \qquad (6.50)$$

so that the range D_i takes the form

$$D_i = \rho_i - B_u - B_i \qquad (6.51)$$

where B_u is the user range clock offset and B_i is the satellite range clock offset. Combining Eqs. (6.50) and (6.51) yields

$$\mathbf{1}_i \cdot \mathbf{R}_u - B_u = \mathbf{1}_i \cdot \mathbf{R}_i - \rho_i + B_i \qquad (i \geq 4) \qquad (6.52)$$

The left-hand side of this equation contains the four unknowns, the three coordinate directions of position error, and the range error due to the user's clock error. The right-hand side contains all known quantities. The satellite knows and transmits its position, pseudorange (ρ_i) is what we actually compute, and the range error due to satellite clock error is transmitted with the satellite position information. This error correction considers a multitude of

6.2. Navaid Sensor Subsystems

corrections that we will treat as known biases. That means we treat them as given and add (or subtract) as appropriate.

For the deterministic case mentioned earlier, the measurement vector z, which consists of the pseudorange measurement errors, can be expressed as $z = Hx$:

$$z_{4 \times 1} = \begin{bmatrix} 1_1^T 1 \\ 1_2^T 1 \\ 1_3^T 1 \\ 1_4^T 1 \end{bmatrix}_{4 \times 4} \begin{bmatrix} R_u \\ B_u \end{bmatrix}_{4 \times 1} \quad (6.53)$$

Also, the least-squares estimate can be obtained from

$$\hat{x} = H^{-1} z \quad (6.54)$$

which solves for R_u and B_u.

A general least-squares formulation solves Eq. (6.54). The solution is obtained by first defining certain vectors and matrices. Using the notation of Ref. 13, let

$$X_{u(4 \times 1)} = [R_{u1} \quad R_{u2} \quad R_{u3} \quad -B_u]^T \quad (6.55)$$

$$G_{u(n \times 4)} = [\Gamma_1 \quad \Gamma_2 \quad \Gamma_3 \quad \Gamma_4]^T \triangleq \begin{bmatrix} \Gamma_1 \\ \Gamma_2 \\ \Gamma_3 \\ \vdots \\ \Gamma_n \end{bmatrix} \quad (6.56)$$

$$A_{u(n \times 4n)} = \text{diagonal}[\Gamma_1 \quad \Gamma_2 \quad \Gamma_3 \quad \Gamma_4]$$

$$\triangleq \begin{bmatrix} \Gamma_1 & 0 & 0 & \cdots & 0 \\ 0 & \Gamma_2 & 0 & & 0 \\ 0 & 0 & \Gamma_3 & & 0 \\ \vdots & \vdots & & \ddots & \vdots \\ 0 & 0 & 0 & \cdots & \Gamma_n \end{bmatrix} \quad (6.57)$$

$$\Gamma_i = [1_{i1}, 1_{i2}, 1_{i3}, 1]$$

where R_{ux} = the components of unknown user position ($x = 1, 2, 3$)
B_u = range offset of the user's clock
(note that for 4 satellites used, the subscript $n = 4$)

The satellite position vector $S_{4n \times 1}$ is given by

$$S_{(4n \times 1)} = [R_{11} \quad R_{12} \quad R_{13} \quad B_1 \quad R_{21} \quad R_{22} \quad R_{23} \quad B_2 \cdots B_4]^T \quad (6.58)$$

and the measured pseudorange matrix by

$$\boldsymbol{\rho}_i = [\rho_1 \quad \rho_2 \quad \rho_3 \quad \rho_4]^T \quad (6.59)$$

Setting up the system of equations, we have

$$\mathbf{G}_u \mathbf{X}_u = \mathbf{A}_u \mathbf{S} - \boldsymbol{\rho} \quad (6.60)$$

This equation can be placed in a least-squares form as follows:

$$\mathbf{G}_u^T \mathbf{G}_u \mathbf{X}_u = \mathbf{G}_u^T [\mathbf{A}_u \mathbf{S} - \boldsymbol{\rho}] \quad (6.61)$$

The least-squares algorithm implements an iterative solution based on the following equation [13]:

$$\hat{\mathbf{X}}_u = [\mathbf{G}_u^T \mathbf{G}_u]^{-1} \mathbf{G}_u^T [\mathbf{A}_u \mathbf{S} - \boldsymbol{\rho}] \quad (6.62)$$

Note that the solution to this equation requires an iterative type solution based on an initial guess as to the user's actual position. On the basis of that guess, the actual position and hence the initial direction cosines to the satellites get updated. Furthermore, the solution iterates until the desired convergence results. Equation (6.62) gives the estimate of the user position and clock offset. Deriving velocity information results from a similar development. For example, in Eq. (6.53), the vector $\mathbf{R}_u$ in the x vector will be replaced by the velocity vector $\mathbf{V}_u$. The delta range or range rate information results from the receiver computing the Doppler shifts of the arriving signals to establish range rates to the particular satellite.

6.2.5.5. Differential GPS

One technique for improving the performance of the GPS position, which has been receiving increasing interest in recent years, is the use of the differential GPS (DGPS) concept. This concept involves the use of data from a GPS reference receiver (RR) in the vicinity of the GPS receiver-equipped users that allows certain errors that are common to both receivers to be removed from the user's position measurements. More specifically, a reference receiver at a known, well-surveyed location measures the range to a satellite. That is, the DGPS reference receiver operates from a well-surveyed antenna, which is capable of tracking all visible satellites. Common-mode bias errors are determined using the GPS reference receiver. Bias error estimates are then used to correct the measurements of other GPS receivers within a specified area around the RR. The measured range includes the actual range to the satellite and associated errors. These errors include (1) satellite clock error, (2) tropospheric delay, and (3) ionospheric delay. Since the reference receiver's coordinates are known with good precision, the satellite's ephemeris message allows the range to be calculated. However, an ephemeris error can exist in the calculated range, so that the difference

6.2. Navaid Sensor Subsystems

between the measured and the calculated ranges includes all errors mentioned above. If this range correction is applied to the measured range of a remote GPS user, a corrected range is obtained that will allow the remote receiver to remove the error in its uncorrected position. Figure 6-21 illustrates the DGPS concept [17].

The true values of the reference receiver's navigation fix are compared against the measured values, and the differences become the differential corrections. These corrections may be transmitted to area user sets in real time, or they may be recorded for postmission use. In either case, the users apply the corrections to their navigation data in order to produce position fixes that are free of GPS-related biases. In Ref. 2, several schemes to accomplish differential GPS navigation are described. Strictly speaking, these methods are combinations of either corrections of pseudoranges, or of the navigation solutions and whether the corrections are made on board the user vehicle or at the RR. The most commonly used or preferred method when a large number of user vehicles must be accommodated is the "uplink of pseudorange corrections for participant onboard processing" [2]. Specifically, this method requires that the position of the RR be known and that the RR measures the pseudoranges to all visible satellites and computes the differences between the calculated range to the satellite and the measured range. These differences are taken after the receiver clock bias has been removed and the calculated range to each satellite uses the satellite's position as defined in the ephemeris message. As mentioned earlier, the user selects the optimum set of four satellites.

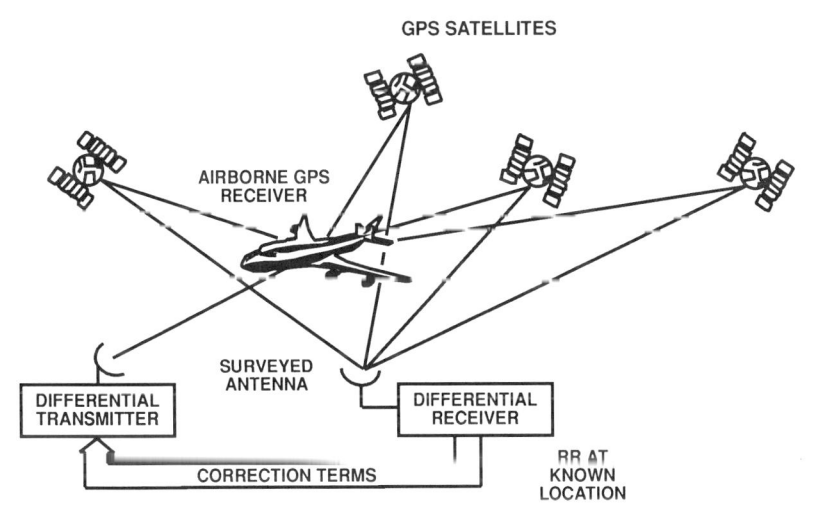

FIGURE 6-21
The differential GPS concept.

The discussion of this section can be summarized by noting that all differential techniques require a reference receiver that is a well-calibrated GPS receiver located at a well-surveyed site. The difference between the RR's navigation solution and the survey coordinates is the "ground truth." The ground truth represents the navigation error in east, north, and vertical components [11]. The difference between a pseudorange calculated using the survey coordinates is the corrected pseudorange. Finally, the DGPS represents a cost-effective method, which significantly improves the reliability and accuracy of GPS navigation.

6.2.5.6. GPS/Inertial Navigation System Integration

As with other navaids, GPS and inertial navigation systems have complimentary features that can be exploited in an integrated system, thus resulting in improved navigation performance. The GPS can operate with sufficient accuracy in a stand-alone configuration (e.g., GPS manpack), or integrated with an INS. In the stand-alone mode of operation, a benign environment is assumed. The INS is able to provide accurate aiding data on short-term vehicle dynamics, while the GPS provides accurate data on long-term vehicle dynamics. Since the GPS receiver provides pseudorange and delta pseudorange data, these data can be used for estimating errors in position, velocity, and other error parameters of the INS and GPS receiver clock. The estimates of the INS error parameters allow GPS/INS navigation with substantially smaller errors than could be achieved with either a GPS stand-alone navigator or an INS operating alone (i.e., free inertial mode). As mentioned earlier, the GPS user equipment consists of a receiver (with one to five channels, depending on the application) that acquires and tracks the satellites and performs pseudorange and delta pseudorange measurements that are processed sequentially in real time to provide the best estimate of user position, velocity, and system time. Figure 6-22 illustrates a simple GPS/INS integration, in which the receiver contains its own internal Kalman filter.

The optimal integration of the GPS-aided INS* is with both INS and GPS jointly providing raw data to a single Kalman filter as shown in Fig. 6-22. Even though the INS and/or GPS may become unstable, the Kalman filter remains stable if the filter estimates all modeled error states in the filter with with a bounded error. The INS provides the GPS with a reference that is corrected by the Kalman filter. Parameters estimated by this filter, such as position, velocity, and attitude corrections, are combined with the INS measured position, velocity, and attitude to generate the best estimates of system position, velocity, and attitude. It is noted that the accuracy of the GPS solution degrades rapidly whenever lock is lost in one or more satellites.

---

*In this scheme, improved performance can be achieved with less than four satellites.

6.2. Navaid Sensor Subsystems

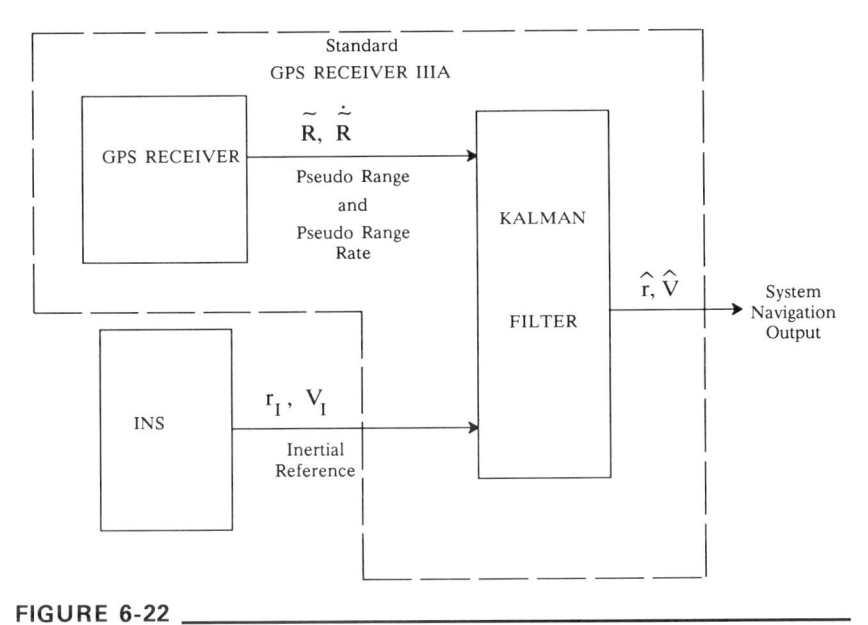

FIGURE 6-22
GPS/INS integration block diagram.

However, this is not a problem if the INS is used in conjunction with the GPS, since the INS is a high-quality extrapolator. Furthermore, the primary benefit of filtering GPS data through the INS is not to improve on the accuracy of the GPS position and velocity but to optimize the system solution in case a GPS outage occurs.

The Standard GPS Receiver Type IIIA* indicated in Fig. 6-22 is commonly designed to be integrated with an INS. Together, the GPS and the INS have reduced sensitivity to jamming and maneuvers in military applications. As will be discussed later, in order to aid the GPS the INS must be aligned. This alignment includes both the initial alignment of the INS and the subsequent in-flight calibration or correction of inertial position, velocity, attitude, and other parameters. The initial alignment is normally performed before an aircraft is ready to take off (ground alignment), thereby providing the best accuracy. Performance of the GPS/INS discussed here can be analyzed by performing a covariance or Monte Carlo simulation. The covariance simulation computes the covariance matrix of errors in the state vector that includes all the random variables that affect the GPS or INS measurements. Two major events executed by the covariance simulation are filter

---

*Note that the Five-channel Standard GPS Receiver Type IIIA is intended for airborne applications while the IIIS type is for shipboard applications.

update and propagation through time (without any update occurring). The covariance simulation is a linear error analysis tool, in the sense that only linear dynamics and linear relationships between sensor measurements and the error state vector are modeled. We will now examine the "truth" model of the GPS and INS.

The INS Truth Model The truth model for an aircraft INS can have as many as 85 states. However, a suboptimal or reduced order truth model may also yield satisfactory results. The basic INS error model consists of nine states (three positions, three velocities, and three platform misalignments). The perturbation equations can be written as nine linear differential equations in which the $n \times 1$ error state vector is defined as

$$\delta \mathbf{x}_I^T = [\delta P_{IN} \quad \delta P_{IE} \quad \delta P_{ID} \quad \delta V_{IN} \quad \delta V_{IE} \quad \delta V_{ID} \quad \psi_N \quad \psi_E \quad \psi_D] \quad (6.63)$$

where $\delta P_{I(\cdot)}$ and $\delta V_{I(\cdot)}$ are the INS position and velocity error components along the north–east–down (NED) navigation coordinate axes. From Eq. (6.63), we note that the error state vector contains all the error parameters, other than white noise. Thus, the error state vector contains all the errors that affect the navigation system to a significant degree. The basic INS truth model can be described by the equation

$$\delta \dot{\mathbf{x}}_I = \mathbf{F}_I \delta \mathbf{x}_I + \mathbf{w}_I \quad (6.64)$$

where $\mathbf{F}_I$ is the $n \times n$ matrix representing the INS linear error dynamics and $\mathbf{w}_I$ is the white driving noise term representing the instrument uncertainties. From Ref. 12

$$\mathcal{E}\{\mathbf{w}\} = 0 \quad \text{and} \quad \mathcal{E}\{\mathbf{w}(t)\mathbf{w}^T(t+\tau)\} = \mathbf{Q}\delta(\tau)$$

where $\mathbf{Q}$ is the strength of the driving noise and $\delta(\tau)$ is the Dirac delta function. In practice, this system is augmented to include errors due to the barometric altimeter measurement and INS instrumentation errors. The nine-state INS error model discussed thus far is the minimum useful configuration for three-dimensional applications and represents the baseline INS error model. In a more complete INS model, the error dynamics [Eq. (6.64)] are driven additionally by gyroscope and accelerometer errors. The gyroscope errors directly drive the platform tilt equations and are modeled as a constant bias plus additional white noise. Refinements, such as slow variations of the bias, scale factor, and g-sensitivity errors are possible, but because of the slow coupling of these errors through the platform misalignments and the short duration of the vehicle maneuvers, if applicable, they are not included in the model. The bias states (gyroscope drift rates) are modeled as random constants that are augmented to the system equations as

$$\dot{D}_N = \dot{x}_{10} = 0, \quad \dot{D}_E = \dot{x}_{11} = 0, \quad \dot{D}_D = \dot{x}_{12} = 0 \quad (6.65)$$

6.2. Navaid Sensor Subsystems

where D_N, D_E, and D_D represent the drift rates due to the north, east, and down gyroscopes, respectively. The accelerometer errors are treated more accurately because of the direct nature in which they drive system errors during the short maneuver duration. The error model contains white noise plus two states for each accelerometer: bias B and scale factor SF. Both are modeled as random constants and augmented to the system equations

$$\dot{B}_N = \dot{x}_{13} = 0$$
$$\dot{B}_E = \dot{x}_{14} = 0$$
$$\dot{B}_D = \dot{x}_{15} = 0$$
$$\dot{SF}_N = \dot{x}_{16} = 0$$
$$\dot{SF}_E = \dot{x}_{17} = 0$$
$$\dot{SF}_D = \dot{x}_{18} = 0$$

The GPS Receiver Error Model Now consider a basic 12-state GPS receiver error model. The state vector δx_G of this basic GPS error model consists of three position error states δr, three velocity errors δV, three accelerometer error states δA, one baroaltimeter aiding state δh, one range error state due to the user clock bias b, and one user clock drift state d. The user position δr and user velocity δV are given in ECEF coordinates. As for the INS error model, the stochastic differential equation for the basic GPS error model can be defined as follows [see also Eq. (6.64)]:

$$\delta \dot{x}_G = F_G \delta x_G + w_G$$

where

$$\delta x_G = \begin{bmatrix} \delta r \\ \delta V \\ \delta A \\ \delta h \\ b \\ d \end{bmatrix}$$

The dynamic equations associated with these states are

$$\dot{x}_i = x_{i+3}$$
$$\dot{x}_j = 0$$
$$\dot{x}_{10} = -\frac{1}{\tau_h} x_{10} + w_{\text{alt}} = \delta \dot{h}' \qquad (6.66)$$
$$\dot{x}_{11} = x_{12}$$
$$\dot{x}_{12} = 0$$

where $i = 1, 2, 3, 4, 5, 6$
$j = 7, 8, 9$
τ_h = correlation time constant = d_{alt}/V
V = speed of the vehicle
d_{alt} = baroaltimeter correlation distance
w_{alt} = baroaltimeter white driving noise

The white driving noise w_{alt} has a strength of

$$Q_{alt} = 2\sigma_{alt}^2/\tau_h \tag{6.67}$$

where σ_{alt} is the standard deviation of the altitude of a constant pressure surface. Since the INS vertical channel is inherently unstable because of the gravity-induced compensation error positive-feedback loop, the basic model [Eq. (6.64)] requires (as a minimum) altitude aiding for stability purposes. A first-order Markov process describing the baroaltimeter error must be augmented to the basic system equation as shown in Eq. (6.66) state $\dot{x}_{10}$. If the GPS receiver system designer wishes to incorporate such errors as atmospheric path delays and code-tracking loop errors, these errors can be modeled as a constant bias and a second-order code-tracking loop [20]. Consequently, these models must be augmented to the basic system dynamics model such that $\dot{x}_{(i+12)} = 0$ ($i = 1, 2, 3, 4$), with second-order code-tracking loops contributing eight additional states. Equations (6.66) represent the baseline 12-state GPS receiver error model dynamics. The system equations for a simple 8-state, single-channel, low-dynamics GPS manpack receiver (Ref. 19 gives the navigation mechanization in an ECEF coordinate frame, which is compatible with the information given on the satellite ephemeris) are as follows:

$$\dot{x}_1 = x_4$$

$$\dot{x}_2 = x_5$$

$$\dot{x}_3 = x_6$$

$$\dot{x}_4 = -\frac{1}{\tau_v} x_4 + \eta_v$$

$$\dot{x}_5 = -\frac{1}{\tau_v} x_5 + \eta_v$$

$$\dot{x}_6 = -\frac{1}{\tau_v} x_6 + \eta_v$$

$$\dot{x}_7 = x_8 + \eta_p$$

$$\dot{x}_8 = -\frac{1}{\tau_f} x_8 + \eta_f$$

6.2. Navaid Sensor Subsystems

where x_1, x_2, and x_3 are the components of the user position, x_4, x_5, and x_6 are the user velocity components, and x_7 and x_8 are the user clock bias and bias rate terms. Furthermore, τ_v and τ_f are the velocity and clock bias drift correlation time constants, η_v is the velocity additive random-noise term, and η_p and η_f are the corresponding random fluctuations in the user clock phase and frequency errors, respectively. In this particular case, the system model is given by

$$\dot{\mathbf{x}}(t) = F\mathbf{x}(t) + \mathbf{\eta}(t)$$

where

$$\mathbf{x}(t) = [x_1 \quad x_2 \quad \cdots \quad x_8]^T$$

$$\mathbf{\eta}(t) = [0 \quad 0 \quad 0 \quad \eta_v \quad \eta_v \quad \eta_v \quad \eta_p \quad \eta_f]^T$$

A typical Kalman filter INS error model will commonly consist of 17–22 filter states. The 17-state Kalman filter includes the following error states: two horizontal positions, two horizontal velocities, three platform tilts, three gyroscope biases, three gyroscope scale factors, two horizontal accelerometer biases, and two horizontal accelerometer scale factors. (Note that the vertical channel is not modeled since it is unstable; the vertical channel information is obtained from a barometric altimeter.)

A 22-state Kalman filter uses approximately 32K words. The Kalman filter structure can be designed in two configuration modes: (1) as a closed loop filter for in-air alignment of the INS by the GPS, and (2) as an open loop filter for the GPS/INS integrated mode.

Before we proceed with the discussion of the integrated system truth model, a few remarks concerning the code-tracking loop error model are in order. A commonly used code-tracking loop in which two states for each receiver channel are implemented is that of a second-order. From Refs. 19, 20, and 24, the equation for describing the code loop error model is

$$\begin{bmatrix} \delta \dot{r}_c \\ \dot{v}_r \end{bmatrix} = \begin{bmatrix} -a_r & 1 \\ -a_r a_v & 0 \end{bmatrix} \begin{bmatrix} \delta r_c \\ v_r \end{bmatrix}$$

$$+ \begin{bmatrix} a_r(r_a + b) & \delta V_{aid} \\ a_r a_v (r_a + b) \end{bmatrix} + \begin{bmatrix} 1 \\ a_v \end{bmatrix} w_c \quad (6.68)$$

where a_r, a_v = constant, predetermined gains
  r_a = error due to uncompensated atmospheric path delays*
  v_r = an internally generated variable
  δr_c = position error due to code loop

*The atmospheric error model for a tropospheric delay is given by $\dot{r}_a = -(1/\tau_a)r_a + v_a$, where w_a is a white Gaussian noise with zero mean.

δV_{aid} = error in the externally supplied aiding velocity
w_c = white driving noise
b = range error due to user clock bias

After specifying the damping ratio ζ and bandwidth B_L of the code loop, the parameters a_r and a_v are determined by the equations

$$a_r = \frac{16\zeta^2 B_L}{1+4\zeta^2}$$

$$a_v = 4B_L - a_r$$

where the nominal value of the damping ratio is taken as $\zeta = 0.7071$. The term δV_{aid} in Eq. (6.68) is

$$\delta V_{aid} = -\mathbf{1}_{LOS} \cdot \delta \mathbf{V} \tag{6.69}$$

where the four vectors of $\mathbf{1}_{LOS}$ and $\delta \mathbf{V}$ are all expressed with respect to the NED navigation coordinate frame. This equation states that the code loop is effected by the projection of the velocity vector onto the user to satellite line-of-sight (LOS) vector. Note that since the GPS position and velocity estimates are in the ECEF frame, the GPS estimates must be transformed into the platform coordinates. The driving noise strength of w_c is determined by solving the Riccati equation

$$\dot{\mathbf{P}} = \mathbf{FP} + \mathbf{PF}^T + \mathbf{Q} \tag{6.70}$$

Setting $\dot{\mathbf{P}} = 0$ results in

$$Q_c = \frac{2a_r^2 \sigma^2}{2a_v + a_r}$$

The value σ^2 is determined from [24]

$$\sigma^2 = \frac{4B_L}{3}$$

while the signal-to-noise ratio (SNR) seen by the GPS antenna is determined from the equation

$$\sigma^2 = \frac{N_0 B_L}{2P_s}\left[1 + \frac{2}{SNR}\right]$$

where N_0 is the thermal noise for the system, B_L is the bandwidth of the code loop, P_s is the strength of the signal, and σ^2 is the variance of the code loop error.

The System Truth Model The truth model is formed by combining common states in the total INS model and the total GPS receiver model,

6.2. Navaid Sensor Subsystems

while eliminating the redundant states. The position, velocity, and barometric altimeter states are duplicates, and these states are eliminated from the GPS truth model. Thus, only one position error vector and one velocity error vector are included in the truth model. The resulting error state vector is given by

$$\delta x_t^T = [\psi^T \quad \delta P^T \quad \delta V^T \quad \delta h \quad D^T \quad B^T \quad SF^T \quad \delta A^T \quad \delta r_a^T \quad \delta r_c^T \quad v_r^T]$$

where all the symbols have been defined previously. The truth model dynamics matrix F_t relating these states is formed as

$$F_t = \begin{bmatrix} F_I & F_{UR} \\ \hline F_{CL} & F_G \end{bmatrix}$$

where F_I is the full INS error truth model, F_G is the GPS truth model minus the redundant states, F_{UR} adds the terms $\delta \dot{V} = \delta A$ due to the acceleration error modeled within the GPS receiver, and F_{CL} adds the terms due to the error in the velocity aiding signal to the code-tracking loop. That is

$$F_{CL} = \delta V_{aid} = -1_{LOS} \delta V$$

where 1_{LOS} represents the unit line-of-sight vectors from the INS position to the satellites. Neglecting time correlation, the best information the GPS receiver Kalman filter has available to simulate the INS velocity aiding error is if δV_{aid} of Eq. (6.69) is treated as an additional white driving noise on the code-tracking loop. This is necessary because the GPS receiver error model does not simulate the dynamics of the INS mechanization. For this reason, it cannot estimate the time-correlated nature of the INS information. These two interacting models are referred to as the "two-filter" full-state system [14].

Now consider a filter that is "external" to the GPS receiver. We now have the outputs of one Kalman filter being used as inputs to a following Kalman filter. Figure 6-23 depicts a simple GPS/INS integration employing two Kalman filters in a filter-driving-filter configuration. The GPS produces a navigation solution and provides its estimate of the vehicle position and velocity to the INS Kalman filter in the form of measurements. At the same time, the INS is providing velocity aiding to the GPS code loop. In designing an integrated GPS/INS system, care must be taken so that system stability is guaranteed. Finally, GPS signals will be used to update the INS at a rate of once per second [5].

In military applications, where a fire control computer (FCC) is used, the fire control computer's Kalman filter determines the best system navigation solution and calculates the errors in the INS estimate of position,

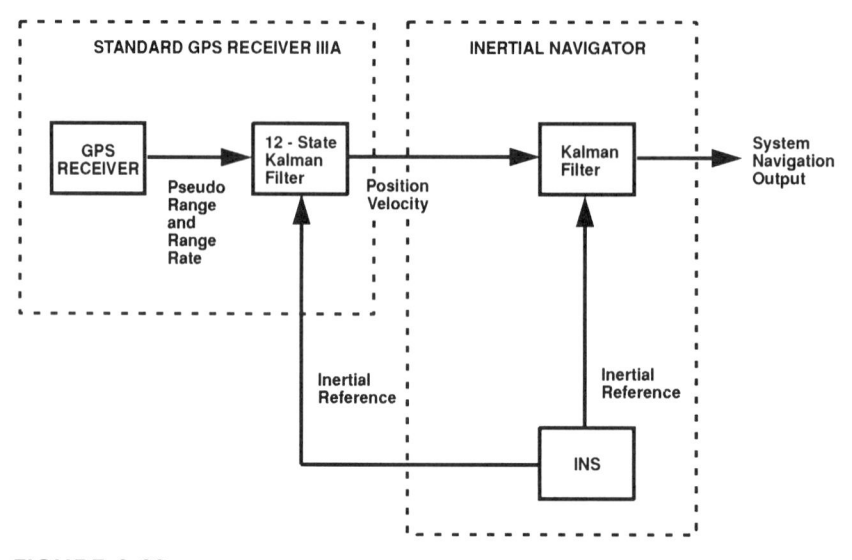

FIGURE 6-23
Two-filter GPS/INS integration.

velocity, and platform alignment. These computations take place in the Kalman filter, which uses position and velocity inputs and mathematically applies values for the earth's curvature, gravity, torquing terms (in the case of gimbaled systems), and random noise to estimate INS misalignment angles, gyroscope bias, and position and velocity vectors. Moreover, the Kalman filter performs a series of calculations that require approximately ten seconds to complete. The filter computes an estimate of how much error exists in the INS. This data is applied to the INS's navigation solution in the 50-Hz range to yield the FCC computed system navigation solution, which more closely approximates the aircraft's "real position." Note that the INS solution differs from the "real" solution by position and velocity errors and platform misalignment angles. As a result, for fighter aircraft using a fire control computer, the GPS is integrated into the fire-control system. Therefore, the fire control computer is responsible for maintaining a navigation solution utilizing data from independent sensors such as GPS, INS, and the Central Air Data Computer (CADC). The inputs to the Kalman filter can be grouped in the following ways:

1. Ground position–velocity fixes
    - Discrete zero velocity
    - Continuous zero velocity

2. Regular position fixes
    - Visual overfly
    - Radar
    - Head-up display (HUD)
3. GPS
    - GPS position only
    - GPS position and velocity

The ground fix modes provide a means of updating the aircraft's position and velocity while on the ground. Regular position fixes are accomplished in the standard manner. Each fix measures true range-to-steerpoint, which the fire control computer compares to the INS computed range-to-steerpoint. Note that a radar fix would be considered more accurate than an overfly fix, if the latter were accomplished at a high altitude where exact position is difficult to ascertain visually. The GPS measurements of position and velocity are available anytime the GPS is powered up and not in an initialization or built-in-test (BIT) mode. The fire control computer Kalman filter estimates are used to correct INS data to give it the accuracy required for fix and weapon delivery functions. Examples of GPS/INS integration can be found presently in the following U.S. Air Force aircraft: (1) AC-130U (Gunship), (2) F-16C/D, (3) F-111F, and (4) T-39 for INS alignment using GPS computed corrections.

In summary, the navigation performance of the integrated GPS/INS system is better than either the GPS stand-alone or the INS stand-alone. Consequently, integrated navigation algorithms can be developed and/or written that produce position data that have the smooth, nonnoisy characteristics of an INS and at the same time do not degrade the position accuracy of the GPS. Also, the velocity accuracy of the integrated system, depending on the complexity of the integrated algorithms, can be much better than either system in the stand-alone mode. The GPS avionics system must be able to function as the aircraft's sole means of navigation for en route and terminal phases of flight and nonprecision approaches.

6.2.5.7. Navstar/GPS Operational Capabilities

When fully implemented, the GPS will provide continuous worldwide coverage, allowing its users to calculate their position and velocity much more accurately than any previous system. Each satellite transmits a spread-spectrum signal encoded with the time of transmission and a set of orbital elements from which the satellite location at the time of transmission can be computed. By simultaneously delay tracking the signals from four satellites, a real-time position estimate with an absolute three-dimensional accuracy on the order of 10 meters RMS in each axis and a velocity of 0.1 m/s can be obtained. This high accuracy is possible because the system can be designed to minimize sources of error and to compensate for error sources that could not be eliminated by design. Recently, however, it has been claimed that, in the determination of position computed from the GPS pseudorange measurements from four satellites ($n=4$), a unique fix is not guaranteed, requiring at least $n \geq 5$ satellites in order to assume a unique fix. Because the GPS system is referenced to a common grid, the WGS-84, civil and military position data can be standardized on a worldwide basis. The user equipment can transform navigation information into other commonly used datums as well. As stated in the previous section, GPS user equipment is capable of operating in an autonomous (i.e., stand-alone) mode, or it can be integrated with other navigation systems such as inertial and Doppler radars. In particular, the GPS bounds INS drift errors and maintains the INS within 10 m of its true position for an entire flight [20].

Typical GPS receiver inputs and outputs are as follows: *inputs*—satellite signals, position, velocity, altitude, acceleration and attitude (from aiding navigation system), waypoints (entered from a CDU or data loader), and aiding navigation system mode and status data; *outputs*—position, velocity and time, altitude (mean sea level or absolute), steering information (i.e., track angle and cross-track error), time and distance to waypoint, groundspeed and groundtrack angle, elevation angle to waypoint, true magnetic heading, magnetic variation, calendar and time of day, and test and status data. Interface characteristics usually are dependent on the type of installation, power, and other requirements.

The real value and/or benefits of the GPS can be summarized as follows:

1. Inertial navigation system
   - Faster and more accurate alignment (including ground, in-flight, and transfer alignment)
   - Reduction of drift
   - Continuous update
2. Weapon systems
   - Better weapon delivery performance through reduction of number of bombs required
   - Better accuracy and probability of kill
   - Passive or blind weapon release
   - Reduction of exposure time
   - Reduction in target acquisition time
3. Navigation
   - Accurate general navigation (en route and terminal)
   - Area navigation and instrument approach and landing
   - Waypoint

- Rendezvous
- Steering for desired track
- Backup steering in the event of INS failure
- Air traffic management (e.g., civil air traffic control and collision avoidance)
- Flight control

When GPS receivers are used in supersonic aircraft, the navigation of such aircraft requires rapid processing of the information transmitted by the satellites. Each GPS receiver contains its own digital computer for the processing of this information. Automatic landing for transport-category aircraft using an integrated GPS/INS guidance system promises to be more cost-effective and rival the accuracy of microwave landing systems (MLS). This type of automatic landing system uses the signals from the satellites as well as inertial navigation data to determine aircraft position. With regard to weapon delivery, an optimally integrated GPS/INS system can provide precision, all-weather navigation for such common weapon delivery modes as (1) continuous computation of impact (target) point (CCIP), that is, compute the point on the ground where the weapon would impact if released at that instant, (2) continuous computation of release point (CCRP), (3) time-to-go weapon release, (4) computer-generated release signal, and (5) limitation of roll command steering. Finally, in a high-jamming–high-dynamics environment, the GPS receiver supplies only range information; for this reason it is desirable to use an INS-supplied rate-aiding signal to the code-tracking loop. Consequently, the INS-supplied rate-aiding signal allows the code-tracking loop to retain track longer than if no aiding signal were present.

6.2.5.8. International Efforts in Space-Based Radio-Navigation Systems

Global satellite radio navigation systems have been under development since the 1970s, notably by the United States and Russia. The U.S. Navstar/GPS system was first launched in 1978, while Russia's Global Navigation Satellite System (GNSS or Glonass) system was inaugurated four years later. Both the Navstar/GPS and the Glonass include a military (P-code) and a civil (C/A) component. Similar to the U.S. Navstar/GPS, the Glonass satellite navigation system foresees an operational configuration of 24 satellites (21 satellites + 3 active spares) with 8 satellites in each of 3 orbital planes separated in right ascension of the ascending node of 120°. The transmission carrier frequencies chosen (3) for the Glonass satellite navigation system operate in the L-band, around 1250 MHz (L_2) and 1600 MHz (L_1). Only the L_1 frequency carries the civil C/A-code. Radio-frequency carriers used by the Glonass are within the bands 1240–1260 MHz and 1597–1617 MHz;

the channel spacing is 0.4375 MHz at the lower frequencies and 0.5625 MHz at the higher frequencies. The carrier frequencies themselves are also multiples of channel spacing, the number of planned channels being 24. Unlike the GPS, in which each satellite is identified by a code in its radiated signal, each Glonass satellite radiates at a slightly different frequency. For the purposes of allowing the user to compute its own position, navigation satellites transmit details of their own positions and a time reference. Each satellite sends data at low-speed from which its own position at any reference time may be calculated. Specifically, this data is transmitted at a 50-bps (bits per second) rate and superimposed on a pseudorandom noise code, which is periodic and longer than a single data bit. The Glonass low-precision code has a length of 511 bits as compared to Navstar's 1023 bits for its equivalent code. A code sequence lasts 1 ms. Bandwidths for the Glonass transmission can be taken at 1 MHz and 10 MHz for the civil and military codes, respectively. These figures compare with 2 MHz and 20 MHz for Navstar's equivalent bandwidths. Civil aviation users equipped with GPS or a combination of GPS/Glonass receivers will be able to acquire the C/A signal with equal accuracy. However, one potential obstacle to a combined GPS/Glonass receiver is that the atomic clocks used in the Russian satellites are synchronized to a slightly different reference than the ones used in the GPS satellites. In satellite navigation systems such as Navstar/GPS and Glonass, the master control stations use an intermediate time reference for the individual space vehicles. This time is known as "system time" and is transmitted as part of the normal data message. Therefore, it is clear that a common reference time is the Coordinated Universal Time (UTC).

Historically, for many years the measurement of time has been coordinated at the Bureau de L'Heure (see also Chapter 2, Section 2.2.6). National laboratories running atomic standards are used to compute the UTC. For example, the UTC (USNO) refers to the atomic master clock at the U.S. Naval Observatory, while the UTC (SU) refers to the hydrogen atomic clock in Moscow. However, the relationship between the Navstar/GPS and Glonass remains unresolved at an accuracy of better than 1 μs, which corresponds to an error in range of about 300 m. Nevertheless, for the purposes of global time comparison, the Glonass system is certain to be linked to the UTC. As in the Navstar/GPS system, Glonass navigation data can be used to update the INS once each second in order to correct for gyroscopic drifts. When the full 24-satellite operational Glonass constellation is deployed in the 1996–1997 time frame, Russia plans to phase out the ground-based navaids.

In addition to the U.S. Navstar/GPS and Russia's Glonass systems, the European Space Agency (ESA) is designing its own space-based radio navigation system, designated "ESA/NAVSAT." At the present time, the ESA/NAVSAT is in the study phase. Table 6-2 presents a comparison of

TABLE 6-2
Summary and Status of the Various Space-Based Radio Navigation Systems

System	Navstar GPS	Glonass	ESA/Navsat
Sponsor	U.S. DoD	Russia	European Space Agency
Coverage	Worldwide	Worldwide	Worldwide
Satellite constellation	21 + 3 spares Inclined, 12-h period	21 + 3 spares Inclined, 12-h period	6 geosynchronous and 12 inclined, 12-h period highly eccentric orbits
	Autonomous	Autonomous	Transponder
Signals	L-band dual frequency spread-spectrum data-modulated	L-band spread-spectrum data-modulated	L-band spread-spectrum data modulated
Accuracy			
Position	SPS: 40-m CEP or 100 m 2-D rms PPS: 10–15 m 3-D SEP	Horizontal : 100 m Vertical: 150 m (predicted)	12.5 m SEP (predicted)
Velocity	0.1 m/s	0.15 m/s	Not known
Time	100 ns	1 μs	Not known
Status	Partial operation Full operation planned by the end of 1993	In development Full operation planned by 1996	In study phase

the space-based radio navigation systems discussed above. Finally, the various radio-navigation aids discussed in this chapter are summarized in Table 6-3. Some of the Navaids listed in Table 6-3 (e.g., Decca, Loran-A) have been phased out or are in the process of being phased out in favor of the Navstar/GPS.

6.2.6. Very-High-Frequency Omnidirectional Ranging (VOR)

The VOR is a ground-based radio system which provides bearing measurements similar to TACAN. Although the accuracies of standard VOR are comparable to TACAN, an improved compatible VOR is capable of providing better accuracy, approximately ±0.4°. VORs operate within the 108–117.95 MHz frequency band and have a power output necessary to provide coverage within their assigned operational service volume. Commonly, the VOR system consists of a transmitter, a VOR antenna, a goniometer, a monitor, and automatic ground check antenna. The function of each of these components will now be briefly described. The transmitter, starting with a precise frequency, low-level output from a temperature compensated crystal oscillator, modulates and amplifies that signal to provide a properly modulated high-level VOR carrier to the antenna bridge. The goniometer also

TABLE 6-3
Performance Comparison of Modern Radio–Navigation Systems

System	Year introduced	Type	Frequency	Range (km)	Accuracy (m) CEP
VOR	1946	Theta	108–118 MHz	370.4	185.2
Doppler[a]-VOR	1960	Theta	108–118 MHz	370.4	See note a
DME	1959	Rho	960–1215 MHz	370.4	914.4
TACAN	1954	Rho–Theta	960–1215 MHz	370.4	400.0
VORTAC	1972	Rho–Theta	L-band	182.5	1,852.0
Decca	1944	Hyperbolic	100 kHz	277.8	3,048.0
Loran-A	1943	Hyperbolic	1.8–2.0 MHz	1,574.2	457.2
Loran-C	1960	Hyperbolic	100 kHz	2,222.4	180.0
Loran-D	1967	Hyperbolic	100 kHz	1,100 (max)	270.1
Omega[b]	1968 (Partial)	Hyperbolic	10–14 kHz	14,816.0	1,828.8
GPS[c]	1970s (1st launch 1978)	LOS	Carrier bands: L_1: 1575.42 MHz P-Code L_2: 1227.60 MHz C/A	Worldwide	Position: SPS: 40-m CEP or 100-m 2-D RMS PPS: 10-15-m 1-D SEP Velocity: 0.1 m/s 1–σ Time: 100 ns 1–σ
ILS[d]	1939	LOS	Localizer: 110 MHz (VHF) Glide slope: 330 MHz (UHF)	9.3–18.5	Category III Vertical error: 1.04 Lateral error: 1.50
MLS[e]	1967	LOS	5 GHz (960–1215 MHz)	37.0	Vertical error: 0.61 Lateral error: 1.50

[a]Doppler accuracy is rated at 0.1–0.25% 1σ of velocity, or 0.5–1.0% of position.
[b]The differential Omega version operates at the same frequency and has a range of about 370 km and an accuracy of 335–579 m.
[c]The differential GPS (DGPS) accuracy is given as 5–10 m RMS in each axis in position, 0.1 m/s RMS in each axis in velocity, and 0.1 μs in time. The Radio Technical Commission for Maritime Service (RTCMS) established a Special Committee 104 on DGPS in November 1983.
[d]In 1943, U.S. government civil and military representatives met at the Pittsburgh Airport and made a decision to standardize the VHF/UHF ILS system. In 1946, the International Civil Aviation Organization (ICAO) also standardized the ILS as the new international standard. This system became officially effective in 1950.
[e]Work on the Microwave Landing System (MLS) to replace the ILS began in 1967. In 1978, the Time Reference Scanning Beam MLS technique was selected as the international standard by the ICAO. The ICAO established the MLS equivalent Precision Distance Measuring Equipment (DME/P) standards in 1985. The MLS is scheduled to replace the ILS for international air carrier operations in 1998.

starts with the crystal oscillator output and provides the antenna bridge with two balanced double-sideband suppressed carrier signals offset ±30 Hz from the carrier. Moreover, these sidebands automatically track the carrier power level through feedback control in order to provide constant modulation percentage. The audio components of the two double-sideband pairs are in phase quadrature and are switched 180° in RF phase on alternate half cycles of the 30-Hz navigation signal. The VOR monitor utilizes a microcomputer design approach and performs the VOR executive monitoring function. The VOR automatic ground check equipment consists of 16 monitor antennas with radomes and a single pole, 16-throw solid state switch. These ground check antennas are switched to the VOR monitor in a synchronized manner, under control of the monitor microcomputer, and an azimuth phase measurement is performed for each antenna radial. The VOR wavelengths are about ten times greater than that of TACAN. VOR is the standard short-range navigation aid in the United States and for other members of the International Civil Aviation Organization (ICAO). Since the equipment operates in the VHF frequency band, it is subject to line-of-sight restrictions, and its range varies proportionally to the altitude of the receiving equipment. However, there is some spillover, and reception at an altitude of 305 m is about 74–83 km. This distance increases with altitude. A variation of the standard VOR is the Doppler VOR (DVOR), which has been designed to overcome the multipath problems created by buildings and local topography.

6.2.7. Distance Measurement Equipment (DME)

Distance Measurement Equipment is the most precise method known for updating airborne inertial navigation systems in real time. Specifically, these systems measure the ranges from the aircraft to a number of fixed, known points on the ground. Using multilateration methods, the aircraft's position relative to these precisely surveyed locations can be determined. The DME range measurement link consists of a cooperative interrogator-transponder pair, one at each end of the range link being measured. Some systems have the DME interrogator in the aircraft and a number of transponders on the ground, while others have a DME transponder in the aircraft and a number of interrogators on the ground. Multilateration DME systems, similar to artificial satellites, are subject to geometric dilution of precision (GDOP). In order to minimize the GDOP, the aircraft should generally fly above and within the boundaries of the ground-based elements of the DME network, and the distance between ground-based elements should be comparable to the altitude of the aircraft. Consequently, under these conditions, the accuracy of the multilateration position fix becomes equivalent to the accuracy of the range measurements themselves, which is about 1.52 m (1σ). When distance measuring capability is added to VOR, it is called a VOR/DME station. VOR/DME is the primary navigation mode for civil aircraft flying in the airspace of many countries in the world. The VOR/DME enable onboard determination of an aircraft's bearing relative to north at the fixed ground station and slant range from the station, respectively. Specifically, the VOR/DME system involves primarily radial navigation, that is, aircraft fly directly to or from the ground beacons. It should be noted that bearing information from VOR/DME is frequently contaminated by aircraft magnetic heading in the aircraft receivers so that range and bearing data results in instrumentation errors.

When used in conjunction with an INS, VOR/DME information can be used to update the inertial navigation system. Since the accuracy of the position information derived from the VOR/DME system depends upon the relative location of the aircraft and the VOR/DME station, as well as the number of VOR and DME stations used, the use of VOR/DME information suggests the possibility of using an error model in a Kalman filter configuration so as to obtain improved navigational accuracy. Such an error model for the VOR/DME has been treated extensively in the literature. For example, in Ref. 3, both VOR and DME errors are separated into two components: a component with a long correlation time that can be modeled as a random bias, and a component with a short correlation time that can be modeled as white noise. Denoting b_V and e_V the VOR error components with a long and a short correlation time, respectively, and similarly b_D and e_D denote the DME error components with a long and a short correlation time, respectively, we have [3]

$$\dot{b}_V = 0 \tag{6.71a}$$

$$\varepsilon\{e_V\} = 0, \qquad \varepsilon\{e_V(t+\tau)e_V(t)\} = 2\sigma_{e_V}^2 T_{e_V}\delta(\tau) \tag{6.71b}$$

$$\dot{b}_D = 0 \tag{6.71c}$$

$$\varepsilon\{e_D\} = 0, \qquad \varepsilon\{e_D(t+\tau)e_D(t)\} = 2\sigma_{e_D}^2 T_{e_D}\delta(\tau) \tag{6.71d}$$

where σ_{e_V} and σ_{e_D} and T_{e_V} and T_{e_D} are the standard deviation and correlation time of e_V, respectively. Consequently, for a system using an inertial navigator, the VOR/DME error state equations that must be added to the system model $\dot{\mathbf{x}} = F\mathbf{x} + \mathbf{w}$ are as follows:

$$\dot{b}_{V_1} = 0$$
$$\dot{b}_{D_1} = 0$$
$$\dot{b}_{V_2} = 0$$
$$\dot{b}_{D_2} = 0$$

A typical value for σ_{e_V} and σ_{b_V}, the standard deviation of e_V and b_V, respectively, is 1°. For an aircraft flying at a speed of 926 km/h (500 knots) and a correlation time of 7.2 s, the correlation distance of e_V is about 1.852 km ($d_c = VT_c$). Similarly, typical values of σ_{e_D} and T_{e_D} are 0.1852 km and 3.6 s, respectively. Finally, a reasonable value for σ_{b_D}, the standard deviation of b_D is 0.259 km.

6.2.8. VOR/TACAN (VORTAC)

A collocated VOR with a TACAN constitutes a VORTAC beacon system. As with VOR and TACAN, VORTAC is a ground-based radio-navigation system operating in the L-band, and is designed to provide the user Rho (range)/Theta (azimuth) information. There are approximately 1200 stations located throughout the National Airspace System that are operated and maintained by the Federal Aviation Administration (FAA). These VORTAC stations include VOR, DME, TACAN, and DVOR, and provide navigation service to civil and military aircraft over the continental United States, Alaska, Hawaii, and other U.S. territories. The equipment is designed for maximum commonality of the VORTAC, VOR/DME, and VOR facilities. For instance, at a VOR/DME, the only significant differences from the VORTAC are that the 5-kW TACAN power amplifier is replaced by a 1-kW DME power amplifier and that the primary ac power is single-phased. The maximum useful range of VORTAC stations is typically 185 km or less. However, because of the VORTAC line-of-sight characteristics and problems with terrain obstacles, minimum en route altitudes (MEAs) are specified by the FAA for airway routes.

Rho–Theta systems have the desirable feature that the lateral position error decreases linearly as the station is approached. On the other hand, the useful maximum range is limited by the ever-increasing lateral error with increasing distance from the station. Furthermore, since these ground-based radio systems operate in the VHF and L-band regions of the spectrum, their line-of-sight propagation characteristics further limit the horizontal and vertical coverage. Note that although the Rho–Theta systems inherently have omnidirectional or area coverage, their application has been largely restricted to the use of radials for routes or airways. VOR and TACAN signal reliability is extremely high. As mentioned above, these systems provide relative range and angle (bearing) data with respect to the station; however, no vertical or velocity data is available from these systems.

6.2.9. Joint Tactical Information Distribution System (JTIDS) Relative Navigation (RelNav)

The Joint Tactical Information Distribution System (JTIDS) is a military, triservice program providing integrated communications, navigation, and identification capabilities. JTIDS was developed to facilitate secure, flexible and jam-resistant data and voice transfer in real time among the dispersed and mobile elements of the military services. More specifically, JTIDS is a high-capacity, synchronous, time-division multiple-access (TDMA), spread-spectrum, integrated communication system which has the inherent capability of furnishing high-accuracy relative navigation (RelNav) with respect to other terminals in the net. The TDMA method contains a unique propagation guard period which assures that the TDMA system will provide data throughput over a 556 km range. JTIDS information is broadcast omnidirectionally at many thousands of bits each second and can be received by any terminal within range. Its TDMA communications architecture not only supports interunit data exchange but also permits measurement of time of arrival (TOA) between cooperative members. When precision TOA measurements are made available to onboard processors of inertial subsystem data, a synthesis is possible to support tactical navigation requirements. The RelNav function provides the synthesis in a software ensemble of modules that were designed to provide geodetic and grid navigation information in terms of position, velocity, and attitude. Basic to the operation of the system is a stable relative navigation grid, which provides compatible interoperation for differing terminal types found in a highly interactive tactical system. The location of the grid origin is defined by one moving user or two stationary users assigned to the role of "navigation controller." Consequently, each relative navigation community selects one of its members as the relative grid origin and identifies this fact to the community by setting its transmitted relative position quality to the highest value allotted. Therefore, the primary intended application of JTIDS RelNav is for tactical relative navigation. The navigation function is obtained by frequent passive ranging and infrequent active (round trip) ranging by means of these TOA measurements. Normally, the RelNav algorithm is implemented in software contained in the unit's terminal processor (see Fig. 6-24). The software is provided with received P-message data and the measured TOA of these messages from the signal processor. From Fig. 6-24, we note three primary software modules used to provide the RelNav function: (1) Kalman filter, (2) source selection, and (3) dead-reckoning data processor. All users continuously update their estimates of position and velocity in the relative grid as well as the location of the grid origin based on the TOA of JTIDS messages called "position and status report" (P-messages) received from other users. Furthermore, by means of a Kalman filter mechanization in the user computer, each terminal passively determines its position, velocity, and time bias with respect to system time on the basis of these sequential TOA measurements. In most applications, dead-reckoning (DR) data from such sensors as inertial platforms, Doppler radar, air data systems, and attitude and heading reference systems (AHRS) are used to extrapolate the TOA-derived data between filter updates. Figure 6-24 illustrates the RelNav function (R/T = receiver–transmitter).

6.2. Navaid Sensor Subsystems

FIGURE 6-24
Basic JTIDS RelNav concept.

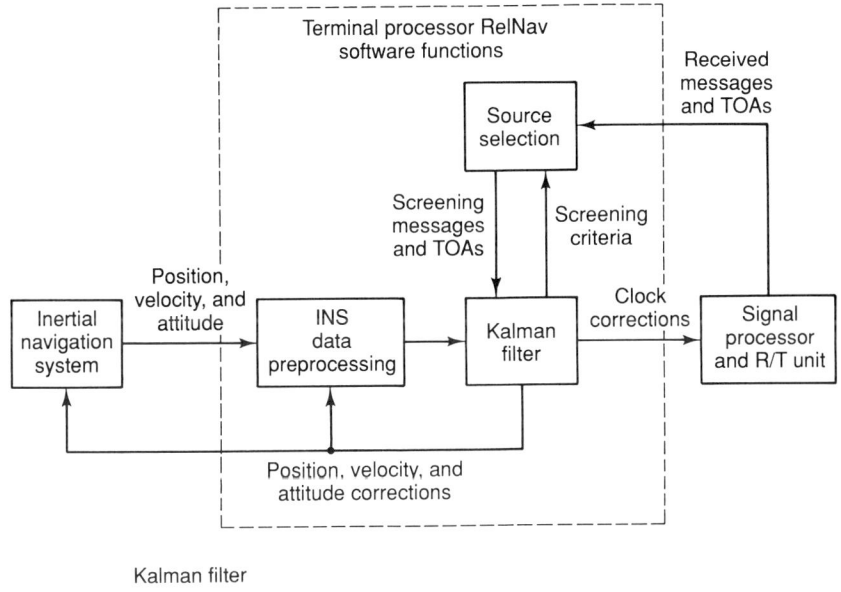

Kalman filter

System: $x_{k+1} = \Phi(k+1, k)x_k + \omega_k \quad x(0) \sim \mathcal{N}(\bar{x}(0), P(0))$

Measurement: $z_k = H_k x_k + \upsilon_k$

JTIDS disseminates its navigation data to the entire tactical community, thereby providing an essential consistency of position location to each of its elements. This consistency applies not only to the positions of each member, but to all data derived in the community from onboard sensors. This permits the acquisition of precise fire-control solutions based on sensor data derived from multiple platforms and stations. It is important to note that the JTIDS RelNav function can be obtained by adding only a software module to the computer program of the basic JTIDS communication terminal computer. However, if some of the terminals are surveyed ground terminals or have independent knowledge of their absolute (geographic) position, accurate* absolute navigation data can also be obtained by the user. Data sources for the JTIDS RelNav function may be both airborne and ground-based.

A simplified model will now be developed by considering an n-member user network attempting to navigate in a horizontal plane by means of JTIDS TOA measurements and round-trip timing measurements. As a

---

*The position accuracy of JTIDS is limited only by its TOA measurement capability, which is a function of the signal bandwidth and SNR.

result, a member of the net who receives a JTIDS broadcast from another member can construct a TOA measurement by subtracting the time of broadcast of the message from the time of reception. This results in a measure of the range between the members. However, errors in their clock phases prevent the immediate use of the TOA as a range measurement. As in the case of the GPS (see Section 6.2.5.2), the TOA is known as a pseudorange measurement. Now, let each member have an INS and a JTIDS clock. Furthermore, let the state of the ith member ($i = 1, 2, \ldots, n$) be modeled as an error state $\delta \mathbf{x}_i$, whose components define the simplified model of the two-dimensional navigation errors of the ith INS. Therefore, the simplified error state vector is given by

$$\delta \mathbf{x}_i^T = [\delta x_i \quad \delta y_i \quad \delta v_{xi} \quad \delta v_{yi} \quad \psi_{zi} \quad \delta t_i \quad \delta f_i]$$

where δx_i = error in the INS indicated x coordinate
δy_i = error in the INS indicated y coordinate
δv_{xi} = error in the INS indicated velocity in the x direction
δv_{yi} = error in the INS indicated velocity in the y direction
ψ_{zi} = platform misalignment about the z axis
δt_i = JTIDS clock phase error with respect to the Greenwich Meridian time
δf_i = JTIDS clock frequency error with respect to the Greenwich Meridian time

The error states δt_i and δf_i are expressed in range equivalent units, so that the speed of light equals unity. Similarly, we can define an error vector, $\delta \mathbf{x}'_{sij}$, which is the set of the "relative" navigation errors of member j with respect to member i. Thus

$$\delta \mathbf{x}'^T_{sij} = [\delta r_{ij} \quad \delta v_{ij} \quad \delta \alpha_{ij} \quad \delta \omega_{ij} \quad \delta \alpha_{ji} \quad \delta t_{ij} \quad \delta f_{ij}]$$

where δr_{ij} = error in the indicated range between members i and j
$\delta v_{ij} = \delta \dot{r}_{ij}$ = error in the indicated LOS velocity
$\delta \alpha_{ij}$ = pointing error of member i to member j
$\delta \alpha_{ji}$ = pointing error of member j to member i
$\delta \omega_{ij}$ = error in the indicated inertial angular velocity of the LOS
$\delta t_{ij} = \delta t_j - \delta t_i$ = relative JTIDS clock phase error
$\delta f_{ij} = \delta \dot{t}_{ij}$ = relative JTIDS clock frequency error

The relative geometry between members i and j can be represented as follows. Let

r_{ij} = the range between members i and j
θ_{ij} = the angle of the LOS with the x axis
α_{ij} = the angle of the LOS with the ith INS platform axis (this is called the *pointing angle*)
$\omega_{ij} = \dot{\theta}_{ij}$ = inertial angular rate of the LOS

6.2. Navaid Sensor Subsystems

Since navigation takes place in the horizontal plane, the relationship of the relative parameters r_{ij}, θ_{ij}, and α_{ij} to the geodetic coordinates of the two members i and j can be expressed by the simple geometric equations [25]

$$r_{ij} = \sqrt{(x_i - x_j)^2 + (y_i - y_j)^2} \tag{6.72a}$$

$$\theta_{ij} = \tan^{-1}\left[\frac{y_j - y_i}{x_j - x_i}\right] \tag{6.72b}$$

$$\alpha_{ij} = \theta_{ij} - \psi_{zi}$$

(Note that the definition θ_{ij} implies that $\theta_{ji} = \theta_{ij} + \pi$.)

As before, the system errors can be modeled by the linear dynamic system $\delta\dot{x}_i = F_i \delta x_i + w_i$ driven by the white noise w_i. References 6, 25, and 26 provide a more detailed description of the JTIDS system.

6.2.10. Visual Flyover

Visual flyover position updating is commonly used in flight test and evaluation of inertial navigation system performance. Flyover updates can be accomplished with an accuracy of less than 30.48 m (100 ft) while flying low-level (76.2 m or 250 ft AGL), the accuracy decreasing with increasing altitude. The visual flyover update mode is generally easy to perform, and there are no serious operator interface problems. Specifically, the flyover (also operator-inserted position updates) position updates require the pilot to maneuver over a known landmark, while the navigator updates the system. Aircraft true latitude, longitude, and altitude can be recorded from test-site range support, which may include time–space–position information (TSPI) from an FPS-16 instrumentation radar, cinetheodolite cameras, and other tracking systems (see Table 6-4). The differences between aircraft INS position update and TSPI position must be compared in order to provide the mission computer update errors.

Flyover updates must be reduced to along-track, cross-track, and radial errors, where the term "track" refers to the aircraft true track. Figure 6-25 shows these errors. Furthermore, for sensor updates such as TACAN, radar, and FLIR (see next section), the data must be reduced to along-track, cross-track, and radial error, where now the term "track" refers to the true bearing from the update navigation reference point (NRP) to the aircraft position. For TACAN and radar updates, slant range error and azimuth angle error must be calculated. Note that the radial error decreases at lower flyover altitudes because it is easier to determine when the aircraft is actually overhead the NRP. Table 6-4 compares the various tracking options that are available. Note that the laser trackers (PATS) are the primary source of trajectory data. At the Yuma Proving Ground six laser trackers are used

TABLE 6-4

Comparison between Various Tracking Systems[a]

System	Accuracy	Constraints	Remarks
Cinetheodolites	<1.52 m (0.1 mrad)	Range <10.0 km Suitable geometry	Position only Weather constraints Tracking/frame speed Terminal area only
CIRIS (or ARS)	3.96 m/axis (0.1–0.2 fps per axis)	Suitable transponder geometry	Performance degradation in turns (available at Holloman AFB)
PATS laser	Az: 0.1 mrad El: 0.1 mrad	Range <30.0 km	Position only Retroreflector shielding in turns
GPS	10.0–15.0 m SEP	Suitable satellite coverage	Requires custom pod in the aircraft Data dropouts in turns Developmental system
FPS-16 radar	Az: 0.2 mrad El: 0.2 mrad R: <4.6 mrms	Range: >277.8 km LOS	Position only
HAMOTS	Real time: 7.98–14.99 m HOR 14.99–29.99 m VER	UTTR range	Position only Antenna shielding in turns
EATS	14.99 m	Point Mugu range	Position only Antenna shielding in turns
High-speed fixed cameras	1.49 m	Terminal area land target only Suitable geometry	Camera pointing and timing Altitude limitations

[a] Abbreviations: PATS = precision automated tracking system; CIRIS = completely integrated reference instrumentation system; ARS = advanced reference system; HAMOTS = high-accuracy multiple object tracking system; UTTR = Utah Test Range; EATS = extended-area tracking system.

to determine the position of an aircraft or ground vehicle fitted with a retroreflector.

6.2.11. Forward-Looking Infrared (FLIR) Updates

The FLIR sensor (discussed briefly in Chapter 2, Section 2.2.11), also known as infrared detection set (IDS), provides the pilot with azimuth and elevation angles to a designated target. In using FLIR for position updating, the operator would normally use the FLIR to locate an IR (infrared) target of known position with the system cursor. FLIR/BARO and FLIR/RALT

6.2. *Navaid Sensor Subsystems* **347**

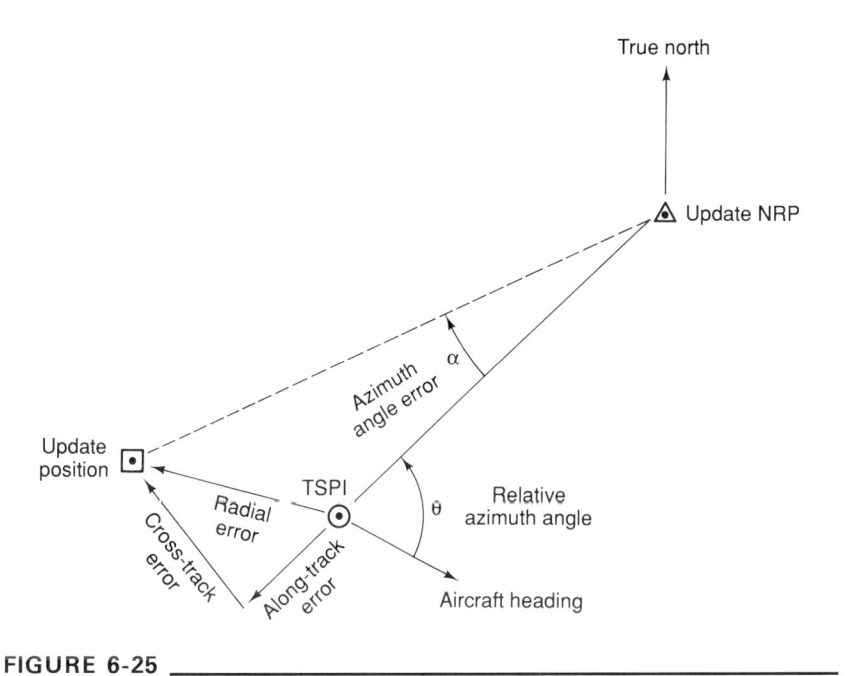

FIGURE 6-25
Flyover update error description.

modes providing barometric altimeter and radar altimeter ranging information, respectively, can be used to achieve a more precise update. As in the case of visual flyover updates, the data can be reduced to along-track, cross-track, and radial errors. Typical position update sighting accuracy for the FLIR position update mode is 3 mrad (RMS) in both azimuth and elevation. The update error increases with decreasing elevation angle. Human factor measurements have shown that the random error associated with cursor placement is approximately 0.127 cm. Higher aircraft speeds and closer target down-range and/or larger cross-range distances will result in larger overall errors due to an increase in operator cursor placement error. These azimuth and elevation angle errors consist of such errors as LOS misalignment, INS altitude, gimbal readout (in the case of gimbaled systems) of LOS to platform misalignment, and operator target cursor placement accuracy. The FLIR narrow field-of-view (FOV) image subtends a 6.7° × 5° FOV, which is presented on a 482 × 300-pixel (interlaced) display. This corresponds to a resolution of 0.3 mrads/pixel. INS time-dependent errors due to such factors as gyroscope drift, accelerometer errors, and gravity anomalies are expected to contribute 1σ bias errors of not more than 1.0 mrad in heading and 0.6 mrad in pitch and roll over a 6-h flight. When conducting an error analysis, errors due to bending of the aircraft structure from the INS to the

FLIR mount must be included. FLIR pointing performance, while performing an en route navigation update, can be evaluated using a simulation of a "fly-by" at, say, 115-m/s groundspeed with the FLIR target offset to the side of the aircraft. Consider, for example, an aircraft that has not performed a position update for the last hour. The resultant navigation error is 457 m CEP. A FLIR/BARO position update at this distance would reduce this error to less than 30 m and enable the pilot to continue on with a greatly reduced navigation error.

6.2.12. Terrain Contour Matching (TERCOM)

TERCOM (TERrain COntour Matching) is a technique for determination of the geographic position location of an airborne vehicle with respect to the terrain over which the vehicle is flying. Reference terrain elevation source data descriptive of the relative elevations of the terrain in the fix-point areas are stored in the vehicle's onboard navigation computer. These data are in the form of a horizontally arranged matrix of digital elevation numbers. As a result, a given set of these numbers describes a terrain profile. Now as the vehicle flies over the matrix area, data describing the actual terrain profile beneath the vehicle are acquired. The actual profile data acquired using a combination of radar and barometric altimeter outputs sampled at specific intervals, and compared against the stored matrix profiles, provides the position location. It should be noted that this navigation system requires extensive computer memory storage for a digitized map of the route to be flown, which is compared against the radar-derived terrain contours.

Historically, the concept of terrain profile uniqueness and the utilization of that uniqueness as a fix-taking technique to update an inertial navigation system was originally developed by the LTV-Electrosystems in 1958. The basic operation of TERCOM is illustrated by the block diagram in Fig. 6-26.

The process of determining vehicle position by the use of terrain contour matching can generally be described as consisting of three basic steps as follows: (1) data preparation, (2) data acquisition, and (3) data correlation. Data preparation consists of selecting a fix-point area large enough to accommodate the along-track and cross-track navigation arrival uncertainties, securing source material that contains contour information, and then digitizing the terrain elevation data into a matrix of cells oriented along the intended flight path. Data is acquired by sampling the altitude of the vehicle above the terrain directly below it, and at an interval equal to the reference map cell size. The last step in the process is the correlation of the data in the terrain elevation file with each column of the reference matrix. Specifically, the reference column, which has the greatest correlation with the

6.2. Navaid Sensor Subsystems

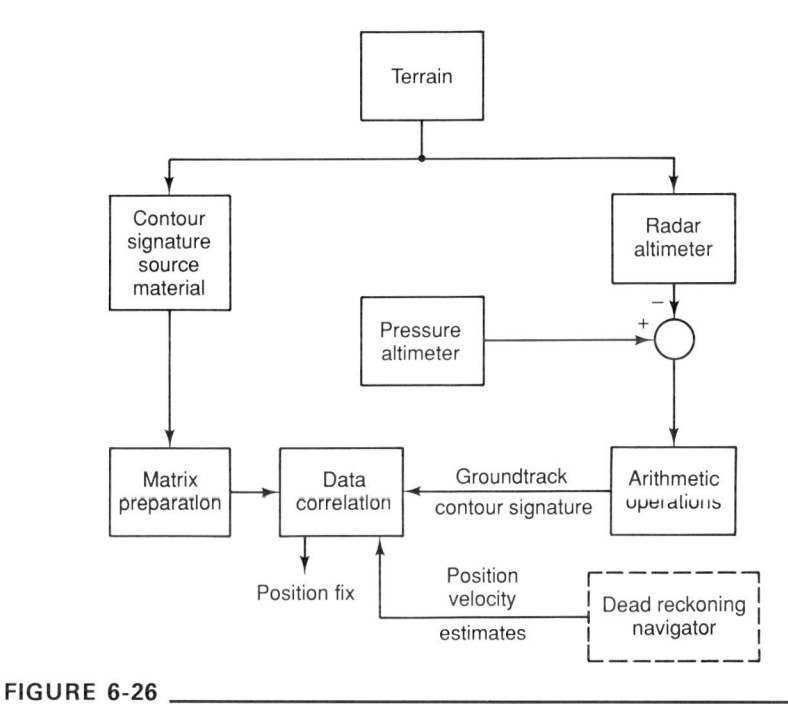

FIGURE 6-26
Basic TERCOM concept block diagram.

terrain elevation file, is the column down which the vehicle has flown. If the navigation error were zero, then the match column would be the center column of the map, since that is the ground track that the navigation system is steering along.

As indicated in Fig. 6-26, the TERCOM system utilizes barometric and radar altimeter measurements to construct a profile of the terrain that is coincident with the ground track of the airborne vehicle. As mentioned above, this measured terrain profile is then compared with a set of prestored reference terrain profiles which correspond to the area over which the vehicle is flying. Once a match is found, the geographic coordinates of the matching reference profile are used by the navigation system to update the vehicle's navigated position. Furthermore, the left side of Fig. 6-26 describes the stored or reference data loop. Source material in the form of survey maps or stereophotographs of the terrain are used to collect the set of altitudes which constitute the reference matrix. The right side of Fig. 6-26 describes the data acquisition loop. The dead-reckoning navigation subsystem supplies velocity information from which position estimates may be extrapolated. This function may be replaced by an airspeed and compass combination or by an inertial navigator. TERCOM maps can be classified as (1) landfall, (2) en route, (3) midcourse, or (4) terminal. These maps differ in length, width, and size. Now consider Fig. 6-27. From the preceding discussion, the TERCOM system yields a fix by comparison of a set of acquired data, in the form of a sequence of terrain elevation measurements, with a set of stored data in the form of a matrix of reference terrain elevations. Note that a matrix is composed of $m \times n$ cells, in which each cell is $d \times d$ meters in size.

The circles in Fig. 6-27 represent points at which the terrain altitude, referred to as the *local mean value*, is determined from contour maps or stereophotographs. The cell size is denoted by d, $L = Nd$ is the length of the profile used for correlation, and N is the number of cells. As the vehicle approaches a fix area, TERCOM begins to acquire two altitude measurements every interval d. One of the two is altitude above mean sea level, whereas the other is altitude above the terrain. Altitude estimates above the terrain are acquired by the radar altimeter. The output is differenced with a barometric altimeter. Acquisition of these measurements is continued until the vehicle is well past the fix area. Each pair is differenced with the result that the sequence of differences yields an estimate of the terrain profile along the vehicle track. Various arithmetic operations (e.g., mean removal and quantization) are then performed on the differenced data.

TERCOM position fixing techniques fall under two categories: the "long sample, short matrix" (LSSM) and the "short sample, long matrix" (SSLM). These two concepts are illustrated in Fig. 6-28. In both cases, the matrix is made wide enough in order to accommodate the navigation cross-track arrival uncertainty.

For the LSSM, the measured terrain elevation file is long enough to accommodate the down-track uncertainty and sampling begins prior to arriving at the map area, while for the SSLM, the stored reference matrix is

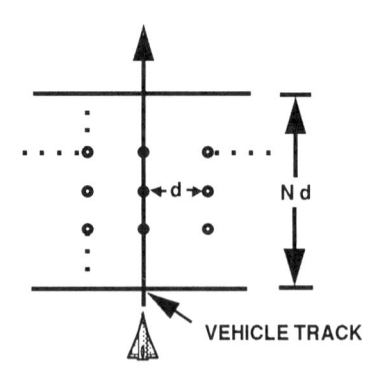

FIGURE 6-27
TERCOM cell designation.

6.2. Navaid Sensor Subsystems

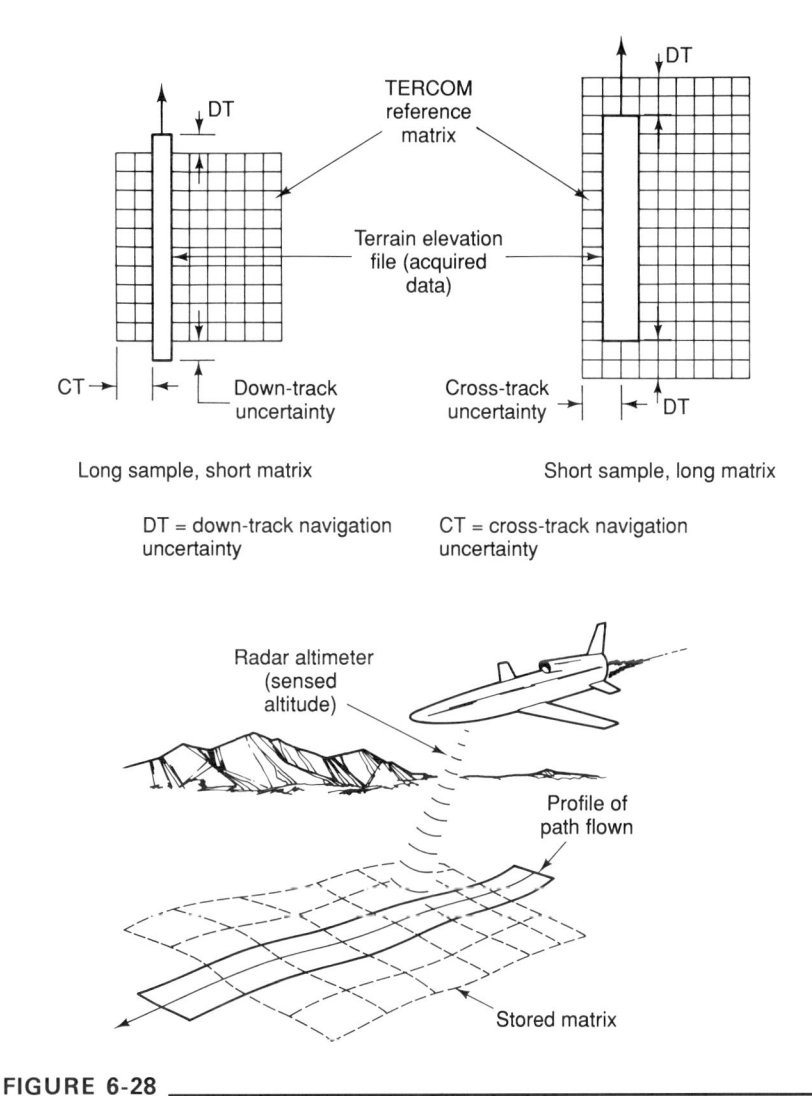

FIGURE 6-28
TERCOM fix concept.

long enough to accommodate the down-track uncertainty and sampling begins and ends while the vehicle is over the map area. Furthermore, the LSSM is used whenever the vehicle arrival uncertainty is relatively large or if only small fix areas are available. The SSLM can be employed during a multiple fix-taking mode. In this mode, faster updating is achieved, provided the search area (i.e., navigation uncertainty) is kept small and enough data is available for the longer reference matrix. The length of the measured terrain profile L can be the same for either mode. A closer examination reveals that in the SSLM technique, the measured terrain elevation samples are sequentially stored in the terrain elevation file until it is filled, at which time no more samples are taken and correlation of the file with the reference matrix begins. In the LSSM technique, sampling begins prior to arrival at the reference matrix area, and once the terrain elevation file has been filled, correlation begins. However, after the file has been correlated with all the reference columns, the first sample S_1 (i.e., the oldest with respect to time) is discarded. The remaining $L-1$ samples are shifted down one location (i.e., $S_i \to S_{i-1}$), and the most recent sample is stored at the top of the file (i.e., S_L). Thus, the updated file is correlated with the reference matrix. This process continues for some distance past the reference map area in order to accommodate the down-track position uncertainty.

There is always, however, a probability that the TERCOM system will fix on a part of the reference matrix that the vehicle did not actually fly over. As a result, this probability of false fix increases as the uniqueness of the terrain in the reference matrix area decreases. Therefore, the reference matrices must be carefully chosen, tested, and screened so that the only ones used are those that exhibit a sufficiently low probability of false fix. This gives rise to the consideration of terrain roughness. Consequently, one factor that is used in selecting an update area is the roughness of the terrain. The variation in terrain elevation provides the TERCOM signal, so that the quality of this signal increases directly with increasing amplitude, frequency, and randomness of the terrain. Usually, terrain roughness is defined as the standard deviation of the terrain elevation samples; that is, it is referred to as "Sigma-T" (σ_T). Mathematically, σ_T can be computed fairly accurately by taking the difference between the highest and lowest terrain elevation points. Thus

$$\sigma_T = \sqrt{\frac{1}{N} \sum_{i=1}^{N} (H_i - \bar{H})^2} \qquad (6.73a)$$

$$\bar{H} = \frac{1}{N} \sum_{i=1}^{N} H_i \qquad (6.73b)$$

where H_i = highest terrain elevation above the mean sea level
$\bar{H}$ = lowest terrain elevation above the mean sea level

Specifically, three parameters are usually used to describe the TERCOM related terrain, and their values can give an indication of the terrain's ability to obtain a successful TERCOM position fix: Sigma-T, Sigma-z, and the terrain correlation length X_T. Sigma-z is defined as the standard deviation of the point-to-point changes in terrain elevation (i.e., the slope), as shown in Fig. 6-29. Similar to Sigma-T, the value of Sigma-z provides a direct indication of terrain roughness and can be an indicator of TERCOM performance. Expressed mathematically, σ_z is given by

$$\sigma_z = \sqrt{\frac{1}{N-1}\left\{\sum_{i=1}^{N-1}(D_i - \bar{D})^2\right\}} \qquad (6.74a)$$

where

$$D_i = |H_i - H_{i+1}| \quad \text{and} \quad \bar{D} = \frac{1}{N-1}\sum_{i=1}^{N-1} D_i \qquad (6.74b)$$

Finally, the two parameters Sigma-T and Sigma-z are related with the third parameter X_T as follows:

$$\sigma_z^2 = 2\sigma_T^2[1 - e^{-(d/X_T)^2}] \qquad (6.75)$$

where d is the distance between elevation samples (i.e., cell size) and where a Gaussian terrain autocorrelation function is commonly assumed. The correlation length X_T represents the separation distance between the rows or columns of the terrain elevation matrix required to reduce their normalized autocorrelation function to a value of e^{-1}. Furthermore, it is usually assumed that parallel terrain elevation profiles, which are separated by a distance greater than X_T, are independent of each other.

As indicated in Fig. 6-29, the TERCOM process involves matching the measured contour of the terrain along the ground track of the vehicle with each down-track column of the reference matrix which is stored in the

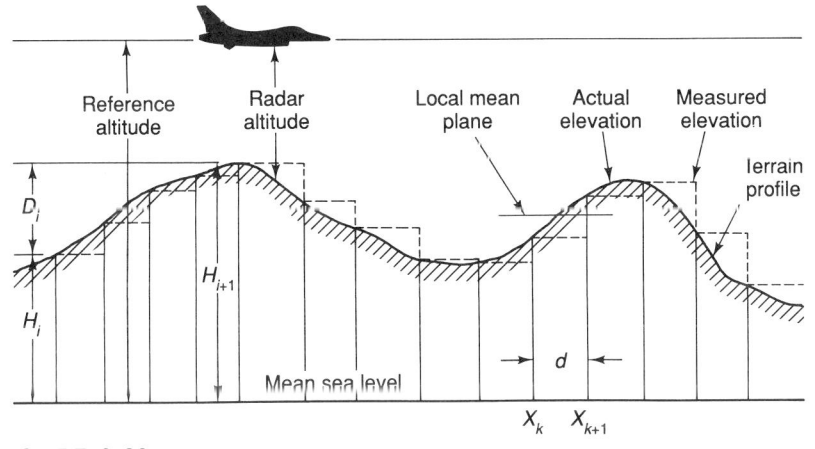

FIGURE 6-29
Definition of Sigma-z (σ_z).

vehicle's digital computer memory prior to flight. However, since the TERCOM system is not noiseless, the terrain profile measured during the flight will probably never exactly match one of the reference matrix profiles. Note that this profile is correlated with all sets of reference profiles in the matrix, in both cross-track and down-track directions. A common assumption to the terrain correlation process is that the geographic distance between the measured terrain elevation profile and the best matching reference matrix column provides an excellent measure of the cross-track and down-track position errors of the vehicle as it flew over the reference matrix area. A number of correlation algorithms of varying complexity and accuracy exist, which could be used to correlate the measured data with the reference data. These algorithms include the following: (1) mean absolute difference (MAD), (2) mean-squared difference (MSD), (3) normalized MAD, and (4) normalized MSD. Of these, the MAD algorithm provides the best combination of accuracy and computational efficiency, for performing real-time terrain contour matching in an airborne digital computer environment, and will be briefly discussed here. The MAD algorithm, which is most commonly used for correlating the measured terrain elevation file with each down-track column of the reference matrix, repeated for each column of stored data, is defined as follows:

$$\text{MAD}_{k,m} = \frac{1}{N} \sum_{n=1}^{N} |h_{k,m} - H_{m,n}| \qquad (6.76)$$

where $\text{MAD}_{k,m}$ = the value of the mean absolute difference between the kth terrain elevation file and the mth reference matrix column

n, m, k = row, column, and terrain elevation file indices

$H_{m,n}$ = the prestored reference matrix data; $1 \leq m \leq M$, $1 \leq n \leq N$

$h_{k,m}$ = the kth measured terrain elevation file; $1 \leq k \leq K$

N = the number of samples in the measured terrain elevation file; also, designates the number of rows in the reference matrix

M = the number of reference matrix columns

K = the number of measured terrain elevation files used in the correlation process (e.g., for the SSLM technique, $K = 1$)

For operation in the SSLM mode, the location of the reference matrix set, which results in the minimum MAD, is the position fix. Therefore, the correlation process consists of taking the one-dimensional measured terrain elevation file containing N elevation samples and computing the MAD value for it and for M reference map columns. If each of the N elevations (i.e., $h_n - H_n$) are equal, then at the end of the correlation, the MAD value will be zero, indicating a perfect match. On the other hand, if the elevation pairs are not identical, then the MAD computation will produce a value or residue greater than zero. Consequently, the magnitude of this MAD residue represents the degree of mismatch between the reference matrix column and the corresponding terrain elevation file.

As stated at the outset of this section, the objective of the TERCOM process is to update the vehicle's inertial navigation system by providing the navigation system with a measured cross-track and down-track vehicle position error. The navigation system then uses the measured position error to update its estimate of the vehicle's true geographic position. Figure 6-30 depicts a typical course correction update process. It is noted in Fig. 6-30 that the planned course of the vehicle is down the center of the fix-taking area, and it is the planned course that the navigation system directs the vehicle along. However, with cross-track errors present in the navigation system, the actual ground track of the vehicle will be either to the left or right of the planned course. Note that down-track position errors must also be considered in a full-scale analysis. Now, as the vehicle approaches the designated map, the radar altimeter scans the map and develops a terrain height profile that is compared with the data in the vehicle's map storage device. Usually, a Kalman filter is used to reduce the navigation system's errors, based on the measured vehicle position error. Furthermore, as discussed earlier, each cell in the matrix specifies a terrain height. The radar altimeter captures the profile of the overflown land and compares it with the cells in the map. The difference between where the computer's memory says the vehicle is and its actual position over the mapped area is fed through the Kalman filter, which makes the necessary corrections. After the navigation system has been updated, position correction commands are sent to the flight-control system, which flies the vehicle back onto the planned course. The navigation system then steers the vehicle toward the next predetermined map area. As the vehicle proceeds toward its intended target, the maps would be of smaller and smaller areas, since the guidance system successively

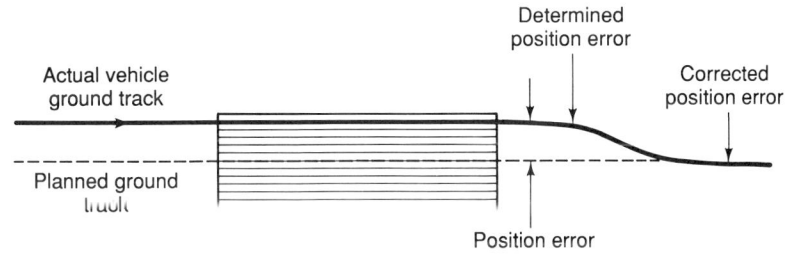

FIGURE 6-30
Course correction after a position update.

removes the errors. That is, each time a terrain correlation position fix is made, the accuracy of the Kalman filter's internal error estimate improved with a resulting decrease in the position error growth rate.

TERCOM system error sources are normally due to (1) vertical measurement errors that directly yield erroneous altitude measurements and (2) horizontal errors that induce vertical errors by causing measurements of terrain elevation to be horizontally displaced from the desired measurement location. More specifically, vertical measurement errors arise from three main sources: (1) vertical inaccuracies in the source data, (2) radar altimeter measurement errors, and (3) barometric-pressure measurement errors. Horizontal errors, on the other hand, induce vertical errors into the terrain correlation process by causing an altitude profile to be displaced horizontally from its desired location. Horizontal errors that affect the TERCOM performance are horizontal velocity and skew errors, vehicle attitude errors, and horizontal quantization (i.e., cell size).

In addition to TERCOM, other position updating navaid techniques have been either used and/or proposed, including (1) Digital Scene Matching Area Correlation (DSMAC), (2) Radiometric Area Correlation (RAC), (3) TERrain PROfile Matching (TERPROM), and (4) MICrowave RADiometer (MICRAD). The DSMAC sensor operates by viewing an area of land, digitizing the view, and comparing it with a stored scene in an onboard digital computer for precise guidance. The RAC system uses reference map data; that is, digital numbers that represent the expected radiometric temperature at each point on the ground are stored and compared reiteratively with real-time data from a scanning radiometer in order to update position, velocity, and other parameters. The radiometric system is passive and fully autonomous and can be used either day or night. TERPROM, developed by British Aerospace, is a computer-based, high-accuracy, terrain-profile navigation system using data from a radar altimeter and a digital map to determine the precise position of the vehicle. MICRAD uses a sensitive receiver that senses thermal microwave radiation emitted and reflected from terrain features. The amount of energy is determined by the emissivity of the object being observed.

As mentioned earlier in this subsection, the concept of terrain contour matching to determine vehicle geographic position was originally developed by the LTV-Electrosystems in 1958. Since that time, other firms entered into research and development of TERCOM, notable among them the E-Systems Inc., and Texas Instruments. Since the early 1960s, research and development of several programs demonstrating the feasibility of the TERCOM concept, and subsequent flight testing, were initiated. Some of these early programs were the Low Altitude COntour Matching (LACOM) developed by LTV-E in the 1963–1965 time frame, the RApid COntour Matching (RACOM) system also developed by LTV-E between 1963 and 1966, and Boeing

Aerospace Company's Subsonic Cruise Armed Decoy (SCAD) in 1968. Also, several programs were sponsored by the U.S. Air Force's SAMSO (Space and Missile Systems Organization) during 1963–1971 to investigate the application of terrain correlation techniques for ballistic missiles such as the Terminal Position Location System (TPLS), TERminal Fix (TERF), and the Terminal Sensor Overland Flight Test (TSOFT). During the late 1960s, LTV-E's original RACOM system was improved and renamed Recursive All Weather COntour Matching (RAWCOM). This new program used smaller cell sizes and a special processing algorithm which markedly improved the terrain correlation position update system accuracy. One more program using the terrain contour matching concept is the Sandia National Laboratories developed SITAN (Sandia Inertial Terrain-Aided Navigation) program. SITAN is a flight computer algorithm that combines the outputs from a terrain elevation sensor, an inertial navigator, and Digital Terrain Elevation Data (DTED) to produce a trajectory whose accuracy is much greater than that produced by the INS alone.

McDonnell Douglas Astronautics and General Dynamics have been conducting TERCOM flight tests since 1973 in support of the U.S. Navy's Submarine Launched Cruise Missile (SLCM) program. McDonnell Douglas also supported the U.S. Air Force TERCOM studies during the same period for the ALCM (Air Launched Cruise Missile), and developed the Ground Launched Cruise Missile (GLCM). The use of TERCOM has been successfully applied, with impressive results, by Boeing Aerospace in the well-known ALCM (AGM-86B) program (the General Dynamics ALCM was designated as AGM-109). Finally, the shipboard launched Tomahawk cruise missile developed by General Dynamics/McDonnell Douglas was used successfully against Iraqi air defense systems during "Operation Desert Storm."

Summarizing the discussion of this subsection, the TERCOM position fixing system is a self-contained, precision navigation–guidance system for drones, aircraft TF/TA (terrain following/terrain avoidance), and cruise missile types. TERCOM operates equally well under ECM (electronic countermeasures) conditions, day night, all-weather, and low/high altitudes.

6.2.13. Star Sightings

Star sightings, used to update an inertial navigation system, imply that a star tracker is available on board the vehicle. Stellar aided inertial navigation, or astroinertial navigation, is based on the use of accelerometer sensed quantities, the directions of which are determined by gyroscopes with updating from star fixes. The heart of an astroinertial system is the star tracker (or stellar monitor). A star tracker will normally consist of the following basic components: (1) a modified Cassegrainian telescope–optics for collecting the electromagnetic energy from the particular star(s) being tracked, (2) a photosensor to convert the electromagnetic energy into an electrical signal, (3) associated electronics for signal detection and processing, and (4) a scanning servomechanism for determining the position of the celestial body in the field-of-view (FOV) and telescope positioning. The instantaneous telescope FOV varies between 30 and 40 arcseconds. Stellar trackers are generally used for azimuth correction and position fixing. Azimuth correction is based on the known relationship of the sun and stars with respect to true north, and by appropriate sightings to selected celestial bodies derives the true heading of the carrier vehicle. Position fixing, on the other hand, employs the known relationship of the stars with a given point on the earth at any specific time. By taking two or more sightings and computing the intersection of the LOPs, an observer's position can be fixed on the earth's surface. This process, however, assumes that a perfect vertical is maintained during the sightings.

Typically, the star tracker is mounted on a gimbaled inertial platform with known azimuth and vertical reference. The telescope has 2 degrees of freedom (azimuth and elevation) and is mounted in such a way that it can view the sky through a suitable window. Furthermore, the telescope system rotates about a vertical axis relative to the inertial platform. Pointing angles to a celestial body are based on time and computed present position. The telescope system measures the error angles between the computed and actual LOS to the star so that, based on them, corrections in platform tilt angle and/or position can be made. In a locally level system, the platform tilts only until the gravity forces sensed by the accelerometer integrate to a value of velocity error that just compensates the gyroscope drift causing the tilt. Thus, it is the result of the onboard navigation computer integrating this velocity error that is corrected by the stellar information. A highly accurate star tracker would provide correction of inertial platform errors by precisely tracking an average of three stars per minute, day or night, and at any altitude. An astroinertial system provides continuous long-term precision heading, position, and attitude reference information, and can be designed to interface with other avionics, such as Air Data Computer, Flight Director, Horizontal Situation Indicator (HSI), Magnetic Compass, and other displays. Moreover, it provides automatic vehicle steering through control of the autopilot, sensor stabilization, and a data reference base. When interfaced with the airspeed and compass units, the following modes of navigation are possible: (1) astroinertial–inertial–airspeed, (2) inertial only, (3) dead reckoning, and (4) attitude heading reference. The star tracker indicates the angular error between the star line and the platform reference line. However, this indication is corrupted with noise. This additive noise is wideband compared with the bandwidth of the overall system, and can be approximated as a white noise. Thus the indicated angular error is given by

$$\theta_{ind} = \theta_s - \theta_p + w(t) \qquad (6.77)$$

where θ_s is the angle of the star line with respect to the inertial reference, θ_p is the platform angular error, and $w(t)$ is a white process noise. Note that the actual star tracker noise is wideband noise rather than a truly white noise. As in the case with other aided systems, stellar–inertial navigation systems are integrated by a real-time Kalman filter. The star LOS data is provided to the Kalman filter to update the state vector estimate of positions, velocities, platform tilts, heading, gyroscope drifts, accelerometer biases, and star tracker elevation bias. The resulting navigation system will be self-contained, independent of ground fixes, with bounded position error and calibration of gyroscope drift rates. Recently, the Northrop-Electronics Systems Division developed a strapdown astroinertial navigator, using holographic lens technology. The holographic lens replaces the gimbaled telescope assemblies with a solid state, optical, wide-angle lens star tracker. More specifically, this system utilizes a single multiexposed holographic optical element to produce three wide-angle, large-aperture lenses. The function of each of these lenses is as follows: one lens is pointed in one direction in space and focuses on a focal plane array; the second lens, superimposed on the first, collects light from another direction in space and focuses on another focal plane array; the third lens, similarly superimposed on the first two, collects light from yet another direction and focuses on a third focal plane array. The appropriate number of focal plane arrays is determined by a vehicle's altitude and mission requirements. References 9 and 15 contain a more detailed account of stellar–inertial systems.

6.2.13.1. Star Tracker

Stellar sensing and tracking devices have been used for several decades now for improving the accuracy and performance of aircraft navigation systems, missile guidance, and reference systems. The U.S. Air Force as well as private industry (e.g., Northrop-Electronics Division, Litton Guidance and Control Systems, and Kearfott-Guidance and Navigation Corporation) have been actively supporting research and development in the field of celestial navigation and guidance. As noted in Section 6.2.13, astroinertial (or stellar–inertial) navigation is an independent, autonomous, day-or-night tracking system capable of providing continuous long-term precision heading, position, attitude reference information, and accurate worldwide navigation, with gravity disturbance being the only limiting error source. Attitude readout of better than 25 arcseconds can be obtained routinely. Other system characteristics are its ability of yielding bounded position error over an extended period of time, it is passive, nonjammable, and uses star sightings as the navigation reference. A single star sighting determines azimuth information and consequently permits corrections of heading misalignment, whereas two star sightings yield position information. Thus, using a star tracker (sensor), one can determine azimuth alignment by measuring the orientation of the inertial platform with respect to the stars. It must be pointed out, however, that in establishing an LOS to a star, a known vertical reference must exist. The vertical reference is obtained by leveling the two horizontal accelerometers during alignment and operation. Precise vertical reference is provided via the star tracker. Note that for in-flight alignment of inertial systems, the vertical reference and initial position must be known.

Present-day star trackers are capable of tracking star magnitudes from -1 to $+4$, and can automatically exclude stars that are within $12.5°$ of the sun, or within $3°$ of other stars. Also, the onboard navigation computer can store permanently in its memory a stellar ephemeris of more than 100 stars. As an example of an astroinertial system, Northrop's NAS-26 astroinertial navigation system is a very accurate, self-contained navigation system, providing continuous position, velocity, heading, and attitude reference information, which bounds position error advertised to better than 305-m CEP independent of flight time. In addition, this system is designed to interface with other avionics systems, such as Doppler radar, ground mapping radar, GPS, TACAN, and altimeter. The benefits of astroinertial navigation for aircraft applications are as follows:

- Navigation
- Waypoint steering
- Ground alignment and calibration
- In-air alignment and calibration
- Continuous real-time Kalman filter mechanization, which can use the following observations (as available): (1) position, (2) radar velocity, (3) true airspeed, (4) stellar, and (5) barometric altitude

6.2.14. Doppler Radar

The Doppler radar system utilizes the Doppler frequency shift phenomenon, predicted by the Austrian physicist Christian Johann Doppler in 1842. Basically, the Doppler radar transmits radio-frequency (RF) energy to the ground and measures the frequency shift in the returned energy to determine groundspeed. Thus, Doppler radar outputs groundspeed V_G and drift angle, δ, which is the angle between the aircraft longitudinal axis and the ground velocity vector. The Doppler output V_G must be resolved into components in the inertial measurements unit's coordinate system using the drift angle and the platform azimuth gimbal angle. Doppler groundspeed error, as it will be discussed later, is characterized by bias and scale factor errors. Also, the drift angle is in error by a bias. For a constant-speed mission, the effects of bias and scale factor errors are indistinguishable. Doppler error is expressed as a percentage of groundspeed. Typically, Doppler speed bias error is on the order of 0.1% of groundspeed. The groundspeed and drift

6.2. Navaid Sensor Subsystems

angle measurements are available as inputs to the onboard navigation computer. Unlike the unaided inertial navigation mode, there is no long-term degradation associated with Doppler radar. Doppler provides good long-term speed information. Therefore, when operating in combination with an inertial system, that is, in Doppler–inertial mode, it results in a system with bounded velocity error that is smaller than that obtained from the inertial system itself. However, the Doppler radar's most significant errors are those that occur during overwater operation. Also, Doppler radar accuracy degrades during high-acceleration maneuvers, such as in a fighter aircraft environment. During this period, reliance on the inertial system is necessary.

Referring to Fig. 6-31, the Doppler principle states that a transmitted radar wave will be received by the aircraft with a Doppler frequency shift [9]

$$\Delta f = (2/\lambda) V \cos \alpha \tag{6.78}$$

where λ is the wavelength of the carrier-frequency radiation, V is the magnitude of the aircraft velocity vector relative to the ground, and α is the angle between the antenna beam and the velocity vector. The constant $(2/\lambda)$ is called the Doppler sensitivity. The Doppler frequency shift Δf is an extremely small fraction of the transmitter frequency f.

The signal from a single beam can provide only the velocity component in the direction of that beam. Complete velocity determination requires the use of at least three beams. The three beam version is called a "lambda" configuration. The radar beamwidth is shown in two orthogonal directions by $\Delta \beta$ and $\Delta \gamma$. Three noncoplanar beams supply velocity information in three orthogonal directions. The fourth beam is used for redundancy to increase both reliability and accuracy. In a two-beam system, vertical velocity is supplied by an external sensor, such as a barometer. Modern Doppler

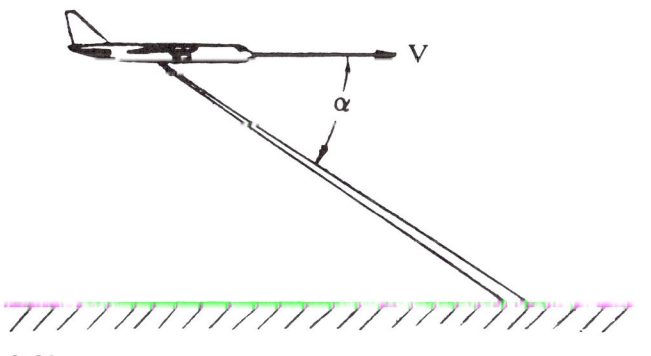

FIGURE 6-31
Basic Doppler beam configuration.

radar systems use a four-beam technique for symmetry (one antenna for each beam). Doppler frequency measurements are obtained sequentially for each of the four beam directions. To relate the beam directions, and hence the measured velocity to an earth oriented coordinate system, a vertical reference is required. Figure 6-32 illustrates the four-beam technique, as designed and developed by the Singer Company—Kearfott Division (such as the AN/ASN-128 Doppler navigation system).

For simplicity, in the Doppler–inertial mode, the INS's dynamic equations will require two additional constant bias terms, one for the Doppler velocity bias b_1 and the other for the Doppler boresight bias b_2, as follows:

$$\frac{db_1}{dt} = 0 \quad \text{and} \quad \frac{db_2}{dt} = 0$$

As stated earlier, the Doppler radar furnishes along-heading velocity and across-heading (drift) velocity. When a suitable reference system (e.g., an inertial platform) is available, the output can be transformed to total ground velocity, V_G) and heading θ. In this case, the measurements are described by the equations

$$V_G = \sqrt{V_N^2 + V_E^2} + b_1 + \eta_v \tag{6.79a}$$

$$\theta = \tan^{-1}(V_E/V_N) + \psi_z + b_2 + \eta_\theta \tag{6.79b}$$

where V_N and V_E are the north and east velocity components, respectively, and η_v and η_θ are random errors. The azimuth misalignment angle ψ_z is included to reflect the fact that the inertial system (that is misaligned from the true north) is the reference system used by the Doppler radar. Note that, in general, the various Doppler radar error sources are modeled as first-order Markov processes of the form

$$\dot{x}(t) = -\frac{1}{\tau} x(t) + w(t) \tag{6.80}$$

where τ is the correlation time and $w(t)$ is a white-noise driving process with strength Q. In the Doppler–inertial mode, Doppler updates are normally made every minute for the duration of the flight.

Typically, a Doppler radar velocity sensor consists of the following basic components: transmitter–receiver, antenna, Doppler frequency measuring device, output signal generators, and displays. The antenna is a very critical element of a Doppler radar since its characteristics determine system performance. Antenna types used in Doppler radar navigation include paraboloids, microwave lenses, and linear–planar arrays. In newer, state-of-the-art systems, the antennas are fixed to the aircraft instead of being gimbaled.

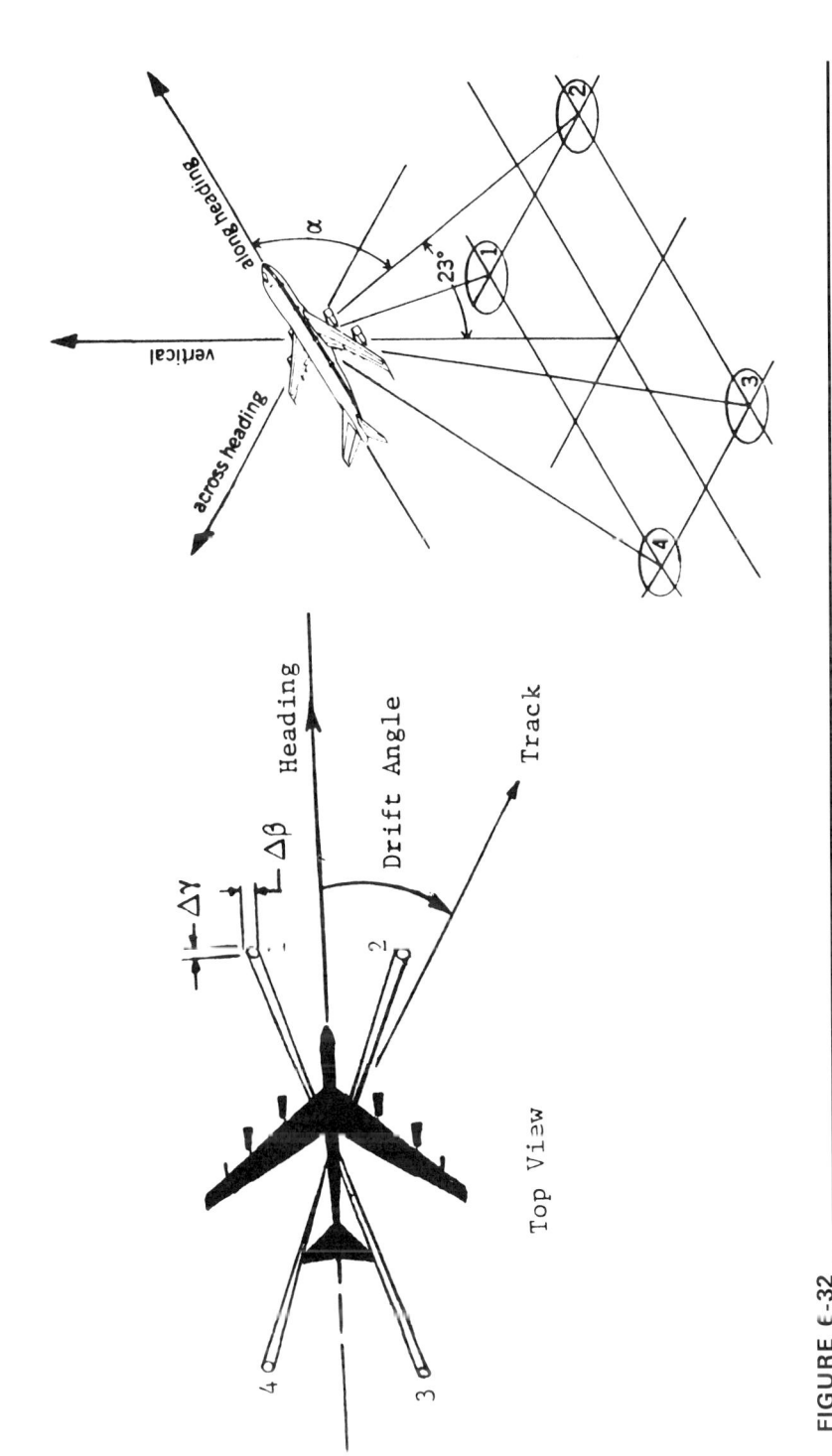

FIGURE 6-32
Typical four beam Doppler radar technique.

In this case, the resulting aircraft body coodinated signals are then transformed into the navigation coordinate system in the onboard navigation computer using the knowledge of aircraft attitude. In order to be used in an integrated Doppler/INS Kalman filter mechanization, a Doppler error model is required. Generally, Doppler radar is modeled to compensate for errors in beam direction, temperature, installation alignment, INS attitude, tracker time constant, surface motion, and beamwidth. Assuming that Doppler errors are dominated by alignment–pitch calibration errors, these errors will appear explicitly in the truth model. A simplification is possible if the aircraft is assumed to fly over land and surface motion errors are neglected. The alignment–pitch calibration errors are modeled as first-order Markov processes.

As stated earlier, Doppler radar navigation is a type of airborne radar system for determining aircraft velocity relative to the earth's surface. Moreover, Doppler radar navigation is a self-contained, dead reckoning system, using aircraft velocity and heading, where the heading information is supplied by the inertial platform. Doppler radar system errors fall under two categories: (1) velocity error from the radar system itself and (2) errors in attitude and heading supplied to the radar system by the attitude–heading system. Velocity errors inherent in the radar system are either random or bias. Random errors vary with time and are independent of the flight path (neglecting the effect of terrain). Bias errors are constant with time but are dependent on hardware and terrain. For the attitude–heading reference, three basic configurations are found in Doppler radars: (1) the antenna is roll/pitch-stabilized, (2) the antenna is only pitch-stabilized, and (3) the antenna is fixed to the body of the aircraft. The first two cases introduce errors due to hardware used for stabilization and the error due to roll in the second case. For the fixed antenna there is error in the inaccuracy of the values of pitch and roll. In all three cases, there is also the error in the value of heading used. In error analyses, the INS velocity indication is compared with the Doppler radar reference velocity indication and the difference is the measurement vector $\mathbf{z}$ ($\mathbf{z} = H\mathbf{x} + \mathbf{v}$). Thus

$$\Delta \mathbf{V} = \mathbf{V}_{\text{diff}} = \mathbf{V}_{\text{INS}} - \mathbf{V}_{\text{ref}} = \mathbf{z} \quad (6.81)$$

Letting the true vehicle velocity be $\mathbf{V}$, an expression involving the INS and reference sensor errors can be obtained. From Chapter 4, the inertial velocity indication is

$$\mathbf{V}_{\text{INS}} = \mathbf{V} + \delta \dot{\mathbf{R}} + \boldsymbol{\rho} \times \delta \mathbf{R}$$

The Doppler radar first obtains the vehicle velocity along the aircraft axes, and then this velocity vector is transformed via the inertial platform resolvers into the platform coordinate system. Thus, the comparison of the velocity can be made in the same coordinate system. Indicating the Doppler radar sensor error by $\delta \mathbf{V}_R$, the reference velocity vector is

$$\mathbf{V}_{\text{ref}} = \mathbf{V} + \delta \mathbf{V}_R \qquad \text{(in vehicle axes)}$$

or

$$\mathbf{V}_{\text{ref}} = \mathbf{V} + \delta \mathbf{V}_R - \mathbf{\psi} \times \mathbf{V} \qquad \text{(in inertial platform axes)}$$

where $\mathbf{\psi}$ is the vector of error angles between the platform axes and the computer axes. Therefore, the velocity difference $\mathbf{V}_{\text{diff}}$ is given by

$$\mathbf{V}_{\text{diff}} = \delta \dot{\mathbf{R}} + \mathbf{\rho} \times \delta \mathbf{R} - \delta \mathbf{V}_R + \mathbf{\psi} \times \mathbf{V}$$

The term $\mathbf{\psi} \times \mathbf{V}$ takes into account the error induced in resolving the Doppler velocity into the inertial platform coordinates. This is the term that causes a delay in airborne azimuth alignment, since the error in the reference velocity, a significant contributor to azimuth misalignment, is limited by this term.

For purposes of illustration, assume a NEU local vertical coordinate system. Furthermore, assume the state vector to consist of the following states: latitude, longitude, north velocity, ground east velocity, three misalignment angles (about the east, north, and vertical or azimuth axes, respectively), three gyroscope drift rates about these axes, two accelerometer biases about the east and north axes, and two Doppler radar bias errors (i.e., velocity and boresight). The nonzero elements of the observation (or measurement) matrix H are given by

$$\frac{\partial V}{\partial V_N} = V_N/V$$

$$\frac{\partial V}{\partial V_E} = V_E/V$$

$$\frac{\partial \theta}{\partial V_N} = -V_E/V$$

$$\frac{\partial \theta}{\partial V_E} = V_N/V$$

$$\frac{\partial \theta}{\partial \psi_z} = 1$$

Then the observation matrix takes the form

$$H = \begin{bmatrix} 0 & 0 & \dfrac{V_N}{V} & \dfrac{V_E}{V} & 0 & 0 & 0 & 0 & 0 & 0 & 1 & 0 \\ 0 & 0 & \dfrac{-V_E}{V} & \dfrac{V_N}{V} & 0 & 0 & 1 & 0 & 0 & 0 & 0 & 1 \end{bmatrix}$$

$$V = \sqrt{V_N^2 + V_E^2}$$

Aircraft inertial navigation systems have numerous built-in equipment redundancies in order to enhance the overall system reliability. For example, four basic navigation modes can be selected, depending, of course, on equipment availability:

Free inertial   In this mode, one uses the IMU, onboard digital computer, and barometric pressure measurements of altitude.

Doppler–inertial   This mode adds Doppler radar measurements to those used in the free inertial mode.

Doppler dead reckoning   This mode can be selected in case of IMU failure, provided Doppler radar and AHRS equipment are available.

True airspeed (TAS) dead reckoning   If both the IMU and the Doppler radar fail, AHRS and airspeed data from the CADC can be used to continue dead reckoning.

In any of these basic modes, position data from Loran, Omega, TACAN, or any other of the navaids discussed in this chapter can also be used to update the best estimate of the state vector.

6.2.15. Indicated Airspeed (IAS)

In Section 4.5.2, indicated airspeed was defined as the speed indicated by a differential-pressure indicator calibrated to the standard formulas relating airspeed to the pitot–static pressures. Specifically, because of the importance of indicated airspeed to safe flight, a direct-reading $p_t - p_s$ gauge is usually placed directly on the instrument panel of the aircraft, in addition to any indication of airspeed driven from the CADC. The direct-reading gauge is calibrated to read IAS or V_{IAS}, by the relation

$$V_{IAS} = \sqrt{\frac{2(p_t - p_s)}{\rho_0}} \tag{6.82}$$

where ρ_0 is the sea-level air density. From the ICAO standard atmosphere published values, the sea-level air density ρ_0 is specified at $t_0 = 15°C$ and has the value $\rho_0 = 1.2250 \text{ kg/m}^3$. The quantity $p_t - p_s$ is the dynamic pressure. However, it should be noted that the IAS is not the true airspeed. For this reason, when the direct-reading gauge is used, a manual correction must be applied in order to obtain the true airspeed. The determinations of minimum and maximum safe flying speed, as well as climb, descent, and cruising speeds, are based on indicated airspeed. IAS is used in autopilots to maintain the same dynamic response over a wide range of airspeeds.

In addition to the IAS discussed in Section 4.5.2, the function of the CADC was discussed. The specific air data measurements which are most useful in an aided inertial navigation system are those of altitude and altitude rate. In both cases, the actual measurement is of static pressure that must be converted to an electrical signal by means of a pressure transducer. As a result, this intermediate device creates a time delay between the input and output signals. In situations where the measurements themselves have dynamics, it is common practice to augment the system state vector by additional variables. Letting h_a and h_r be the auxiliary state variables to account for the dynamics in barometric altitude and barometric altitude rate, respectively, then the dynamic equations governing these variables are given by

$$\frac{dh_a}{dt} = -\frac{1}{\tau_a(h)}(h_a - h) \tag{6.83a}$$

$$\frac{dh_r}{dt} = -\frac{1}{\tau_r(h)}(h_r - V_z) \tag{6.83b}$$

where $\tau_a(h)$ and $\tau_r(h)$ are known functions of altitude and the physical characteristics of the specific pressure transducer and V_z is the vertical velocity. The barometric altitude and altitude rate measurements can be expressed mathematically in the form

$$y_a = h_a + k_a h_a + b_a \tag{6.84a}$$

$$y_r = h_r + k_r h_r + b_r \tag{6.84b}$$

where k_a and k_r are the scale factor errors associated with the pressure transducers and b_a and b_r are the bias errors. The terms b_a and b_r include the bias error present in the instrument itself as well as bias that is present in a datum (commonly established before takeoff). Although the instrument bias may be considered constant, random variations in sea-level temperature and pressure suggest that the bias term reflects this random variation. Therefore

$$\frac{db_a}{dt} = \xi_a \tag{6.85a}$$

$$\frac{db_r}{dt} = \xi_r \tag{6.85b}$$

where ξ_a and ξ_r are random disturbances.

6.2.16. EM-Log or Speedlog

The previous sections emphasized that, in order to reduce the Schuler errors, an external measurement of velocity is often used. For aircraft, as we have seen, radar measurements of groundspeed are used. In marine integrated navigation systems, an EM (electromagnetic)-Log or Speedlog may be used to measure waterspeed. Specifically, since the undamped Schuler loop has errors that grow unbounded with time, it is desired to estimate these errors in order to be able to compensate for them. This may be accomplished by using the EM-Log measurement. The EM-Log consists of a probe that extends in front of the vehicle (e.g., a ship) in the water. An ac magnetic field is maintained between two vertical poles on the probe. As the probe moves, the salt water moves through this field. Since salt water is an electrical conductor, an electric field is induced in it, which is picked up by two horizontal poles. The resulting electric field is proportional to the speed of the conductor (i.e., salt water) through the magnetic field, and hence, proportional to the speed of the vehicle with respect to the water. Furthermore, since the electric field is only induced perpendicular to the magnetic field, it is the waterspeed only along the desired axis of motion that is measured. The EM-Log has been found to give a very accurate measurement of waterspeed.

For a straight-steaming vehicle, the EM-Log gives the waterspeed V_w in the direction of motion as follows:

$$V_{\text{EM-Log}} = V - V_w \qquad (6.86)$$

where V is the velocity of the vehicle relative to the earth. However, it should be noted that there may be a small bias calibration error associated with the EM-Log. As in any optimally integrated INS, an EM-Log model is necessary. Before the EM-Log model is developed, we note that the EM-Log sensor model provides x, y-velocity components to the INS for velocity damping. Thus, the EM-Log velocity outputs are

$$V_{\text{EM-Log }x} = V_{\text{true-}x} + \varepsilon_b \cos \psi_T + \varepsilon_w \qquad (6.87\text{a})$$

$$V_{\text{EM-Log }y} = V_{\text{true-}y} + \varepsilon_b \sin \psi_T + \varepsilon_w \qquad (6.87\text{b})$$

where ψ_T is the vehicle's true heading, ε_b is the EM-Log bias error, and ε_w is the ocean current bias error. This velocity is compared with the INS velocity creating a ΔV_{ref}. Now let V_{um} and V_{vm} be random velocity errors in the EM-Log data, whereby these errors include the effects of ocean currents in addition to instrument errors. Therefore, the EM-Log can be modeled as a stochastic differential equation (first-order Markov) in the form

$$\frac{dV_{\text{um}}}{dt} = -\beta_{V_{\text{um}}} V_{\text{um}} + \sqrt{2\sigma^2 \beta_{V_{\text{um}}}}\, \eta \qquad (6.88a)$$

$$\frac{dV_{\text{vm}}}{dt} = -\beta_{V_{\text{vm}}} V_{\text{vm}} + \sqrt{2\sigma^2 \beta_{V_{\text{vm}}}}\, \eta \qquad (6.88b)$$

where η is a white-noise forcing function and $\beta = 1/\tau$ ($\tau =$ correlation time), and the σ values are the a priori covariances associated with the EM-Log instrument noise in the along-track and cross-track directions, respectively.

6.3. THE UPDATING PROCESS

In an inertial navigation system, the onboard navigation computer integrates the accelerometer outputs to determine vehicle position and velocity outputs, by utilizing initial vehicle position and velocity at the time of IMU turn-on. However, the position and velocity computations are in error due mainly to accelerometer and gyroscope imperfections and initial position and velocity uncertainties. Therefore, in designing inertial navigation systems, it is important to determine the influence of sensor and initialization errors on system position and velocity accuracy. The best method of combining position fix information and inertial navigation data in real time is through a centralized Kalman filter mechanization. In essence, the Kalman filter is a statistical weighting and error propagation device. When an external position or velocity measurement is made, the filter updates the inertial navigation system's position and velocity on the basis of statistically weighting the estimated accuracy of the external measurement against the estimated position or velocity accuracy of the system itself. Between updates, the filter propagates in real time the best estimate of the system's position, velocity, attitude, and sensor errors as well as the statistical uncertainty (i.e., covariance matrix) associated with these errors. Moreover, the Kalman filter provides a fully rational approach to the design of aided navigation systems, not only because it supplies an optimal estimate of navigation system errors that are directly observable, such as position and velocity, but because it also provides estimates of all modeled navigation sensor error sources which have significant correlation times.

For example, in the mechanization mentioned above, the navigation computer computes the aircraft's position and compares it with the aircraft position as supplied by the updating system (e.g., radar, VOR/DME, GPS). As a result, the difference between the computed and measured positions is an error signal. This error signal is multiplied by a set of filter gains and added in at various modes or states in the inertial computation chain to correct or reset those computational states. Furthermore, the Kalman filter computes the gains as a function of (1) the relative accuracies of the measured and computed positions, (2) the relative geometry between the measured and computed position (e.g., range measurements vs. latitude–longitude), and (3) the correlation between the measurements and the state being updated, that is, velocity states being updated with position measurements. In general, the Kalman gains vary with time and with aircraft dynamics.

At this point, let us look more closely at the position and velocity update modes.

Position update mode   In this mode, updates to the system state are made by feedback of measurements of, say, range, range rate, azimuth, and elevation angles, to stored checkpoints (or waypoints). These measurements are used by the Kalman filter to compute and apply corrections to the system state being propagated by the navigation equations. The effect is to minimize the free-inertial growth of position, velocity, and attitude errors in all channels. In addition, as long as position updates are available, no position errors accumulate. There are two reasons for this: (1) the position update measurements minimize attitude error, thereby minimizing position error buildup between updates; and (2) the position errors are removed at the time of a position update.

Velocity update mode   In the velocity update mode, the velocity measurement of, say, range rate is used by the Kalman filter to update the range rate predicted by the navigation equations. As in the position update mode, the effect is to prevent buildup of position errors that would otherwise be seen in the free-inertial mode. The velocity update alone cannot remove position errors; it can only minimize the buildup of such errors. Initial position errors will remain with the system state after velocity updates.

In order to address the updating problem, a simulation program must be written that evaluates the effectiveness of external measurements in the aiding of an inertial navigation system. Specifically, the inertial system designer must construct a simulation program that will update periodically the system error covariance matrix on the basis of external position and/or velocity measurements. The effect of incorporating external measurement information via the Kalman filter is to reduce, in an optimal manner, the statistical covariance uncertainty associated with system errors. Initial errors (including specified navigation sensor errors), a flight mission profile, and an update sequence are input to the program. A propagation routine propagates input errors into system position, velocity, and attitude errors on the basis of linear acceleration and platform torquing rate information (in the case of gimbaled platforms). The errors modeled in the simulation program are treated as

6.3. The Updating Process

initially uncorrelated, normally distributed random variables having zero means. An initial covariance matrix position, velocity, attitude, and modeled sensor errors is propagated along a reference trajectory, which can be printed out when desired. This covariance matrix is updated with a Kalman filter whenever position or velocity observations are available. These updates can be made at regular intervals over any part of the flight. In order to keep track of the relative accuracies of the measured and computed positions, the Kalman filter includes error models of both the updating device and the inertial system. Figure 6-33 shows the error behavior in an inertial navigation system using external information sources. As stated in Section 6.1, the effect of the position fixes is to reset the position error in the inertial system to the same level of accuracy inherent in the position fixing technique. In between position fixes, the inertial error grows at a rate equal to the velocity error in the inertial system. This results in the well-known sawtoothed position error curve as indicated in Fig. 6-33.

Kalman filtering techniques also generate an improvement in the inertial system's velocity accuracy by virtue of the strong correlation between position error (i.e., the updated quantity) and velocity error; the former is essentially the direct integral of the latter. Furthermore, one can estimate the velocity accuracy of the updated system by dividing the position update accuracy by the available smoothing time. (Note that the available smoothing time is defined as the time it takes an acceleration error to create a position error equivalent to the position updating error.)

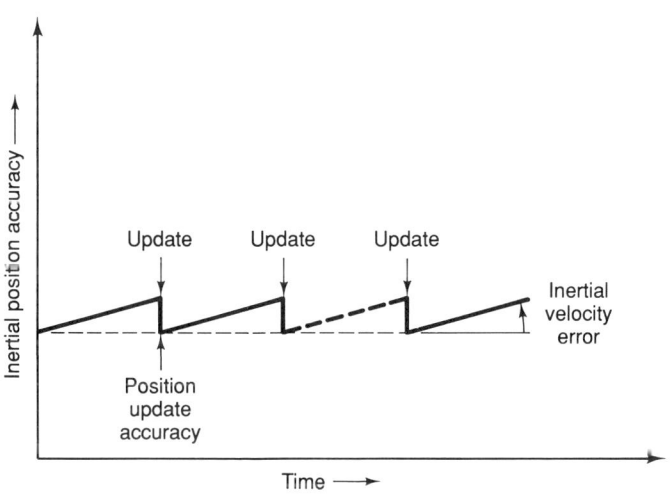

FIGURE 6-33

Navigation error behavior for a typical INS using updates from navaids.

The typical filter states mechanized in an inertial navigation system are (1) position, (2) velocity, and (3) attitude. Although not mechanized as true states, the effects of instrument error sources such as bias noise, scale factor errors, and instrument nonorthogonalities are modeled as noise input driving terms. Random errors utilized by the program may be essentially constant (unknown biases), slowly varying (correlated effects or "colored" noise), or rapidly varying ("white" noise) relative to a given mission time. For the reader's convenience, the discrete-time Kalman filter equations needed for the measurement update and time update step are given below.

System model:

$$\mathbf{x}(k) = \Phi(k, k-1)\mathbf{x}(k-1) + \mathbf{w}(k), \quad \mathbf{w}(k) \sim N(0, \mathbf{Q}(k)), \quad x(0) = x_0 \quad (6.89)$$

$$k = 1, 2, \ldots \qquad \qquad \mathcal{E}\{x_0\} = \mu_x$$

Measurement:

$$\mathbf{z}(k) = H(k)\mathbf{x}(k) + \mathbf{v}(k), \quad \mathbf{v}(k) \sim N(0, \mathbf{R}(k)) \quad (6.90)$$

$$k = 1, 2, \ldots$$

Initial conditions:

$$\mathbf{x}(0) \sim N(\mu_x, \mathbf{P}(0))$$

[Gaussianly distributed, with known mean μ_x and covariance matrix $\mathbf{P}(0)$].

Measurement update:

Filter gain: $\quad \mathbf{K}(k) = \mathbf{P}(k/k-1)\mathbf{H}^T(k)[\mathbf{H}(k)\mathbf{P}(k/k-1)\mathbf{H}^T(k)$

$$+ \mathbf{R}(k)]^{-1} \quad (6.91)$$

Filter: $\quad \hat{\mathbf{x}}(k/k) = \hat{\mathbf{x}}(k/k-1) + \mathbf{K}(k)[\mathbf{z}(k) - \mathbf{H}(k)\hat{\mathbf{x}}(k/k-1)] \quad (6.92)$

Covariance: $\quad \mathbf{P}(k|k) = \mathbf{P}(k|k-1) - \mathbf{K}(k)\mathbf{H}(k)\mathbf{P}(k|k-1) \quad (6.93)$

Time update:

Filter: $\quad \hat{\mathbf{x}}(k/k-1) = \Phi(k, k-1)\hat{\mathbf{x}}(k-1/k-1) \quad (6.94)$

$$\mathbf{x}(0|0) = \mu_x$$

Covariance: $\quad \mathbf{P}(k/k-1) = \Phi(k, k-1)\mathbf{P}(k-1/k-1)\Phi^T(k, k-1)$

$$+ \mathbf{Q}(k-1) \quad (6.95)$$

$$k = 1, 2, \ldots$$

Combining both the measurement and time update steps for the calculation of $\mathbf{P}(k+1/k)$ yields the following single expression for the discrete-time

Riccati equation:

$$\begin{aligned}\mathbf{P}(k+1|k) &= \mathbf{\Phi}(k+1,k)\{\mathbf{P}(k|k-1) - \mathbf{P}(k|k-1)\mathbf{H}^T(k)+\mathbf{R}]^{-1}\mathbf{H}(k) \\ &\quad \times \mathbf{P}(k|k-1)\}\mathbf{\Phi}^T(k+1,k) + \mathbf{Q}(k) \\ &= \mathbf{\Phi}(k+1,k)\{[\mathbf{I}-\mathbf{K}(k)\mathbf{H}(k)]\mathbf{P}(k|k-1)[\mathbf{I}-\mathbf{K}(k)\mathbf{H}(k)]^T \\ &\quad + \mathbf{K}(k)\mathbf{R}\mathbf{K}^T(k)\}\mathbf{\Phi}^T(k+1,k) + \mathbf{Q}(k). \end{aligned} \qquad (6.96)$$

This single equation can be used to provide $\mathbf{P}(k/k-1)$ for all k's of interest.

Current research in optimal control and estimation theory indicates that it may be advantageous to use a federated Kalman filter architecture, which is applicable to decentralized sensor systems with parallel processing capabilities. This architecture provides significant benefits for real-time multisensor applications, such as integrated inertial navigation systems. In a federated architecture, each subsystem operates independently and has its own link with the central computer. Furthermore, distributed architecture subsystems have a common link, which distributes information according to defined priorities, destinations, and processing procedures.

REFERENCES

1. Bird, M. W., Wierenga, R. D., and TenCate, J. V.: "Kalman Filter Design and Performance for an Operational F-4 Loran Inertial Weapon Delivery System," *Proceedings of the NAECON'76* (Dayton, Ohio, May 18–20, 1976), pp. 331–340.
2. Blackwell, E. G.: "Overview of Differential GPS Methods," papers published in *Navigation*, Vol. III, 1986, pp. 89–100.
3. Bobick, J. C., and Bryson, A. E., Jr.: "Updating Inertial Navigation Systems with VOR/DME Information," *AIAA Journal*, 11(10), 1377–1384, October 1973.
4. Brown, A. K.: "NAVSTAR/GPS Operational Principles and Applications," Short Course, Wright-Patterson AFB, Ohio, May 3–5, 1989.
5. Cox, D. B., Jr.: "Integration of GPS with Inertial Navigation Systems," *Navigation: Journal of the Institute of Navigation*, 25(2), 236–245, 1978.
6. Fried, W. R.: "Principles and Simulation of JTIDS Relative Navigation," *IEEE Transactions on Aerospace and Electronic Systems*, **AES-14**(1), 76–84, January 1978.
7. Fried, W. R.: "A Comparative Performance Analysis of Modern Ground-Based, Air-Based, and Satellite-Based Radio Navigation Systems," *Navigation: Journal of the Institute of Navigation*, 24(1), 48–58, 1977.
8. Jorgensen, P. S.: "Navstar/Global Positioning System 18-Satellite Constellations," *Navigation: Journal of the Institute of Navigation*, 27(2), 89–100, 1980.
9. Kayton, M., and Fried, W. R. (eds): *Avionics Navigation Systems*, Wiley, New York, 1969.
10. Kerr, T. H.: "Decentralized Filtering and Redundancy Management for Multisensor Navigation," *IEEE Transactions on Aerospace and Electronic Systems*, **AES-23**(1), 83–119, January 1987.
11. Lee, Y. C.: "Ground Monitoring Schemes for the GPS Integrity Channel," paper presented at the 6th National Technical Meeting of the Institute of Navigation, San Mateo, Calif., January 23–27, 1989.

12. Maybeck, P. S.: *Stochastic Models, Estimation, and Control*, Vol. I, Academic Press, New York, 1979.
13. Milliken, R. J., and Zoller, C. J.: "Principle of Operation of NAVSTAR and System Characteristics," *Navigation: Journal of the Institute of Navigation*, **25**(2), 95–106, 1978.
14. Schlee, F. H., Toda, N. F., Islam, M. A., and Standish, C. J.: "Use of an External Cascaded Kalman Filter to Improve the Performance of a Global Positioning System (GPS) Inertial Navigator," *Proceedings of the NAECON'88* (Dayton, Ohio, May 23–27, 1988), pp. 142–147.
15. Siouris, G. M.: "Navigation, Inertial," *Encyclopedia of Physical Science and Technology*, Vol. 8, Academic Press, San Diego, 1987, pp. 668–717.
16. Sturza, M. A., Brown, A. K., and Kemp, J. C.: "GPS/AHRS: A Synergystic Mix," *Proceedings of the NAECON'84*, Vol. I (Dayton, Ohio, May 21–25, 1984), pp. 339–348.
17. Teasley, S. P., Hoover, W. M., and Johnson, C. R.: "Differential GPS Navigation," *Proceedings of the IEEE PLANS* (Position, Location, and Navigation Symposium), December 1980, pp. 9–16.
18. Thomin, D. W.: "Global Positioning System—A Modification to the Baseline Satellite Constellation for Improved Geometric Performance," M. S. thesis, U.S. Air Force Institute of Technology, December 1984.
19. Upadhyay, T. N., and Damoulakis, J. N.: "A Real-Time Sequential Filtering Algorithm for GPS Low-Dynamics Navigation System," *Proceedings of the NAECON'79* (May 15–17, 1979, Dayton, Ohio), pp. 739–749.
20. Upadhyay, T. N., et al.: "Benefits of Integrating GPS and Inertial Navigation Systems," The Institute of Navigation: *Proceedings of the Thirty-Eighth Annual Meeting* (Washington, D.C., 1982), pp. 120–132.
21. Van Dierendonck, A. J., Russell, S. S., Kopitzke, E. R., and Birnbaum, M.: "The GPS Navigation Message," *Navigation: Journal of the Institute of Navigation*, **25**(3), 147–165, 1978.
22. Van Etten, J. P.: "Navigation Systems: Fundamentals of Low and Very-Low Frequency Hyperbolic Techniques," *Electrical Communication*, **45**(2), 192–212, 1970.
23. Vass, E. R.: "Omega Navigation System: Present Status and Plans 1977–1980," *Navigation: Journal of the Institute of Navigation*, **25**(1), 40–48, 1978.
24. Widnall, W. S.: "Stability of Alternate Designs for Rate Aiding of Non-Coherent Mode of a GPS Receiver," Contract F3361577C1004, Intermetrics, Inc., Cambridge, Massachusetts, September 25, 1978.
25. Widnall, W. S., and Gobbini, G. F.: "Stability of the Decentralized Estimation in the JTIDS Relative Navigation," *IEEE Transactions on Aerospace and Electronic Systems*, **AES-19**(2), 240–249, March 1983.
26. Widnall, W. S., and Kelley, J. F.: "JTIDS Relative Navigation with Measurement Sharing: Design and Performance," *Proceedings of the IEEE PLANS* (Position, Location, and Navigation Symposium), San Diego, Calif., November 1984.
27. U.S. Coast Guard Academy: "Loran-C Engineering Course," New London, Conn., 1988.
28. U.S. Department of Transportation, FAA, Systems Research and Development Service: "Loran-C, Omega, and Differential Omega Applied to Civil Air Navigation Requirement of CONUS, Alaska, and Off-Shore," Vol. II: *Analysis*, Washington, D.C., July 28, 1978.
29. U.S. Department of Transportation, FAA, Office of Systems Engineering Management: "Evaluation of Various Navigation System Concepts," Washington, D.C., July 1981.

7
Steering, Special Navigation Systems, and Modern Avionics Systems

7.1. INTRODUCTION

Present-day commercial and military navigation systems are designed to facilitate functional interfacing with existing and future avionic systems such as flight directors, autopilots, air data systems, weather radars, and other navigation systems. In addition, they have comprehensive input/output capabilities, both digital and analog. Advances in inertial sensors, displays, and VLSI/VHSIC technologies made possible the use of navigation systems to be designed for commercial aviation aircraft to use all-digital inertial reference systems (IRS). The IRS interfaces with a typical transport aircraft flight management system. The primary outputs from the system are linear accelerations, angular rates, pitch/roll attitude, and north–east–vertical velocity data used for inputs to a transport aircraft flight-control system. All-digital laser IRSs are now flying aboard commercial aviation aircraft such as the Boeing 757/767. In modern avionics systems, for instance, the interfacing of navaids with an inertial navigation system is accomplished by using a Kalman filter based on an error model of the system elements that accurately characterizes the significant system errors for the specified dynamic environment in which performance capabilities must be achieved. Furthermore, modern avionics systems provide flexibility and growth potential for future requirements. In addition, the avionics system is partitioned to take advantage of autonomous subsystem capabilities and fault detection. For example, fighter aircraft avionic systems support all-weather air-to-air and air-to-surface combat missions. All-weather air-to-surface weapon delivery, for instance, can be achieved by providing a precision navigation capability and a precision targeting capability using high-resolution synthetic aperture radar (SAR). Also, in weapon delivery and aircraft flight-control computations, the following inertially derived data are required: (1) linear acceleration, (2) velocity, (3) position, (4) attitude, and (5) attitude rate. The inertial instruments sense directly the linear acceleration, attitude, and attitude rate; consequently, the inertially sensed data is used to compute velocity and position. In a gimbaled inertial system, for example, the vehicle attitude is obtained from the gimbal angle measurements, providing the vehicle pitch, roll, and yaw in relation to the geodetic local-level, north-pointing (or other) reference frame. Vehicle attitude rate data can be obtained from gyroscopes that are part of the flight-control system. Formatted displays and master mode design enhance one-switch selection of optimized text and video in easily discernible displays. Recent multifunction display sets represent an advanced concept in fighter aircraft controls and displays. In essence, this set provides the main human–machine interface between all of the aircraft weapons and sensors and allows integrated or multifunction operations to be accommodated specifically within a very small cockpit. Current-generation multifunction display sets provide high contrast images of radar, electrooptical (EO) weapon, and TV video information on two independent multifunction displays simultaneously, each with a different symbology overlay. In addition, dual video recorder outputs are provided from the set to allow the operator to record the display scene that is being presented on either multifunction display. Future avionics systems will impose increasingly stringent requirements on effectiveness and life-cycle costs of navigation, avionic, and weapon systems.

In order to meet these requirements, new systems must incorporate advanced integration concepts with greater complexity and sophistication. New approaches based on modern system, control, and estimation theory need to be applied effectively to achieve both improved effectiveness and life-cycle cost (LCC). As mentioned above, new developments in inertial sensor technology, estimation theory, and guidance computer technology will be used to the fullest extent in achieving future requirements. In the implementation of multisensor integrated navigation systems, models must be developed for establishing INS requirements that ensure maximum capability with external navaids such as GPS, Omega, JTIDS, and Loran. The ultimate integration of complete avionics systems under a unifying architecture and information structure will enhance system performance, increase the reliability, and eliminate the costs associated with separately designed systems.

7.2. STEERING

Basically, steering involves determination of the change in vehicle motion required to obtain a desired result. The steering program provides steering commands and related information to the displaying instruments, and steering errors to the automatic flight-control system (AFCS). The aircraft can be flown manually using the information displayed on the instruments or automatically by engaging the AFCS. In the case of an automatic steering program, advantages accrue through a fully automated interface with the autopilot that reduces the pilot and/or navigator workload, offers opportunities for fuel savings, and increases the probability of obtaining the desired position fixes. The steering software converts the navigation equations output from inertially referenced quantities into earth-referenced and course-referenced quantities. The initial operation is to transfer vehicle velocity and position of this inertially referenced coordinate system into a local-vertical system. The second calculation is to compute the desired local-vertical parameters for a defined course. Therefore, the difference between the actual and desired values are used to compute the display values for the control-and-display unit (CDU). These display values include (1) distance-to-go, (2) time-to-go, (3) cross-track position error, and (4) track-angle error. (Note that time-of-arrival control is sometimes implemented by automatically computing the true airspeed required to achieve a user-specified time-of-arrival at a selected waypoint.) The FROM and TO displays on the CDU (control-and-display unit) identify which prestored endpoints are being used.

The objectives of the steering task are as follows:

- Provide autopilot steering control parameters
- Develop intercept steering algorithm(s) that account for anticipated coordinated turns in subsequent course changes
- Modify the CDU displays in the steering software

It should be pointed out that the steering functions are enabled only after the system modes NAV AUTO and NAV MAN are selected. For this reason, navigation is not operative until the alignment is complete. Thus, this procedure and/or program will have the capability to take the aircraft through preselected flight paths. Steering endpoints or waypoints can be specified by a destination number, and can be loaded into the CDU. As mentioned earlier, the pilot enters present position in terms of latitude/longitude and destination into the CDU prior to departure. Figure 7-1 illustrates the INS units used in steering.

The CDU illustrated in Fig. 7-1b is of an earlier design. CDUs of modern design incorporate either LED (light-emitting diode) or CRT (cathode-ray-tube) alphanumeric displays, thereby providing several advanced features to the flight crew. Leading companies in the area of CDU design and manufacturing are Litton Aero Products, Honeywell, and Collins Avionics Division (Rockwell International). The Litton design employs an LED matrix display, which displays five lines and 16 characters per line of data entered via data-entry pushbuttons. The top line gives the title heading while

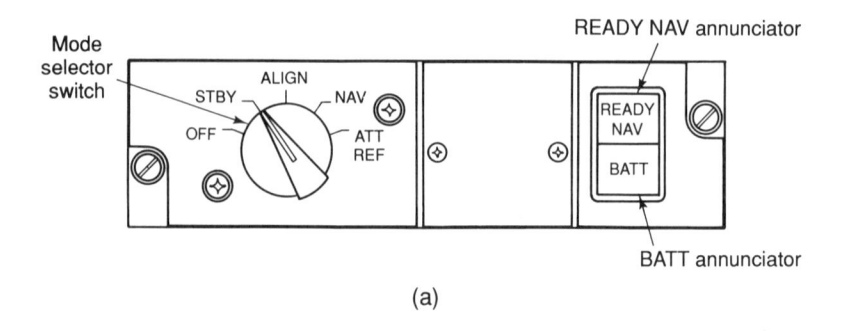

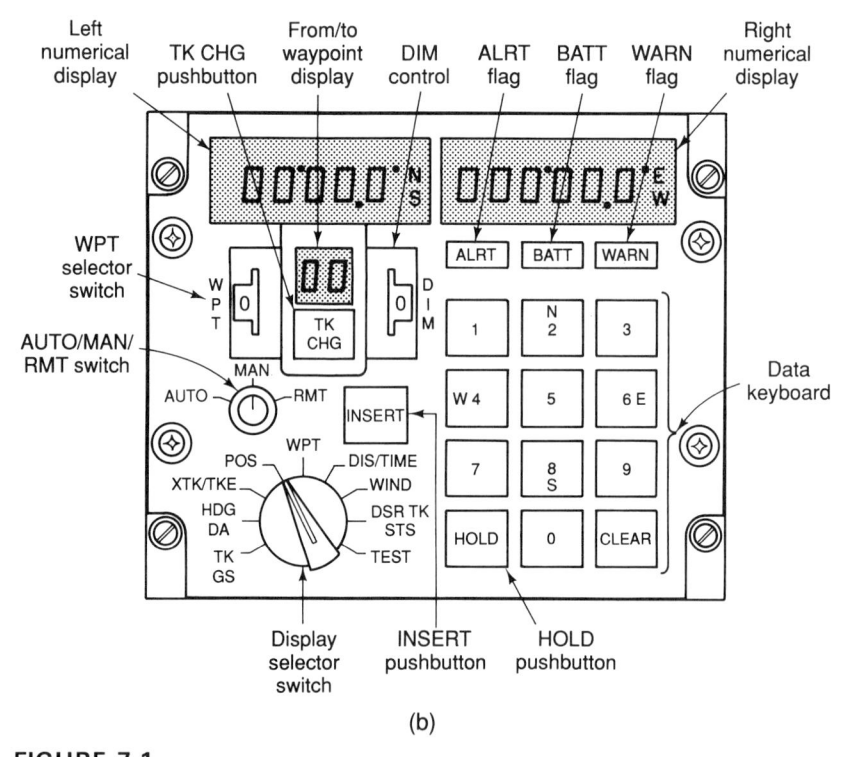

FIGURE 7-1

Inertial navigation system functional units: (a) mode selector unit (MSU); (b) control-and-display unit (CDU).

the remaining four lines are data entered or INS output data. Figure 7-2 depicts the Litton CDU design.

The Honeywell CDU design features a CRT data display consisting of a 26-element alphanumeric keyboard, control keys, and system-operation annunciators. An integral mode-select switch eliminates the need and space

7.2. Steering

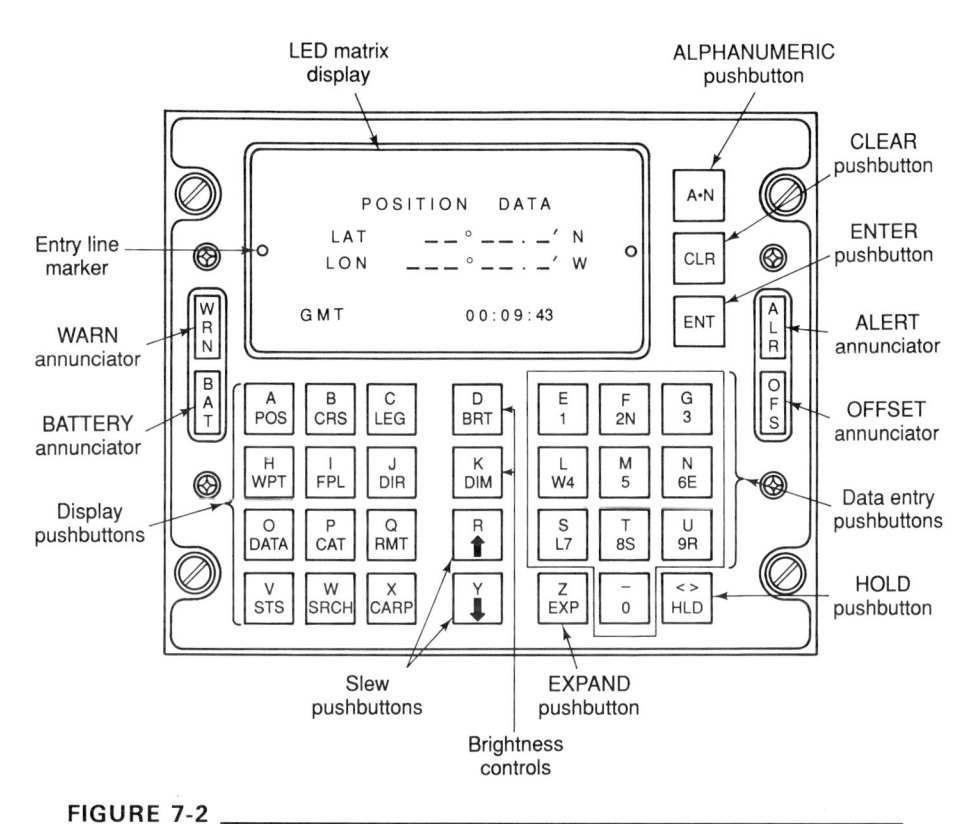

FIGURE 7-2
LED/CRT control-and-display unit.

required for a separate mode selector unit (MSU). The high-contrast CRT display, readable in sunlight, depicts alphanumeric flight data in rows of 14 characters and standard notations. Moreover, the display data are grouped into pages that can be called by depressing a given function key. This self-prompting interactive display guides the flight crew through data entry and selection of available features and modes, providing a logical and versatile operation. A message key-light signals the presence of a message from the INS. As in the Honeywell design, the Collins CDU design also employs a CRT display. In this design, a scratchpad line lets the operator touch-in data and correct it without losing the currently displayed data. All systems offer automatic and direct great circle flight planning when departure point and destination are entered into the CDU. The CDU can display the following navigation data when the navigation (NAV) mode is initiated: (1) true heading (T-HDG), (2) magnetic heading (M-HDG), (3) true track (T-TK), (4) magnetic track angle (M-TK), (5) track angle error (TKE), (6) true desired track (T-DTK), (7) cross-track distance (XTK), (8) groundspeed (GS), (9) wind direction and velocity (WD), (10) true airspeed (TAS), (11) drift angle (DA), (12) present position (e.g., Lat/Lon updated and nonupdated), (13) aircraft altitude (ALT), (14) distance and time to waypoints on the flight plan, (15) Greenwich Mean Time (GMT), (16) position of selected waypoints, and (17) present position in terms of radial–slant range to closest TACAN station when available. Other information, depending on mission requirements, is also possible. It should be noted that the wind direction and speed are calculated by the navigation computer from known heading, true airspeed, and aircraft velocity in relation to a local vertical, free or wander–azimuth frame, and from present direction cosine values. The navigation function supports "en route" navigation and "terminal approach" navigation. These navigation functions are defined as follows:

En route navigation   The en route navigation function is required to operate in course "fly-to" mode. Course mode navigates between two points, either directly or radially. If the radial option is selected, the aircraft will be steered along an operator "fly-in" radial. Fly-to mode navigates between the aircraft present position and a selected waypoint. Fly-to guidance will be performed only when course mode is not available.

Terminal navigation   The terminal navigation function provides relative position for autonomous landing guidance to a selected touchdown point. The terminal navigation function is required to operate in the following modes: (1) radar, (2) dead reckoning, and (3) forward-looking infrared (FLIR).

Waypoints are geographic points, given in terms of latitude and longitude, that may be defined as points along the mission profile or trajectory near which the aircraft is commanded to fly. The waypoints (sometimes also referred to as "action points") are used to define each change of state or flight mode.

A waypoint is commonly defined as a geographic position in terms of latitude/longitude or bearing/distance from a known location defined by a latitude/longitude. For waypoints described by latitude/longitude, the resolution capability for both values must be 0.01 minute or better, whereas for waypoints described as a bearing and distance from a latitude/longitude, the resolution capability of the system must be $1.0°$ and 0.1852 km or better, respectively.

Each turn, change of speed, change of altitude, or any other mission-dependent parameter is performed by changing the appropriate command at the desired waypoint. These points may be identified visually, via electrooptical sensors, or via radar to provide the pilot with assurance relative to navigation accuracy. Navigating from, say, waypoint 1 to waypoint 2 can be visualized by referring to Fig. 7-3.

7.2. Steering

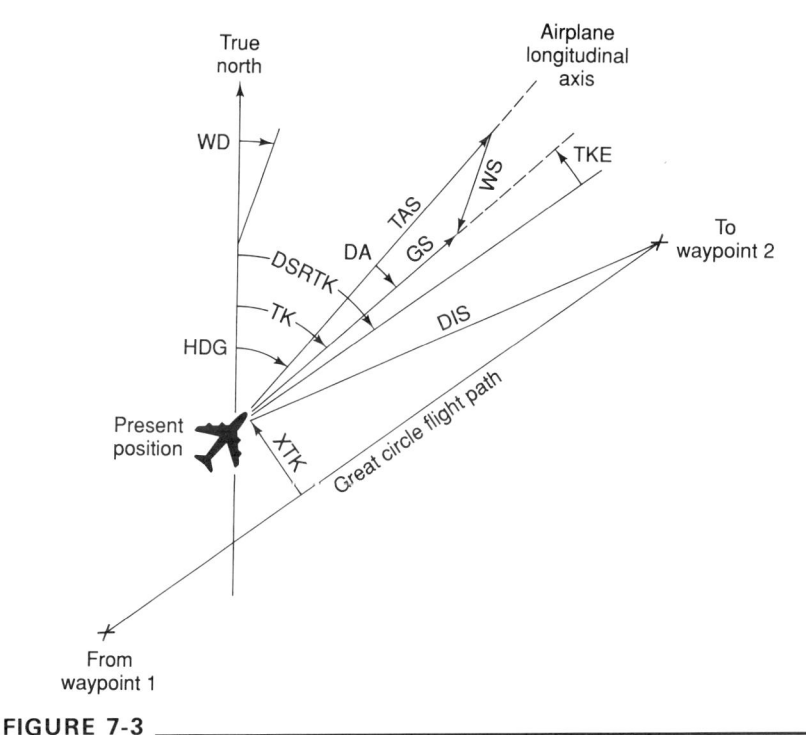

FIGURE 7-3

Definition of aircraft present position and waypoints.

When the data selector switch on the CDU is set to POS, the latitude and longitude of the aircraft's present position are displayed. The computation of latitude and longitude is accomplished by using the direction cosine values that define the present position vector. Furthermore, when the data selector is set to POS, the pilot may also load present position using the keyboard. Depression of the first key in the loading process locks the display to the keyboard. That is, the display will reflect the keys pressed by the operator and will be updated as successive keys are pressed. The data will not be provided to the onboard navigation computer until the ENTER key is pressed. After the pilot enters the data and initiates the initial track leg (note that this can be done on the ground or during flight), the INS will automatically navigate from waypoint to waypoint and continuously provide navigation data for the aircraft's autopilot or flight director control, and attitude data for the aircraft's instrumentation. All these data are available for cockpit display. In addition, the pilot can accomplish waypoint bypassing, and waypoint position change, as well as several methods of track change to allow for flight-plan changes. Flight instructions can be commanded or changed only at waypoints. When the data selector switch is set to WPT (waypoint), the same type of operation is commanded. This time, the navigation computer supplies latitude and longitude of the waypoint selected. Loading of waypoints latitude and longitude is accomplished essentially the same as loading of present position.

Coordinates of more than 25 waypoints can be entered into the INS navigation computer either during the self-alignment sequence while the aircraft is on the ground or after takeoff; that is, more than 25 destinations (latitude and longitude) can be stored in the onboard navigation computer. The latitude and longitude of the highest stored destination number will appear in the display in degrees, minutes, and tenths of minutes. It should be pointed out, however, that waypoint coordinates cannot be entered until after the present position coordinates have been entered. Once entered, waypoints will remain in the INS until new waypoints are entered. (In some INSs all waypoints are automatically cleared when the mode selector switch is set to the OFF position.) Additional waypoint coordinates can be entered sequentially into subsequent waypoint storage locations. Figure 7-4 shows navigation along several waypoints.

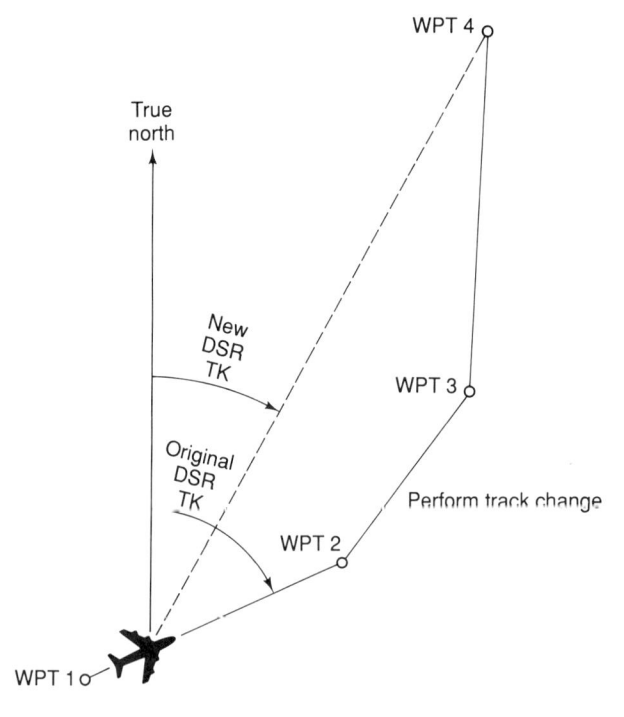

FIGURE 7-4
Multiple waypoint navigation.

In military applications, such as strategic missiles, more than 250 waypoints can be stored in the onboard navigation computer. In the automatic leg change mode, the track leg change occurs prior to arrival at the waypoint in order to permit a smooth turn onto the new track. Consequently, the aircraft speed and new track angle will determine the time when the track leg change occurs. Therefore, the aircraft will be steered along the extension of the previous desired track. This situation is evident in cruise missile-type applications, where the vehicle is commanded to fly a large number of legs stored on the onboard computer, and determined in advance by the mission planner. For instance, in flying two legs (AB and BC) of a great circle, that is, from WP_i to WP_{i+2}, a turn angle can be executed whereby the vehicle flies close to the intermediate waypoint WP_{i+1} (see Fig. 7-5).

The bypassing of a waypoint can be intentional as in the case of a commercial aircraft where the pilot can change course, or if the vehicle is having a position error such as in a cruise missile. In this case, the turn position should occur before waypoint WP_{i+1} and at a distance equal to the tangent of one-half the turn angle multiplied by the turn radius. Mathematically, the distance at the point of turn initiation is $AB - R * \tan(\Delta\psi/2)$. Note that $R * (\Delta\psi/2)$ can only be determined after $\Delta\psi$ is determined.

The following steering modes are commonly used from the preceding discussion:

Direct steering to waypoint   Direct steering in the great circle guidance mode results in the aircraft flying a great circle path directly from its current position (e.g., waypoint 1) to its destination (e.g., waypoint 2).

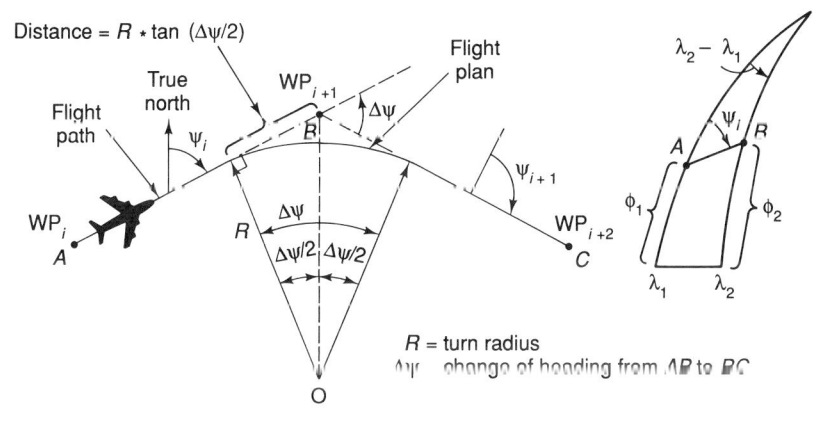

FIGURE 7-5
Geometry of waypoint bypassing.

Centerline recovery steering Centerline recovery steering in the great circle guidance mode results in the aircraft achieving and maintaining a desired trajectory, defined by the great circle path between its initial position and its destination.

Turn-short steering In certain cases, it is desired to perform an anticipatory turn, instead of overflying a waypoint. The turn-short or waypoint bypassing geometry is illustrated in Fig. 7-5.

In transport applications, especially in military applications with more than two crew members, an automatic flight-control system (AFCS) is used to reduce the crew workload and aid in precision navigation tasks. Modern transport AFCSs consist of flight controls and enhanced guidance and flight director system processing contained within the onboard navigation computer (or mission computer). This system uses inputs from the inertial navigation system, radar, and other navaids, to aid in precision navigation. Basically, the AFCS consists of a flight director computer and an autopilot. The flight-control modes are (1) pitch hold, (2) bank hold, (3) heading hold, and (4) en route R-NAV.

The autopilot system maintains the aircraft at a predetermined attitude, heading, and altitude with minimum pilot effort, and automatically coordinates turns and provides for roll and pitch adjustments. The autopilot system normally functions in two modes: (1) engaged and (2) coupled. In the engaged mode, the operation of the elevators, ailerons, and rudder are controlled by three servos. The yaw damper system is connected to the rudder and can be independently engaged. Furthermore, when the autopilot is engaged, the yaw damper engages automatically, and the elevator trim tabs are controlled automatically to compensate for out-of-trim conditions in pitch. When the autopilot is coupled, lateral and vertical steering commands are accepted from the flight director computer. In this case, aerodynamic surfaces are positioned in response to computed steering commands during en route (i.e., nonterminal), heading, ILS approach, and altitude hold modes. The autopilot cannot be coupled to the flight director unless a lateral mode has been selected.

The lateral aircraft motion resulting from the deflection of the rudder and ailerons is measured by the navigation sensors. This information (i.e., position, velocity, attitude, attitude rates) is sent to the navigation computer, flight director computers, and yaw damper. The lateral navigation measurements used in the navigation computer during autopilot operation are the aircraft latitude, longitude, altitude, east velocity, north velocity, true airspeed (TAS), and roll angle. Moreover, the aircraft groundspeed V_g and ground track angle (GTA), η, are computed from the east and north velocity components by applying the transformation from rectangular to polar coordinates. Assuming no wind and turns are coordinated (i.e., zero side slip),

7.2. Steering

the lateral aircraft dynamics are given by the following first-order, nonlinear differential equations:

$$\frac{d\eta}{dt} = \frac{g}{V_a} \tan \phi$$

$$\frac{dx}{dt} = V_g \sin(\eta - \eta_0)$$

where V_a = true airspeed
 $V_g = \sqrt{V_E^2 + V_N^2}$
 $\eta = \tan^{-1}(V_E/V_N)$

These equations describe how roll angle and aircraft velocity translate into GTA and cross-track position. The above equations can be linearized by making small-angle approximations. Thus

$$\frac{d\eta}{dt} = \frac{g}{V_a} \phi$$

$$\frac{dx}{dt} = V_g(\eta - \eta_0)$$

This is a linear model for the lateral aircraft dynamics.

In summary, the steering problem is the determination of the ground track required at the present aircraft position to reach a given destination by the shortest route. This is known as *great circle steering*. The following two subsections will discuss in some detail great circle and rhumb-line steering.

7.2.1. Great Circle Steering

In order to minimize fuel, time, and distance, the best course to take from point to point on the earth's surface is a great circle route, since a great circle is the shortest distance between two points on a sphere. Current methods of determining great circle routes and distances use tables and equations that limit the accuracy to locations less than 10,013.7 km (5,406.9 NM) apart. This is due to the use of trigonometric equations that originated from celestial navigation and the appropriate tables from the Hydrographic Office. Historically, great circle calculations relied on astronomic navigation tables, which were simply an extension of celestial navigation. Many of the operational inertial navigation systems discussed in this book are particularly suited for great circle steering, whereby the steering is accomplished along a great circle in the plane determined by the center of the earth, aircraft present

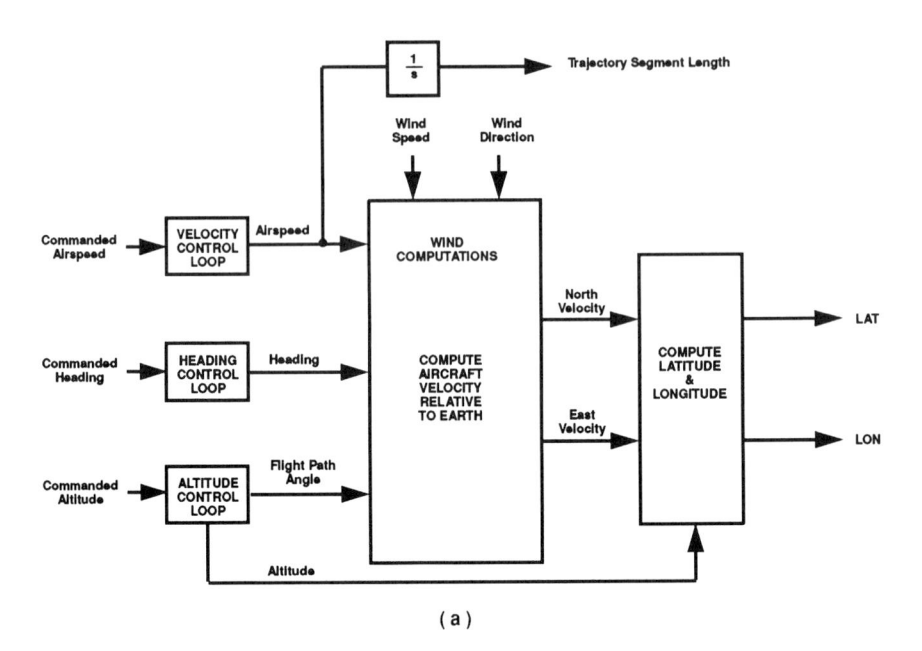

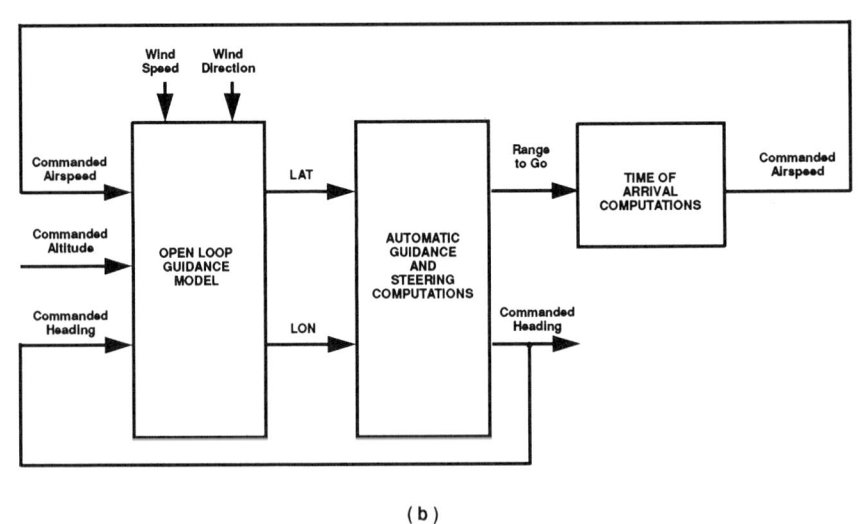

FIGURE 7-6
Open-loop (a) and closed-loop (b) guidance mode block diagram.

position, and the desired destination. That is, a great circle is defined as a circle on the surface of a sphere, of which the plane of the circle passes through the center of the sphere. Between two points on a sphere, there is only one great circle route, unless the points are exactly opposite each other, and then there are an infinite number of great circles (all the same distance). However, only one great circle occurs, since three points (the current position, destination, and the center of the sphere) determine a plane, and that plane forms only two arcs between the points on the surface of the sphere: a long arc (the long way around a sphere) and the desired great circle arc. Lateral steering, if desired, can be accomplished by nulling out track position and velocity or relocated latitude and north velocity. Depending on the mission requirements, great circle steering can also be mechanized as a space-stabilized azimuth system, analytic system, or similar. Transcontinental and intercontinental transport aircraft equipped with inertial reference systems and/or inertial navigation systems use their capability of flying a great circle route between departure and destination points routinely. The great circle segment (or leg) is commonly defined to be from top of the climb to top of descent, that is, roughly from release by the departure control to pickup by the approach control center. However, since for transcontinental flights it is not required that commercial IRSs or INSs be designed to indicate geographic coordinates along the great circle track line (except present position and inserted waypoints), problems can arise in obtaining flight clearances because the air traffic control (ATC) on these routes requires reporting points designated in advance by the flight crew (the flight crew knows only the departure and arrival points of the intended great circle track). Before the equation for a great circle flight path is developed, a brief discussion of the trajectory generator is in order. In applications where a trajectory generator is required, the trajectory generator can operate in either an open-loop or closed-loop guidance mode as depicted in Fig. 7-6.

In the open-loop guidance mode, the inputs to the flight control loops are determined by the user in the form of altitude, velocity, and heading commands. In the closed-loop guidance mode, heading commands are computed automatically by solving either great circle or rhumb-line (see Section 7.2.2) guidance equations and associated steering equations. The great circle guidance equations that are used in the trajectory generator are commonly expressed in an earth-centered inertial (ECI) coordinate frame. For the present analysis, we will assume a unit sphere defined by the unit vectors $(\mathbf{i}, \mathbf{j}, \mathbf{k})$ as illustrated in Fig. 7-7, where the unit vectors $\mathbf{i}$ and $\mathbf{k}$ pass through the Greenwich Meridian and the earth's polar axis, respectively, and $\mathbf{j}$ is a unit vector completing the right-hand coordinate system.

In the coordinate frame shown in Fig. 7-7, the present position, initial position, and destination of an aircraft can be defined by three unit vectors $\mathbf{1}_0$, $\mathbf{1}_1$, and $\mathbf{1}_2$, respectively. Furthermore, define the position unit vector in

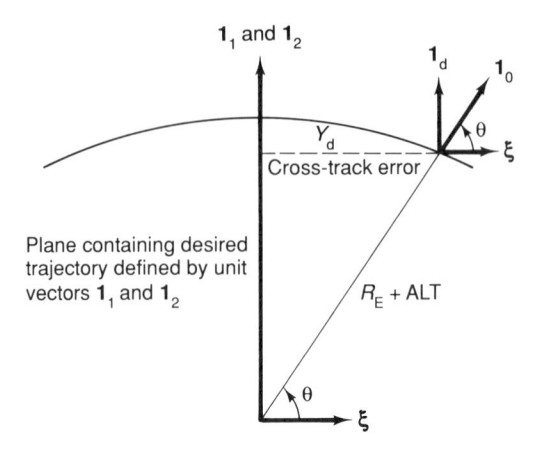

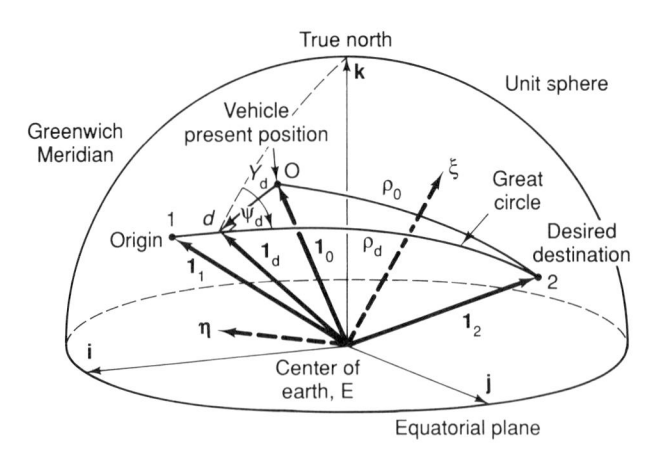

388 CHAPTER 7 • Steering, Special Navigation, and Modern Avionics

FIGURE 7-7
ECI coordinate frame for great circle guidance.

terms of spherical coordinates (i.e., latitude and longitude) as [7]

$$\mathbf{1}_n = C_{31} * \mathbf{i} + C_{32} * \mathbf{j} + C_{33} * \mathbf{k}$$
$$= \cos \phi \cos \lambda * \mathbf{i} + \cos \phi \sin \lambda * \mathbf{j} + \sin \phi * \mathbf{k} \quad (7.1)$$

where $\mathbf{i}, \mathbf{j}, \mathbf{k}$ are unit vectors along the earth-fixed coordinate axes defined above. Note that the magnitude of this vector $\mathbf{1}_n$ is the mean radius of the spherical earth plus altitude above the earth. Therefore, the unit vectors $\mathbf{1}_0$,

7.2. Steering

$\mathbf{1}_1$, and $\mathbf{1}_2$ can be written as

$$\mathbf{1}_0 = [\cos \phi_0 * \cos \lambda_0] * \mathbf{i} + [\cos \phi_0 * \sin \lambda_0] * \mathbf{j} + [\sin \phi_0] * \mathbf{k} \quad (7.2a)$$

$$\mathbf{1}_1 = [\cos \phi_1 * \cos \lambda_1] * \mathbf{i} + [\cos \phi_1 * \sin \lambda_1] * \mathbf{j} + [\sin \phi_1] * \mathbf{k} \quad (7.2b)$$

$$\mathbf{1}_2 = [\cos \phi_2 * \cos \lambda_2] * \mathbf{i} + [\cos \phi_2 * \sin \lambda_2] * \mathbf{j} + [\sin \phi_2] * \mathbf{k} \quad (7.2c)$$

In general, the great circle distance can be found by defining two points on the sphere with the unit vectors $\mathbf{1}_1$ (a vector from the center of the earth to the point 1, the origin or vehicle present position 0) and $\mathbf{1}_2$ (a vector from the center of the earth to point 2, the destination); the inner product can be used to find the angular distance μ between the two positions on the surface of the sphere. Thus

$$\mathbf{1}_1 * \mathbf{1}_2 = |\mathbf{1}_1||\mathbf{1}_2| \cos \mu$$

Since $\mathbf{1}_1$ and $\mathbf{1}_2$ are unit vectors, solving for μ yields

$$\mu = \cos^{-1}(\mathbf{1}_1 \cdot \mathbf{1}_2) \quad (7.3)$$

Therefore, the arc length (i.e., $s = r\theta$) can be determined from the following relationship:

$$\rho_d = r \cos^{-1}(\mathbf{1}_1 \cdot \mathbf{1}_2) \quad (7.4)$$

where $r = R_E + h$ (R_E is the radius of the earth and h is the aircraft altitude). Similarly, the distance from the vehicle present position ρ_0 to the destination is given by

$$\mathbf{1}_0 * \mathbf{1}_2 = |\mathbf{1}_0||\mathbf{1}_2| \cos \rho_0$$
$$\rho_0 = r \cos^{-1}(\mathbf{1}_0 \cdot \mathbf{1}_2) \quad (7.5)$$

The commanded heading and range-to-go can be readily obtained from Fig. 7-7 using vector notation. Let ξ be a unit vector perpendicular to the geodesic defined by the unit vectors $\mathbf{1}_1$ and $\mathbf{1}_2$. Note that here we assume that the distances ρ_0, ρ_d, and Y_d are normalized. Thus

$$\xi = \frac{\mathbf{1}_1 \times \mathbf{1}_2}{|\mathbf{1}_1 \times \mathbf{1}_2|} \qquad -180° \leq \xi \leq 180° \quad (7.6)$$

Furthermore, the angle θ can be formed by the unit vectors ξ and $\mathbf{1}_0$ expressed in terms of the vector dot product as follows:

$$\cos \theta = \frac{\mathbf{1}_0 \cdot \xi}{|\mathbf{1}_0||\xi|} \quad (7.7)$$

The unit vector $\mathbf{1}_d$, directed at the desired position, can be expressed as

$$\mathbf{1}_d = \mathbf{1}_0 - [\mathbf{1}_0 \cdot \xi]\xi \quad (7.8)$$

while the cross-track error Y_d between the aircraft and the present position $\mathbf{1}_0$ is given by

$$Y_d = (R_E + h) \cos \theta = (R_E + h) \left\{ \frac{\mathbf{1}_0 \cdot \boldsymbol{\xi}}{|\mathbf{1}_0||\boldsymbol{\xi}|} \right\} \tag{7.9}$$

where R_E and h have been defined above. The desired, or commanded, heading angle ψ_d is given by

$$\psi_d = \tan^{-1} \left[\frac{\boldsymbol{\eta} \times \boldsymbol{\xi}}{\boldsymbol{\eta} \cdot \boldsymbol{\xi}} \right] \qquad -180° \leq \psi_d \leq 180° \tag{7.10}$$

where $\boldsymbol{\eta}$ is a unit vector perpendicular to the meridian plane containing the desired trajectory $\mathbf{1}_1$ and $\mathbf{1}_2$. Therefore

$$\boldsymbol{\eta} = \mathbf{1}_d \times \mathbf{k} \tag{7.11}$$

Note that the angle ψ_d is also the desired azimuth from the true north. True course is an angle between 0° and 360°, with 0° and 360° the true north. However, since $|\boldsymbol{\eta} \times \boldsymbol{\xi}|$ is always positive, the following tests are needed to determine the proper quadrant for ψ_d:

If (**k** component of $\boldsymbol{\xi}$) > 0, the desired course is easterly and $\psi_d = \psi_d$.
If (**k** component of $\boldsymbol{\xi}$) < 0, the desired course is westerly and $\psi_d = -\psi_d$.
If (**k** component of $\boldsymbol{\xi}$) = 0, the desired course is 0° or 180°.
If $\boldsymbol{\eta} * \boldsymbol{\xi} > 0$, then $\psi_d = 0°$.
If $\boldsymbol{\eta} * \boldsymbol{\xi} < 0$, then $\psi_d = 180°$.

Note that these angles apply only for departing the present position, and as a result the heading must be updated. The range-to-go, that is, the distance from the present position to the destination, can be computed from the expression

$$R_{TG} = \rho_0 = \gamma [R_E + h] \tag{7.12}$$

where γ is the angle between $\mathbf{1}_0$ and $\mathbf{1}_2$, which is given by

$$\gamma = \tan^{-1} \left[\frac{\mathbf{1}_0 \times \mathbf{1}_2}{|\mathbf{1}_0 \cdot \mathbf{1}_2|} \right] \tag{7.13}$$

Finally, the time-to-go can be computed from the relationship

$$t_0 = \rho_0 / V_g \tag{7.14}$$

where V_g is the groundspeed. The sign conventions commonly used for latitude and longitude in the great circle equations are as follows:

ϕ = latitude: north from equator is positive $\qquad 0° \leq \phi \leq 90°$
south of the equator is negative $\qquad -90° \leq \phi \leq 0°$
λ = longitude: east from Greenwich is positive $\qquad 0° \leq \lambda \leq 360°$
west from Greenwich is negative $\qquad -360° \leq \lambda \leq 0°$

7.2. Steering

As a navigation system starts from a known position and travels a distance along the great circle, the navigation system must determine its new position and update the heading. Assuming for simplicity that the platform does not drift and depart from the great circle track, the new position is found by adding the angular distance traveled on the great circle from the last known position. The next heading is calculated between the new position and the destination. The navigation computer will sample data such as the current position and velocity at some sampling rate $1/t$. The heading should be updated during each data sample. Unfortunately, if wind, Coriolis effect, instrument errors, and modeling approximations are not corrected, the platform will drift off the great circle track. Finding the true course for a great circle route is difficult because almost all great circle routes require a constant heading change along the route. Only a route along the equator or a line of constant longitude will have courses that remain constant. For example, a great circle route from New York to London begins in a northeasterly direction, but arriving in London would be a southeasterly direction. The course changes because the great circle route crosses every line of longitude at a slightly different angle. Note that, as stated earlier, the great circle equations are not valid if the great circle distance between destinations or between present position and a destination is greater than $(\pi R_E/2) = 10{,}013.675$ km (or $5{,}406.952$ NM); for this computation, the value of the earth's radius used was from the WGS-84 system, i.e., $R_E = 6{,}378{,}137$ m.

Working with vectors can, at times, be cumbersome. Commonly, the vector calculations are a two-step process. First, the position is converted to latitude and longitude in degrees (to 16 decimal places), and then these are used to compute the position vectors. The position vectors are geocentric unit vectors pointing toward their respective locations on the earth. For this reason, in the discussion that follows, the equations for the great circle distance and initial heading will be developed using trigonometric relations. An important aspect of spherical triangles is to be aware that the sides are measured in degrees of arc. This is a significant fact because the magnitude of the angle for a constant length side will not be constant from one sphere to another but will depend on the radius of the sphere. The spherical triangles that are used in great circle navigation are subject to the following conditions: (1) the sum of the sides is less than $360°$, (2) the sum of the angles is between $180°$ and $540°$, (3) equal sides are opposite equal angles, (4) the greater of two unequal sides is opposite the greater angle and conversely, and (5) the sum of two sides is greater than the third side. As is the case for the planar triangle, obtaining solutions to spherical triangle problems entails using the equations known as the Law of Sines and the Law of Cosines. However, these equations are not those used in planar trigonometry and are referred to as the Law of Sines and Cosines for Spherical Triangles. The solution to the spherical triangle can be achieved in much the same manner as for the planar one. Given the right combination of three of the quantities, the remaining three can be solved for by using the Law of Sines and Cosines for Spherical Triangles. Obtaining the great circle path distance requires the spherical triangle to be positioned on the sphere as shown in Fig. 7-8. The sphere in this case is the earth with its appropriate radial dimension.

Suppose now it is desired to navigate from point 1 to point 2 as defined in Fig. 7-8. It is clear that this is the shortest distance to be navigated; this is called the *great circle distance*. Using the known parameters in conjunction with the Law of Sines and/or Cosines of Spherical Trigonometry will result in determining the great circle path distance in degrees. To convert this result from, say, nautical miles to degrees, simply multiply by 60.76 NM/deg. (Note that 1 NM = 1.852 km.) Now, in Fig. 7-8, let a be defined as the

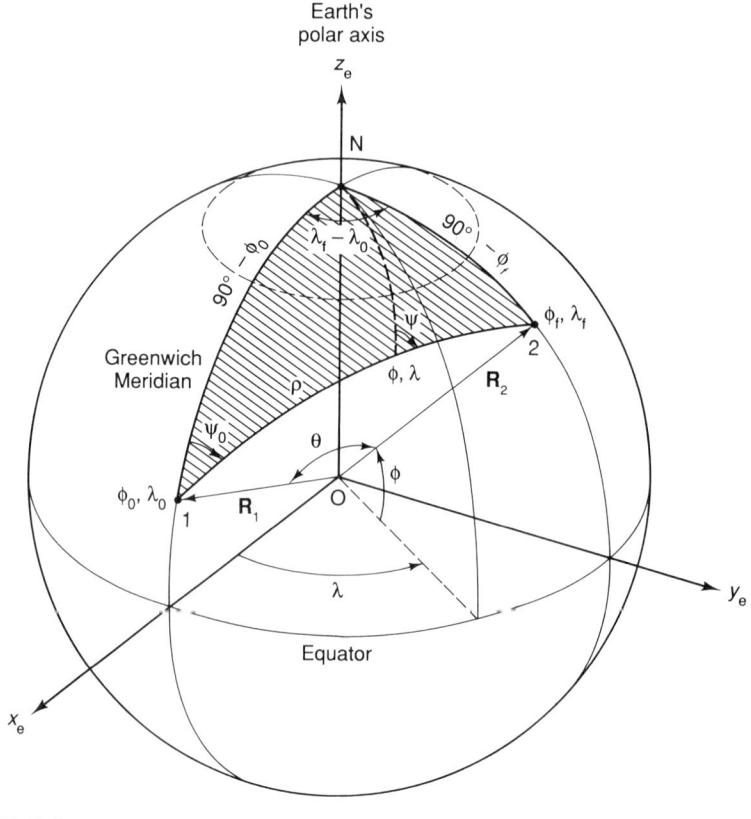

FIGURE 7-8
Spherical triangle for great circle computations.

7.2. Steering

average radius of the earth. Then

$$a = \frac{R_1 + R_2}{2} \tag{7.15}$$

For small angles, $\rho = a\theta$ or

$$\cos\theta = \cos(\rho/a) \tag{7.16}$$

From the Law of Cosines for Spherical Triangles, we obtain for the spherical triangle 1N2, for travel from point 1 to point 2 [1,8]:

$$\cos(\rho/a) = \cos(90° - \phi_1)\cos(90° - \phi_2) + \sin(90° - \phi_1)$$
$$\times \sin(90° - \phi_2)\cos(\lambda_1 - \lambda_2) \tag{7.17}$$

or

$$\cos(\rho/a) = \sin\phi_1 \sin\phi_2 + \cos\phi_1 \cos\phi_2 \cos(\lambda_1 - \lambda_2)$$
$$\rho/a = \cos^{-1}\{\sin\phi_1 \sin\phi_2 + \cos\phi_1 \cos\phi_2 \cos(\lambda_1 - \lambda_2)\}$$
$$\rho = a * \cos^{-1}\{\sin\phi_1 \sin\phi_2 + \cos\phi_1 \cos\phi_2 \cos(\lambda_1 - \lambda_2)\} \tag{7.18}$$

This is the great circle path distance between waypoints 1 and 2. The initial heading angle ψ_0 can similarly be obtained using the Law of Cosines for Spherical Triangles as follows:

$$\cos\psi_0 = \frac{\sin\phi_2 - \sin\phi_1 \cos\left(\dfrac{\rho}{a}\right)}{\cos\phi_1\left[\sin\left(\dfrac{\rho}{a}\right)\right]} \tag{7.19}$$

or, since $\sin(\rho/a) = \sqrt{1 - \cos^2(\rho/a)}$, we obtain the heading angle in the form

$$\psi_0 = \cos^{-1}\left\{\frac{\sin\phi_B - \sin\phi_A[\sin\phi_B \sin\phi_A + \cos\phi_B \cos\phi_A \cos(\lambda_B - \lambda_A)]}{\cos\phi_A\sqrt{[1 - (\sin\phi_B \sin\phi_A + \cos\phi_B \cos\phi_A \cos(\lambda_B - \lambda_A)^2]}}\right\} \tag{7.20}$$

where (ϕ_A, λ_A) is the lat/long of the current position and (ϕ_B, λ_B) is the destination. The time required to travel along a great circle path from waypoint 1 (or from point j) to waypoint 2 (or to point $j+1$) is computed as follows:

$$\Delta t_j = (a/V_j) * \cos^{-1}\{\sin\phi_{j+1} \sin\phi_j$$
$$+ \cos\phi_j \cos\phi_{j+1} \cos(\lambda_j - \lambda_{j+1})\} \tag{7.21}$$

$$V_j = \sqrt{V_x^2 + V_y^2}$$

where Δt_j = time in seconds
V_x, V_y = velocity components of the velocity vector lying along the great circle path between points 1 and 2

In the preceding development, a spherical earth model was used for the derivation of the great circle distance. The purpose of the following discussion is to derive and incorporate into a computer program the necessary mathematical equations for calculating the great circle distance between two given points on the surface of the earth and for estimating the down-range (X_d) and cross-range (Y_d) errors expected if an initial incorrect heading is flown from the first (or initial) position. The program also includes calculations for the corrected initial and final heading between these same two points and allows the user to select the number of intermediate headings desired. Since the basis for the mathematics involved in the calculations is from spherical trigonometry, a spherical earth model is utilized in all derivations, except for the calculations of the earth's radius, which utilizes the "ellipsoidal" model. Furthermore, since the earth's radius is not constant between the two points of interest, this ellipsoidal model is used to calculate the radius at each of the two points and the average of these two calculations is utilized in the subsequent calculations employed in the program. As a result of this approximation, the accuracy of the information provided by the program is degraded as the distance between the two points selected increases, although not significantly. We can now write the x and y coordinates for a point on the ellipse as follows [1]:

$$x = \frac{a_e \cos L}{\sqrt{1-e^2 \sin^2 L}} \qquad y = \frac{a_e(1-e^2) \sin L}{\sqrt{1-e^2 \sin^2 L}}$$

where a_e is the earth's equatorial radius, L is the geodetic latitude, and e^2 is the eccentricity. For a point that is an elevation H above the ellipsoid (assumed to be the mean sea level), the x and y components of the elevation normal to the adopted ellipsoid are $\Delta x = H * \cos L$ and $\Delta y = H * \sin L$. Therefore, in terms of the geodetic latitude, elevation above the mean sea level, and eccentricity, the x and y components assume the form

$$x = \left[\frac{a_e}{\sqrt{1-e^2 \sin^2 L}} + H\right] \cos L \qquad \text{and} \qquad y = \left[\frac{a_e(1-e^2)}{\sqrt{1-e^2 \sin^2 L}} + H\right] \sin L$$

Therefore, it follows directly that $R_i = (x^2 + y^2)^{1/2}$, where R_i is the computed earth's radius at the ith point. Similarly, a radius can be computed for the second point, R_{i+1}, so that the average radius is calculated as

$$R_{\text{avg}} = \frac{R_i + R_{i+1}}{2}$$

Referring to Fig. 7-9, the initial heading A between the two points can be obtained from the Law of Cosines as follows:

$$\cos A = \frac{\cos a - \cos b \cos c}{\sin b \sin c}$$

7.2. Steering

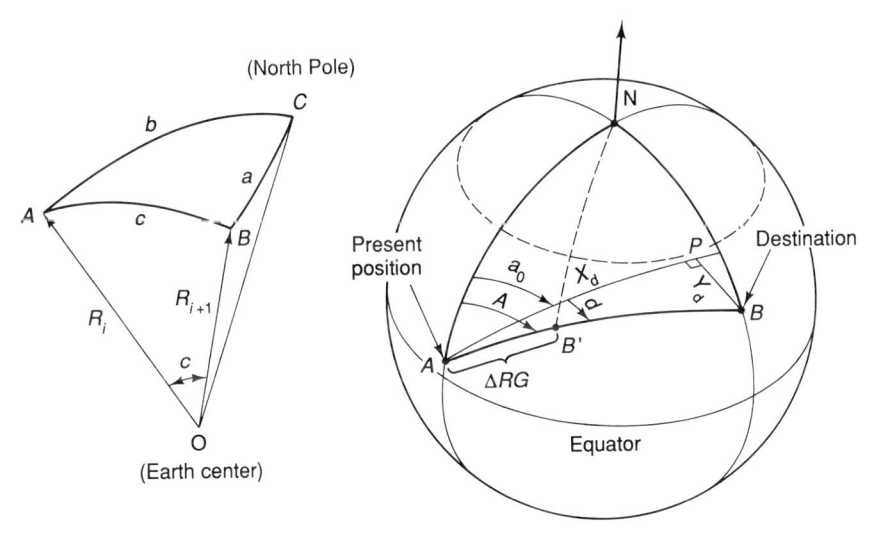

FIGURE 7-9
Geometry for calculating the down-range and cross-range.

or

$$A = \text{HDG}$$
$$= \cos^{-1}\left[\frac{\sin(\text{Lat 2}) - \sin(\text{Lat 1})\cos(RG/R_{\text{avg}})}{\cos(\text{Lat 1})\sin(RG/R_{\text{avg}})}\right] \quad (7.22)$$

In the program, the initial heading A is called ZETA. Similarly, the angle B can be calculated from the expression [4]

$$B = \cos^{-1}\left[\frac{\sin(\text{Lat 1}) - \sin(\text{Lat 2})\cos(RG/R_{\text{avg}})}{\cos(\text{Lat 2})\sin(RG/R_{\text{avg}})}\right] \quad (7.23)$$

from which we can obtain the final heading (HDGF) as

$$\text{HDGF} = 180° - B \quad (7.24)$$

Intermediate headings and great circle distances from the intermediate point to the destination can similarly be derived using the Laws of Sines and Cosines. In this case, the great circle distance remaining will be the difference between the total great circle distance (A to B) minus the distance from point A to the point in question being computed.

In certain situations, it may be desirable to calculate the down-range and cross-range errors expected from an initial great circle and an initial heading selected from the first point (A). Defining the cross-range as the perpendicular distance from the point to the initial great circle as shown in Fig. 7-9, then the down-range is the distance along the initial great circle from the initial point to point P, at which the cross-range is measured. Solving the right spherical triangle APB using spherical trigonometry, the down-range X_d and cross-range Y_d are obtained from the relations

$$X_d = R_{avg} \cos^{-1}\left[\frac{\cos(RG/R_{avg})}{\cos\{\sin^{-1}[\sin(RG/R_{avg})\sin d]\}}\right] \qquad (7.25)$$

$$Y_d = R_{avg} * \sin^{-1}[\sin(RG/R_{avg})\sin d] \qquad (7.26)$$

where $d = A -$ nominal initial heading $= A - a_0$ and RG is the great circle distance from A to B (i.e., $c = RG/R_{avg}$). Finally, and as stated above, in order to compute the intermediate headings between the two points A and B, the spherical triangle ANB (see Fig. 7-9) can be divided into two other spherical triangles. Again, using the Laws of Sines and Cosines for spherical triangles, we obtain the following relationships:

$\cos a' = \cos b' \cos c' + \sin b' \sin c' \cos A'$
$\cos a' = \cos(90° - \text{Lat } B')$ or $\text{Lat } B' = \sin^{-1}(\cos a')$
$\cos b' = \cos(90° - \text{Lat } 1) = \sin(\text{Lat } 1)$
$\cos c' = \cos(\Delta RG/R_{avg})$
$\sin c' = \sin(\Delta RG/R_{avg})$
$\cos c'' = \cos\{(RG - \Delta RG)/R_{avg}\}$

where Lat B' represents the latitude of the intermediate point. From these relations, the intermediate heading A_{int} is given by

$$\cos A_{int} = \cos A''$$
$$= \frac{\sin(\text{Lat } 2) - \cos a' \cos((RG - \Delta RG)/R_{avg})}{\cos(\text{Lat } B') \sin((RG - \Delta RG)/R_{avg})} \qquad (7.27)$$

If we desire to calculate the longitude at the intermediate point (B'), we can similarly use the Laws of Sines and Cosines once again to compute the angle C'. Since C' represents the change in longitude (DELONG), we can easily obtain the new longitude at the intermediate point using the relationship

$$\text{Long } B' = \text{Long } 1 + \text{DELONG} \qquad (7.28)$$

7.2. Steering

A FORTRAN program* for these above calculations is as follows:

```
C       THIS PROGRAM CALCULATES THE GREAT CIRCLE DISTANCE BETWEEN TWO
C       POINTS, IN ADDITION TO THE DOWNRANGE AND CROSSRANGE DISTANCE
C       FROM AN INITIAL GREAT CIRCLE DETERMINED FROM A NOMINAL
C       HEADING FROM THE FIRST POSITION. IT THEN COMPUTES THE
C       CORRECTED INITIAL HEADING, FINAL HEADING, AND TIME OF FLIGHT
C       BETWEEN THESE TWO POINTS FOR A GIVEN SPEED. SHOULD THE USER
C       DESIRE INTERMEDIATE HEADINGS ENROUTE, THE PROGRAM WILL
C       CALCULATE THEM AS WELL.
        REAL LAT1,LAT2,LONG1,LONG2,H,H1,H2,LAT,SI,X,Y,R,R1,R2,HDG,AE
        REAL ECTY,RAVG,PI,PARTX,RG,XD,ZETA,SPEED,TIME
        REAL DIST,DELONG,LATM,HDGD,COSAP,COSAPP,STEP,B,HDGF,LONGM
        INTEGER M,N,L,NBR
10      PRINT *,'ENTER 1 TO ENTER DATA, 0 TO TERMINATE PROGRAM'
        READ *,M
        IF (M.NE.1) GO TO 700
        PRINT *,'ENTER (DEGREES) LAT1,LONG1,LAT2,LONG2,HDG'
        READ *,LAT1,LONG1,LAT2,LONG2,HDG
        PRINT *,'ENTER ELEV1(FT MSL),ELEV2,SPEED(KNOTS),',
     1  'NUMBER OF STEPS'
        READ *,H1,H2,SPEED,NG4BR
        PI=3.1415926
        AE=3443.922786
        ECTY=.08182
        PRINT 40, LAT1,LONG1
40      FORMAT(/'LAT1 =',F10.2,' DEGREES'/,'LONG1 =',F10.2,' DEGREES')
        PRINT 50,LAT2,LONG2
50      FORMAT('LAT2 =',F10.2,' DEGREES'/,'LONG2 =',F10.2,' DEGREES')
        PRINT 60,HDG,SPEED
60      FORMAT('ESTIMATED HEADING = ',F10.2,' DEGREES'/,
     1  'SPEED(NO WIND) = ',F10.2' KNOTS')
        LAT 1=LAT1*PI/180.0
        LONG1=LONG1*PI/180.0
        IF (LONG1.LT.0.0) LONG1=2*PI+LONG1
        LAT2=LAT2*PI/180.0
        LONG2=LONG2*PI/180.0
        IF (LONG2.LT.0.0) LONG2=2*PI+LONG2
        HDG=HDG*PI/180.0
        H=H1*.3048/1852.0
        LAT=LAT1
        L=0
        N=0
400     PARTX=1-ECTY**2*(SIN(LAT))**2
        PARTX=AE/SQRT(PARTX) + H
        X=PARTX*COS(LAT)
        PARTX=(PARTX-H)*(1-ECTY**2) + H
        Y=PARTX*SIN(LAT)
        R=SQRT(X**2 + Y**2)
        N=N+1
        IF (N.EQ.2) GO TO 600
```

*This program was written by David W. Thomin, Captain, U.S. Air Force.

```
              R1=R
              LAT=LAT2
              H=H2*.3048/1852.0
              GO TO 400
600           R2=R
              RAVG=(R2+R1)/2
              RG=RAVG*ACOS(SIN(LAT2)*SIN(LAT1)+COS(LAT2)*COS(LAT1)*
             1 COS(LONG2-LONG1))
              ZETA=ACOS((SIN(LAT2)-SIN(LAT1)*COS(RG/RAVG))/(COS(LAT1)*
             1 SIN(RG/RAVG)))
              HDGF=ACOS((SIN(LAT1)-SIN(LAT2)*COS(RG/RAVG))/(COS(LAT2)*
             1 SIN(RG/RAVG)))
610           SI=ABS(ZETA)-ABS(HDG)
              XD=RAVG*ACOS(COS(RG/RAVG)/COS(ASIN(SIN(RG/RAVG)*SIN(SI))))
              YD=RAVG*ASIN(SIN(RG/RAVG)*SIN(SI))
              HDG=ZETA
              TIME=RG/SPEED
              PRINT 70,RG
70            FORMAT(//'GREAT CIRCLE DISTANCE =',F10.3,' NM')
              PRINT 80,TIME
80            FORMAT('TIME TO TRAVERSE =',F10.2,' HOURS')
              PRINT 90,XD,YD
90            FORMAT('XD =',F10.3,' NM'/'YD =',F10.3,' NM')
              PRINT 100,RAVG
100           FORMAT('AVERAGE RADIUS OF EARTH =',F10.3,' NM')
105           HDG=HDG*180.0/PI
              HDGF=HDGF*180.0/PI
107           IF (LONG2.LT.LONG1) GO TO 650
              GO TO 680
110           HDGD=HDG
              PRINT 115,HDG,HDGF
115           FORMAT('INITIAL HEADING = ',F10.2,' DEGREES'/,
             1 'FINAL HEADING = ',F10.2,' DEGREES'//)
              PRINT 118
118           FORMAT(/'        LATITUDE         LONGITUDE     '
             1 ,'   HDG UPDATE        DIST REMAINING')
120           IF (L.GE.(NBR-1)) GO TO 10
              L=L+1
              STEP=RG/NBR
              B=STEP*L/RAVG
              COSAP=SIN(LAT1)*COS(B)+COS(LAT1)*SIN(B)*COS(ZETA)
              COSAPP=(SIN(LAT2)-COSAP*COS(RG/RAVG-B))/(SIN(RG/RAVG-
             1 B)*COS(ASIN(COSAP)))
              HDG=ACOS(COSAPP)
              LATM=ASIN(COSAP)
              DELONG=ACOS((COS(B)-SIN(LATM)*SIN(LAT1))/(COS(LATM)*
             1 COS(LAT1)))
              HDG=HDG*180.0/PI
              LATM=LATM*180.0/PI
              GO TO 107
130           IF (HDGD.LT.180) LONGM=LONG1+DELONG
              IF (HDGD.GE.180) LONGM=LONG1-DELONG
              IF (LONGM.GT.PI) LONGM=-1*(2*PI-LONGM)
              LONGM=LONGM*180.0/PI
              DIST=RG-STEP*L
              PRINT 135,LATM,LONGM,HDG,DIST
```

7.2. Steering　　　　　　　　　　　　　　　　　　　　　　　　　　　　　　　　　**399**

```
135     FORMAT(F15.2,F15.2,F15.2,F15.2)
        GO TO 120
650     IF ((LONG1-LONG2).LT.PI) HDG =360.0-HDG
        IF ((LONG1-LONG2).LT.PI) HDGF=180.0+HDGF
        IF ((LONG1-LONG2).GE.PI) HDGF=180.0-HDGF
        IF (L.NE.0) GO TO 130
        GO TO 110
680     IF ((LONG2-LONG1).GT.PI) HDG=360.0-HDG
        IF ((LONG2-LONG1).GE.PI) HDGF=180.0+HDGF
        IF ((LONG2-LONG1).LT.PI) HDGF=180.0-HDGF
        IF (L.NE.0) GO TO 130
        GO TO 110
700     END


SAMPLE CASE

ENTER (DEGREES) LAT1,LONG1,LAT2,LONG2,HDG
39.75,-84.25,51.50,-.12,50     DAYTON, OHIO TO LONDON, UK
ENTER ELEV1(FT MSL),ELEV2,SPEED(KNOTS), NUMBER OF STEPS
0,0,800,25

LAT1  =   39.75 DEGREES
LONG1 =  -84.25 DEGREES
LAT2  =   51.50 DEGREES
LONG2 =    -.12 DEGREES
ESTIMATED HEADING =    50.00 DEGREES
SPEED (NO WIND)   =   800.00 KNOTS

GREAT CIRCLE DISTANCE = 3400.834 NM
TIME TO TRAVERSE      =    4.250 HOURS
XD = 3399.698 NM
YD = -108.997 NM
AVERAGE RADIUS OF THE EARTH = 3438.049 NM
INITIAL HEADING =  47.83 DEGREES
FINAL HEADING   = 113.75 DEGREES
```

7.2.2. Rhumb-Line Navigation

In Section 7.2.1 we saw that the shortest distance between two points on the surface of the earth is the distance measured along a great circle that passes through both points. For example, the shortest distance between two points, both of which are on the equator, is the distance measured along the equator. Similarly, the shortest distance between two points on a meridian is the distance measured along that meridian. Thus, in the first case, the shortest course for an aircraft (or a ship) flying between the two points will make the same angle (i.e., 90°) with every meridian crossed, while in the second case, the course will maintain a constant angle (i.e., 0°) with the meridian. For any other case, that is, for two points not on the equator or the same meridian, the great circle that joins the two points will cut each succeeding meridian at a different angle. From the preceding discussion, it is clear that if a vehicle is moving from one place to another by the shortest route, the track must be the arc of a great circle, so that the direction of such a track is constantly changing. Consequently, for distances that are not great, the saving in time by following a great circle track is very small; in addition, this involves considerable calculation and requires frequent changes of course. For this reason, it is generally more convenient to steer a constant course, or to follow a *rhumb line*.

A rhumb line (or *loxodrome*) is defined as a line on the earth's surface which intersects all meridians at the same angle. Any two places may be connected by such a line. Rhumb-line courses may be approximated by assuming that the distances involved are short enough that a flat earth assumption is satisfactory. Therefore, rhumb-line guidance and/or steering is based on traversing a straight-line course on a mercator chart that yields a "constant heading" command over the trajectory flown. Under the above definition, the meridians, the equator, and the parallels are rhumb lines. Since the rhumb-line course is cutting all the meridians at the same angle, it should be noted that special rhumb-line courses are also great circle courses; this angle is $\pm 90°$ for an equatorial flight and $0°$ for meridional flights. The parallels are a special type of rhumb line, as they are at right angles to all the meridians and are everywhere equidistant from the poles. All other rhumb lines are loxodromic curves (i.e., spirals that approach the poles but never reach them). It should be noted that in the lower latitudes near the equator, the loxodromic curve is fairly flat, so that the rhumb line is very nearly the shortest distance between two points on the earth's surface; while as the pole is approached, the spiral becomes more pronounced as it must cross the converging meridians at the same angle and the difference between the great circle (shortest) distance and rhumb-line distance becomes greater. However, as mentioned above, unless the distance between the point of departure and destination is great, the convenience of using a constant course outweighs the disadvantage of a longer run, and so the rhumb line may be used more frequently than the great circle track.

Rhumb-line steering can be used with any inertial navigation system configuration, if position and velocity are transformed to latitude and longitude and north and east velocities. Both a loxodrome and a great circle are important geometric curves in navigation. Historically, the loxodrome was used in marine navigation to which the track of a ship was made to conform, when steaming moderate distances, for the convenience of maintaining a constant course.

In order to derive the rhumb-line distance and heading angle between the current position and the desired destination, consider Fig. 7-10. A spherical earth is assumed for the derivation.

From Fig. 7-10 it is noted that even though the right triangle ABC is not spherical (only $R\,\Delta\phi$ lies on the great circle), it can be considered as a

7.2. Steering

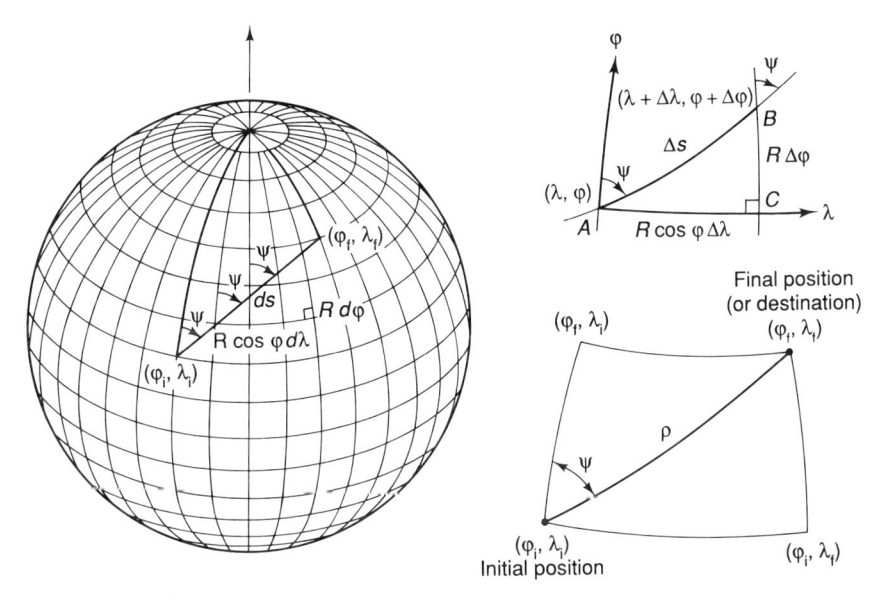

FIGURE 7-10
Loxodrome (rhumb line) geometry.

spherical triangle only when $\Delta\phi$ and $\Delta\lambda$ are chosen to be sufficiently small. Therefore, two differential relationships in terms of the latitude and longitude angles (ϕ, λ), respectively, for points on the path can be written as follows:

$$\tan \psi = \frac{\Delta\lambda \cos \phi}{\Delta\phi} \tag{7.29a}$$

and

$$\Delta s \cos \psi = R \Delta\phi \tag{7.29b}$$

where R is the radius of the earth. In the limit, as $\Delta\phi \to 0$, we obtain two differential equations:

$$\frac{d\lambda}{d\phi} = \frac{\tan \psi}{\cos \phi} \tag{7.30a}$$

and

$$\frac{ds}{d\phi} = \frac{R}{\cos \psi} \tag{7.30b}$$

Since the rhumb heading (or bearing) angle ψ is by definition constant, we obtain by integrating Eq. (7.30a) the expression

$$d\lambda = \tan \psi \, \frac{d\phi}{\cos \phi}$$

$$\int d\lambda = \tan \psi \int_{\phi_1}^{\phi_2} \frac{d\phi}{\cos \phi}$$

$$\lambda = \tan \psi \left[\ln \tan \left(\frac{\pi}{4} + \frac{\phi}{2} \right) + C \right]$$

$$\lambda_2 - \lambda_1 = \tan \psi \left[\ln \tan \left(\frac{\pi}{4} + \frac{\phi_2}{2} \right) - \ln \tan \left(\frac{\pi}{4} + \frac{\phi_1}{2} \right) \right] \quad (7.31)$$

or by solving for ψ yields

$$\psi = \tan^{-1} \left[\frac{\lambda_f - \lambda_i}{\ln \tan \left(\frac{\pi}{4} + \frac{\phi_f}{2} \right) - \ln \tan \left(\frac{\pi}{4} + \frac{\phi_i}{2} \right)} \right]$$

or (7.32)

$$\psi = \tan^{-1} \left[\frac{\lambda_f - \lambda_i}{\ln \left[\tan \left(\frac{\pi}{4} + \frac{\phi_f}{2} \right) \Big/ \tan \left(\frac{\pi}{4} + \frac{\phi_i}{2} \right) \right]} \right]$$

If λ_f and λ_i are given in degrees and ϕ_f and ϕ_i in radians, then

$$\psi = \tan^{-1} \left[\frac{\pi (\lambda_f - \lambda_i)}{180 \left\{ \ln \tan \left(\frac{\pi}{4} + \frac{\phi_f}{2} \right) - \ln \tan \left(\frac{\pi}{4} + \frac{\phi_i}{2} \right) \right\}} \right]$$

The equation for a rhumb line true course (or bearing) over the surface of an ellipsoidal earth is given by

$$\psi = \tan^{-1} \left[\frac{\Delta \phi}{\frac{180}{\pi} [\tanh^{-1}(\sin \phi) - e \tanh^{-1}(e \sin \phi)]} \right]$$

while the range along the rhumb line track (from the equator to the point

7.2. Steering

where it intercepts the parallel of latitude) is

$$\rho = a(1-e^2)[(1+\tfrac{3}{4}e^2)\phi - \tfrac{3}{8}\sin 2\phi]\sec\psi$$

where $\Delta\phi = \phi_2 - \phi_1 =$ change in latitude

$e^2 = 1 - \dfrac{b^2}{a^2}$

$a =$ semimajor axis of the ellipse
$b =$ semiminor axis of the ellipse
$e =$ eccentricity of the ellipse

Finally, integration of Eq. (7.30b) results in

$$s = \rho = \frac{R}{\cos\psi}(\phi_2 - \phi_1) \qquad (7.33)^*$$

From Eq. (7.29b), for example, it can be noted that a flight path to the North Pole [$\phi = (\pi/2)$] with a constant angle $\psi = 60°$, results in

$$s = \left\{R*\left(\frac{\pi}{2}\right)\right\}\bigg/0.5 = \pi R \qquad (7.34)$$

while the shortest path along a meridian ($\psi = 0°$, and $\cos\psi = 1$) is

$$s' = R*\left(\frac{\pi}{2}\right) \qquad (7.35)$$

indicating that the path along the rhumb line is twice as long as that along a great circle. The following is a FORTRAN program** for computing the rhumb-line distance and heading:

```
CCC   PROGRAM:  L.FOR

CCC   THIS PROGRAM COMPUTES THE DISTANCE ALONG A RHUMB LINE, GIVEN
C     THE INITIAL AND FINAL POSITIONS; USING THE INITIAL AND FINAL
C     COORDINATES, THE PROGRAM COMPUTES FIRST THE NAVIGATION HEADING
C     ANGLE. THE PROGRAM QUERIES FOR NECESSARY INPUTS: INITIAL
C     LATITUDE AND LONGITUDE, AND FINAL LATITUDE AND LONGITUDE.

C     THE PROGRAM IMPLEMENTS THE FOLLOWING EQUATIONS:

C         ARGUMENT = TAN(PI/4. + PHIf/2.)/TAN(PI/4. + PHIi/2.)

C         PSI = ATAN((LAMBf - LAMBi)/ALOG(ARGUMENT))

C     AND

C         RHO = Req*(PHIf - PHIi)/COS(PSI).
```

*Note that in the preceding equations, the subscripts 1, i, 2, and f are used interchangeably.
**This program was written by Dr. Gerald G. Cano, Lt. Col., U.S. Air Force Reserves.

```
CCC     VARIABLE DEFINITION TABLE:
C       INPUTS -
C               PHIi  :  INITIAL LATITUDE
C               PHIf  :  FINAL LATITUDE
C               LAMBi :  INITIAL LONGITUDE
C               LAMBf :  FINAL LONGITUDE

C       EQUATION OUTPUTS -
C               RHO   :  DISTANCE ALONG RHUMB LINE
C       CONSTANTS -
C               Req   :  EQUATORIAL RADIUS IN METERS*
CCC     BEGIN PROGRAM CODE

        IMPLICIT REAL (L)
        CHARACTER*64 FNAME
        CHARACTER*1 TOFILE, ANOTHER
        DATA PI/3.1415927/,Req/6378137./
        RADIANS = PI/180.0

        WRITE(*,5000)
        READ(*,5001) TOFILE
        IF(TOFILE.EQ.'Y'.OR.TOFILE.EQ.'Y')THEN
                WRITE(*,5002)
                READ(*,5003) FNAME
                OPEN(1,FILE=FNAME,STATUS='NEW')
        ENDIF

CCC     ENTER INITIAL AND FINAL COORDINATES

50      WRITE(*,7000)
        READ(*,*) PHIi
        WRITE(*,7002)
        READ(*,*) LAMBi
        WRITE(*,7004)
        READ(*,7006)
        READ(*,*) LAMBf

CCC     CONVERT ANGLES FROM DEGREES TO RADIANS

        PHIi  = PHIi*RADIANS
        LAMBi = LAMBi*RADIANS
        PHIf  = PHIf*RADIANS
        LAMBf = LAMBf*RADIANS

CCC     TO COMPUTE THE DISTANCE ALONG A RHUMB LINE, THE NAVICATION
C       HEADING ANGLE IS NEEDED; THEN COMPUTE THE DISTANCE.
C       THE DISTANCE EQUATION DOES NOT HOLD WHEN THE INITIAL, AND
C       FINAL LATITUDE ARE THE SAME - DISCONTINUITY.
C       COMPUTE ARGUMENT AND USE AS A TEST - IS EQUAL TO ONE
C       AT DISCONTINUITY.
```

---

*The units of R_{eq} are given in meters. Although the radius in the program is the equatorial radius of the earth, this need not be the case. Any value of R can be used.

7.2. Steering

```
              ARGUMENT = TAN(PI/4. + PHIf/2.)/TAN(PI/4. + PHIi/2.)
              IF(ABS(ARGUMENT).NE.1)THEN
C             FOR CONDITIONS INDICATED, 0<=PSI<90
              IF(PHIf-PHIi.GE.0..AND.LAMBf-LAMBi.GE.0.)THEN
              PSI = ATAN((LAMBf - LAMBi)/ALOG(ARGUMENT))
              RHO = Req*(PHIf - PHIi)/COS(PSI)
              ENDIF

C             FOR CONDITIONS INDICATED, 90<PSI<=180

              IF(PHIf-PHIi.LT.0..AND.LAMBf-LAMBi.GE.0.)THEN
              PSI = ATAN((LAMBf - LAMBi)/ALOG(ARGUMENT))
              PSI = PI + PSI
              RHO = Req*(PHIf - PHIi)/COS(PSI)
              ENDIF

C             FOR CONDITIONS INDICATED, 180<PSI<270

              IF(PHIf-PHIi.LT.0..AND.LAMBf-LAMBi.LT.0.)THEN
              PSI = ATAN((LAMBf - LAMBi)/ALOG(ARGUMENT))
              PSI = PI + PSI
              RHO = Req*(PHIf - PHIi)/COS(PSI)
              ENDIF

C             FOR CONDITIONS INDICATED, 270<PSI<360

              IF(PHIf-PHIi.GE.0..AND.LAMBf-LAMBi.LT.0.)THEN
              PSI = ATAN((LAMBf - LAMBi)/ALOG(ARGUMENT))
              PSI = 2.*PI + PSI
              RHO = Req*(PHIf - PHIi)/COS(PSI)
              ENDIF

              ELSE

C             ACCOUNT FOR DISCONTINUITIES AT PSI = 90 AND PSI = 270

              PSI = PI/2.
              IF(LAMBf-LAMBi.LT.0.) PSI = 3.*PI/2.
              RHO = Req*ABS(LAMBf - LAMBi)
              ENDIF

C             CONVERT ANGLES FROM RADIANS TO DEGREES

              PHIi  = PHIi/RADIANS
              LAMBi = LAMBi/RADIANS
              PHIf  = PHIf/RADIANS
              LAMBf = LAMBf/RADIANS
              PSI   = PSI/RADIANS

              WRITE(*,8000) LAMBi,PHIi
              WRITE(*,8002) LAMBf,PHIf
              WRITE(*,8004) PSI, RHO
              IF(TOFILE.EQ.'Y'.OR.TOFILE.EQ.'Y')THEN
              WRITE(1,4000)
              WRITE(1,8000) LAMBi,PHIi
              WRITE(1,8002) LAMBf,PHIf
```

```
              WRITE(1,8004) PSI, RHO
              WRITE(1,4000)
              ENDIF
              WRITE(*,6000)
              READ(*,5001) ANOTHER
              IF(ANOTHER.EQ.'Y'.OR.ANOTHER.EQ.'Y')THEN
              GO TO 50
              ENDIF
              STOP

       4000   FORMAT(' ')

       5000   FORMAT(' DO YOU WANT OUTPUT ALSO TO FILE (Y/N) >: ')
       5001   FORMAT(A)
       5002   FORMAT(' ENTER FILENAME >: ')
       5003   FORMAT(A64)

       6000   FORMAT(' DO YOU WANT TO CONTINUE (Y/N) >: ')

       7000   FORMAT(' ENTER INITIAL LATITUDE IN DEGREES >: ')
       7002   FORMAT(' ENTER INITIAL LONGITUDE IN DEGREES >: ')
       7004   FORMAT(' ENTER FINAL LATITUDE IN DEGREES >: ')
       7006   FORMAT(' ENTER FINAL LONGITUDE IN DEGREES >: ')

       8000   FORMAT(' INITIAL POSITION: ',F8.3,' DEGREES LONGITUDE ',
              1F8.3,' DEGREES LATITUDE')
       8002   FORMAT(' FINAL POSITION: ',F8.3,' DEGREES LONGITUDE ',
              1F8.3,' DEGREES LATITUDE')
       8004   FORMAT(' HEADING IS ',F8.3,' DEGREES,'
              1'DISTANCE ALONG RHUMB LINE IS: ',F12.1,' METERS')
              END
```

The computation for distance RHO is somewhat approximate. PSI is a function of PHI_f and PHI_i. This is ignored in doing an integration to derive RHO. Thus, there are discontinuities at headings of 90° and 270°. However, the program corrects for these. Regardless of whether an aircraft is flying a great circle route or a loxodrome, the commanded heading angle between two waypoints may be represented as shown in Fig. 7-11.

7.3. SPECIAL NAVIGATION SYSTEMS

7.3.1. Dead-Reckoning Navigation

Dead reckoning (DR) is a technique of computing the position of a vehicle from measurements of velocity. More specifically, it is a method of navigating in the absence of position fixes, and consists of calculating the position of the vehicle by integrating the estimated or measured groundspeed. Therefore, the dead-reckoning navigation mode provides navigation capability and serves as backup to the aided or unaided inertial mode. In these modes,

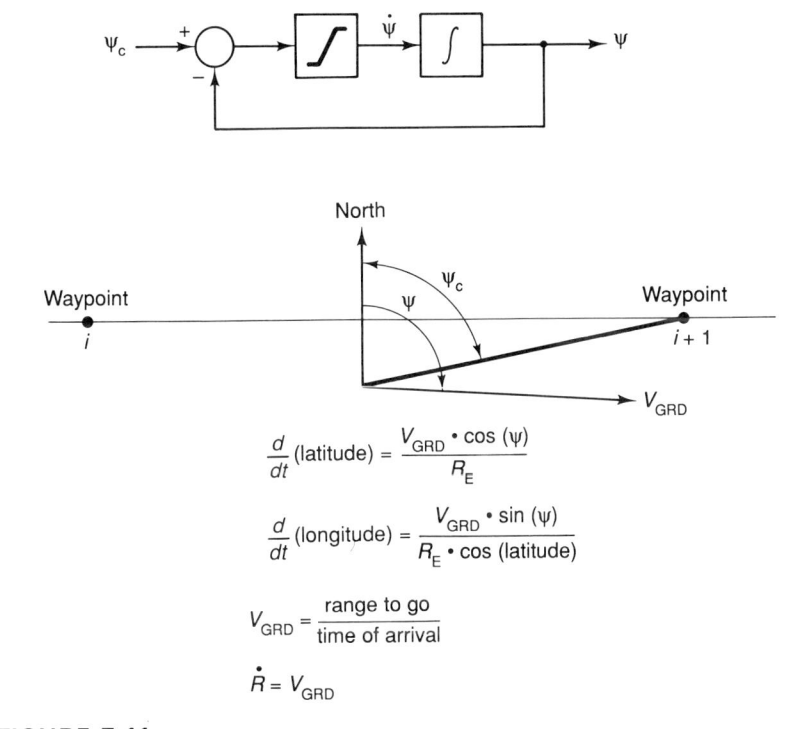

7.3. Special Navigation Systems

FIGURE 7-11
Commanded heading for navigation between two waypoints.

$$\frac{d}{dt}(\text{latitude}) = \frac{V_{GRD} \cdot \cos(\psi)}{R_E}$$

$$\frac{d}{dt}(\text{longitude}) = \frac{V_{GRD} \cdot \sin(\psi)}{R_E \cdot \cos(\text{latitude})}$$

$$V_{GRD} = \frac{\text{range to go}}{\text{time of arrival}}$$

$$\dot{R} = V_{GRD}$$

the Doppler or Central Air Data Computer (CADC) and the gyroscope stabilized subsystem (GSS), when such a system is used, provide velocity and heading angle data to the avionics control unit. Thus, in the event of inertial system failure, heading data from the auxiliary flight reference unit and velocity data from the Doppler radar or from true airspeed (TAS) plus winds are used for dead-reckoning navigation. In the dead reckoning mode, two basic pieces of information are needed: (1) the magnitude of the vehicle's velocity and (2) the direction of this velocity relative to the geographic coordinates.

The dead-reckoning (DR) position can be computed from the following expressions:

$$\phi_d = \phi_0 + \left(\frac{V_g \cos\theta_{TC} \cdot \Delta t}{60}\right)$$

$$\lambda_d = \lambda_0 - \left(\frac{V_g \sin\theta_{TC} \cdot \Delta t}{60 \cos\phi_0}\right) \quad \text{if } \theta_{TC} = 90°, 270°$$

$$\lambda_d = \lambda_0 - \frac{180}{\pi} \left\{ \tan \theta_{TC} \left[\ln \tan \left(45° + \frac{\phi_d}{2} \right) \right. \right.$$
$$\left. \left. - \ln \tan \left(45° + \frac{\phi_0}{2} \right) \right] \right\} \quad \text{otherwise}$$

where ϕ_0 = initial latitude
  ϕ_d = destination latitude
  λ_0 = initial longitude
  λ_d = destination longitude
  θ_{TC} = true course angle
  V_g = groundspeed
  Δt = time between positions

(Note that the flight path may not cross the North Pole.)

In the Doppler DR mode, the Doppler supplies the magnitude of the vehicle's velocity (or groundspeed) and the orientation of the velocity vector relative to, say, the aircraft centerline (i.e., drift angle). The GSS supplies part of the missing directional information, that is, the aircraft centerline orientation relative to the GSS reference axes. The computer supplies the remaining information (e.g., a correction for magnetic variation from true north). Furthermore, in the airspeed DR mode, the CADC supplies TAS (the magnitude of the aircraft's velocity relative to the airmass). In case the Doppler radar fails, the airspeed DR mode provides aircraft navigation data. In addition, the navigation computer must supply a velocity correction for wind, or the motion of this airmass relative to the earth. Normally, the GSS has two basic modes: (1) slaved magnetic and (2) free azimuth. Thus, when the GSS operates in the slaved magnetic mode, its indicated aircraft heading (or yaw angle) is basically magnetic heading. In this case the navigation computer supplies a correction for magnetic variation based on tabular interpolation. In the free-azimuth mode, the GSS heading angle changes in time with respect to some initial value. Therefore, in this case the navigation computer computes the change using stored heading (or corrected magnetic heading) as an initial value [2,6].

Dead-reckoning navigation computations are quite simple. After having computed the aircraft heading relative to the north, and knowing the Doppler radar's drift angle, the navigation computer performs the following functions: (1) resolves the groundspeed into north and east navigation coordinates, (2) computes latitude and longitude rates, and (3) updates the aircraft latitude and longitude via a simple integration procedure. Note that position fix data may be used to correct the indicated latitude and longitude. In the discussion that follows, we will discuss the airspeed and Doppler radar dead-reckoning modes. In using airspeed DR, the indicated airspeed (in "airmass" coordinates) must first be resolved in aircraft body axes, and then

7.3. Special Navigation Systems

transformed into the navigation axes. Therefore, the transformation from "airmass" coordinates to airframe coordinates is given by the equations

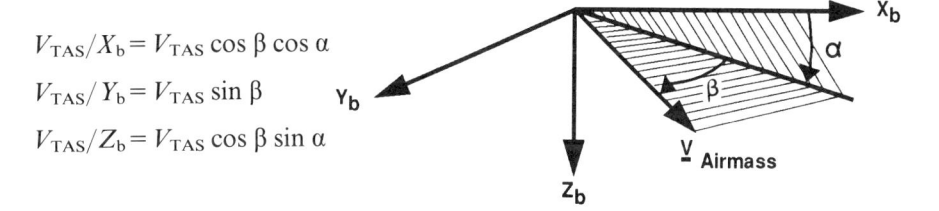

$$V_{TAS}/X_b = V_{TAS} \cos \beta \cos \alpha$$
$$V_{TAS}/Y_b = V_{TAS} \sin \beta$$
$$V_{TAS}/Z_b = V_{TAS} \cos \beta \sin \alpha$$

where α is the angle of attack and β is the sideslip angle. For the present discussion, it is assumed that the angle is made available by the CADC. Furthermore, it is also assumed that the sideslip angle is less than 5°, so that $\sin \beta = 0$ and $\cos \beta = 1$ in the transformation equations. The airmass velocity components in navigation coordinates must be corrected with the last best wind estimate, expressed in navigation coordinates, before the aircraft position can be determined. As a result, the wind correction equations are given in the form

$$V_E = V_{TAS/E} + W_E \quad (7.36a)$$
$$V_N = V_{TAS/N} + W_N \quad (7.36b)$$

where V_E and V_N are now the estimated aircraft velocity components in east–north navigation axes and W_E and W_N are the wind estimates. Airspeed dead reckoning can now proceed as does Doppler DR; that is, latitude and longitude are updated on the basis of integrating the relative aircraft angular rates about the east and north axes. Figure 7-12 illustrates in block diagram form the airspeed DR mode.

In the Doppler radar DR mode, the Doppler-indicated ground-track speed and drift angle information must be converted into velocity components into the GSS-related east–north (or west–north) navigation coordinates. Therefore, the velocity transformation from Doppler to navigation coordinates (i.e., east–north coordinates) can be expressed simply as

$$V_N = V_g \cos(\alpha_g + \gamma + \delta) \quad (7.37a)$$
$$V_E = V_g \sin(\alpha_g + \gamma + \delta) \quad (7.37b)$$

where V_g is the ground-track speed, γ is the GSS yaw angle (projected aircraft centerline from the reference GSS axis), α_g is the GSS reference axis heading from north, and δ is the Doppler radar drift angle. The latitude and longitude equations are then given by the following expressions:

Latitude

$$\phi = \phi_0 + \int \dot{\phi} \, dt, \quad \dot{\phi} = \frac{V_N}{(R_e + h)} \quad (7.38)$$

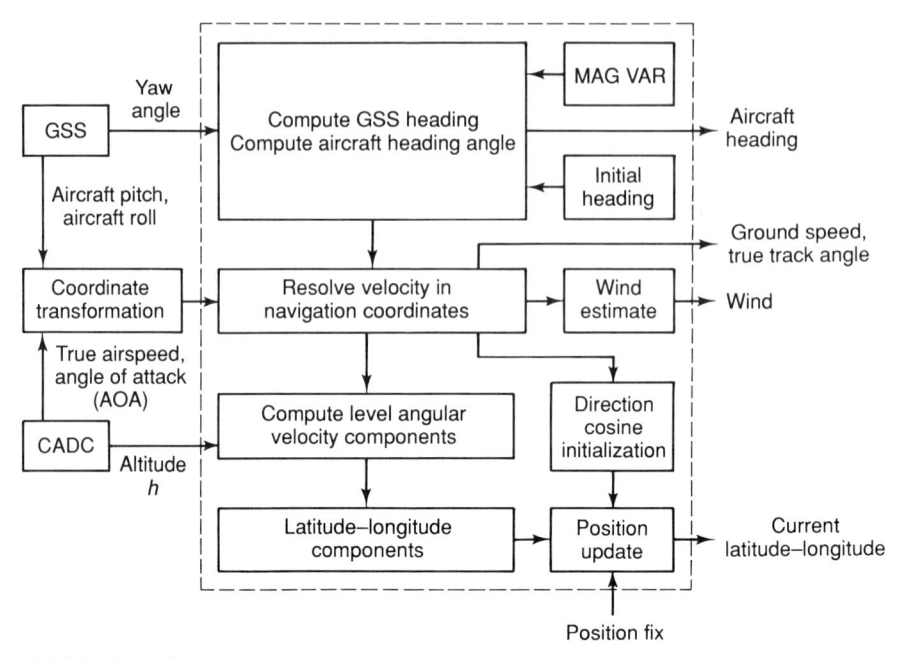

FIGURE 7-12
Airspeed dead-reckoning mode.

Longitude

$$\lambda = \lambda_0 + \int \dot{\lambda} \, dt, \qquad \dot{\lambda} = \frac{V_E}{(R_e + h) \cos \phi} \qquad (7.39)$$

where R_e is the radius of the earth and h is supplied from the CADC. Finally, in the GSS slaved magnetic mode, the angle α_g is the correction angle supplied by the MAG VAR table for magnetic variation. Moreover, in the GSS free azimuth mode, α_g is the system wander angle, assuming the form

$$\alpha_g = \alpha_{g_0} + \int (\dot{\lambda} + \Omega_e) \sin \phi \, dt \qquad (7.40)$$

where α_{g_0} is an initial value, such as stored heading or corrected magnetic heading. The Doppler radar block diagram is similar to Fig. 7-11, with the exception that the coordinate transformation block is replaced by the Doppler radar, and the outputs from the Doppler block are the groundspeed and drift angle.

As stated earlier, the airspeed sensor or CADC senses (scalar) the velocity of the aircraft relative to the airmass. In equation form, we can

7.3. Special Navigation Systems

write the following relationship:

$$\mathbf{V}_{air} = \mathbf{V}_g - \mathbf{W}, \qquad |\mathbf{V}_g| = (V_N^2 + V_E^2)^{1/2} \qquad (7.41)$$

where $\mathbf{V}_g$ is the groundspeed and $\mathbf{W}$ is the wind vector. However, because of various errors existing in the airspeed hardware, the airspeed output can be modeled more completely by the equations

$$\hat{V}_{air} = (1 + \Delta SF) V_{lag} + V_{noise} + V_{bias} \qquad (7.42)$$

$$\frac{d}{dt} V_{lag} = \frac{1}{\tau_{CADC}} (V_{air} - V_{lag}) \qquad (7.43)$$

where ΔSF is the scale factor error and τ_{CADC} is the dynamic lag time in seconds. Therefore, the true airspeed is modified by a dynamic lag of the sensor, and subsequently the output process adds a scale factor, noise, and bias errors. Figure 7-13 shows the mechanization equations that constitute the Doppler dead-reckoning mode.

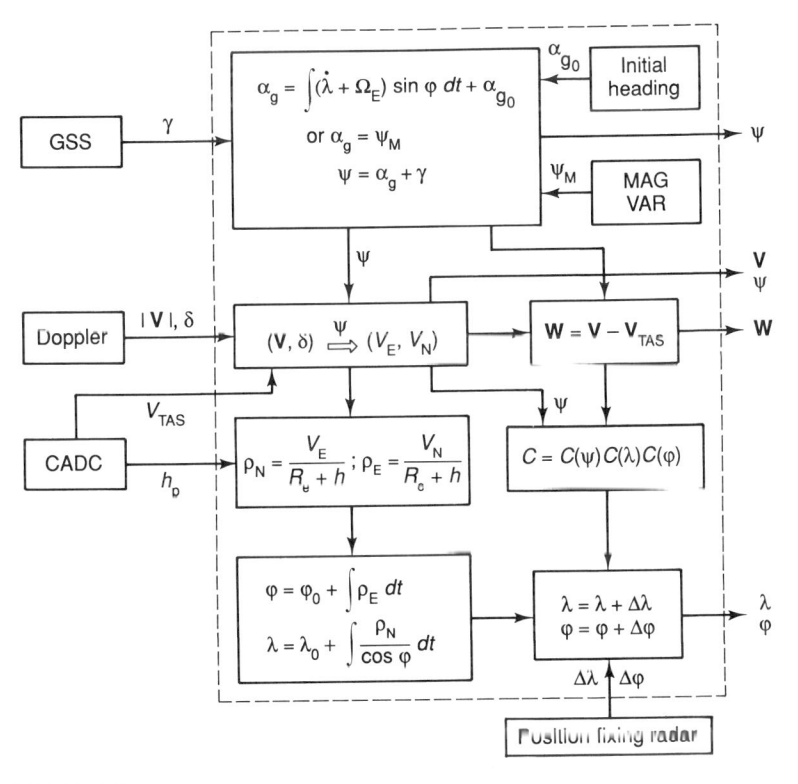

FIGURE 7-13
Doppler dead-reckoning mode.

A simplified version for a DR error model in terms of the velocity, and latitude–longitude errors, that can be used in connection with a covariance analysis simulation program satisfies the following linear stochastic differential equations:

$$\frac{d}{dt}\phi(t) = V_N(t) \tag{7.44a}$$

$$\frac{d}{dt}\lambda(t) = V_E(t) \tag{7.44b}$$

$$dV_N(t) = -\frac{1}{\tau_c} V_N(t)\, dt + dw_1(t) \tag{7.44c}$$

$$dV_E(t) = -\frac{1}{\tau_c} V_E(t)\, dt + dw_2(t) \tag{7.44d}$$

where ($\tau_c > 0$) denotes the correlation time, V_N and V_E denote the north and east components of the velocity error, and ϕ and λ denote the latitude and longitude errors, respectively. Furthermore, the process noise $\mathbf{w} \equiv [w_1 \ w_2]^T$ is a Wiener process with independent components. Note that this model does not represent the real world but is used here for the purposes of illustration. In today's airspace environment, acceptable short-range navigation system accuracy can be accomplished using VOR/DME. In an integrated VOR/DME-DR system that uses true airspeed (V_{TAS}), heading, and estimated wind velocity, the measurements of the Kalman filter are distance differences (i.e., distance from the DR position to the DME ground station used, minus the DME distance). Table 7-1 lists the various parameters that a DR system outputs.

TABLE 7-1

Dead Reckoning Outputs

Parameter	Symbol
Latitude	ϕ
Longitude	λ
Altitude	h_0
North velocity	V_N
East velocity	V_E
Vertical velocity (up)	V_Z
North wind	W_N
East wind	W_E
Groundspeed	V_g
Roll	ϕ
Pitch	θ
True heading	ψ_T
Velocity heading	ψ_V
Drift angle	δ

7.3.2. Area Navigation

Area navigation (R-NAV) is a flight technique for straight-line and point-to-point parallel-route air navigation used by commercial and military transport category aircraft as well as general aviation aircraft. A typical flight, however, is not a straight line directly between the point of departure and destination but is a sequence of straight-line segments, defined by waypoints. Area navigation permits aircraft operations on any desired course within the coverage of station referenced navigation signals, or within the limits of self-contained system capability. Moreover, area navigation utilizes the airspace or corridors removed from the basic VOR/DME ground routes, thus eliminating convergence at the en route VORTAC stations. A typical inertial based R-NAV system, such as the Litton Aero Products LTN-72RL system, offers accurate, worldwide all-weather area navigation, flight and fuel management, true or magnetic heading reference, position in Lat/Long and bearing to a VORTAC ground station, Greenwich mean time/ETA (estimated time of arrival), and automatic radio updating and tuning capability. Above all, this system is a state-of-the-art navigation system and is compatible with the ARINC*561 INS installations. As noted in Chapter 6, system position in latitude and longitude is computed from the inertial navigation system position and the latitude and longitude update of either radio navaids or manual entry. In the R-NAV mode, however, the updates are corrected by sequential measurements of DME range and VOR omnibearing. Ground-based systems applicable to R-NAV include the following: (1) VOR/DME, (2) DME/DME, (3) TACAN, (4) Omega, (4) Loran, and (5) GPS. When using the VOR/DME mode (using range and bearing, i.e., Rho–Theta), the R-NAV system provides highly accurate horizontal navigation, independent of time and inertial navigation system drift errors. It might be pointed out that use of VOR/DME is extensive, especially in high-traffic areas.

From the preceding discussion, it is seen that when provided with signals from ground-based navaids, the inertial navigation system functions as an area navigation system. In this mode, the system can automatically control the channel or frequency selection of, say, VOR and TACAN receivers aboard the aircraft, so that sequential data from multiple stations is available for use in the system update Kalman filter. More specifically, the Kalman filter uses the varying aircraft–station geometry, caused by the aircraft moving relative to the station, to execute system position updates. Following a period of radio updates, the pure inertial navigation system error propagation is reduced by the corrections to the INS error sources estimated during the update process. With regard to accuracy, the ground-based aids provide angle accuracy data as a function of range to the station. In the direction of

---

*Aeronautical Radio, Inc.

the station, the accuracy is about 0.1852 km (0.1 NM). On the other hand, in the perpendicular direction the accuracy is directly proportional to the bearing measurement error and range. Typically, a catalogue of ground-based navaids or stations is stored in the aircraft's navigation computer memory for use during en route navigation, as well as in executing an R-NAV approach to an airport. The aforementioned list of ground stations as well as geographic and operational data on individual stations is available for display on the control and display unit (CDU). Stations can be tuned automatically by the system "autotune" feature, or manually by the operator.

Self-contained inertial navigation systems operating independently of ground radio aids also offer area navigation capability. These systems compute distances, bearings, and command guidance signals relative to preselected waypoints using the present position obtained from DR. Typically, the R-NAV system can guide an aircraft over a complete flight, with a minimum of pilot operations. R-NAV features the following: (1) select desired flight plan (note that the desired flight plan is selected while the aircraft is at the gate), (2) select departure runway and procedure, (3) enter en route changes as required by ATC (air traffic control), and (4) select arrival and approach. Finally, the R-NAV can be used to provide continuous lateral and vertical guidance without the necessity of operation on conventional airways. In addition, the system can automatically compute the vertical angle to fly from one waypoint to another, with the selected along-track offset. The following diagram illustrates the concept of area navigation in the lateral mode.

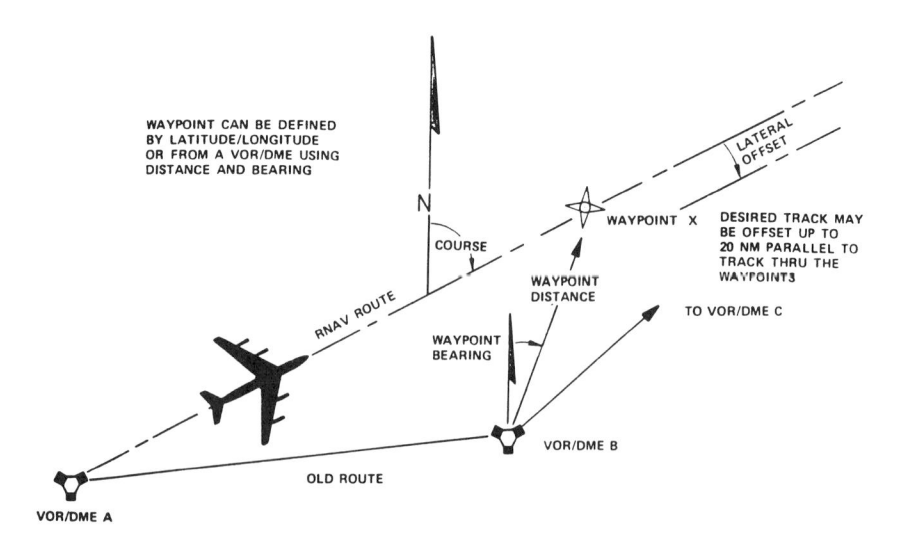

7.3.3. Attitude-and-Heading Reference Systems

An attitude–heading reference system (AHRS) provides roll, pitch, and heading information to aircraft indicators, flight directors, flight controls, and other avionics subsystems. Strictly speaking, an AHRS is a dead-reckoning system which provides continuous attitude, heading, position, and velocity information, and fills the gap between slaved directional gyrocompass systems and vertical gyroscopes, as well as inertial navigation systems used in the most demanding applications. In the unaided mode, the AHRS provides low-accuracy position and velocity with unbounded errors. On the other hand, an integrated GPS/AHRS system combines the best properties of both systems, resulting in a continuous navigation system with bounded errors [10]. Because of the widespread use of ring laser gyro (RLG) and/or fiber-optic technology, most AHRS systems in use today are of the strapdown type. For this reason, in the ensuing discussion we will consider a strapdown AHRS configuration. An attitude–heading reference system implies the availability of heading and velocity reference information. For the AHRS system under consideration, magnetic heading data is assumed to be available for use as the heading reference, while the airspeed, as provided by the CADC, constitutes the velocity reference. Therefore, a typical modern strapdown system will consist basically of a strapdown inertial measurement unit (IMU), an air data system, and a flux valve. The inertial package will normally consist of three single-degree-of-freedom gyroscopes to measure angular motion in three axes, and a triad of three accelerometers to detect linear acceleration. This inertially derived information is processed via the appropriate computational algorithms and then combined with reference data in a Kalman filter to generate optimal estimates of aircraft attitude and heading. Note that other navigational data, such as velocity, position, and body rates, and accelerations are also available for use by other aircraft subsystems. The air data system provides temperature and pressure data for computing the true airspeed and barometric altitude. In addition, the true airspeed (TAS) is employed in order to control the roll and pitch, as well as the level velocity errors of the system (i.e., AHRS), while the baro altitude provides a reference for damping the system's vertical channel. Figure 7-14 shows a generic AHRS.

If a Doppler radar is available for aiding instead of the airspeed sensor, then the system performance will only be limited by the Doppler radar performance and the dynamical response of the strapdown system itself. Body angular rate and acceleration data are transferred to the high-speed digital computer in the form of angle and velocity increments, where they are processed by attitude propagation and navigation algorithms to compute aircraft roll, pitch, heading, velocity, and position. Magnetic heading reference information, as mentioned above, is provided by a flux valve or other

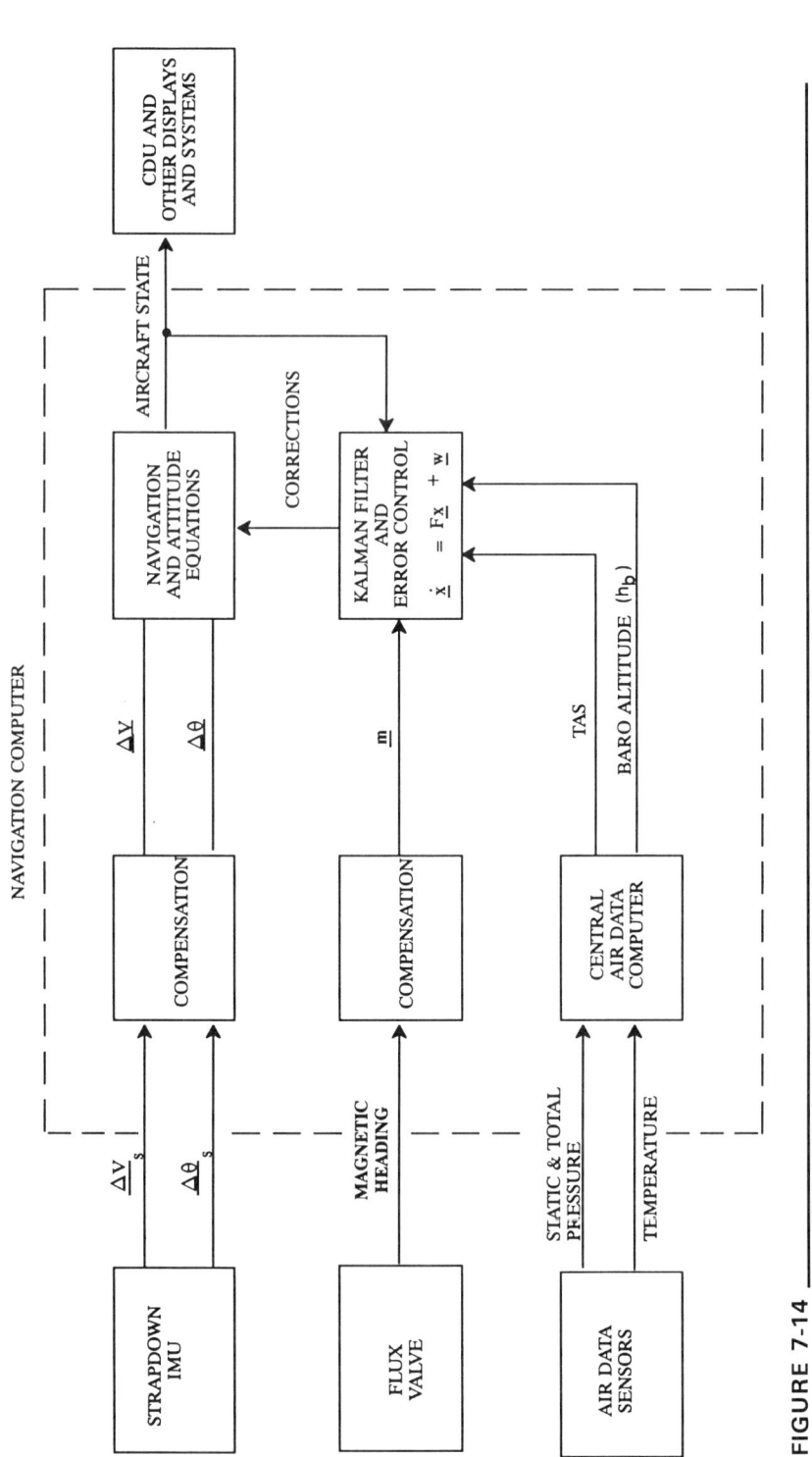

FIGURE 7-14
Strapdown AHRS block diagram.

magnetic sensor. Corrections for magnetic variation (MAG VAR) can be considered to be generated in the computer via an appropriate position-dependent model. The attitude and heading can be generated via a third-order quaternion algorithm; velocity and position are obtained in north, east, and down (NED) coordinates by integration of the transformed incremental velocity data. Furthermore, an attitude and heading reference system is designed to determine the orientation of the carrying vehicle relative to a locally level, north-pointing system. This orientation or attitude is defined by the conventional Euler angles (ϕ, roll; θ, pitch; ψ, yaw or azimuth).

As with all navigation systems, a Kalman filter can be employed to estimate the AHRS error state from redundant sensor data. Estimated errors are fed back into the system using a combination of continuous and impulsive control methods in order to bound the inertial velocity, attitude, and heading errors, as well as to reduce the rate of position error growth. As discussed in Chapter 6, the Kalman filter provides statistically optimal estimates of all variables in a dynamical system based on noisy observations of some of these variables and their interrelations. In particular, inertially derived velocity, altitude, attitude, and heading are combined with barometric altitude, TAS, and magnetic field data in the Kalman filter to obtain minimum-variance estimates of the system time-correlated errors. Consequently, these estimates are fed back into the system via an error-control algorithm, which resets the inertial velocity, position, and attitude integrators every filter cycle, and compensates measured acceleration and angular rate data for estimated gyroscope and accelerometer bias errors [10].

AHRS system outputs include the following parameters: (1) roll (ϕ) and pitch (θ), (2) true (ψ) and magnetic heading (ψ_M), (3) V_N, V_E, V_D velocity components, (4) altitude (h), (5) W_N, W_E wind components, and (6) body accelerations and angular rates. It should be noted that the proper utilization of the magnetic heading and airspeed data is necessary for effective operation of the attitude system.

7.3.3.1. System Error Model

The development of the AHRS system error model is similar to the error models developed for an INS in Chapters 5 and 6. For this reason, the development of the AHRS error model will be brief. Since for the present discussion a strapdown system has been assumed, the compensated gyroscope outputs are used to propagate the direction cosine matrix C_N^B, which transforms the aircraft body axes (B) into a local-level north–east–down (NED) navigation coordinate frame (N). The Euler angles (roll, pitch, and yaw) can be computed from this direction cosine matrix. From Chapter 2, Section 2.2.10, we know that the direction cosine entries may be derived by successive applications of rotations about the Z direction through the azimuth angle (ψ), then a rotation through the pitch angle (θ) about an intermediate axis, and a final rotation through the roll angle (ϕ) about a second intermediate axis. Therefore, the transformation from the navigation (N) body coordinate frame to the body (B) coordinates is given by [see also Eq. (2.35)]

$$C_N^B = \begin{bmatrix} 1 & 0 & 0 \\ 0 & c\phi & s\phi \\ 0 & -s\phi & c\phi \end{bmatrix} \overset{\text{Roll}}{} \begin{bmatrix} c\theta & 0 & -s\theta \\ 0 & 1 & 0 \\ s\theta & 0 & c\theta \end{bmatrix} \overset{\text{Pitch}}{} \begin{bmatrix} c\psi & s\psi & 0 \\ -s\psi & c\psi & 0 \\ 0 & 0 & 1 \end{bmatrix} \overset{\text{Azimuth}}{}$$

$$= \begin{bmatrix} c\theta c\psi & c\theta s\psi & -s\theta \\ -s\psi c\phi + s\theta c\psi s\phi & c\phi c\psi + s\phi s\theta s\psi & s\phi c\theta \\ s\psi s\phi + s\theta c\psi c\phi & -s\phi c\psi + c\phi s\theta s\psi & c\phi c\theta \end{bmatrix}$$

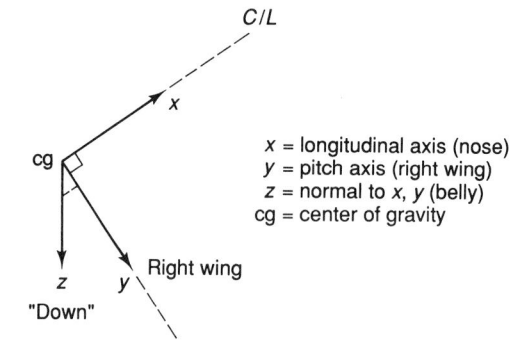

x = longitudinal axis (nose)
y = pitch axis (right wing)
z = normal to x, y (belly)
cg = center of gravity

where $c(\cdot)$ denotes the cosine function and $s(\cdot)$ denotes the sine function. The angles (ϕ, θ, ψ) are the usual Euler angles. From the transformation matrix C_N^B, we obtain the Euler angles as follows:

$$\theta = -\sin^{-1}(C_{13}) = -\tan^{-1}\left[\frac{C_{13}}{\sqrt{1-C_{13}^2}}\right] \quad (7.45a)$$

$$\phi = \tan^{-1}\left[\frac{C_{23}}{C_{33}}\right] \quad (7.45b)$$

$$\psi = \tan^{-1}\left[\frac{C_{12}}{C_{11}}\right] \quad (7.45c)$$

When the accelerometer outputs are transformed into the navigation frame, then $\Delta \mathbf{V}^N = C_B^N \Delta \mathbf{V}^B$. Note that the AHRS must be aligned before the attitude algorithms are initialized; this is done by calculating the starting value of

C_B^N. The Kalman filter used for processing the AHRS sensor data is a variable-dimension model of the system error behavior. In particular, this model defines the way in which the system errors propagate in time. The mathematical model (or plant) is of the standard form

$$\dot{\mathbf{x}} = F\mathbf{x} + \mathbf{w} \tag{7.46}$$

$$\mathbf{z} = H\mathbf{x} + \mathbf{v} \tag{7.47}$$

where $\mathbf{x}$ ($n \times 1$) is the state vector, F ($n \times n$) is the system error dynamics matrix, $\mathbf{w}$ ($n \times 1$) is a white process noise vector with spectral density Q, $\mathbf{z}$ ($m \times 1$) is the measurement vector, H ($m \times n$) is the measurement matrix, and $\mathbf{v}$ ($m \times 1$) is a white measurement noise sequence with covariance R. A typical dynamical error model that can be used for error propagation in the analysis of the strapdown attitude system will consist of sixteen states: three velocity errors, three angular (or attitude) errors, three gyroscope bias errors, three accelerometer errors, two wind error equations, one magnetic heading reference error, and one airspeed sensor bias error. The number of states can be reduced by noting that, in general, the vertical axis error equations are not strongly coupled to the error equations of the horizontal axes. That is, the vertical errors do not exhibit the interaction between the states in an AHRS that the horizontal errors do. Consequently, the vertical velocity and position do not have to be included in the plant dynamics. These two parameters can be updated in a separate loop. Furthermore, if a constant altitude flight profile is assumed, the cross-coupling is further weakened. The state vector $\mathbf{x}$ can be partitioned into three subvectors as follows: (1) those states describing the unforced dynamical behavior of the unaugmented plant, (2) the states describing the dynamical characteristics of the plant error sources (i.e., gyroscope drift and accelerometer errors), and (3) the states describing the errors in the references (i.e., errors in the magnetic heading and airspeed) [9]. The discussion in Chapter 5 can be used to set up the appropriate inertial error equations.

The AHRS makes an excellent backup system for commercial as well as military transport aircraft. Thus, the AHRS systems serve an important function, namely, to guide pilots if the primary (e.g., INS) aircraft navigation system fails. Typical AHRS performance characteristics achieved with the present state-of-the-art inertial sensors are as follows:

Attitude accuracy (rms):
    Heading: $0.50°$ (true)
                                     $0.80°$ (magnetic)
    Pitch/roll: $\leq 0.20°$
Angular rates (rms): 0.1 deg/s
Groundspeed: 4.0 kt
Linear acceleration: 0.1 m/s^2

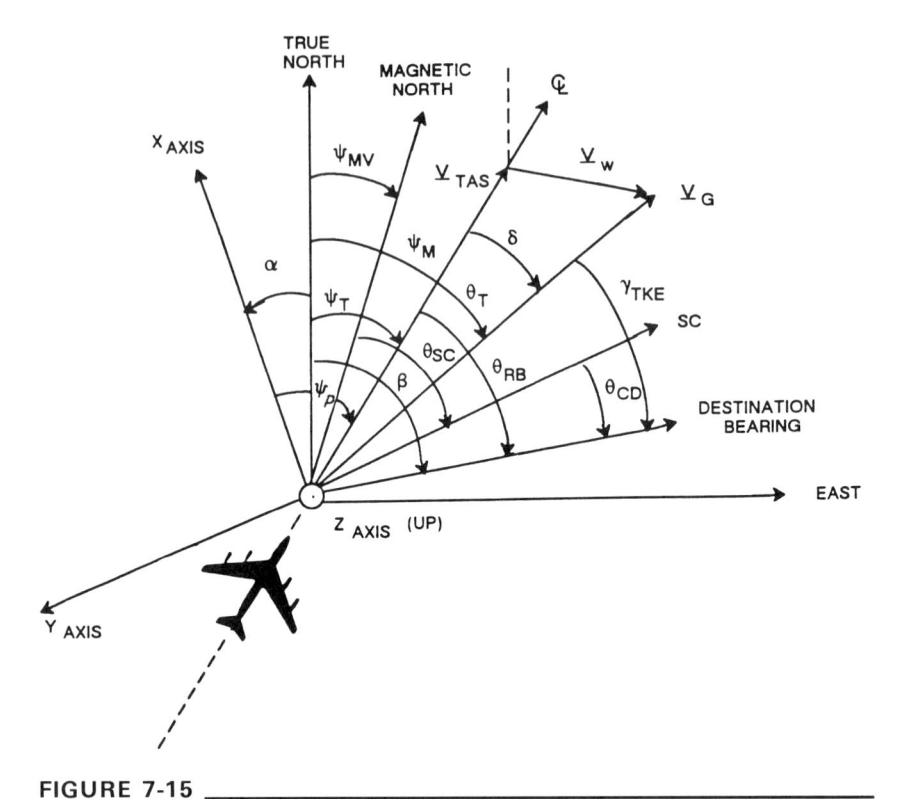

FIGURE 7-15
Navigation angles.

The various aircraft angular relationships (in the horizontal plane) that are used for steering, dead reckoning, and navigation from one waypoint to another are summarized in Fig. 7-15.

The angles in Fig. 7-15 are defined as follows:

V_G = aircraft groundspeed = $\sqrt{(V_E)^2 + (V_N)^2}$
V_w = wind speed
V_{TAS} = true airspeed
α = azimuth (wander) angle
β = great circle bearing to destination
γ_{TKE} = great circle bearing error
δ = drift angle = $\theta_T - \psi_T$
θ_T = aircraft ground (true) track angle = $-[\alpha + \tan^{-1}(V_E/V_N)]$
ψ_P = platform heading
ψ_T = true heading = $\psi_M + \psi_{MV} = \psi_P - \alpha$
ψ_M = magnetic heading = $\psi_T - \psi_{MV}$

ψ_{MV} = magnetic variation
θ_{RB} = relative bearing
θ_{SC} = selected magnetic course (selected via the HSI knob)
θ_{CD} = course deviation = $\beta - (\theta_{SC} + \psi_{MV})$
SC = selected course (magnetic)
$\text{\textcentiline}$ = aircraft centerline

Some of the definitions, such as course and heading, may vary in aircraft and marine applications. The definitions for the appropriate application are given in the following lists.

Aircraft Applications

Course—the intended direction or path of travel over the surface of the earth with respect to north; that is, a planned route or direction of flight with reference to a line on the surface of the earth that gives the desired track. Course may be expressed as magnetic or true.

Heading—the angle formed between the longitudinal axis of the aircraft and a north reference measured clockwise from the north in the horizontal plane. Heading may be true or magnetic. True heading is with respect to the true north; magnetic heading is with respect to magnetic north.

Track—the actual direction or path of travel over the ground with respect to true north.

Bearing—the direction of a point from an aircraft, measured in a clockwise direction in the horizontal plane from a line of reference. Bearing may be true, magnetic, or relative.

Azimuth bearing in the horizontal plane, usually expressed as an angle, measured clockwise from true north or magnetic north from 0° to 360°.

Drift angle—the angular difference between the true heading and ground track.

Marine Applications

Course—the direction prescribed for the ship's movement or progress; also defined as the direction in which the ship sails from one place to another.

Heading—the direction in which the ship actually points or heads at any particular moment. "Heading" should not be confused with "course." A ship frequently is off course, but is never off the heading. There are three kinds of headings: (1) true, (2) magnetic, and (3) compass.

Relative bearing—the direction of an object from the ship, relative to the ship's head; refers to the heading of the ship, and equals the angle between fore-and aft line of the ship and the bearing line of the object, measured clockwise from 0° to 360°.

7.3.4. Electronic Combat Systems and Techniques

In Section 6.2.9, it was mentioned that the JTIDS system is designed to be jam-resistant. During war operations, it becomes necessary to jam enemy radars and/or communications systems. This section describes briefly the jamming methods presently available. Jamming techniques have been used by the USAF with considerable success during the Persian Gulf War ("Operation Desert Storm"). The USAF made a large investment in electronic warfare technology, particularly in radar-jamming protection pods that have been used since the Vietnam War. (The external radar jammer pod is normally carried below the fuselage on a weapon pylon). Specifically, the USAF's tactical aircraft (e.g., A-10s, F-15C/E, F-16C/D) are equipped with a variety of electronic countermeasures (ECM) systems. For instance, radar warning receivers (RWR) provide warning and classification of radar threats and rough directional information. As stated above, radar jammers are carried internally or in external pods. These radars are designed to confuse threat radars as to the aircraft's present position. Expendable decoy dispensers eject chaff to divert radar-guided surface-to-air and air-to-air missiles and flares in order to divert infrared (IR)-guided missiles. In addition to the above equipment, aircraft equipped with "radar threat warning" systems can detect, identify, and determine the azimuth bearing of the various threats, either land-based or ship-based. The system can also indicate SAM (surface-to-air missile) weapon status, whether the aircraft is being tracked or whether a weapon has been launched. The azimuth indicator is a CRT that displays threats as alphanumeric or geometric symbols relating to each threat, and is commonly located to the left of the head-up display. The prime indicator control panel contains six pushbutton indicators that allow the pilot to select the system modes of operation.

The USAF's attack and fighter aircraft are also protected by electronic combat support aircraft, such as the EF-111A Raven radar jamming aircraft, the EC-130H Compass Call communications jamming aircraft, and the F-4G Wild Weasel defense suppression aircraft. The F-4G aircraft is armed with the Texas Instruments' High-Speed Antiradiation Missile (HARM), which homes in on enemy radars. The Compass Call is a specially configured aircraft built and electronically integrated by the Lockheed Company. Compass Call aircraft flew numerous sorties in Operation Desert Storm, supporting attack as well as air superiority missions by jamming enemy command, control, and communications (C^3) systems. Furthermore, USAF fighters were heavily supported during their missions by the EF-111 and the Navy's EA-6B dedicated jamming aircraft and by HARMs fired at enemy air defense radars by F-4G Wild Weasels and Navy A-7s and F/A-18s. At the present time, the F-4G is the only aircraft that can autonomously detect, identify, and accurately locate and destroy enemy target tracking and acquisition radars with HARMs.

7.4. MODERN AVIONICS SYSTEMS

Present-day commercial and military aircraft avionics systems, such as multiple radios, navigation systems, weapons, aircraft survivability systems, and flight management systems, are ever-increasing in complexity. Each has dedicated controls that require the pilot's attention, particularly during critical phases of the mission. Moreover, the task is compounded when the pilot's accessibility to dedicated controls is limited by cockpit space restrictions (such as in a fighter aircraft). Advanced flight management systems have already been incorporated into the Boeing 757/767 and 747. These newer flight management systems include an advanced flight management computer with bubble memory, providing twice the memory of current generation systems. Sperry, Honeywell, Rockwell-Collins, and Litton are some of the companies in the United States that are in the forefront in the design of advanced flight-control systems. SEXTANT Avionique in France is also a leader in this field. Heading into the twenty-first century, cockpit management systems, using integrated avionics systems, will reduce a multiple of controls into essentially one panel-mounted CDU. Some of the benefits of such a system include the following: (1) reduced cockpit workload, (2) simplified future avionic upgrades, (3) simplified and consistent pilot procedures, and (4) fully redundant avionic control. The cockpit management system will combine internal navigation inputs from the INS, GPS, Loran, Doppler radar, and AHRS to provide a simple determination of position at any given time. Furthermore, it will estimate the error states in the individual navigation subsystems to improve the quality of navigation following the failure of any subsystem. Above all, using intelligent management software, the cockpit management system will provide graceful degradation, which will automatically reconfigure responses to failures or changes in system status, with minimal impact to the mission.

Even today, as we proceed into the 1990s, it is certain that the 1990s will usher in a new era of advanced avionics systems for commercial airliners, which will make the pilot's task easier. For instance, it is anticipated that the new generation of commercial aircraft will use integrated modular avionics, which will become an integral part of the avionics architecture for these aircraft. The integrated modular avionics suite will enable the integrated avionics to share such functions as processing, input/output (I/O), memory, and power supply generation. The flight decks of these new generation of airliners will incorporate advanced features such as flat-panel screens instead of cathode-ray tubes (CRTs), which will display flight, navigation, and engine information. Another important area is that of "fault-tolerant avionics," in which no single failure can cause the loss of an essential avionics function. This level of performance can be achieved by building redundancy into the system. Seen from another point of view, such a system must fail active (e.g., no change in functionality or safety) on the first fault, fail active on the second fault, and fail passive (with no degradation in flight safety) on the third fault. When fully developed, fault-tolerant avionics may not even alert the flight crew that a failure has occurred, unless the crew must take steps to do something about the failure, or the system is a failure away from serious operational degradation requiring their immediate attention. In addition, since airline operations are driven by seat-mile economics, payload and range factors that enable the airlines to offer services more economically, commercial aircraft manufacturers designing new aircraft for the 1990s and beyond, such as the Boeing 777, will incorporate into the avionics suite an airplane information management system (AIMS), as well as an air data/inertial reference system (ADIRS). Such combined air data/inertial reference systems have been developed by Honeywell and Delco. The AIMS system will provide the flight-deck crews with pertinent information on the condition of the aircraft, its maintenance requirements, as well as key operating functions such as flight, thrust, and communications management. The ADIRS navigation system, being developed by Honeywell for the Boeing 777, will consist of a six ring laser gyro (6-RLG) package, in a skewed-axis configuration for fault-tolerant operation. The ADIRS will replace the conventional air data, inertial reference system.

Many of the new generation jetliners will be designed using "fly-by-wire" flight-control systems. In this design, advanced electronic mechanization and control theory are used to eliminate the conventional mechanical linkages and control cables in all axes. As a result, no mechanical linkages or control cables are used between the cockpit controllers (side-stick, rudder pedals, etc.) and the control surface integrated servo-actuators, the leading-edge flap power drive unit, and the speed brake. Fly-by-wire will save weight and lower maintenance costs by means of improved reliability and fewer wires and connectors.

7.4.1. Flight Instruments and Displays

This section will be devoted to the flight instruments and displays that are part of commercial and/or military aircraft avionics suites, which will help the flight crew perform their duties in the most efficient manner and complete the mission safely. Displays associated with the navigation system are on the data-entry displays of the up-front controls. In particular, the magnetic variation display will show the automagnetic variation or it can be entered via a manual mode. The INS memory display shows the contents of the memory locations entered by the operator. Furthermore, the INS alignment mode shows the alignment status, time into alignment, latitude, longitude, system altitude, and true heading. The various flight instruments and displays, some of which are conventional while others are modern, are described below. As mentioned above, in today's modern airliners that are using flat panel color displays, replacing current-generation CRTs as primary flight and engine instrument indicators, many of the avionics functions (e.g., airspeed, attitude display, altimeter, vertical speed indicator) have been combined into one instrument. To improve readability, the present units have photometers that read light striking the surface and adjust the intensity of the displays. The rate of intensity change is timed to the eye's adjustment speed. Other innovations include a Category 3B Automated Flight Control and Augmentation System (AFCAS). The Category 3B system, which allows a decision height of 4.57 m (15 ft) and runway visual range of 150 m (492 ft), includes three flight-control computers, a dual-flight augmentation computer, and two sets of servos. In case one of the components fails, the AFCAS system is still certified for Category 3A landings [the Category 3A autoland system allows a decision height of 15.24 m (50 ft) and runway visual range of 200 m (656.2 ft)]. In addition, the autoland system is designed to be compatible with microwave landing systems (MLS) and the global positioning system, as these systems become operational. All control inputs are fed through the flight control computer, so that commands that would exceed the aircraft's design limits are ignored. As is the case with current generation aircraft, the flight management system and computer allow the pilot to insert the route numbers. Consequently, the computer plots all headings, flight times, and other information and directs the pilot to fly the designated route. Thus, an aircraft can be flown automatically from immediately after takeoff to landing. In the case of wind shear, the flight management system can take corrective action automatically as the aircraft enters such a region. Finally, the advanced displays and/or integrated flight management systems are an outgrowth of airline–aircraft manufacturer cooperation.

Control-and-Display Unit (CDU) The CDU contains the primary navigation display and data-entry controls for the operation of the INS, and it is the interface unit that permits the operator to communicate with the onboard navigation computer. The CDU provides the flight with the following capabilities: (1) loading the flight plan into the navigation computer, (2) monitoring navigation data derived from the computer, (3) loading position update information during flight in order to correct position errors, and (4) providing a means of reading malfunction codes supplied by the computer. The keyboard provides the capability of loading flight information, that is, initial position coordinates, waypoint-destination position coordinates, and the desired performance index number. The information is encoded and routed to the computer I/O section. In addition, indicators show the system state of readiness, requested data displays, malfunctions, and other numerical, discrete display, and word messages. For example, the power supply is enabled by a discrete from the navigation unit when the INS is placed in the standby mode, while the WARN and BAT lamps are illuminated by discretes received from the navigation unit.

Radar Altimeter Some aircraft and/or cruise missiles are equipped with a combined altitude radar altimeter (CARA). The purpose of the CARA is to provide the pilot and other avionics subsystems with accurate altitude information from 0 to 15,240 m (50,000 ft). Altitude information can be displayed on a head-up display. In this case, five different scales are used to display the aircraft's altitude. The altitude information, which is height above sea level, comes from the CADC. The radar altitude is displayed in a dedicated window that is preceded by the letter R when in the barometric altitude mode, and AR when in the autoradar altitude mode. The interface between the CARA and other avionics subsystems consists of data blocks of serial digital words to the head-up and display unit that are an analog value of the aircraft's altitude, and discretes.

Head-up Display (HUD) The head-up display is an electronic and optical instrument that presents, in symbolic form, such essential functions as aircraft performance information and attack (in the case of a fighter) and navigation or landing guidance on a single display. The symbology is focused to infinity, projected on a transparent combiner, and displayed in front of the pilot at eye level. The HUD receives computed attack, navigation, and landing data from the computer, aircraft performance data from aircraft flight sensors, and discrete signals from various aircraft systems. The HUD operates in navigation, attack, landing, and manual modes with appropriate indicators and "fly-to" steering commands for each. The flight path marker, representing the point toward which the aircraft is flying (i.e., velocity vector), and flight path angle lines are constantly displayed. Scales indicating airspeed, altitude, magnetic heading, and vertical velocity are displayed or can be turned off at the pilot's discretion.

Horizontal Situation Indicator (HSI) The horizontal situation indicator provides the pilot with heading and navigation information. Navigation information is derived from the INS, instrument landing system (ILS), or TACAN, depending on the position selected on the mode switch on the instrument mode select coupler. The HSI provides the following displays:

Aircraft magnetic heading This heading is displayed by a rotating compass card that is read against a fixed lubber line at the top of the dial.

Selected heading A heading may be selected using the HDG knob. A split double bar on the outer periphery of the compass card rotates to indicate the selected heading.

Selected course Here, a TACAN course can be selected using the CRS knob. The selected course is indicated by an arrow, which is read against the compass card. The course selection is also displayed on a three-digit course display in the upper right corner of the instrument.

Bearing Bearing to or from the TACAN station selected is indicated by a bearing pointer located on the outer periphery of the compass card. Also, two small arrows on the dial face indicate whether the aircraft is to or from the selected station. If an arrow points to the head of the course arrow, the aircraft is heading toward the station; if the arrow points to the tail, the aircraft is heading away from the station.

Distance Distance to or from the TACAN station or INS computed range to destination (in nautical miles) is indicated by a three-digit display in the upper left corner of the instrument face.

Figure 7-16 shows a conventional (analog or digital) HSI.

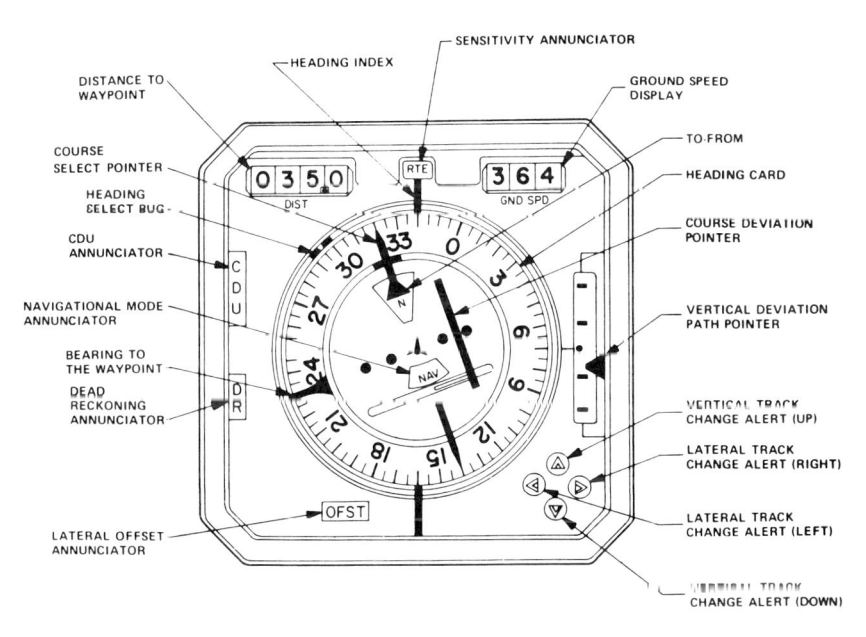

FIGURE 7-16
Conventional horizontal situation indicator.

Attitude Direction Indicator (ADI) The conventional ADI provides pitch and roll attitude, ILS glide slope and localizer deviation, as well as aircraft turn rate and slip information. The ADI consists of the following displays:

Attitude display The attitude display consists of an attitude sphere that is read against a fixed aircraft symbol for pitch and pointers on the top and bottom of the sphere, which are read against fixed roll indices along the outer periphery of the indicator face. The attitude display has operating ranges of 0° to 87.5° in pitch and 0° to 360° in roll.

Flight director display This display consists of horizontal and vertical steering bars across the front of the indicator and a glide slope pointer on the left of the attitude display. These two pointers indicate the following information:

- The vertical pointer displays the aircraft horizontal position with respect to the proper glide path.
- The horizontal and glide slope pointers display the aircraft vertical position with respect to the proper glide path.

Alarm flags Four alarms indicate malfunctions in the indicator or to indicator inputs. During normal operation, all flags are out of view. The flags indicate the following:

- Glide slope (GS) alarm flag: indicates loss of flight director information to the horizontal pointer.
- Localizer (LOC) alarm flag: indicates loss of flight director information to the vertical pointer.
- Auxiliary attitude (AUX) alarm flag: indicates that the ADI is displaying attitude information from a backup source.
- Attitude warning (OFF) flag: indicates one or more of the following conditions—loss of 115 V to ADI, failure of the roll and pitch displays to track attitude input signals, and loss of external attitude data validity from the attitude reference source.

Primary Flight-Control System The principal function of the primary flight-control system (PFCS) is to provide aircraft three-axis flight path control. This is accomplished through the use of minimum displacement-type force-sensing control stick and rudder pedals and through pilot initiated trim commands in each of the three axes. The PFCS includes stability augmentation to provide static stability in pitch, in addition to continuous automatic damping about the three control axes. Command augmentation is used to provide the pilot with superior three-axis control via the following hydraulically actuated control surfaces: (1) *pitch*—symmetric deflection of the all-movable horizontal tail, (2) *roll*—asymmetric deflection of flaperons on each wing and asymmetric deflection of the all-movable horizontal tail,

7.4. Modern Avionics Systems

and (3) *yaw*—deflection of the rudder on the trailing edge of the vertical stabilizer. The PFCS interfaces directly with the secondary flight-control system and the air data system. For example, in the F-16 fighter aircraft, the pilot achieves longitudinal (i.e., pitch) control through the fore-and-aft forces applied to the side-stick controller. The longitudinal (pitch) axis employs both command augmentation and stability augmentation for achieving precise flight path control. The PFCS uses an angle-of-attack limiter function to limit the maximum obtainable angle-of-attack, enabling the pilot to maneuver the aircraft to its maximum usable angle-of-attack without loss of control.

Instrument Mode Select Coupler The instrument mode select coupler (IMSC) is a panel-mounted control unit that provides switching functions to select and control the displays of the HSI and ADI. The mode select control knob is used to select one of four input combinations: (1) TACAN, (2) ILS/TACAN, (3) NAV, and (4) ILS/NAV.

Magnetic Compass The magnetic compass is used as a standby to the HSI compass card. It contains a rotating compass card with magnets attached so that the card continuously aligns with the magnetic north. Normally, the compass card is graduated in north, south, east, and west indications with $50°$ graduations following each indication.

Angle-of-Attack Indicator The angle-of-attack (AOA) displays true AOA using a moving tape read against a fixed lubber line. The indicating range is from minus $5°$ to $\pm 40°$ true AOA (fighter aircraft applications).

Airspeed Mach Indicator The airspeed Mach indicator is pneumatically operated by Pitot and static pressure. Indicated airspeed is displayed by a moving pointer against a fixed dial. The Mach number is used in the weapon delivery computations and in the automatic flight control computations. It should be pointed out that weapon delivery calculations require frequent velocity information.

Vertical Velocity Indicator The vertical velocity indicator is used to provide rate-of-climb or descent information.

Fire-Control System The fire-control system works in conjunction with the communication, navigation, and identification and survivability avionics to penetrate defenses and locate, acquire, and deliver air-to-surface and air-to-air weapons. Key elements include the following subsystems: (1) fire-control radar (FCR), (2) HUD, (3) multifunction display set, (4) INS, (5) fire-control computer (FCC), (6) stores management set, and (7) radar altimeter.

Forward-Looking Radar The forward-looking radar's (FLR) primary function is to provide slant range to target information to the tactical computer for computed updating of HUD information; it also sends required pitch command signals to the ADI, and presents information in ground-map or PPI (plan position indicator) display on the radarscope, when the fighter aircraft is so equipped. The FLR provides the pilot with nine modes of operation as follows:

1. *Air-to-ground ranging*—provides slant-range information to the Navigation/Weapon Delivery Computer (NWDC) and supplies range information to the tactical computer for use by the automatic weapon release system and activates the "in-range" light on the HUD when the aircraft achieves the desired range from the target.
2. *Terrain following (TF)*—displays terrain following information in a 18.52-km (10-NM) range; enables the pilot to fly at a present altitude above the ground and terrain obstacles and controls HUD pullup command symbology.
3. *Terrain avoidance (TA)*—displays all terrain at or above the aircraft altitude track in 9.26-km (5-NM) or 18.52-km (10-NM) range and enables maneuvering around terrain obstacles.
4. *Ground mapping (shaped beam)*—provides high-altitude ground mapping.
5. *Ground mapping (pencil beam)*—provides higher-resolution mapping at lower altitudes.
6. *Beacon*—interrogates airborne (e.g., tanker) beacons and the reply is displayed as coded pulse on the scope to indicate range and bearing of the target for rendezvous.
7. *Cross-scan terrain avoidance*—combined TF/TA scanning.
8. *Cross-scan ground map (pencil beam)*—presents simultaneous terrain following commands with ground-map pencil display.
9. *TV*—provides display on closed-circuit monitor for aiming weapons.

Instrument Landing System (ILS) The instrument landing system has been in use since the 1940s. The main functions of the ILS are to provide guidance displays and to allow the pilot to make a safe, all-weather approach to a suitably equipped runway. In essence, the ILS consists of three subsystems: (1) *glide slope*—provides vertical to align the aircraft on a descent path that will intersect the runway at a point after the threshold has been crossed; (2) *localizer*—provides lateral guidance to allow the aircraft to the center of the runway; and (3) *marker beacon*—provides the positional guidance and is used to determine the approximate distance from the aircraft to the runway threshold. Furthermore, each subsystem has a related ground transmitter that must be operational in order to make an ILS approach. The glide slope beam requires reflection from the airport surface to form the guidance signal. This requires that a 457-m (1500-ft) area in front of the antenna be flat and free from obstacles. Each marker beacon radiates a fan-shaped vertical beam that is approximately ±40° wide along the glide path by ±85° wide perpendicular to the path. Markers are placed as follows: *outer marker* (OM)—distance varies from 7.4 to ~13 km (4–7 NM) as specified in the approach charts for each airport; *middle marker* (MM)—placed where the glide slope is 61 m (200 ft) above the runway or 1067 m (3500 ft) from the threshold; *inner marker* (IM)—required only at airports certified for Category II landings where the glide slope is 30.48 m (100 ft) above the runway [2]. The localizer beam is a wide beam (6°–10°) that should be free from interference from nearby structures. The ILS cannot be installed in sites with irregular terrain because its usefulness will be limited. Even though not fully operational, a GPS/ILS integration is possible. The following is a list of ILS characteristics:

Power requirements	28 V dc
Frequency range	
Localizer	108.10–111.95 MHz
Glide slope	329.15–335.00 MHz
Marker beacons	75 MHz
Channels	40

Microwave Landing System (MLS) Basically, the microwave landing system configuration consists of an approach azimuth (AZ) signal, approach elevation signal, a set of basic data words, and a "precision DME" (DME/P) signal capable of being transmitted over a 200-channel range. This basic configuration can be expanded to include a back azimuth signal and a set of auxiliary words. The approach azimuth antenna is located at the stop end of the runway, as is the ILS localizer, and scans a beam up to ±60° in azimuth to either side of the runway centerline (most sites in the U.S. will scan ±40°). The elevation (EL) antenna is located approximately 305 m (1000 ft) from the runway threshold at the glide path intercept point (GPIP). The EL antenna has a scan coverage of 0.9°–30°. The DME/P transponder has a 360° coverage up to 41 km (22 NM), and is usually collocated with the AZ antenna. By using the three signals (AZ, EL, and DME/P), the three-dimensional position of the aircraft, in relation to the runway, can be determined. The addition of a back azimuth (BAZ) antenna to the approach end of the runway will provide guidance to the aircraft for takeoff and missed approaches. Two MLS formats have been proposed: (1) a "scanning beam" MLS and (2) a "Doppler radar" MLS. An International Civil Aviation Organization (ICAO)-compatible MLS calls for a single accuracy standard to be implemented worldwide. There are several operational advantages of using the MLS. Specifically, it provides a distinct advantage over the ILS by providing command guidance in a much greater coverage volume.

Consequently, the pilot has much more flexibility in approaches to landing, missed approaches, and takeoffs. At the present time, many approaches can be flown only to nonprecision decision heights because of the complexity of the approach and the lack of positive guidance. Therefore, complex approaches are required when conducting noise abatement and obstacle clearance maneuvers and when flying in restricted airspace. The MLS also provides advantages for special operations such as STOL (short takeoff and landing) and decelerated approaches. MLS is also advantageous to the air traffic controllers. For instance, once an aircraft is in the terminal and MLS coverage area, the pilot will fly a prescribed approach to landing. The controller will no longer need to provide continuous radar vectors to the pilot for interception of the ILS localizer beam or aircraft separation, thus reducing the controllers workload. The U.S. Air Force's updated C-130s were the first aircraft equipped with an airborne MLS. The Federal Aviation Administration (FAA) continues research and development on the MLS, working toward the goal of having the system fully operational in 1998. However, if it is determined before that date that the Category II and III approaches can be flown with the GPS, it is conceivable that work on the MLS will be stopped. An integrated MLS/GPS system is also being investigated for commercial aviation.

Synthetic Aperture Radar (SAR)   As is the case with traditional (real-beam) radars, the synthetic aperture radar forms images by resolving the target reflectivity in the range (LOS) and azimuth directions. The resolution in the range is determined by the accurate timing of pulses, sometimes spread in time during transmission via chirping or a similar technique. Therefore, SAR is an image-making radar that is generally used to produce a two-dimensional map of an array of radar-wave scatterers. Specifically, the SAR achieves high resolution of the photograph-like image by coherently processing the Doppler histories of the backscattered signals. Consequently, there is a requirement that there exist relative motion between the SAR and the signal scatterer. For the present discussion, it will be assumed that the SAR is mounted in an aircraft, and is used in a ground mapping mode. There are three SAR modes of operation: (1) the strip map mode, (2) the spotlight mode, and (3) the Doppler beam-sharpening (DBS) mode. In the strip map mode, the area to be mapped traverses through the real beam of the antenna, which has a fixed squint angle θ_s. The antenna in this case is not steered during the mapping process, and if the return signal is suitably recorded, the image can be easily constructed in non-real time. The strip map mode is generally used in military applications such as reconnaissance. In the spotlight mode, an image is generated about specific map coordinates, so that continuous real-beam coverage of the area to be mapped is required. Moreover, if the real beam is too narrow to provide the continuous coverage over

7.4. Modern Avionics Systems

the SAR processing time, then real-beam steering is required. The spotlight mode is used for navigation system updates or weapon delivery, so that all signal processing is done in real time. Finally, the DBS mode is similar to the strip map mode, except that real-beam steering is provided to allow a variable squint angle. Coherent processing then provides enhanced resolution, while at the same time allowing a PPI display. The way in which the SAR processing differs from previous methods is its handling of resolution in the azimuth direction. Generally speaking, the angular resolution of a real array antenna is proportional to the wavelength of the signal and inversely proportional to the diameter of the antenna. To a rough approximation, for an X-Band radar ($\lambda = 3$ cm) and a 3-m antenna, the resolution will be 10 mrad. In order to achieve better resolution with smaller antennas, SAR, as mentioned above, uses the motion of the aircraft to synthesize an antenna of larger aperture. Figure 7-17 illustrates the SAR operation. Consider the azimuth (cross-range) resolution δ_a of a real aperture radar for an airborne system. The 3-dB beamwidth of a conventional antenna of dimension D operating at a wavelength λ is nominally $\theta = \lambda/D$ [5].

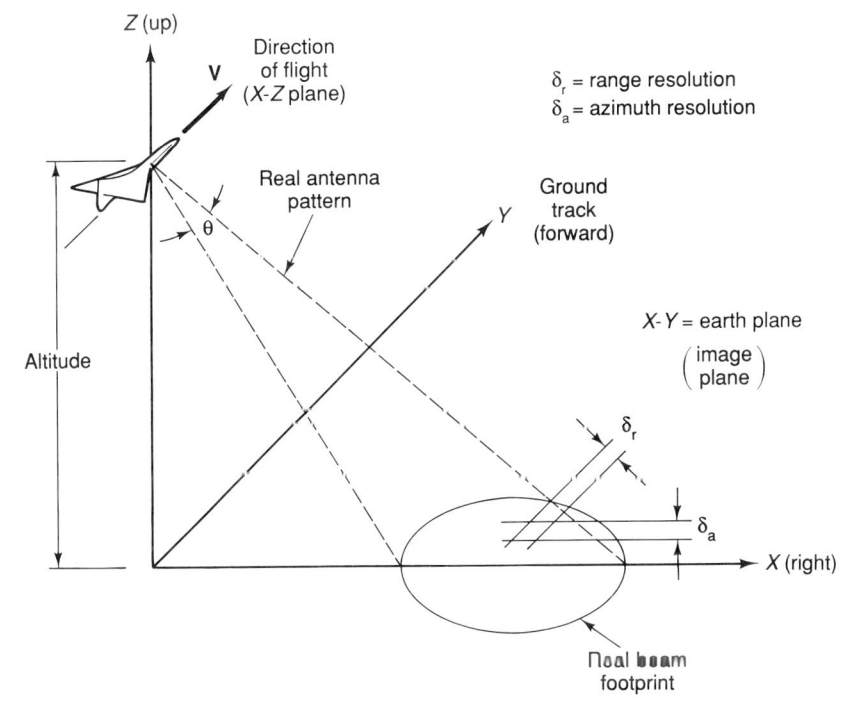

FIGURE 7-17
Broadside SAR geometry.

For the SAR system, the corresponding angular resolution at a distance R is given by

$$\delta_a = 2R * \tan\left(\frac{\theta}{2}\right) \cong R\theta = \frac{R\lambda}{D} \qquad (7.48)$$

This equation indicates that, for a given operating frequency, one must increase the aperture size D in order to improve the angular resolution. From the preceding discussion, it is noted that the range resolution δ_r and the azimuth resolution δ_a define the resolving power (i.e., the ability to distinguish between two closely spaced scatterers) of SAR, but do not by themselves totally define the quality of the image produced. Specifically, the range resolution is determined by the effective length of the transmitted pulse while the azimuth resolution is determined by the effective antenna beamwidth. For a pencil beam antenna with $\theta < 20°$, the maximum effective aperture size D_{SAR} is approximately

$$D_{SAR} \cong R\theta \qquad (7.49)$$

Therefore, in SAR the 3-dB beamwidth is expressed by the equation

$$\theta_{SAR} = \frac{\lambda}{2D_{SAR}} = \frac{\lambda}{2R\theta} \qquad (7.50)$$

so that the ideal SAR azimuth (cross-range) resolution δ_a, is given by

$$\delta_a = \frac{D}{2} \qquad (7.51)$$

In high-resolution SAR systems, it is important to know the precise position of the antenna during the integration time. Typically, the position of the antenna is calculated from the navigation data of the INS onboard the aircraft. A number of factors can cause errors in the calculated position. For instance, an error can be caused by the flexible modes of the antenna lever arm. In order for the proper phase compensation to be made so as to account for the motion of the radar, the true value of the LOS velocity must be available and correctly used. There are a number of reasons why this velocity may not be correct: (1) quantization of accelerometer and attitude outputs, (2) differential motion between the motion sensor and the radar antenna that is not observable at the sensor, (3) sensor noise, and (4) insufficiently high data rates. In designing a SAR system, a motion compensation study must be carried out in order to identify the principal error sources for motion compensation, evaluate their impact at the output of both processor and display, and investigate methods for reducing the impact of such errors on the image produced.

The synthetic aperture radar requires accurate inertial data in order to track signal phase and to subsequently perform coherent processing used to obtain an image. For instance, a high-resolution synthetic array requires that relative target–antenna motion be tracked to within a fraction of a wavelength over the entire synthetic aperture, which can translate into a subcentimeter-level accuracy requirement in position over kilometers of flight. However, as stated earlier, motion compensation errors are important limiting factors for SAR performance. On the other hand, it should be noted that SAR is relatively insensitive to weather.

SAR-aided navigation modeling has been proposed and/or used, in which the SAR image is used to update the navigator; that is, the amount of image displacement is the basic observation used to update the navigator. Here, the basic approach is to augment the navigation Kalman filter with two additional "image states" that account for the integrating properties of coherent SAR processing. These additional states are essentially the coordinates, in a two-dimensional "image plane," of a coherently processed point source image. Thus, this procedure accurately models the integrated effects of navigation error as it enters the processing of a SAR image. Moreover, the two SAR image states are reinitialized at the start of each synthetic aperture. Through integration over the time span of the aperture, the two image states are highly correlated with position and velocity, so that the image measurement serves to optimally update the navigated estimates of position and velocity. As noted earlier, the Kalman filter provides a precise means of estimating the amount of image smear attributable to the navigator.

In addition to SAR, in which a stationary target is illuminated, an inverse synthetic aperture radar (ISAR) is used to account and/or track targets that are in motion. As its name implies, ISAR utilizes the inverse techniques used in an airborne SAR, which uses the different Doppler shifts in echoes from fixed terrain that result from the motion of an airborne radar antenna. The ISAR image is a range–Doppler display of radar-return intensity. The Doppler resolution results from continuous Fourier transform processing of the coherent radar signal that shows Doppler effects caused by target motion (i.e., roll, pitch, and yaw). The image is stabilized by precision range–Doppler tracking of a dominant scatterer on the target. Consequently, the displayed information is of a photographic nature, resembling an isometric view of the target. Developed by the Naval Research Laboratory and Texas Instruments [designated as AN/APS-137(V)], ISAR will enable airborne surveillance radars to display the profile contours (e.g., two-dimensional profile image) of a surface ship, facilitating its identification, and/or classification. However, in order for ISAR to be effective, the Doppler shift caused by the ISAR's own aircraft velocity relative to the ship, as well as the ship's own forward motion, must be canceled out so that the radar senses only the Doppler shifts due to the relatively slow roll, pitch, and yaw motions of different parts of the ship.

Computed Air-Release Point (CARP) The computed air-release point is a standard tactical drop system. Mathematical in nature, a CARP is based on average parachute ballistics and fundamental dead-reckoning principles. The precision air-drop function requires the capability to compute an air-release point between sequential waypoints [viz., the initial point (IP) and the drop zone (DZ)]; the IP should be located as close as possible to the axis of the DZ. Specifically, the precision air-drop function uses stored and/or manually inserted ballistic data, programmed or sensor derived point of impact location, and OAPs (offset aimpoints), when available. The pilot and navigator jointly confirm the offset distance for the CARP. (The pilot assumes the responsibility for maintaining the offset distance and required track, while the navigator is responsible for the actual solution of the CARP.) A portion of the precision air-drop processing is the CARP computation. The CARP system computation performs the following functions:

1. *Dead reckoning*—using geodetic coordinates and aircraft velocity.
2. *Radar relative*—to DZ or OAP.
3. *Visual relative*—to initial point.
4. *FLIR relative*—to DZ or OAP.

From the preceding discussion, the CARP system supports the following functions:

1. Ballistic computation of the air-release point, giving along-track and cross-track offsets of the CARP relative to the point of impact. The formulas must incorporate corrections for nonstandard atmospheric conditions.
2. Annunciation of arrival at the CARP and end of the drop zone can be done via the HSI as well as numeric display of the distance and time to CARP.
3. Supply of CARP steering signals to the autopilot and cockpit instruments.

Fire Control Computer (FCC) In essence, the FCC is the principal component of the weapons-control subsystem, whereby its related terminals are integrated into a highly flexible and accurate means of computing automatic air-to-air and air-to-surface weapon deliveries, navigation functions (e.g., GPS integration), and energy management information. Note that weapon delivery includes bombing, air-to-air, air-to-surface, and missile launching. As such, the most stringent inertial reference system (IRS) requirements occur with unguided weapons. In this case, the approach is to establish unguided weapon error budgets in which velocity errors are commensurate with other system errors. For example, on the basis of these budgets, the most stringent IRS velocity requirement is about 0.154 m/s (0.5 ft/s) during bomb delivery. The most important functions of the FCC are to serve as the primary bus controller for the serial digital buses, such as the avionics multiplex and the display multiplex buses, and to provide storage for centralized fault gathering and reporting of weapons control terminals for self-test information. Data on the MUX buses is normally transmitted and received at a 1-MHz rate. The central processing unit (CPU) of an FCC normally uses 16-, 32-, and 48-bit data words for single- and double-precision fixed-point, and single- and extended-precision floating-point calculations. Naturally, these values may change with higher-speed digital computers. The CPU accomplishes all arithmetic computations for the FCC as dictated by the operational flight program (OFP), which is stored, for instance, in the 64K (64-kilobyte) core memory. Commonly, the FCC has six major addressing modes that enable it to address 64K words of memory, and it uses a 16-bit general register file to accomplish this function. The I/O system is the means by which the FCC communicates with external sources. (The I/O system consists of an analog-to-digital converter card, a discrete I/O card, and the serial digital interface.) The FCC OFP, when loaded into the fire-control computer, will provide the required real-time computations for the following functions: air-to-air combat, air-to-ground attack, master mode select, stores (weapons) data select, system data management, display control, energy management, navigation, fix-taking, self-test, and multiplex (i.e., time-sharing) bus control.

Multiplex Databus Systems In an inertial navigation system, internal and external interface is accomplished via a multiplex (MUX) MIL-STD-1553/1553B databus. Specifically, the state information of modern INSs is transmitted to the mission (or onboard) computer via the MIL-STD-1553B system databus. For example, this information may be used to control sensor (radar, FLIR, etc.) stabilization and pointing, and to support various flight director and guidance functions. The INS accepts the following digital inputs in serial format via the MIL-STD-1553B databus: (1) position updates (e.g., latitude–longitude), (2) velocity updates, (3) angular updates, and (4) other inputs as necessary. Outputs provided by the INS over the MUX databus include pitch, roll, heading (both true and magnetic), present position, velocity, and steering information. Messages on the 1553 databus are typically grouped into subframes and frames. Within a subframe, messages tend to be grouped in time, close to one another. In particular, subframes are transmitted at regular intervals, known as the subframe time or subframe rate. For instance, in a typical aircraft, it's on the order of milliseconds, measured from the start of one subframe to the start of the next. A remote terminal (RT) must validate command and data words received from the bus controller. This involves detection by a remote terminal of electrical and protocol errors in a message. Words and messages with errors are divided into two classes: (1) invalid and (2) illegal. An invalid word is any command or data word containing an improper sync character or a bit with an abnormal Manchester II code (i.e., something other than 16 bits plus parity or incorrect parity). An invalid message is any message containing an invalid word. The 1553 databus protocol has one mode command that is used for polling the terminals only. The difference between the 1553 and 1553B databus protocols is the mode commands. MIL-STD-1553B bus protocol can accomplish up to 10 different functions with or without an associate data word. In military aircraft, all onboard navigation aids are integrated into a self-contained navigation system through the MIL-STD-1553B databus.

Software It was mentioned above that the central processing unit (CPU) accomplishes all arithmetic computations for the FCC by the operational flight program (OFP). In the existing systems, the OFP is programmed in the JOVIAL J73 high-order language (MIL-STD-1589B). The minimum CPU processing speed, based on average operation times and a typical instruction mix, is 120 KOPS (kilo-operations per second). Future avionics systems will use high-order languages such as Ada. Ada (MIL-STD-1815) is a high-order language developed for use throughout the life cycle of DoD embedded systems. Specifically, Ada will be used in the design, development, implementation, and maintenance stages of the software life cycle. Combined with modern avionics software design, Ada encourages good programming practices by incorporating software engineering principles such as modularity, portability, reusability, and readability. With modularity, Ada organizes code into self-contained units (i.e., structured) that can be planned, written, compiled, and tested separately. This feature allows programs to be written in portions by software teams working in parallel before being integrated into the final product. In the portability feature, Ada developed for one system can easily be recompiled and ported to other systems, since Ada compilers are validated upfront and standardized internationally by MIL-STD-1815A, ANSI (American National Standards Institute), and ISO (International Standards Organization).

With regard to reusability, Ada's package concept allows users to develop software components that may be retrieved, used, and/or changed without affecting the rest of the program. Moreover, Ada's generic program units allow programmers to perform the same logical function on more than one type of data. Reliability is another key feature of Ada. Thus, Ada's exception-handling mechanism supports fault-tolerant applications by providing a complete and portable way of detecting and gracefully responding to error conditions. Finally, with regard to maintainability, Ada's program structuring based on modularity and high level of readability make it easier for one programmer to modify or improve software written by another programmer. Modularity also allows package modification without affecting other programs. Consequently, these features reduce costs in software development, verification, debugging, and maintenance associated with software development. Ada's increasing acceptance in government, scientific, and commercial sectors demonstrates its value as a desirable tool for large and complex systems. In fact, the Ada computer programming has been accepted by the DoD and the NATO nations. On the negative side, Ada carries an overhead penalty into the object code that affects both memory and throughput performance. Above all, since Ada is a relatively new language, cross-compiler maturity will remain a critical issue in the design of Ada embedded software [3].

Cockpit Instrumentation As part of an avionics modernization program, the U.S. Air Force is replacing analog instrument panels in an upgraded C-130 Hercules cockpit with the new generation of flat-panel, liquid-crystal displays (LCDs) discussed earlier. The upgraded C-130 was first flown early in 1991 in a 4-h test; testing continued through October 1991. In these flight tests, six full-color, active-matrix LCD displays, each measuring 6 × 8 in., replaced more than 60 analog-type cockpit instruments on an operational C-130. Five displays were on the main instrument panel and one in the navigation station. The purpose of the flights was to study LCD reliability and maintainability over electromechanical instruments and cathode-ray tubes (CRTs). These new displays present flight attitude, navigation, weather radar, and engine-operating data on the screens.

In the near future, the LCDs are expected to succeed the CRTs as the next generation of electronic displays for both military and commercial aircraft. Several airlines have already begun studies for modernizing existing jetliners and/or new generation jetliners (e.g., Boeing 777, MD-11/12) with these next-generation cockpit displays. The Air Force's C-17 transport will also have an all-new flight deck with the most advanced avionics instruments. The LCDs offer volume, weight, and power savings over the conventional displays, and will probably cost less with increased production. The LCDs use digital processing for lower power consumption, increased durability, and reduced weight. In addition, they are more reliable. Whereas only parts of an LCD screen fail, the entire display is lost when a CRT goes out.

Today's modern avionics cockpit consists of a flight management system (FMS), such as the Collins Avionics, Rockwell International Corporation's FMS-800. The FMS-800 automates navigation and routine cockpit functions to reduce aircrew workload, thereby improving mission precision. Specifically, the FMS-800 outputs dynamic guidance data to the flight instruments and automatic flight-control system using the MIL-STD-1553B or ARINC 429/561 signals. In particular, the Collins FMS-800 integrates the functions of communication, navigation, and IFF (Identification/Friend or Foe) control, GPS/INS navigation, flight instruments and controls, autopilot, stores and radar for transport, tanker, and utility aircraft. The FMS-800 is also designed to take maximum advantage of color multifunction displays such as the Collins digital Electronic Flight Instrument System (EFIS-85) display system. Figure 7-18 illustrates the Collins Avionics high-resolution,

FIGURE 7-18

The Collins EFIS-700 display units: (*top*) the Electronic Attitude Director Indicator (EADI); (*bottom*) the Electronic Horizontal Situation Indicator (EASI). (Courtesy of Collins Avionics, Rockwell International Corporation.)

full-color, EFIS-700 system consisting of the Electronic Attitude Director Indicator (EADI) display unit and the Electronic Horizontal Situation Indicator (EHSI) display unit.

The EADI display includes groundspeed, radio altitude, localizer, and glideslope deviation information on approach as well as speed command and basic roll and pitch data. In addition, autopilot and thrust management

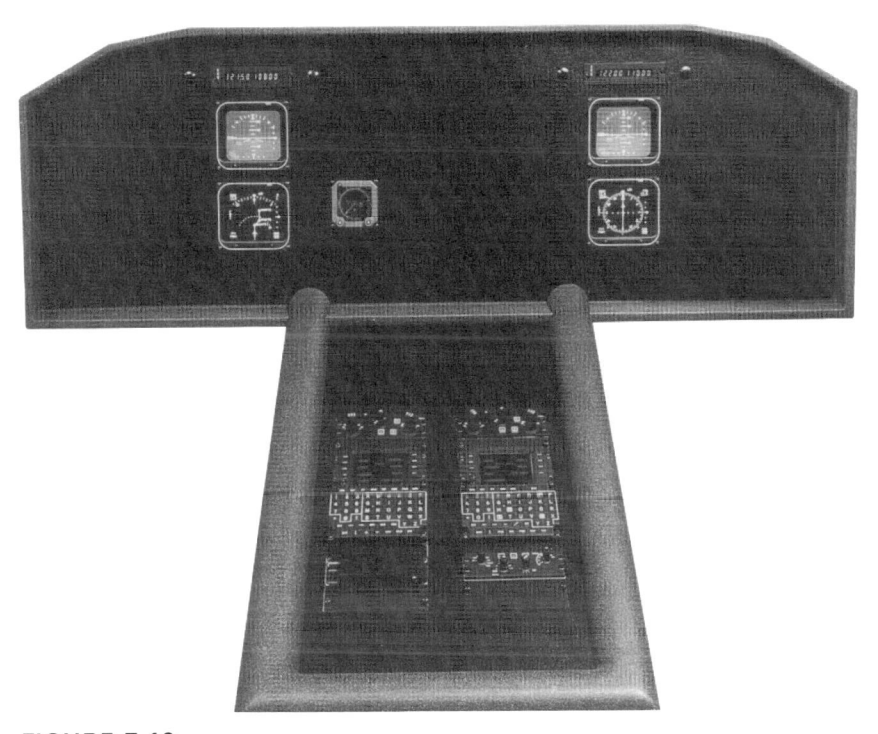

FIGURE 7-19
Modern cockpit instrumentation compatible with the USAF T-1A program. (Courtesy of Collins Avionics, Rockwell International Corporation.)

mode annunciations are also displayed. Five pilot-selectable EHSI modes include the traditional compass rose, expanded ILS or VOR formats, a map mode showing the aircraft's position relative to specific waypoints, and a north–up mode showing the flight plan. Weather radar displays may be superimposed over the map, VOR, or ILS modes. The Collins EFIS systems have been selected for such Boeing aircraft as the 757/767-300, 737-300, and 747-400, and Fokker 100.

Figure 7-19 illustrates a partial cockpit instrumentation for modern U.S. Air Force aircraft. Specifically, the Collins FMS-800 can drive conventional electromechanical flight instruments. However, the FMS-800 is also designed to take maximum advantage of color multifunction displays such as the Collins EFIS-85 display system and APS-85 autopilot used on the U.S. Air Force Tanker Transport Training System (TTTS).

REFERENCES

1. Bate, R. R., Mueller, D. D., and White, J. E.: *Fundamentals of Astrodynamics*, Dover, New York, 1971

2. Kayton, M., and Fried, W. R. (eds): *Avionics Navigation Systems*, Wiley, New York, 1969.
3. Lin, C. F.: *Modern Navigation, Guidance, and Control Processing*, Prentice-Hall, Englewood Cliffs, N.J., 1991.
4. Macomber, G. R., and Fernandez, M.: *Inertial Guidance Engineering*, Prentice-Hall, Englewood Cliffs, N.J., 1962.
5. Morris, G. V.: *Airborne Pulsed Doppler Radar*, Artech House, Norwood, Mass., 1988.
6. Pitman, G. R., Jr. (ed.): *Inertial Guidance*, Wiley, New York, 1962.
7. Siouris, G. M.: "Inertial Navigation," *Encyclopedia of Physical Science and Technology*, Vol. 8, Academic Press, San Diego, 1987, pp. 668–717.
8. Smart, W. M.: *Text-Book on Spherical Astronomy*, 5th ed., Cambridge University Press, Cambridge, U.K., 1962.
9. San Giovanni, C., Jr.: "Performance of a Ring Laser Strapdown Attitude and Heading Reference for Aircraft," AIAA Guidance and Control Conference, Palo Alto, Calif., August 7–9, 1978, Paper No. 78-1240, pp. 12–19.
10. Sturza, M. A., Brown, A. K., and Kemp, J. C.: "GPS/AHRS: A Synergistic Mix," *Proceedings of the NAECON'84*, Vol. 1 (Dayton, Ohio, May 21–25, 1984), pp. 339–348.

Appendix A
System Performance
Criteria

The objective of this appendix is to provide the basic formulation of system performance measurement methods, namely (1) circular error probable (CEP), also known as "circle of equal probability"; (2) spherical error probable (SEP); and (3) radial position error (RER).

A.1. CIRCULAR ERROR PROBABLE

This section develops the basic mathematical concepts of CEP leading to the currently used simplified formulas that an engineer can immediately employ for a given situation. Above all, by understanding what CEP means, how it was generated, why it is used, and what limitations it has, the engineer can then apply it as a useful tool in measuring system performance with some level of confidence in the numbers produced. Basically, the CEP concept comes from the general bivariate (two-variable) Gaussian distribution [3].

"Circular error probable" is simply that radius of a circle (centered at the mean point of impact) that encloses 50% of the probability of a hit; that is, a 50% probability of a hit (i.e., success) is equivalent to the area enclosed by the circle. In addition to its use in evaluating inertial navigation system performance, this is a very important military concept relative to the accurate delivery of a weapon (e.g., bombs, missiles, ordnance) onto a selected target. Therefore, in the analysis of the accuracy of a given delivery system, this CEP parameter is often used. The circular description arises because of the two-dimensional aspect of the problem of hitting any target. The error in the position of the earth's surface has two dimensions: one along the flight direction called the "down-range" or x direction, and the other perpendicular to the down-range called the "cross-range" or y direction. Therefore, any planar targeting problem must contend with errors in these two directions. Since CEP has the error distribution information in its formulation, we can state that the smaller the CEP, the smaller the errors, and hence, the more accurate the system. Note that quoting the performance of an inertial navigation or weapon guidance and control system using this CEP parameter can relate a whole wealth of information about the overall performance of the system; however, it states nothing about the quality or accuracy of the data used in computing the location of the target. This is an important concept for the engineer to understand. CEP is independent of where the actual target is, since it is a measure of dispersion and not a measure of central tendency. This means that CEP is a function of the total delivery system error.

As stated earlier, the derivation of the CEP is based on statistical concepts. Errors in a weapon delivery system (e.g., biases and standard deviations), for example, are sum totals of the error contributions of a number of subsystems and operational error sources and can be characterized as unpredictable and, therefore, random in nature. Irrespective of the probability distributions of the individual error sources, if the number of error sources is large enough, the total will approximate the Gaussian distribution (via the central limit theorem). The Gaussian density function can therefore be used to study the CEP concept. The general bivariate Gaussian distribution function (also referred to as the "joint distribution of two variables") expressed in planar Cartesian coordinates x and y is given by the following relationship [3]:

$$f(x, y) = \frac{1}{2\pi\sigma_y\sigma_x\sqrt{1-\rho_{xy}^2}} \exp\left\{\frac{-1}{2\sqrt{1-\rho_{xy}^2}}\left[\left(\frac{x-\mu_x}{\sigma_x}\right)^2 - 2\rho_{xy}\frac{(x-\mu_x)(y-\mu_y)}{\sigma_x\sigma_y} + \left(\frac{y-\mu_y}{\sigma_y}\right)^2\right]\right\} \quad (A.1)$$

where

$\rho_{xy} = \frac{\sigma_{xy}}{\sigma_x\sigma_y}$ = the error correlation coefficient between x and y: range is $-1 \leq \rho \leq 1$

σ_{xy} = covariance
σ_x = standard deviation in range direction
σ_y = standard deviation in deflection direction
x = range coordinate
μ_x = mean of range
$x - \mu_x$ = range error

A.1. Circular Error Probable

y = deflection coordinate
μ_y = mean of deflection
$y - \mu_y$ = deflection error

This function is called an "elliptical distribution" because the lines of equal densities form concentric ellipses. In order to evaluate the ellipticity for conversion to CEP, it is very desirable to treat the down-range and cross-range errors separately. Therefore, in order to do this, they must be statistically independent, which requires the correlation coefficient in Eq. (A.1) to equal zero ($\rho = 0$). In general, the probability that a hit will be scored within a region R can be expressed as [2]

$$P(R) = \int_R f(x, y) \, dx \, dy \tag{A.2}$$

To simplify the integration of Eq. (A.2), a coordinate transformation or rotation is necessary to transform x and y into normally distributed, uncorrelated random variables. If there were no correlation ($\rho = 0$) between the down-range and cross-range variables (i.e., a change in x does not affect y), then Eq. (A.2) would simplify to one having fewer variables. It is important to note that the correlation can be removed via a rotation so that we can achieve statistical independence. Coordinate rotation takes care of the correlation, while it does not change the absolute distribution. This rotation then simplifies Eq. (A.2) into a more manageable form as follows:

$$P(x, y) = \iint_R \frac{1}{2\pi\sigma_y\sigma_x} \exp\left\{-\frac{1}{2}\left[\left(\frac{x-\mu_x}{\sigma_x}\right)^2 + \left(\frac{y-\mu_y}{\sigma_y}\right)^2\right]\right\} dx \, dy \tag{A.3}$$

Now, if the bias can be removed so that the mean μ is zero in each direction (i.e., cross-range and down-range), a simpler integral to work with results. A simple translation would accomplish this by shifting and collocating the center of the distribution at the center of the target. Thus

$$P(x, y) = \iint_R \frac{1}{2\pi\sigma_y\sigma_x} \exp\left\{-\frac{1}{2}\left[\frac{x_x^2}{\sigma_x^2} + \frac{y^2}{\sigma_x^2}\right]\right\} dx \, dy \tag{A.4}$$

Transforming Eq. (A.4) into polar coordinates yields the following equation:

$$P(R) = \frac{1}{2\pi\sigma_y\sigma_v} \int_0^R \int_0^{2\pi} \left\{\exp -\frac{r^2}{2}\left[\frac{\cos^2\theta}{\sigma_x^2} + \frac{\sin^2\theta}{\sigma_y^2}\right]\right\} r \, dr \, d\theta \tag{A.5}$$

where

$$R = [x^2 + y^2]^{1/2}$$

From the preceding discussion, the CEP is that value of R which yields a value of 0.5 from the integral of Eq. (A.5).

A special case of Eq. (A.5) is the "circular case," which is formed when $r = R$ and $\sigma_x = \sigma_y = \sigma_c$, where the subscript c denotes the circular case. For this special case, we have

$$P(R) = P_c = 1 - \exp\{-(R^2/2\sigma_c^2)\} \tag{A.6}$$

where P_c = the circular probability distribution function, a special case of $P(R)$
σ_c = the circular standard deviation, a special case of σ_r when $\sigma_r = \sigma_x = \sigma_y$
R = the radius of the probability circle

The CEP for a truly circular (i.e., symmetric, $\sigma_x = \sigma_y$) distribution is computed by the following equations:

$$P(R) = 0.5 = 1 - \exp\{-(R^2/2\sigma_c^2)\}$$

$$\exp\{-(R^2/2\sigma_c^2)\} = \ln(0.5)$$

$$R^2 = 0.69315(2\sigma_c^2)$$

$$R = 1.1774\sigma_c$$

or

$$\text{CEP} = 1.1774\sigma_c \tag{A.7}$$

Similarly, for 75% probability, $R = 1.665\sigma_c$. Another equation for the CEP is the following:

$$\text{CEP} = 1.1774\sqrt{\sigma_x \sigma_y} \tag{A.8}$$

This equation produces an error of as much as 14%; therefore, it is seldom used in practice. Figure A-1 presents the circular normal distributions.

For the "noncircular" case, $\sigma_x \neq \sigma_y$, Eq. (A.5) must be solved in terms of Bessel functions. With some algebraic manipulation, Eq. (A.5) takes the form

$$P(R) = \frac{1}{\sigma_y^2 \rho} \int_0^R I_0\left[\frac{r^2(1-\rho^2)}{4\sigma_y^2 \rho^2}\right] \exp -\frac{r^2(1-\rho^2)}{4\sigma_y^2 \rho^2} \, dr \tag{A.9}$$

where I_0 is a modified Bessel function of zeroeth order and $\rho = \sigma_x/\sigma_y (\rho < 1)$. Obviously, this equation can be plotted, resulting in a curve that is tedious and frustrating for the engineer to work with. Figure A-2 shows, in addition to a plot of Eq. (A.9), two straight-line approximations that the engineer could readily use:

$$\text{CEP} = 0.589[\sigma_x + \sigma_y] \tag{A.10}$$

$$\text{CEP} = 0.615\sigma_x + 0.562\sigma_y ; \quad \sigma_y > \sigma_x \tag{A.11}$$

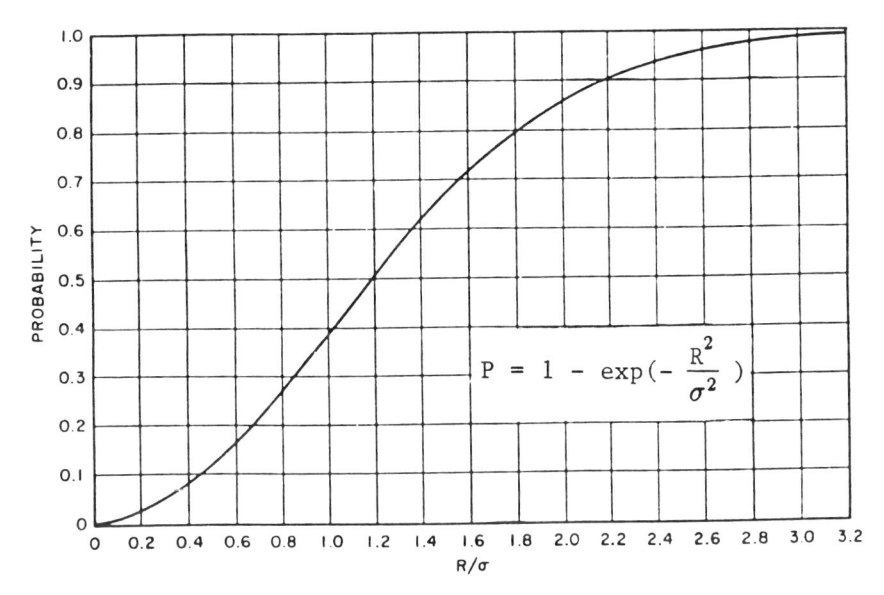

A.1. Circular Error Probable

FIGURE A-1
The circular normal distribution curve.

Now consider two variables with input variances σ_x^2 and σ_y^2. A more general equation for the CEP that takes into account correlation is given by

$$\text{CEP} = 0.589[(\sigma_x^2 \cos^2 \theta + \sigma_y^2 \sin^2 \theta)^{1/2} + (\sigma_x^2 \sin^2 \theta + \sigma_y^2 \cos^2 \theta)^{1/2}]$$

where

$$\theta = \frac{1}{2} \tan^{-1} \left[\frac{2\sigma_x \sigma_y}{\sigma_x^2 - \sigma_y^2} \right]$$

If x and y are independent, and there is no correlation, this equation reduces to Eq. (A.10).

The engineer should become inquisitive when "quoted" the CEP of a particular INS or weapon delivery system. This CEP value, whether in meters, feet, or whatever units, says something of the accuracy of the system. However, without information about the variances and the means, the engineer should be suspicious about the distribution. Table A-1 summarizes the equations available for computing the CEP, for varying degrees of accuracy.

Often, the engineer is interested in relating the 50% circular error probability to other standard error probabilities, such as 1-sigma or 2-sigma (that is, $1\sigma = 68.269\%$, $2\sigma = 95.449\%$, $3\sigma = 99.730\%$, $3.5\sigma = 99.980\%$, $4\sigma = 99.993\%$, $5\sigma = 99.999\%$). This linear conversion is comparatively illustrated in Fig. A-3.

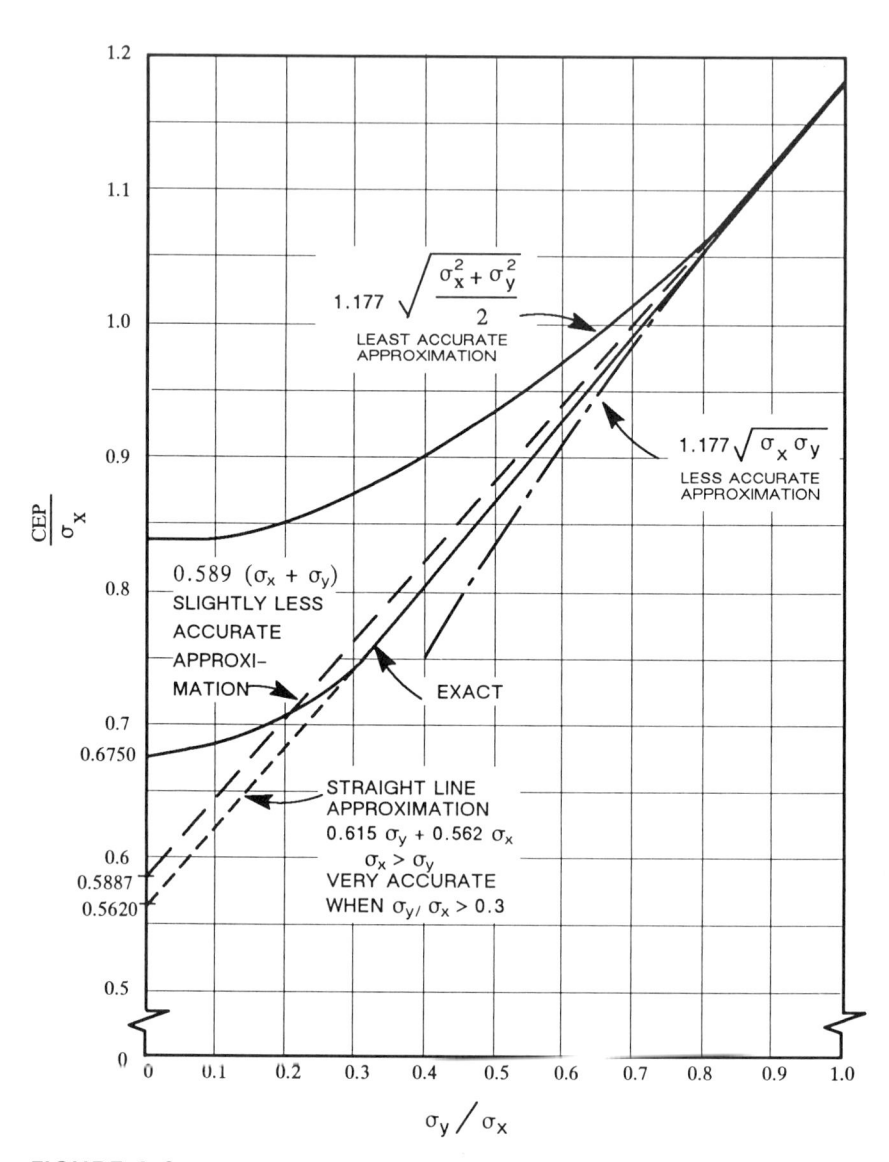

FIGURE A-2
CEP curves for elliptical error distribution approximations.

A.1. Circular Error Probable

TABLE A-1
Summary of CEP Equations[a]

Method	Formula	Accuracy	Remarks
1.A	$CEP = 1.774 \sqrt{\dfrac{\sigma_S^2 + \sigma_L^2}{2}}$	Error < 17%	Straight-line approximation for $0.2 \leq w \leq 1.0$
1.B	$CEP = 1.774 \sigma_c$		Special case: when $\sigma_x = \sigma_y = \sigma_c$ (circular case)
2.A	$CEP = 0.5887 (\sigma_S + \sigma_L)$	Error ≤ 3%	Straight-line approximation for $0.154 \leq w \leq 1.0$
2.B	$CEP = 0.6745 \sigma_L$	Error ≤ 3%	Straight-line approximation for $0.0 \leq w \leq 0.154$
3.A	$CEP^e = [0.6152 \sigma_S + 0.5620 \sigma_L]$	Error < 0.26%	Straight-line approximation for $0.3 \leq w \leq 1.0$
3.B	$CEP^e = (0.8200w - 0.0070)$ $\sigma_S + 0.6745 \sigma_L$	Error < 0.49%	Parabolic curve for $0.0 \leq w \leq 0.3$

[a] The following definitions are basic to the six equations given above: $\sigma_L = \sigma_{large}$, $\sigma_S = \sigma_{small}$, $w = \sigma_{small}/\sigma_{large} = \sigma_x/\sigma_y$.

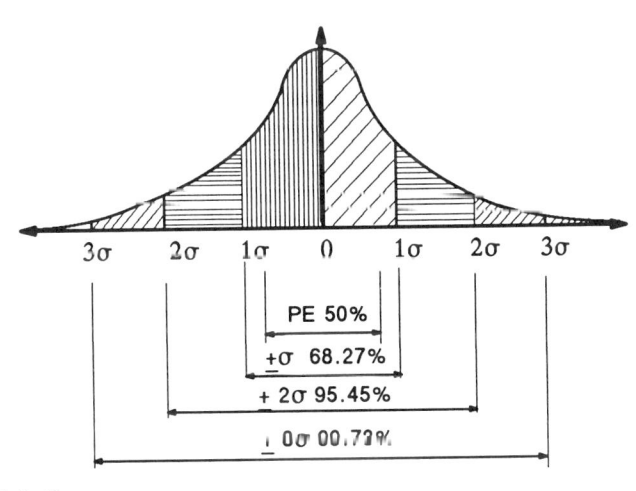

FIGURE A-3
Normal linear distribution.

TABLE A-2
Linear Error Conversion Factors

From	To 50%	68.27%	90%	99.73%
50%	1.0000	1.4826	2.4387	4.4475
68.27%	0.6745	1.0000	1.6449	3.0000
90%	0.4101	0.6080	1.0000	1.8239
99.73%	0.2248	0.3333	0.5483	1.0000

The circular error probable represents 50% of the area under the Gaussian curve. Table A-2 shows the factors for converting from one probability level to another.

A program for computing the CEP equations given in Table A-1 is presented here:

```
CEP FORMULAE: THIS PROGRAM CONTAINS THE SIX MOST COMMONLY USED
ESTIMATES FOR CEP CALCULATIONS. THE FOLLOWING DEFINITION IS BASIC
TO THE SIX FORMULAE GIVEN BELOW.

SIGMAx=SIGL=SIGMA LARGE, SIGMAy=SIGS=SIGMA SMALL, w = SIGS/SIGL
```

METHOD	FORMULA	ACCURACY	REMARKS
1.A	CEP = 1.1774* SQRT(SIGS^2 + SIGL^2)/2)	Error<17%	Straight line approximation for 0.2<w<1.0
1.B	CEP = 1.1774*SIGC		Special case: SIGS=SIGL=SIGC (circular)
2.A	CEP = 0.5887* (SIGS + SIGL)	Error<3%	Straight line approximation for 0.154<w<1.0
2.B	CEP = 0.6745*SIGL	Error<3%	Straight line approximation for 0.154<w<1.0
3.A	CEP = 0.6152*SIGS + 0.562*SIGL	Error<0.26%	Straight line approximation for 0.3<w<1.0
3.B	CEP = (0.82*w - 0.007) *SIGS + 0.6745*SIGL	Error<0.49%	Parabolic curve for 0<w<0.3

```
ENTER VALUES FOR SIGMA SMALL AND SIGMA LARGE WHICH WILL REPRESENT
```

A.1. Circular Error Probable 451

```
THE STANDARD DEVIATION OF THE CROSSRANGE AND DOWNRANGE ERRORS:

SIGS =    20.0000       SIGL =     100.0000

1.A     CEP =    84.903520
2.A     CEP =    80.644000
2.B     CEP =    67.450000
3.A     CEP =    68.504000
3.B     CEP =    70.590000


      PROGRAM CEP
      CHARACTER*15 FNAME
      WRITE(*,3)
3     FORMAT(' WHAT IS THE DESIRED OUTPUT FILENAME ? ',/,
     1' FOR EXAMPLE: A:CEP.DAT ',/)
      READ(*,4)FNAME
4     FORMAT(A)
      OPEN(9,FILE=FNAME,STATUS='UNKNOWN')
      WRITE(*,1)
      WRITE(9,1)
1     FORMAT(5x,' CEP FORMULAE: THIS PROGRAM CONTAINS THE SIX MOST '
     1,/,' COMMONLY USED ESTIMATES FOR CEP CALCULATIONS. THE '
     1,/,' FOLLOWING DEFINITION IS BASIC TO THE SIX FORMALAE BELOW.'
     1,/,/,
     1' SIGMAx=SIGL=SIGMA LARGE, SIGMAy=SIGS=SIGMA SMALL, w=SIGS/SIGL'
     ,/,/,
     1' METHOD    FORMULA                 ACCURACY    REMARKS ',/,
     1'  1.A      CEP = 1.1774*           Error<17%   Straight line'
     1,/,6x,'    SQRT(SIGS^2+SIGL^2)/2)               approx. for'
     1,/,6x'                                          0.2<w<1.0'
     1,/,
     1'  1.B      CEP = 1.1774*SIGC                   Special case:'
     1,/,6x,'                                         SIGS=SIGL=SIGC'
     1,/,6x,'                                         (circular)')
      WRITE(*,7)
      WRITE(9,7)
7     FORMAT(
     1'  2.A      CEP = 0.5887*           Error<3%    Straight line'
     1,/,6x,'         (SIGS+SIGL)                     approx. for'
     1,/,6x,'                                         0.154<w<1.0'
     1,/,
     1'  2.B      CEP = 0.6745*SIGL       Error<3%    Straight line'
     1,/,6x,'                                         approx. for'
     1,/,6x,'                                         0<w<0.154'
     1,/,
     1'  3.A      CEP = 0.6152*SIGS       Error<.26%  Straight line'
     1,/,6x,'         +0.562*SIGL                     approx. for'
     1,/,6x,'                                         0.3<w<1.0'
     1,/,
     1'  3.B      CEP   (0.82*w - 0.007)  Error<.49%  Parabolic'
     1,/,6x,'         *SIGS+0.6745*SIGL               curve for'
     1,/,6x,'                                         0<w<0.3')
      WRITE(*,5)
      WRITE(9,5)
```

```
      5 FORMAT(/,' ENTER VALUES FOR SIGMA SMALL AND SIGMA LARGE',/,
     1' WHICH WILL REPRESENT THE STANDARD DEVIATION OF THE',/,
     1' CROSSRANGE AND DOWNRANGE ERRORS:',/)
        READ(*,*)SIGS, SIGL
        WRITE(*,9)SIGS,SIGL
      9 FORMAT(' SIGS = ',F15.4,'     SIGL = ',F15.5,/,/)
C       DEFINE W = SIGS/SIGL
C
C       STRAIGHT LINE APPROXIMATION FOR (.2.LE.W.LE.1.0)
C       ERROR .LT. .17%
        CEP1A=1.1774SQRT((SIGS*SIGS + SIGL*SIGL)/2.)
        WRITE(*,10)CEP1A
        WRITE(9,10)CEP1A
     10 FORMAT(' 1.A    CEP = ',F14.6)
C
C       SPECIAL CASE: WHEN SIGS = SIGL = SIGC (CIRCULAR CASE)
C       TEST IF THEY ARE REALLY CLOSE.
        EPS = SIGL - SIGS
        IF (EPS .LT. .0001) THEN
        SIGC = SIGL
        CEP1B = 1.1774*SIGC
        WRITE(9,15)CEP1B
        WRITE(9,15)
     15 FORMAT(' 1.B    CEP = ',F14.6)
        ENDIF
        CEP2A = 0.5887*(SIGS + SIGL)
        WRITE(*,20)CEP2A
        WRITE(9,20)CEP2A
     20 FORMAT(' 2.A    CEP = ',F14.6)
        CEP2B = 0.6745*SIGL
        WRITE(*,30)CEP2B
        WRITE(9,30)CEP2B
     30 FORMAT(' 2.B    CEP = ',F14.6)
        CEP3A = 0.6152*SIGS + 0.5620SIGL
        WRITE(*,40)CEP3A
        WRITE(9,40)CEP3A
     40 FORMAT(' 3A     CEP = ',F14.6)
        CEP3B = (0.82*SIGS*SIGS/SIGL) - 0.007*SIGS + 0.6745*SIGL
        WRITE(*,50)CEP3B
        WRITE(9,50)CEP3B
     50 FORMAT(' 3.B    CEP = ',F14.6)
        END
```

A.2. SPHERICAL ERROR PROBABLE

The above results can be extended to the three-dimensional case. Therefore, by extending our coordinates to three dimensions, we can apply the theory of CEP to the analysis a more general case of weapon system performance. In particular, the spherical error probable (SEP) is an integral of the trivariate (three-variable) Gaussian probability density function over a sphere, which is centered at the mean. In spherical coordinates, the SEP assumes the form [compare with Eq. (A.5)]

$$P(R) = \frac{1}{(\sqrt{2\pi})^3 abc} \int_0^R \int_0^\pi \int_0^{2\pi} \rho(r, \phi, \theta) r^2 \sin \phi \, d\theta \, d\phi \, dr \quad \text{(A.12)}$$

$$\rho(r, \phi, \theta) = \exp\left[-\frac{1}{2}\left(\frac{r^2 \sin^2 \phi \cos^2 \theta}{a^2} + \frac{r^2 \sin^2 \phi \sin^2 \theta}{b^2} + \frac{r^2 \cos^2 \phi}{c^2}\right)\right]$$

where r is the radius of the sphere, $a^2 = \sigma_x^2$, $b^2 = \sigma_y^2$, and $c^2 = \sigma_z^2$. [Note that in Eq. (A.12) ϕ is the colatitude and θ is measured along the longitude.] The above integral can be solved by means of an infinite series expansion. The SEP finds application in such areas as determination of the three-dimensional position error of orbiting satellites (i.e., GPS), and weapon delivery system performance in space. Working with trivariate Gaussian distributions, we would be looking at error volumes (e.g., ellipsoids and spheres). The best of the analytical approximations to the SEP, for the unbiased case, is given by the following equations [1]:

$$\text{SEP} \approx [\sigma_T^2 (1 - V/9)^3]^{1/2} \quad \text{(A.13)}$$

$$\sigma_T^2 = \sigma_N^2 + \sigma_E^2 + \sigma_U^2$$

$$V = \frac{2(\sigma_N^4 + \sigma_E^4 + \sigma_U^4)}{\sigma_T^4}$$

where σ_T = total standard deviation
σ_N = north standard deviation
σ_E = east standard deviation
σ_U = vertical (or up) standard deviation

A FORTRAN program listing for computing the SEP is presented here:

```
PROGRAM SPERPR
WRITE(*,*)'ENTER SIGMA NORTH, SIGMA EAST, SIGMA UP'
READ(*,*)SIGN,SIGE,SIGU
SIGT = SQRT(SIGN**2 + SIGE**2 + SIGU**2)
V = 2*(SIGN**4 + SIGE**4 + SIGU**4)/SIGT**4
SIGI = SIGT**2*(1.0 - V/9.0)**3
SEP = SQRT(SIGI)
WRITE(*,*)'SEP = ',SEP
END
```

A.3. RADIAL POSITION ERROR

In certain cases, the engineer is called on to determine the radial position error (RER) of an inertial navigation system. The radial position error can be determined from the relation

$$\text{RER} = \sqrt{(\phi - \phi_0)^2 R^2 + (\lambda - \lambda_0)^2 R^2 \cos^2 \phi_0} \tag{A.14}$$

where ϕ_0 = initial latitude
 λ_0 = initial longitude
 R = radius of the earth

The radial position error can also be expressed in terms of the CEP by assuming that $\sigma_x = \sigma_y = \sigma$. Thus

$$\text{RER} = (\sigma_x^2 + \sigma_y^2)^{1/2} = \sqrt{2}\,\sigma = \sqrt{2}\left(\frac{\text{CEP}}{1.1774}\right)$$

$$= 1.2\,\text{CEP (rate)} \tag{A.15}$$

A program for computing the radial position error, given the initial position in terms of latitude and longitude, is presented as follows:

```
      PROGRAM RADERR
C
C***************************
C   REAL DEFINITION(S)
C***************************
      REAL*8 COS_PHI_NOT
      REAL*8 LATITUDE
      REAL*8 INIT_LAT
      REAL*8 LONGITUDE
      REAL*8 INIT_LONGITUDE
      REAL*8 DELTA_PHI
      REAL*8 DELTA_LAMBDA
      REAL*8 RADER
      REAL*8 R
C
C***************************
C   INITIALIZE VARIABLE(S)
C***************************
      R = 6378137. m
      R = R*R
C***************************
C   PROMPT USER VALUES
C***************************
      PRINT *
      PRINT '(a,$)',' PLEASE ENTER INITIAL LATITUDE (IN DEGREES):'
      READ(5,*) INIT_LATITUDE
      PRINT '(a,$)',' PLEASE ENTER INITIAL LONGITUDE (IN DEGREES):'
      READ(5,*) INIT_LONGITUDE
      PRINT '(a,$)',' PLEASE ENTER FINAL LATITUDE (IN DEGREES):'
      READ (5,*) LATITUDE
      PRINT '(a,$)',' PLEASE ENTER FINAL LONGITUDE (IN DEGREES):'
      READ (5,*) LONGITUDE
```

A.3. Radial Position Error

```
C*****************************************************
C    CONVERT INPUT VALUES FROM DEGREES TO MINUTES
C*****************************************************
      LATITUDE = LATITUDE*60.
      INIT_LATITUDE = INIT_LATITUDE*60.
      LONGITUDE = LONGITUDE*60.
      INIT_LONGITUDE = INIT_LONGITUDE*60.
C*****************************************************
C    COMPUTE RADER
C*****************************************************
      COS_PHI_NOT = COS(INIT_LATITUDE) * COS(INIT_LATITUDE)
      DELTA_PHI = LATITUDE - INIT_LATITUDE
      DELTA_PHI = DELTA_PHI * DELTA_PHI
      DELTA_LAMBDA = LONGITUDE - INIT_LONGITUDE
      DELTA_LAMBDA = DELTA_LAMBDA * DELTA_LAMBDA
      RADER = SQRT((DELTA_PHI * R) + (DELTA_LAMBDA*R*COS_PHI_NOT))
C*************************
C    OUTPUT RADER
C*************************
      PRINT *
      WRITE (5,11) RADER
 11   FORMAT(' RADER = 'F16.5)
      END
```

In summary, the following statistical parameters are presented as a convenience to the reader [2]:

1. Sample mean:

$$m_x = \frac{1}{n} \sum_{i=1}^{n} x_i$$

where x_i = individual sample at time t_i
       n = number of samples

2. Geometric mean (GM):

$$\text{GM} = n\sqrt{\left(\prod_{i=1}^{n} r_i\right)}$$

where n = sample size
       r = radial error from target (in meters or feet)

3. Root mean square (rms):

$$\text{rms} = \sqrt{\left(\frac{1}{n} \sum_{i=1}^{n} r_i^2\right)}$$

If $GM/rms \leq 0.6$: $\quad CEP = rms[0.7(GM/rms) + 0.3]$

If $GM/rms > 0.6$: $\quad CEP = rms[0.7(GM/rms) + 0.4\sqrt{GM/rms}]$

4. Standard deviation:

$$\sigma_x = \sqrt{\frac{1}{n-1} \sum_{i=1}^{n} (x_i - m_x)^2}$$

5. Cross-correlation coefficient:

$$\rho_{xy} = \left(\frac{1}{n-1} \sum_{i=1}^{n} (x_i - m_x)(y_i - m_y)\right) / \sigma_x \sigma_y$$

6. Autocorrelation function:

$$\phi_{xx}(\tau) = \frac{1}{n} \sum_{i=1}^{n} (x_i - m_x)(x_j - m_x)$$

where $\tau = t_i - t_j$, for discrete values of $\tau = 0, 1, 2$ s.

7. It is sometimes preferred to assess the navigation system's performance in terms of 90-percentile error predictions. Specifically, the 90-percentile value of the radial error, for successive static and mobile tests, can be computed as follows. Let γ be the ratio of the geometric mean (GM) to the rms. Then

$$\gamma = \frac{GM}{rms}$$

Therefore

$$R_{90} = \begin{cases} rms[\gamma + 1.6(1 - \gamma^2)] & \text{for } \gamma \leq 0.6 \\ rms[1 + \sqrt{1 - \gamma}] & \text{for } \gamma > 0.6 \end{cases}$$

REFERENCES

1. Childs, D. R., Coffey, D. M., and Travis, S. P.: "Error Measures for Normal Random Variables," *IEEE Transactions on Aerospace and Electronic Systems*, **AES-14**(1), 64–67, January 1978.
2. Papoulis, A.: *Probability, Random Variables, and Stochastic Processes*, 2nd Ed., McGraw-Hill, New York, 1984.
3. Pitman, G. R., Jr. (ed.): *Inertial Guidance*, Wiley, New York, 1962, Appendix E.

Appendix B
The World
Geodetic System

The Department of Defense World Geodetic System (DoD WGS) geoid is defined by a spherical harmonic expansion, which is based on an equipotential ellipsoid of revolution. An equipotential ellipsoid is an ellipsoid defined to be an equipotential surface on which all values of the potential are equal. Therefore, the equipotential ellipsoid serves not only as a reference surface or geometric figure of the earth, but leads to a closed formula for normal gravity at the ellipsoid surface. The WGS-84 coordinate system is defined as follows [1]:

1. The origin of the coordinate system is the earth's center of mass.
2. The Z axis is parallel to the direction of the conventional international origin (CIO) for polar motion as defined by the Bureau International de L'Heure (BIH) on the basis of the latitudes adopted for the BIH stations.
3. The X axis is defined by the intersection of the WGS-84 reference meridian plane and the plane of the mean astronomic equator, the reference meridian being parallel to the zero meridian defined by the BIH on the basis of the longitudes adopted for the BIH stations.
4. The Y axis completes a right-handed, earth-fixed orthogonal coordinate system, measured in the plane of the mean astronomic equator 90° east of the X axis.

Historically, the DoD in the late 1950s generated a geocentric reference system to which different geodetic networks could be referred. Furthermore, because of the failure of local systems to provide intercontinental geodetic information, a unified world system became essential [1]. Consequently, all the military services participated in developing the first DoD WGS-60 sys-

TABLE B-1
Reference Ellipsoid Constants

Reference ellipsoids	Equatorial radius a(m)	Flattening f	Where used
Krasovskii (1940)	6,378,245.000	1/298.30	Russia
Clarke (1866)	6,378,206.400	1/294.9786982	North America
Clarke (1880)	6,378,249.145	1/293.465	France, Africa
International (1924)	6,378,388.000	1/297.000	Europe
Bessel (1841)	6,377,397.155	1/299.1528128	Japan
Everest (1830)	6,377,276.345	1/300.8017	India
Australian National	6,378,160.000	1/298.25	Australia
South American (1969)	6,378,160.000	1/298.25	South America
Airy (1830)	6,377,563.396	1/299.3249646	United Kingdom
Fischer (1960)	6,378,155.000	1/298.30	South Asia
Fischer (1968)	6,378,150.000	1/298.30	South Asia
WGS-72	6,378,135.000	1/298.26	Worldwide
WGS-84	6,378,137.000	1/298.257223563	Worldwide

tem. In order to do this, it was necessary to consider all available observed data to determine the best absolute reference system to provide a good fit for the entire earth. In 1966, a World Geodetic System (WGS) Committee was charged with the responsibility of developing a WGS needed to satisfy mapping, charting, and geodetic requirements. Since 1960, additional surface

TABLE B-2
Geodetic and Geophysical Parameters of the WGS-84 Ellipsoid

Parameters	Notation	Value
Semimajor axis	a	6378137 m
Flattening (ellipticity)	f	1/298.257223563 (0.00335281066474)
Second degree zonal harmonic coefficient of the geopotential	$\bar{C}_{2,0}$	$-484.16685 \times 10^{-6}$
Angular velocity of the earth	Ω	7.292115×10^{-5} rad/s
The earth's gravitational constant (mass of earth's atmosphere included)	GM	3986005×10^{8} m^3/s^2
Mass of earth (includes the atmosphere)	M	5.9733328×10^{24} kg
Theoretical (normal) gravity at the equator (on the ellipsoid)	γ_e	9.7803267714 m/s^2
Theoretical (normal) gravity at the poles (on the ellipsoid)	γ_p	9.8321863685 m/s^2
Mean value of theoretical (normal) gravity	$\bar{\gamma}$	9.7976446561 m/s^2

TABLE B-3
Associated Constants of the WGS-84 Ellipsoid

Constants	Notation	Formula	Value
Semiminor axis	b	$b = a(1-f)$	6,356,752.3142 m
Major eccentricity (first eccentricity)	ε	$\varepsilon = [f(2-f)]^{1/2}$	0.0818191908426
Major eccentricity squared	ε^2	$\varepsilon^2 = f(2-f)$	0.00669437999013
Minor eccentricity (second eccentricity)	ε'	$\varepsilon' = \varepsilon/(1-f)$	0.0820944379496
Axis ratio	b/a	$b/a = 1-f$	0.996647189335

gravity observations, results from the extension of triangulation and trilateration networks, and data from Doppler radar and satellites became available. The WGS 66 gravitational model was thus developed. In 1970 the WGS Committee began work for a replacement of the WGS-66, and in 1972 completed the DoD WGS-72 system. In developing the WGS-72 system, the WGS Committee adhered to the approach and methods used by the International Union of Geodesy and Geophysics (IUGG) in establishing the Geodetic Reference System-1967 (GRS-67). Since 1972, extensive gravity data has been collected and analyzed, giving rise to a more accurate world geodetic system, the WGS-84 system. Table B-1 lists the various reference ellipsoid constants, while Tables B-2 and B-3 give the various parameters and constants of the WGS-84 system, respectively.

B.1. SUMMARY OF SELECTED FORMULAS

Reference Ellipsoid

$$r = a[1 - f \sin^2 \phi_c - \tfrac{3}{8} f^2 \sin^2 \phi_c - \cdots] \tag{B.1}$$

where a = equatorial radius
  ϕ_c = geocentric latitude
  f = flattening

International Gravity Formula

Jeffreys:
$$g = g_0[1 + 0.0052891 \sin^2 \phi + 0.0000059 \sin^2 2\phi] \tag{B.2}$$

Airy:
$$g = g_0[1 + 0.0052884 \sin^2 \phi - 0.0000059 \sin^2 2\phi] \tag{B.3}$$

where g_0 = gravity at the equator = 9.780373 m/s^2
  ϕ = geodetic latitude

Normal Gravity Formula

$$g = \begin{bmatrix} 0 \\ 0 \\ g_z \end{bmatrix}$$

$$g_z = g_0 \left[1 - 2A \left(\frac{h}{a}\right) + B \sin^2 \phi \right] \tag{B.4}$$

$A = 1 + f + m$
$B = \frac{5}{2} m - f$
$m = \dfrac{\Omega^2 a^2 b}{GM}$
g_0 = equatorial gravity
h = altitude
ϕ = latitude
M = mass of earth
Ω = angular rate
g = gravitational constant = 66.7×10^{-9} cm^3 g^{-1} s^{-2}
a = ellipsoid semimajor axis
b = ellipsoid semiminor axis
f = flattening of ellipsoid

REFERENCE

1. "Department of Defense World Geodetic System 1984: Its Definition and Relationship with Local Geodetic Systems," DMA TR 8350.2, September 30, 1987.

Index

Acceleration, 135, 141, 196
   centripetal, 140, 143–144
Accelerometer, 3, 265
   bias, 215, 218, 241
   misalignment, 265
   output, 141
   random constant, 266
   scale factor, 231
Ada, 438–439
Aeronautical Radio, Inc. (ARINC), 198, 413, 439
Airmass, 408–409
Air traffic control (ATC), 387, 414
Altimeter, 197–199, 216
Altitude, 7, 197–199
   absolute, 198
   barometric, 199
   calculation, 224
   divergence, 223–226
   pressure, 199, 204
   system, 199
   true, 199
Analytic inertial navigation system (INS), 36, 41, 48, 130, *see also* Strapdown inertial navigation system
Angle-of-attack indicator (AOA), 429

Angstrom, 85
Angular velocity, 24
Area navigation (R-NAV), 280, 413–414
Atmosphere, 197–198
   error model, 329
   exponential, 203
   nonstandard, 199, 219–221
   standard, 198, 200
Attitude direction indicator (ADI), 20, 206, 428
Attitude error equations, 249–252
Attitude and heading reference system (AHRS), 86, 342, 415–419
Autocorrelation function, 353, 456
Automatic flight-control system (AFCS), 377, 384

Bearing, 421, 427
   relative, 277, 421
Boltzmann's constant, 95
   equation, 95
Bureau International de L'Heure (BIH), 28, 457

Caley–Hamilton theorem, 19

461

Centerline recovery steering, 384
Central Air Data Computer (CADC), 197–199, 205–206, 366–367
Centripetal acceleration, 140, 143, 152, 154
Circular error probable (CEP), 5, 267, 443–452
Combined altitude radar altimeter (CARA), 426
Computed air-release point (CARP), 269, 436
Computed airspeed (CAS), 208
Control and Display Unit (CDU), 69, 377–379, 381, 425–426
Coordinated Universal Time (UTC), 28
Coordinate frame, notation, 7–13
  accelerometer, 12
  body, 10
  earth, 9–10
  geocentric, 10
  geographic, 10
  gyroscope, 13
  inertial, 9
  navigation, 10
  nonorthogonal, 12–13
  platform, 12
  tangent plane, 156
  wander–azimuth, 11
Coordinate transformation, 13–51
Coriolis acceleration, 140, 143
Coriolis law, 143
Course, 421
Covariance, 230, 267, 311–312, 318–319, 370–371
Covariance matrix, 230, 311
Craft rate, 157–158, 160, 166–167, 241, 256

Dead-reckoning (DR) navigation, 1, 342, 349, 364, 366, 406–412, 415, 436
Dilution of precision (DOP), 311, 313–314
Direction cosine matrix (DCM), 14–17
  differential equation, 17–19
  propagation, 175

  time derivative, 18–19, 168–169
  update, 19, 169
Distance Measurement Equipment (DME), 339–341, 412
Doppler radar, 39, 360–366

Earth rate, 26–27, 29, 141
Eccentricity, 149, 191
Electromagnetic Log (EM-Log), *see* Speedlog
Electronic countermeasures (ECM), 422
Ellipticity, 40, 75, 149, *see also* Flattening
En route navigation, 380, 414
Ephemeris Time (ET), 28–29, 328
Error analysis, 229–231
Error angles, 232–233
Error budget, 229, 266–268
Error equations, 241–245
  local level, 237–238
  wander–azimuth, 254–260
Error model, 259–264
Error sources, 230–231, 267
Error state vector, 252, 259, 344
Euclidean norm, 15
Euler angles, 14, 20–22, 61–63, 67

Fabry–Pérot interferometer, 85, 99
Faraday rotator, 114–115, 117
Federal Aviation Administration (FAA), 277, 341, 432
Fiber-optic gyro (FOG), 118–123
  Interferometric fiber-optic gyro (IFOG), 119, 121–123
Fire control computer, 219, 331–333, 429, 436–437
First point of Aries, 29
Flattening, 40, 75, 149, 458–460
Fly-by-wire, 424
Force, specific, 48, 141–142, 188, 195, 216, 223, 248
Forcing function, 242, 252
Forward-looking infrared (FLIR), 38, 272, 346–348
Forward-looking radar (FLR), 430

Free air temperature (T_{FAT}), 197
Free azimuth system, 41, 45, 47, 162, 238
Free inertial mode, 2, 270–271

Gaussian distribution, 443–444, 453
Gauss–Markov process, 215
Geometric dilution of precision (GDOP), 285–286, 312, 317
Glide slope, 428, 430
Global Positioning System (GPS), 296–324
   differential GPS, 322–324
   geometry, 314–315
   system outage, 310
Glonass (GNSS), 335–337
GPS/INS integration, 257, 324–332
Gravitation, 224
   gradient, 146
Gravitational acceleration, 142, 195
Gravitational potential, 145
Great circle navigation, 254, 385–399
Greenwich Mean Time (GMT), 11, 28
Gyrocompassing, 46, 132
Gyroscope, 3, 88, 123
   bias, 161
   drift, 157, 187, 234, 254, 265
   fiber-optic, 118–123
   laser, 84, 86–107

Heading, 420–421
Head-up-display (HUD), 206, 426
Horizontal dilution of precision (HDOP), 313
Horizontal situation indicator (HSI), 206, 358, 426–427

Indicated airspeed (IAS), 205, 207, 366–367
Inertial navigation system (INS)
   concept, 2–4, 136–137, 141–144, 197–198, 229–236, 241–260
   free-azimuth, 41–42, 45–47
   local-level, 10, 253–254
   space-stabilized, 187–196

Inertial reference system (IRS), 85, 375
Innovation, 289–290
Instrument landing system (ILS), 269, 426, 430–431
International Civil Aviation Organization (ICAO), 200, 205, 222
International gravity formula, 459
Inverse synthetic aperture radar (ISAR), 435

Joint distribution function, 444
Joint Tactical Information Distribution System (JTIDS), 272, 341–345
   RelNav, 341–343

Kalman filter, 189, 230, 270, 369–370, 372–373
Kinematic equations, 135–140

Lapse rate, 200, 202–203
Laser gyro, 83–86
   beat frequency, 90–91, 103, 115
   bias, 124–125, 127
   coherence, 100
   dither, 107, 110–112, 128–129, 132
   energy levels, 95–97
   error model, 123–125
   error sources, 107–113
   gain medium, 89, 95, 105
   lasing action, 93–94
   lock-in, 109, 111
   metastable, 95–97
   mode pulling, 112–113
   multioscillator, 104, 113
   null shift, 108–109
   oscillation modes, 97–98
   scale factor, 90, 92–94, 107, 115, 121, 125–127
Latitude, 7, 11, 147, 390, 391
   computation, 150, 165, 172, 191, 409
   geodetic, 147, 150, 191
Least-squares estimate, 309, 316–322
Legendre polynomials, 145

Lever-arm correction, 434
Line of position (LOP), 282–287, 289, 291, 293
  error ellipse, 287
Localizer, 428, 430–431
Local level inertial navigation system, 41
Longitude computation, 150, 165, 172, 190, 410
Loran C, 281–287
  Loran–inertial system mechanization, 289–290
  Loran pulse, 287
Loxodrome, see Rhumb line

Mach number, 205–207, 429
Magnetic compass, 429
Magnetic heading, 415–417
Magnetic-variation (Mag Var), 410, 417, 421
Marker beacon, 430–431
Markov model, 215, 262, 362
Mass attraction, 141, 144, 153–154, 176
Maximum dilution of precision (MDOP), 313–314
Mean sea level (MSL), 199, 216, 353
Mechanization equations, 155
  latitude–longitude, 157–161
  space-stabilized, 187–196
  wander–azimuth, 161–176
Meridian convergence, 77–78
Mesosphere, 202
Microwave landing system (MLS), 335, 425, 431–432
Multimode radar, 272, 274, 276
Multiplex databus, 137, 302, 436–438

Navaids, 272–274, 295
Navigation, 1, 135, 147, 155–157, 375
  angles, 420–421
  hyperbolic, 282–287
  inertial, 2–4
  rho–rho, 293
Newton, I., 9, 135, 142
Normal gravity formula, 460
Normal gravity vector, 146
North-slaved system, 10, 41, 45, 47

Omega, 290
  differential, 294–296
  lane ambiguity, 291
  lane slip, 291, 294
Oscillation modes, see Laser gyro, oscillation modes

Perturbation, 195, 235
Pinson error model, 233, 238, 241
Platform, 48
  angular rate, 172–173, 175
  drift rate, 240
  inertial, 3–4, 10
  misalignment, 48, 50
  torquing, 175
Polarization, 93, 105, 115–117
Position dilution of precision (PDOP), 313–314
Position vector, 188
  geocentric, 188
Primary flight-control system (PFCS), 428
Pseudorange, 302–308
Psi equation, 233–235

Quaternions, 51–68

Radar, 274–277
Radar altimeter, 227, 426
Radial error, 345, 347
Radial position error (RPE), 453–456
Random bias, 124
Random constant (or bias), 124, 260
Random drift, 130
Random walk, 91, 111, 127–129, 218
Ranging mode, 293–294
Reaction time, 5, 131–132
Reference ellipsoid, 10, 40–41, 43–44, 242, 459
Relative bearing, 2, 277, 421
Relative Navigation (RelNav), 341–343
Rho–rho navigation, 293
Rho–theta, 280, 341, 413
Rhumb line, 399–406
Rotating coordinates, 137–140

Rotation matrix, 13, 16, 21
Rotation vector, 79–81

Sagnac, G., 84, 87
  effect, 88, 117–118, 120–121
  interferometer, 87–89
Schuler frequency, 144, 210, 241, 270
Sidereal day, 10, 27
  hour angle, 26–28, 188
  time, 28
Skew–symmetric matrix, 18–19, 50, 248
Slant range, 276, 430
Space-stable system, 41, 46–47, 187–195
Specific force, 48, 141–142, 188, 195, 216, 223, 248
Speedlog, 273, 368–369
Spherical error probable (SEP), 452–453
Spherical harmonics, 145, 457
Standard deviation, 287, 444, 447
Star tracker, 357, 359–360
State–space notation, 242
State transition matrix, 237
Steering, 375, 383–385
Stored heading, 408
Stored magnetic variation, 408
Strapdown inertial navigation system, 10, 12, 19, 36, 41, 48, 63, 156
Stratosphere, 200, 202
Synthetic aperture radar (SAR), 432–435
System altitude, 199

Tactical Air Navigation (TACAN), 277–278
Terminal navigation, 380
Terrain Contour Matching (TERCOM), 348–357
Terrain following/terrain avoidance (TF/TA), 315, 357, 430
Terrain Profile Matching (TERPROM), 356
Tesseral harmonics, 145
Thermosphere, 202
Time dilution of precision (TDOP), 313–314

Time-division multiple-access (TDMA), 342
Time of arrival (TOA), 342–344
Time–space–position information (TSPI), 345
Transformation, coordinate, 7–13, 13–41
Transport rate, 46, 144, 150
Troposphere, 200
True airspeed (TAS), 206–207, 385, 415, 417
True equinox, 27
Truth model, 259, 326, 330–331

Unipolar system, 41, 45, 47
Unit vector, 14, 17, 32
Universal Polar Stereographic (UPS) system, 69
Universal Transverse Mercator (UTM), 68–72
  convergence, 77–78
  false easting, 73–75
  false northing, 73–75

Variance, 260, 312, 319
Velocity, 1, 162–165, 170
  angular, 24–26, 154
  computation, 148, 152–153, 158, 165, 169–170, 173–174
  error equations, 245–247
Vernal equinox, 29
Vertical channel, 197–198, 212–225, 270–271
  damping, 209–210
Vertical dilution of precision (VDOP), 313–315
Very-High-Frequency Omnidirectional Ranging (VOR), 337, 413
Visual flyover, 333, 345–346
VOR/DME, 412–413
VOR/TACAN (VORTAC), 341, 413

Wander angle, 11, 32–33, 44, 161–165, 171–172, 256–257, 410
Wander–azimuth system, 41, 46–47

Waypoint, 304, 334, 380–383
White noise, 125, 196, 215, 372
World Geodetic System (WGS), 40, 210, 457–459

Yaw axis, 10, 20–22, 33–34, 36–37, 39

Zonal harmonics, 145

ISBN 0-12-646890-7